# Systems & Control: Foundations & Applications

Series Editor

Christopher I. Byrnes, Washington University

Associate Editors

S.-I. Amari, University of Tokyo
B.D.O. Anderson, Australian National University, Canberra
Karl Johan Åström, Lund Institute of Technology, Sweden
Jean-Pierre Aubin, EDOMADE, Paris
H.T. Banks, North Carolina State University, Raleigh
John S. Baras, University of Maryland, College Park
A. Bensoussan, INRIA, Paris
John Burns, Virginia Polytechnic Institute, Blacksburg
Han-Fu Chen, Academia Sinica, Beijing
M.H.A. Davis, Imperial College of Science and Technology, London
Wendell Fleming, Brown University, Providence, Rhode Island
Michel Fliess, CNRS-ESE, Gif-sur-Yvette, France
Keith Glover, University of Cambridge, England
Diederich Hinrichsen, University of Bremen, Germany
Alberto Isidori, University of Rome
B. Jakubczyk, Polish Academy of Sciences, Warsaw
Hidenori Kimura, University of Osaka
Arthur J. Krener, University of California, Davis
H. Kunita, Kyushu University, Japan
Alexander Kurzhanski, Russian Academy of Sciences, Moscow
Harold J. Kushner, Brown University, Providence, Rhode Island
Anders Lindquist, Royal Institute of Technology, Stockholm
Andrzej Manitius, George Mason University, Fairfax, Virginia
Clyde F. Martin, Texas Tech University, Lubbock, Texas
Sanjoy K. Mitter, Massachusetts Institute of Technology, Cambridge
Giorgio Picci, University of Padova, Italy
Boris Pshenichnyj, Glushkov Institute of Cybernetics, Kiev
H.J. Sussman, Rutgers University, New Brunswick, New Jersey
T.J. Tarn, Washington University, St. Louis, Missouri
V.M. Tikhomirov, Institute for Problems in Mechanics, Moscow
Pravin P. Varaiya, University of California, Berkeley
Jan C. Willems, University of Gröningen, The Netherlands
W.M. Wonham, University of Toronto

Xunjing Li
Jiongmin Yong

*Optimal Control Theory for Infinite Dimensional Systems*

Birkhäuser
Boston • Basel • Berlin

Xunjing Li
Department of Mathematics
Fudan University
Shanghai 200433
China

Jiongmin Yong
Department of Mathematics
Fudan University
Shanghai 200433
China

### Library of Congress Cataloging-in-Publication Data

Li, Hsün-ching
   Optimal control theory for infinite dimensional systems / Xunjing
Li, Jiongmin Yong.
      p.    cm. - (Systems & control)
   Includes bibliographical references and index.
   ISBN (invalid) 0-8176-3722-2 (acid-free)
   1. Control theory.  2. Mathematical optimization. 3. Linear
systems.    I. Yong, J. (Jiongmin), 1958-   . II. Title.
III. Series.
QA402.3.L4878    1994                              94-37168
515'.64--dc20                                       CIP


Printed on acid-free paper
© Birkhäuser Boston 1995

*Birkhäuser*

Copyright is not claimed for works of U.S. Government employees.
All rights reserved. No part of this publication may be reproduced, stored in a retrieval system, or transmitted, in any form or by any means, electronic, mechanical, photocopying, recording, or otherwise, without prior permission of the copyright owner.

Permission to photocopy for internal or personal use of specific clients is granted by Birkhäuser Boston for libraries and other users registered with the Copyright Clearance Center (CCC), provided that the base fee of $6.00 per copy, plus $0.20 per page is paid directly to CCC, 222 Rosewood Drive, Danvers, MA 01923, U.S.A. Special requests should be addressed directly to Birkhäuser Boston, 675 Massachusetts Avenue, Cambridge, MA 02139, U.S.A.

ISBN 0-8176-3722-2
ISBN 3-7643-3722-2
Reformatted by Texniques, Inc. from authors' disks.
Printed and bound by Quinn-Woodbine, Woodbine, NJ.
Printed in the U.S.A.

9 8 7 6 5 4 3 2 1

# Contents

Preface ................................................................. ix

## Chapter 1. Control Problems in Infinite Dimensions ............ 1

§1. Diffusion Problems ............................................. 1
§2. Vibration Problems ............................................. 5
§3. Population Dynamics ........................................... 8
§4. Fluid Dynamics ................................................ 12
§5. Free Boundary Problems ....................................... 15
Remarks ............................................................. 22

## Chapter 2. Mathematical Preliminaries ........................ 24

§1. Elements in Functional Analysis ............................... 24
  §1.1. Spaces ..................................................... 24
  §1.2. Linear operators ........................................... 27
  §1.3. Linear functionals and dual spaces ......................... 28
  §1.4. Adjoint operators .......................................... 31
  §1.5. Spectral theory ............................................ 32
  §1.6. Compact operators .......................................... 33
§2. Some Geometric Aspects of Banach Spaces ....................... 36
  §2.1. Convex sets ................................................ 36
  §2.2. Convexity of Banach spaces ................................. 41
§3. Banach Space Valued Functions ................................. 45
  §3.1. Measurability and integrability ............................ 45
  §3.2. Continuity and differentiability ........................... 47
§4. Theory of $C_0$ Semigroups ..................................... 49
  §4.1. Unbounded operators ........................................ 49
  §4.2. $C_0$ semigroups ........................................... 52
  §4.3. Special types of $C_0$ semigroups .......................... 55
  §4.4. Examples ................................................... 57
§5. Evolution Equations ........................................... 63
  §5.1. Solutions .................................................. 63
  §5.2. Semilinear equations ....................................... 66
  §5.3. Variation of constants formula ............................. 68
§6. Elliptic Partial Differential Equations ....................... 71
  §6.1. Sobolev spaces ............................................. 71
  §6.2. Linear elliptic equations .................................. 75
  §6.3. Semilinear elliptic equations .............................. 78
Remarks ............................................................. 80

## Chapter 3. Existence Theory of Optimal Controls ............... 81

§1. Souslin Space .................................................. 81

§1.1. Polish space............................................... 81
§1.2. Souslin space.............................................. 84
§1.3. Capacity and capacitability ............................... 86
§2. Multifunctions and Selection Theorems....................... 89
§2.1. Continuity................................................ 89
§2.2. Measurability............................................. 94
§2.3. Measurable selection theorems...........................100
§3. Evolution Systems with Compact Semigroups ................109
§4. Existence of Feasible Pairs and Optimal Pairs ................106
§4.1. Cesari property..........................................106
§4.2. Existence theorems ......................................110
§5. Second Order Evolution Systems ............................113
§5.1. Formulation of the problem .............................113
§5.2. Existence of optimal controls............................118
§6. Elliptic Partial Differential Equations
    and Variational Inequalities ................................121
Remarks.........................................................129

## Chapter 4. Necessary Conditions for Optimal Controls
     — Abstract Evolution Equations ................. 130

§1. Formulation of the Problem..................................130
§2. Ekeland Variational Principle ...............................135
§3. Other Preliminary Results ..................................137
§3.1. Finite codimensionality .................................137
§3.2. Preliminaries for spike perturbation .....................143
§3.3. The distance function ...................................146
§4. Proof of the Maximum Principle ............................150
§5. Applications ................................................159
Remarks.........................................................165

## Chapter 5. Necessary Conditions for Optimal Controls
     — Elliptic Partial Differential Equations.........168

§1. Semilinear Elliptic Equations...............................168
§1.1. Optimal control problem and
      the maximum principle .................................168
§1.2. The state constraints....................................171
§2. Variation along Feasible Pairs ..............................175
§3. Proof of the Maximum Principle ............................179
§4. Variational Inequalities.....................................183
§4.1. Stability of the optimal cost............................184
§4.2. Approximate control problems ..........................185
§4.3. Maximum principle and its proof........................188
§5. Quasilinear Equations.......................................191
§5.1. The state equation and the optimal control problem ......191

§5.2. The maximum principle .................................. 196
§6. Minimax Control Problem ................................... 197
   §6.1. Statement of the problem ............................. 197
   §6.2. Regularization of the cost functional ................. 199
   §6.3. Necessary conditions for optimal controls ............. 200
§7. Boundary Control Problems .................................. 207
   §7.1. Formulation of the problem ........................... 207
   §7.2. Strong stability and the qualified maximum principle .. 209
   §7.3. Neumann problem with measure data .................... 212
   §7.4. Exact penalization and a proof of
         the maximum principle ................................ 214
Remarks ....................................................... 220

## Chapter 6. Dynamic Programming Method for Evolution Systems .... 223

§1. Optimality Principle and Hamilton-Jacobi-
    Bellman Equations ......................................... 223
§2. Properties of the Value Functions .......................... 227
   §2.1. Continuity ............................................ 228
   §2.2. $B$-continuity ........................................ 231
   §2.3. Semi-concavity ........................................ 234
§3. Viscosity Solutions ........................................ 239
§4. Uniqueness of Viscosity Solutions .......................... 244
   §4.1. A perturbed optimization lemma ........................ 244
   §4.2. The Hilbert space $X_\alpha$ .......................... 248
   §4.3. A uniqueness theorem .................................. 250
§5. Relation to Maximum Principle and Optimal Synthesis ........ 256
§6. Infinite Horizon Problems .................................. 264
Remarks ....................................................... 272

## Chapter 7. Controllability and Time Optimal Control ........... 274

§1. Definitions of Controllability ............................. 274
§2. Controllability for linear systems ......................... 278
   §2.1. Approximate controllability ........................... 279
   §2.2. Exact controllability ................................. 282
§3. Approximate controllability for semilinear systems ......... 286
§4. Time Optimal Control — Semilinear Systems .................. 294
   §4.1. Necessary conditions for time optimal pairs ........... 294
   §4.2. The minimum time function ............................. 299
§5. Time Optimal Control — Linear Systems ...................... 302
   §5.1. Convexity of the reachable set ........................ 303
   §5.2. Encounter of moving sets .............................. 308
   §5.3. Time optimal control .................................. 315
Remarks ....................................................... 317

## Chapter 8. Optimal Switching and Impulse Controls ......... 319

§1. Switching and Impulse Controls ............................. 319
§2. Preliminary Results ........................................... 322
§3. Properties of the Value Function ............................. 328
§4. Optimality Principle and the HJB Equation .................. 331
§5. Construction of an Optimal Control ......................... 334
§6. Approximation of the Control Problem ...................... 338
§7. Viscosity Solutions ........................................... 344
§8. Problem in Finite Horizon .................................... 352
Remarks ........................................................... 359

## Chapter 9. Linear Quadratic Optimal Control Problems ..... 361

§1. Formulation of the Problem .................................. 361
  §1.1. Examples of unbounded control problems ................ 361
  §1.2. The LQ problem ......................................... 366
§2. Well-posedness and Solvability ............................... 371
§3. State Feedback Control ...................................... 379
  §3.1. Two-point boundary value problem ...................... 379
  §3.2. The Problem $(LQ)_t$ .................................... 382
  §3.3. A Fredholm integral equation ........................... 386
  §3.4. State feedback representation of optimal controls ........ 391
§4. Riccati Integral Equation .................................... 395
§5. Problem in Infinite Horizon .................................. 401
  §5.1. Reduction of the problem ............................... 401
  §5.2. Well-posedness and solvability .......................... 405
  §5.3. Algebraic Riccati equation .............................. 407
  §5.4. The positive real lemma ................................ 408
  §5.5. Feedback stabilization .................................. 412
  §5.6. Fredholm integral equation and
        Riccati integral equation ............................... 414
Remarks ........................................................... 415

**References** ...................................................... **419**

**Index** ........................................................... **443**

# Preface

Infinite dimensional systems can be used to describe many phenomena in the real world. As is well known, heat conduction, properties of elastic-plastic material, fluid dynamics, diffusion-reaction processes, etc., all lie within this area. The object that we are studying (temperature, displacement, concentration, velocity, etc.) is usually referred to as the *state*. We are interested in the case where the state satisfies proper differential equations that are derived from certain physical laws, such as Newton's law, Fourier's law etc. The space in which the state exists is called the *state space*, and the equation that the state satisfies is called the *state equation*. By an infinite dimensional system we mean one whose corresponding state space is infinite dimensional. In particular, we are interested in the case where the state equation is one of the following types: partial differential equation, functional differential equation, integro-differential equation, or abstract evolution equation. The case in which the state equation is being a stochastic differential equation is also an infinite dimensional problem, but we will not discuss such a case in this book.

It is known by now that the *Pontryagin maximum principle*, the *Bellman dynamic programming method*, and the *Kalman optimal linear regulator theory* are three milestones of modern (finite dimensional) optimal control theory (see Fleming [1]). The study of optimal control theory for infinite dimensional systems can be traced back to the beginning of the 1960's. A main goal of such a theory is to establish the infinite dimensional version of the above-mentioned three fundamental theories. In the past 30 years, many mathematicians and control theorists have made great contributions in this research area.

In recent years we and some of our colleagues have been involved in the study of optimal control theory for infinite dimensional systems. Compared with the works of other mathematicians, we find that ours have their own flavor, and some of the methods might interest other people working in this area or in some related fields. Also, in the past few years, we have taught several courses entitled "Optimal Control for Distributed Parameter Systems" to graduate students at the Institute of Mathematics, Fudan University. The materials were taken from our recent works; many new results developed by other mathematicians in recent years were also adopted. We decided that it would be better to write a book to give a unified presentation of these theories.

The main feature of this book is the following. First of all, we have given a unified presentation of optimal control theory for infinite dimensional systems. This includes the existence theory, the necessary conditions (Pontryagin type maximum principle), the dynamic programming method (involving the viscosity solution of infinite dimensional Hamilton-Jacobi-

Bellman equations), and the linear-quadratic optimal control problems. Second, we have made efforts to provide self-contained proofs for many preliminary results that were not contained in previous control theory books. This will help many graduate students or scholars from other fields understand the theory. Among these proofs, let us mention two such efforts: (i) We have spent a very reasonable number of pages introducing the theory of Souslin space and general selection theorems for measurable multifunctions, which are essential in the presentation of existence theory. (ii) We present a perturbed optimization lemma resulting from the work of Ekeland and Lebourg, which is closely related to the well-known Ekeland's variational principle and plays an important role in the proof of the uniqueness of viscosity solutions. Third, instead of just making some negative statements, we have presented or cited a number of counterexamples, such as nonclosable operators, continuous functions not attaining an infimum on the closed unit ball in infinite dimensional spaces, nonconvexity of the reachable set in infinite dimensional spaces, etc. This will help the readers understand some basic features of infinite dimensional spaces. Finally, to keep the book at a reasonable length, we had to leave out much important material. To compensate for this, we have made some brief bibliographic remarks at the end of each chapter to survey some of the related works.

From the above, we see that this book is something between a monograph and a textbook. It is our desire that the book be useful for graduate students and researchers in the areas of control theory and applied mathematics. People from other fields such as engineering and economics might also find the book valuable.

There is a vast amount of literature devoted to the study of optimal control theory for infinite dimensional systems and related problems. We have not attempted to give a complete list of references. What we have cited at the end of the book are those that we find closely related to our presentation. We admit that very many important works might be overlooked. Fortunately, while preparing the book, we were informed that Professor H.O. Fattorini (of UCLA) was writing a book relevant to ours. The readers are suggested to consult that book for some related works on the topic and a possibly better list of references.

The book is organized as follows. We begin with many examples of control problems in infinite dimensions in Chapter 1. Chapter 2 discusses some very basic elements of preliminaries such as functional analysis, semigroups, evolution equations, and elliptic partial differential equations. The rest of the book is divided into five "parts." The first part is Chapter 3, in which the existence theory for optimal controls is presented. The systems discussed contain elliptic, parabolic, and hyperbolic partial differential equations. Results for elliptic variational inequalities are also presented. The second part consists of Chapters 4 and 5. In these two chapters, we present necessary conditions of the Pontryagin maximum principle type. The results cover abstract evolution equations, elliptic partial differential equations, and variational inequalities. Distributed control, boundary control, and minimax problems, etc. are discussed. The main tools are the Ekeland variational principle and the spike variation technique for different systems. The third part consists of Chapters 6 and 8. In this part, we present the Bellman dynamic programming method for optimal control of abstract evolution equations. A viscous solution in infinite dimensions is the main topic in these two chapters. The relation between the Pontryagin maximum principle and the viscosity solution to the Hamilton-Jacobi-Bellman equation is studied. The optimal switching and the impulse control problem are also discussed. The fourth part is Chapter 7, in which we discuss the controllability and the time optimal control problem. Linear as well as semilinear systems evolutions are treated. In the last part, Chapter 9, we discuss the linear-quadratic (LQ for short) optimal control problem. We concentrate on the parabolic problem with unbounded control. The general problem (not necessarily parabolic) with bounded control is also covered. The operators appearing in the cost functional are allowed to be indefinite. For the finite horizon case, it is shown that, in some sense, the solvabilities of the LQ problem, an operator Riccati equation and a Fredholm integral equation are equivalent. Then the infinite horizon case is briefly discussed. The stabilization of the system is naturally contained in this part.

Some words about the numbering convention. All the *heads* (by this we mean definitions, examples, lemmas, propositions, remarks, theorems, etc.) are numbered consecutively within each section of each chapter with the first number being the section number. For example, Theorem 3.2 is the second *head* in Section 3. When a *head* of another chapter is cited, the number of that chapter will be indicated each time. The equations are also numbered consecutively within each chapter with the first number being the section number. Thus, (4.5) refers to the fifth equation in Section 4.

The authors would like to acknowledge support from the following: the NSF of China, the Chinese State Education Commission NSF, and the Fok Ying Tung Education Foundation. For many years the authors benefited from their colleagues at Fudan University, to whom we would like to extend our thanks. Also, the following colleagues deserve special acknowledgment: Shuping Chen and Kangsheng Liu (Zhejiang University); Zuhao Chen and Shige Peng (Shandong University); Fulin Jin, Liping Pan, Shanjian Tang, and Yunlong Yao (Fudan University); and Jingyuan Yu (Beijing Institute of Information and Control). During the preparation of the book, the second author spent a year participating in the Control Year held at the Institute for Mathematics and Its Applications, University of Minnesota, U.S.A. He would like to thank Professor Avner Friedman for his kind invitation and partial support. During that period the second author benefited from many stimulating conversations, discussions and collaborations with other visitors. Among them, the following deserve special acknowledgment: Gang Bao, Eduardo Casas, Hector O. Fattorini, Scott W. Hansen, Suzanne M. Lenhart, Walter Littman, Wensheng Liu, Zhuangyi Liu, Jin Ma, Srdjan

Stojanovic, and Bing Yu Zhang. In his mathematical career, the second author has been deeply influenced mathematically by Professor Leonard D. Berkovitz, to whom he owes sincere gratitude and would like to dedicate the book. Finally, the authors would like to thank their family members for their understanding and patience during the long and tedious preparation of the book.

Xunjing Li and Jiongmin Yong<br>
Fudan University, China<br>
August 1994

# Chapter 1
# Control Problems in Infinite Dimensions

In this chapter we present some typical control problems in infinite dimensional spaces. Our purpose is to make the point that many physical problems need to be modeled in the infinite dimensional space framework, and to study these problems, many sophisticated mathematical tools are necessary. On the other hand, we will see that these problems make the infinite dimensional control theory very rich. In what follows, we do not attempt to exhaust all possible situations — this is of course not possible. Instead, we only present some combinations of different control systems with different types of control problems. It is important that the control actions be physically realizable. Thus, in the presentation of control problems, we need to specify the meaning of the particular control actions. To this end, some derivations for the state equations are necessary. Those who are already familiar with these derivations can skip some parts of this chapter.

Let us tell the readers that among the problems we present below, only some will be studied in great detail in the later chapters. Many other problems actually still lack complete theory for a solution and are awaiting further investigation.

## §1. Diffusion Problems

The main feature of this class of problems is that the governing equations will be parabolic or elliptic partial differential equations.

We begin with the heat transfer problem. Let $\Omega \subset \mathbb{R}^3$ be a domain in which we are considering the temperature distribution $\theta(t, x)$, which is a function of the time $t$ and the position $x$. Here, we let $t \in [0, T]$, a fixed time duration. The difference of the temperature at different points creates a heat flow (or heat flux):

$$(1.1) \qquad \mathbf{q}(t, x) = -K(t, x) \nabla \theta(t, x), \qquad (t, x) \in [0, T] \times \Omega,$$

where $K$ is a matrix-valued function, that represents the thermal conductivity of the material occupied in $\Omega$. The above (1.1) is usually referred to as *Fourier's law*. The negative sign basically means that the direction of the heat flow is opposite to the gradient of the temperature $\theta$. Thus, the heat flow moves from the position with a higher temperature to the position with a lower temperature. Now, we take a surface element $dS$ at some point $x \in \Omega$. Suppose its unit normal is $\nu$. Then, the amount of heat flowing through this piece of surface in the direction $\nu$, within the time duration $dt$, is given by (note (1.1))

$$(1.2) \qquad dQ = \mathbf{q} \cdot \nu \, dS \, dt = -(K \nabla \theta) \cdot \nu \, dS \, dt.$$

Next, we suppose the specific heat of the material in $\Omega$ is $c(x)$. Then, for any given small ball $B$ centered at $x_0 \in \Omega$, the total heat content in $B$ is given by $\int_B c(x)\theta(t,x)\,dx$. This quantity (as a function of $t$) is changed, in the time duration $dt$, by a flow of heat through the boundary $\partial B$ of $B$ and some *source* (or *sink*) $f$ inside of $B$. Thus, the total increase of the heat content in $B$ is given by

$$(1.3) \qquad dQ = \left( -\int_{\partial B} K\nabla\theta \cdot \nu\, dS + \int_B f\, dx \right) dt,$$

where $\nu$ stands for the inward normal of $\partial B$ with respect to $B$. Comparing with (1.2), by the conservation of the heat content, we obtain

$$(1.4) \qquad \begin{aligned} \int_B c\frac{\partial \theta}{\partial t}\,dx &= -\int_B K\nabla\theta \cdot \nu\, dS + \int_B f\, dx \\ &= \int_{\partial B} \nabla \cdot (K\nabla\theta)\, dx + \int_B f\, dx. \end{aligned}$$

Here, we have used *Green's formula*. Since the above holds for all small balls $B$, we obtain the following equation for the temperature $\theta$:

$$(1.5) \qquad c\theta_t - \nabla \cdot (K\nabla\theta) = f, \qquad \text{in } \Omega \times (0,T).$$

In the case where $c \equiv 1$, $K = I$, the identity matrix, and $f = 0$, we end up with the usual *heat equation* $\theta_t - \Delta\theta = 0$. Next, let us look at boundary conditions. There are several types of conditions. If the temperature is specified on the boundary $\partial\Omega$, we have

$$(1.6) \qquad \theta\big|_{\partial\Omega} = \varphi.$$

This is called the *Dirichlet (boundary) condition*. If the heat flux is given on the boundary, then we have

$$(1.7) \qquad \frac{\partial \theta}{\partial \nu}\bigg|_{\partial\Omega} = \varphi.$$

This is called the *Neumann (boundary) condition*. We may have more complicated boundary conditions. For example, if the heat flux on the boundary is proportional to the local temperature, we have the so-called *Robin (boundary) condition*:

$$(1.8) \qquad \left( \frac{\partial \theta}{\partial \nu} + \sigma\theta \right)\bigg|_{\partial\Omega} = 0.$$

If we are further given the initial temperature distribution $\theta_0(x)$, i.e.,

$$(1.9) \qquad \theta\big|_{t=0} = \theta_0(x), \qquad x \in \Omega,$$

then under proper conditions, we can solve equation (1.5) with the *initial condition* (1.9) and with one of the boundary conditions (1.6), (1.7), or

## §1. Diffusion Problems

(1.8). The solution $\theta(t,x)$ gives the temperature distribution at any point $x \in \Omega$ at time $t \in [0,T]$.

Now, let us consider some control problems related to the above equation. For definiteness, we first consider equation (1.5) with (1.9) and (1.6). Suppose we are able to change the source (or sink) $f$ in the domain $\Omega$. This means we can change the right-hand side of (1.5). Then, a different $f$ will give us a different solution $\theta$. Hence, for a desired temperature distribution $\bar{\theta}(t,x)$, we might try to choose a suitable $f$, so that the solution $\theta(t,x)$ is close to $\bar{\theta}(t,x)$ in some sense. This is the situation that results when we put some heat source in the house to raise the temperature in the winter or put an air conditioner in the house to lower the temperature in the summer. We usually refer to the $\theta$ as the *state* and to the $f$, which one can change, as the *control*. Equation (1.5) is called the *state equation*. In the above described situation, the control appears in the right-hand side of the state equation, or the control acts inside of the domain $\Omega$. We call such a control a *distributed control*. Sometimes we can change the temperature on the boundary $\partial\Omega$; namely, the $\varphi$ in (1.6) can be manipulated. In this case, $\varphi$ is called a *boundary control*. We can also define the same notions for the other boundary conditions.

It is seen that, say, equation (1.5) with conditions (1.6) and (1.9) gives a unique way (under proper conditions) of determining $\theta$ by $f$. We refer to such an object that specifies a relation between the control and the state as a *control system*. Thus, (1.5), (1.6), and (1.9) is a control system; likewise, (1.5), (1.7), and (1.9) is another control system.

Sometimes we want to achieve our goal in an optimal way, say at minimum time, or with minimum energy, etc. In this case, we have an *optimal control problem*. For an optimal control problem, a criterion to measure the performance of the control system should be specified. Such a criterion is referred to as a *cost functional*. Let us consider the above temperature control problem (1.5), (1.6), and (1.9), with the control being $f$. We want the temperature $\theta(t,x)$ to be close to $\bar{\theta}(t,x)$. Then, one may set the following cost functional:

$$(1.10) \quad J(f) = \int_0^T \int_\Omega |\theta(t,x) - \bar{\theta}(t,x)|^2 \, dxdt + \int_0^T \int_\Omega |f(t,x)|^2 \, dxdt.$$

Our goal is to minimize the above $J(f)$ by choosing $f$ properly. In the above, the first term requires the $\theta$ to be "close" to $\bar{\theta}$ and the second term requires that the energy spent should not be too much. We may have some other types of cost functionals. If $\varphi$ is also a control, we may have term $\int_0^T \int_{\partial\Omega} |\varphi(t,x)|^2 \, dxdt$ appearing in the cost functional, to penalize the "cost" spent on the boundary.

If instead, we want the temperature to be close to $\bar{\theta}(x)$ at some given moment $T$, then the cost functional could be

$$(1.11) \quad J(f) = \int_\Omega |\theta(T,x) - \bar\theta(x)|^2 \, dx + \int_0^T \int_\Omega |f(t,x)|^2 \, dxdt.$$

Now, suppose we start with $\theta_0(x)$ and want the temperature $\theta(t,x)$ to be close to $\bar\theta(x)$ as soon as possible. In this case, we can set the cost functional as follows. First, suppose we measure the closeness of the temperature by $L^2(\Omega)$ norm and let $\varepsilon > 0$ be a given accuracy. Let

$$(1.12) \quad Q = \{ y \in L^2(\Omega) \mid \|y - \bar\theta\|_{L^2(\Omega)} \leq \varepsilon \}.$$

Next, for any control $f$, let the corresponding temperature distribution be $\theta(t,x;f)$, indicating the dependence on $f$. Then, we define the cost functional to be

$$(1.13) \quad T(f) = \inf\{ t \geq 0 \mid \theta(t,\cdot\,;f) \in Q \}, \qquad (\inf \phi = +\infty).$$

Our goal is to minimize this cost functional. This kind of problem is referred to as a *time optimal control problem*.

Sometimes it is necessary to use some other ways of measuring the closeness. For example, if the temperature $\theta(T,x)$ needs to be close to $\bar\theta(x)$ pointwise, then we can pose the following cost functional:

$$(1.14) \quad J(f) = \max_{x \in \Omega} |\theta(T,x) - \bar\theta(x)|^2.$$

This is a maximum type cost functional that we want to minimize. Thus, we call such a problem a *minimax control problem*.

For many diffusion-reaction situations, for example, some chemical reaction processes, we may have similar state equations, in which the source term $f$ may depend on the state and/or the gradient of the state. The diffusion coefficient $K$ may also depend on the state and/or the gradient of the state. Hence, we will have general semilinear or quasilinear (even possibly fully nonlinear) *parabolic partial differential equations*. Similarly, the boundary conditions can also be nonlinear.

Next, let us look at the stationary case. For definiteness, we again consider (1.5), (1.6), and (1.9). Suppose the source (or sink) $f$ and the boundary temperature specification $\varphi$ are time independent. Then, intuitively (or under some conditions), as time $t \to \infty$, the temperature $\theta(t,x)$ should go to some equilibrium state $\widetilde\theta(x)$. This function, then, should satisfy the following equation:

$$(1.15) \quad -\nabla \cdot (K \nabla \widetilde\theta) = f, \qquad \text{in } \Omega.$$

This is an *elliptic equation*. We note that in the situation of heating the house with some heat source, the temperature distribution $\theta(t,x)$ is approximately time independent (actually, it is approximately a constant most of the time), after a sufficiently long time duration. Thus, the design of a heating system for a house can be approximately regarded as a problem with an elliptic equation like the above. Now, (1.15) together with some boundary conditions (like (1.6), (1.7) or (1.8)) will give us the control systems (i.e., relations between the control and the state). We can pose control problems for these control systems just as before. Of course, as these are time independent problems, we will not pose time optimal control problems for elliptic equations.

Again, in many other situations, $K$ and $f$ can depend on the state and/or the gradient of the state. This will give us semilinear or quasilinear elliptic state equations.

To conclude this section, let us mention another important issue. In the temperature control problem, we actually have some constraints for the state. In fact, while designing a heating system, we have to consider that the highest temperature over the whole house should not exceed a certain value, for example. This means, we have the following type of *state constraint* for $\widetilde{\theta}$:

$$(1.16) \qquad g(x, \widetilde{\theta}(x)) \leq 0, \qquad x \in \Omega,$$

for some function $g$. Sometimes, we need constraints like

$$(1.17) \qquad |\widetilde{\theta}(x_i) - b_i| \leq \varepsilon_i, \qquad 1 \leq i \leq m,$$

for some points $x_i \in \Omega$, the desired states $b_i \in \mathbb{R}$, and some accuracies $\varepsilon_i > 0$. We refer to these constraints as *pointwise state constraints*. In the discussion of the time optimal control problem above, we already had some kind of state constraint, namely $\theta(T, \cdot) \in Q$. This is usually referred to as a *terminal state constraint* because it imposes a restriction on the state at time $T$, the final moment.

## §2. Vibration Problems

Vibration phenomena are commonly seen in the real world. Easy examples are *strings*, *membranes*, *beams*, and *plates*. In many situations, we would like to control the behavior of these objects. The governing state equations for these problems will be *wave equations, beam equations, plate equations*, etc. In what follows, we will just write these equations with some illustrations; complicated derivations will be omitted. Corresponding possible control problems will be discussed.

We start with the following wave equation:

$$(2.1) \qquad y_{tt} - y_{xx} = f, \qquad (t, x) \in (0, T) \times (0, 1).$$

It is well known that this equation describes the displacement of a string under vibration. Here, $y(t, x)$ represents the displacement of the point $x$ on the string at time $t$, and $f(t, x)$ is the external force acting on $x$ at time $t$. The initial condition is usually given by

$$(2.2) \qquad y\big|_{t=0} = y_0(x), \qquad y_t\big|_{t=0} = y_1(x), \qquad x \in (0, 1).$$

The boundary conditions can have several forms. For example,

(2.3) $$y\big|_{x=0} = y\big|_{x=1} = 0, \qquad t \in [0,T],$$

(2.4) $$y\big|_{x=0} = 0, \qquad y_x\big|_{x=1} = 0, \qquad t \in [0,T].$$

Condition (2.3) means both ends of the string are fixed (clamped), and condition (2.4) means one end is clamped and the other is free. We may have other kinds of boundary conditions. Now, for definiteness, let us consider (2.1) with conditions (2.2) and (2.3). This is a control system if we take $f$ as a control. A basic control problem for such a system is the *stabilization*. Namely, we would like to choose a suitable control $f$ so that the displacement $y(t,x)$ together with its velocity $y_t(t,x)$ will go to zero as $t$ goes to infinity. Furthermore, we would like to see if it is possible to find a control so that at some time $t = T$, it actually holds that $y(T,x) = y_t(T,x) = 0$, $\forall x \in (0,1)$. If this happens, we say that the control system is *exactly null controllable*. We may also pose some optimal control problems. For example, if we want to stabilize the vibration with as little energy as possible (in some sense), then one can try to minimize a cost functional like the following:

(2.5) $$J(f) = \int_0^\infty \int_0^1 \{|y(t,x)|^2 + |y_t(t,x)|^2 + |f(t,x)|^2\} dx\, dt.$$

This is a quadratic type cost functional and the state equation is linear. Thus, such an optimal control problem is referred to as a *linear-quadratic* (LQ for short) problem.

Now, if we want the vibration to be close to a desired one, for example, in an instrument like a violin, then we may define the cost functional as

(2.6) $$J(f) = \int_0^T \int_0^1 \{|y(x,t) - \varphi(x,t)|^2 + |y_t(t,x) - \psi(t,x)|^2\}\, dx\, dt.$$

We may also impose some state constraints like the parabolic and elliptic cases. Again, it is possible to consider more general hyperbolic second order partial differential equations (PDEs) with more general boundary conditions. Boundary controls are also often considered.

Next, let us look at some other equations. An important class is the beam equations. There are several standard models. We briefly list them here. Consider a beam of unit length. Different models will be compared by their *kinetic energy* $K$ and *strain energy* $U$.

1. Euler–Bernoulli beam.

In this model, one takes the following forms of the kinetic energy $K$ and the strain energy $U$:

(2.7) $$K = \frac{1}{2}\int_0^1 \rho\{y_t\}^2\, dx, \qquad U = \frac{1}{2}\int_0^1 EI\{y_{xx}\}^2\, dx.$$

## §2. Vibration Problems

Here, $y(t,x)$ is the displacement of the beam at location $x$ (on the beam) at time $t$, $\rho = \rho(x)$ is the mass density of the beam, $E = E(x)$ is the Young's modulus of the beam material, and $I = I(x)$ is the cross-sectional moment of inertia. Now, suppose there is no externally applied force and there is also no work done on the beam by the internal dissipative mechanisms. Then, under proper boundary conditions, by *Hamilton's principle*, we can obtain the following *beam equation*:

$$\rho y_{tt} + \left(EIy_{xx}\right)_{xx} = 0. \tag{2.8}$$

As far as the boundary conditions are concerned, we have the following types of conditions (take the endpoint $x = 0$ as an example):

(i) *Clamped endpoint*:

$$y(t,0) = y_x(t,0) = 0. \tag{2.9}$$

(ii) *Hinged endpoint*:

$$y(t,0) = y_{xx}(t,0) = 0. \tag{2.10}$$

(iii) *Free endpoint*:

$$EIy_{xx}(t,0) = \left(EIy_{xx}\right)_x(t,0) = 0. \tag{2.11}$$

In the case where there is an external force $f(t,x)$ distributedly acting on the beam, equation (2.8) is replaced by the following nonhomogeneous one:

$$\rho y_{tt} + \left(EIy_{xx}\right)_{xx} = f. \tag{2.12}$$

Also, if we have lateral force $\varphi$ and lateral moment $\mu$ acting on the endpoint $x = 0$, then (2.11) is changed to

$$EIy_{xx}(t,0) = \varphi; \qquad \left(EIy_{xx}\right)_x(t,0) = \mu. \tag{2.13}$$

Clearly, $f, \varphi$, and $\mu$ can serve as the control variables; $f$ is a distributed control and $\varphi, \mu$ are boundary controls. It is possible to impose all the above-mentioned control problems for the Euler–Bernoulli beam.

2. *Rayleigh beam.*

This is a modification of the Euler–Bernoulli beam. In this case, one takes the $U$ as in (2.7) and $K$ as follows:

$$K = \frac{1}{2}\int_0^1 \rho\{y_t\}^2 + \frac{1}{2}\int_0^1 I_\rho\{y_{xt}\}^2, \tag{2.14}$$

where $I_\rho$ is the density of the mass moment of inertia about the neutral axis. As before, we may derive the following equation:

$$\rho y_{tt} - \left(I_\rho y_{ttx}\right)_x + \left(EIy_{xx}\right)_{xx} = f. \tag{2.15}$$

Boundary conditions similar to those for the Euler–Bernoulli beams can be imposed at the endpoints with one difference — the condition corresponding to the second condition in (2.13) should be changed to the following:

(2.16) $$\left(EIy_{xx}\right)_x(t,0) - \left(I_\rho y_{xt}\right)_t(t,0) = \mu.$$

### 3. Timoshenko beam.

In this case, the shearing is taken into account. The kinetic energy $K$ and the strain energy $U$ are taken as the following forms:

(2.17) $$\begin{cases} K = \dfrac{1}{2}\int_0^1 \{\rho\{y_t\}^2 + I_\rho\{\psi_t\}^2\}\,dx, \\ U = \dfrac{1}{2}\int_0^1 \{EI\{\psi_x\}^2 + k\{y_x - \psi\}^2\}\,dx. \end{cases}$$

Here, $\psi(t,x)$ is the sum of the bending angle and the shearing angle at location $x$ at time $t$ and $k = k(x)$ is the modulus of elasticity in shear. By Hamilton's principle, we can obtain the following Timoshenko equations:

(2.18) $$\begin{cases} \rho y_{tt} - \left(k\{y_x - \psi\}\right)_x = 0, \\ I_\rho \psi_{tt} - \left(EI\psi_x\right)_x + k\{\psi - y_x\} = 0. \end{cases}$$

Clearly, the above is a coupled system of two wave equations. If there are lateral force $\varphi$ and moment $\mu$ applying at $x = 0$, then

(2.19) $$k\{y_x - \psi\}\big|_{x=0} = \varphi, \qquad \left(EI\psi_x\right)\big|_{x=0} = \mu.$$

At this moment, we point out that all the control problems discussed before can be posed for this model of beam equations.

We may consider much more complicated equations, such as the Kirchhoff plate equation, the Mindlin–Timoshenko plate equation, and the von Karman plate equation. We omit the details here because of their complexity. Furthermore, it seems possible to study general control systems for elastic bodies. To our knowledge, many control problems for these equations are still open.

## §3. Population Dynamics

In this section, we look at population dynamics. Let us begin with some simple models.

Let $y(t)$ be the total population at time $t$ of a certain species. The most naive model for the growth of the population is given by

(3.1) $$y'(t) = \alpha y(t), \qquad t \in [0, \infty),$$

where $\alpha > 0$ is called the *growth rate*, which is the difference between the *birth rate* and the *death rate*. This model, known as the *Malthus model*, was

## §3. Population Dynamics

posed by Malthus in 1798. From (3.1), we see that the total population $y(t) = y_0 e^{\alpha t}$ will go to infinity exponentially as time $t$ goes to infinity, which is not possible because the living space and the food resources are limited. A more realistic model was posed by Verhulst in 1836. This is called the *logistic model* and it reads as follows:

(3.2) $$y'(t) = \alpha(1 - y(t)/K)y(t), \qquad t \in [0, \infty).$$

Here, $K > 0$ is called the *carrying capacity*. Clearly, if $y(t) > K$, then $y'(t) < 0$ and if $y(t) < K$, then $y'(t) > 0$. Thus, this model takes the environment limitation into account. It is intuitive that if $y(0) > 0$, then the total population $y(t)$ will approach $K$ as $t \to \infty$. This is actually the case from the following explicit form of the solution:

(3.3) $$y(t) = \frac{y(0)e^{\alpha t}}{1 + (y(0)/K)(e^{\alpha t} - 1)}, \qquad t \geq 0.$$

We note that in model (3.2), the term $\alpha(1 - y(t)/K)$ serves as the growth rate and the change of the total population instantaneously changes it. But, in many real situations, such a *biological self-regulator action* responds only after a certain time lag $r > 0$. Thus, a further modified model should be the following:

(3.4) $$y'(t) = \alpha(1 - y(t-r)/K)y(t), \qquad t \in [0, \infty).$$

This is a *delay equation*. Now, suppose that certain control actions $u(t)$ are applied, like putting in (or harvesting) the species. Then (3.4) becomes a control system

(3.5) $$y'(t) = \alpha(1 - y(t-r)/K)y(t) + u(t), \qquad t \in [0, \infty).$$

At this point, we may pose many interesting optimal control problems. For example, we may define the following cost functionals:

(3.6) $$J(u(\cdot)) = \int_0^T \{c_1|y(t) - z(t)|^2 - c_2 y(t) + c_3 u(t)^2\} \, dt.$$

To minimize the above amounts means we would like to keep the total population $y(t)$ close to some desired one $z(t)$ in a certain sense, and at the same time, we want to maximize the revenue from harvesting the species and minimize the total "energy" or cost in harvesting.

We point out that although the above problem is formulated in $\mathbb{R}$, it is actually an infinite dimensional problem. The reason is that the solution $y(\cdot)$ of (3.5) can only be determined if the value of $y(\cdot)$ is specified on $[-r, 0]$. It turns out that the state space for this problem should be $C([-r, 0])$ or $L^p(-r, 0) \times \mathbb{R}$. We will see this point later.

Delay equations appear in many other applications. As a matter of fact, delay represents some kind of "memory effect." In the above, the previous biological self-regulator action $\alpha(1 - y(t-r)/K)$ is "memorized" at the present time $t$. Clearly, we may think of very many situations in which a similar "memory effect" exists. One example is a process of mixing different solutions in a tank by pouring a specified solution into it. In these situations, one will end up with a control system governed by delay equations. We omit the details here.

Next, we consider the age and space dependent population models. These will give us partial differential equations.

Consider a domain $\Omega \subset \mathbb{R}^3$ in which a species is distributed. Let $\rho(t, x, a)$ be the age dependent density at location $x \in \Omega$ and at time $t \in [0, T]$. Here $a$ is the age. We then have the following dynamics for the evolution of the density:

$$\begin{aligned} \rho_t + \rho_a = \Delta \rho - m(t, x, a)\rho(t, x, a) - f(t, x, a), \\ t \in (0, T), \quad x \in \Omega, \quad a \geq 0. \end{aligned} \tag{3.7}$$

In the above, the left-hand side represents the total rate of change of the density with respect to the time and the age; on the right-hand side, the first term represents the diffusion, the second term represents the rate of change due to death ($m(t, x, a)$ is the *death rate*), and the third term represents the rate of change due to the exterior efforts, say, harvesting. The boundary conditions are given as follows:

$$\rho\big|_{a=0} = \int_0^\infty \beta(t, x, a)\rho(t, x, a) \, da, \qquad (t, x) \in (0, T) \times \Omega, \tag{3.8}$$

$$\frac{\partial \rho}{\partial \nu}\bigg|_{\partial \Omega} = 0. \tag{3.9}$$

Condition (3.8) gives the density of the species at age 0, which is nothing but the birth of the species ($\beta$ is the *birth rate*) and condition (3.9) means that the species is confined in $\Omega$ (no flux). The initial condition is given by

$$\rho\big|_{t=0} = \rho_0(x, a). \tag{3.10}$$

Then, under proper conditions, (3.7)–(3.10) admit a unique solution $\rho$. Regarding $f$ as a control, we obtain a control system, and we may now pose some control problems.

Suppose this species is a kind of harmful insect and we want to control it and eventually extinguish it. Then, we can try to choose $f$ so that

$$\lim_{t \to \infty} \rho(t, x, a) = 0. \tag{3.11}$$

Clearly, this is a *stabilization* problem. Further, we can try to find a control so that $\rho(T, x, a) = 0$ for some $T > 0$. This is then an *exact null-controllability* problem. Suppose our power is limited and we would like to

## §3. Population Dynamics

control the size of the species in some optimal way. Then we may define a cost functional as follows:

$$(3.12) \qquad J(f) = \int_0^\infty \int_\Omega \int_0^\infty \{|\rho(t,x,a)|^2 + |f(t,x,a)|^2\} da\, dx\, dt.$$

Next, suppose that we are growing this species by ourselves and we want to get a profit from harvesting it. Our goal is to maximize the total profit. In this case, we may define the following reward functional:

$$(3.13) \qquad J(f) = \int_0^T \int_\Omega \int_0^\infty \{-|f(t,x,a)|^2 + c(t,x,a)\rho(t,x,a)\} da\, dx\, dt,$$

and we want to maximize the above. In (3.13), the first term is the cost and the second term is the total income. Thus, $J(f)$ gives the total profit. If we are supposed to keep the resource of such a species at a certain level for the balance of the ecosystem, then, in the integrand of the reward functional, we should add a term like $-\alpha(t,x,a)\rho(t,x,a)^2$. This will prevent too much harvesting.

It is also possible to consider *multispecies systems*. In these cases, the state equation is a system of coupled partial differential equations. The coupling comes from the competition or predator-prey behavior among the different species. To illustrate the idea, let us just look at a two-species system. Consider the following system of equations:

$$(3.14) \qquad \begin{cases} y_t = \Delta y + (\alpha_1 - \beta_1 y)y + \gamma yz + f, \\ z_t = \Delta z + (\alpha_2 - \beta_2 z)z - \gamma yz + g. \end{cases}$$

This is a model of a *predator-prey system*. $y$ is the density of the predator and $z$ is the density of the prey. $\alpha_i, \beta_i, \gamma$ are supposed to be positive constants. Roughly speaking, we have the following feature for the system (3.14): An increase in the density for the predator $y$ will result in a decrease in the rate of the density of prey $z$; and an increase in the density of the prey $z$ will result in an increase in the rate for the density of the predator $z$. The interaction between the two species is represented by the term $\gamma yz$. Now, functions $f$ and $g$ can be regarded as the control (harvesting or put-into actions). Clearly, the above state equation is a system of semilinear parabolic equations. We may consider the problem in a bounded region $\Omega$ with proper boundary conditions, or in the whole space $\mathbb{R}^3$. As before, many kinds of control problems can be posed.

Now, let us look at a two-species *competition system*:

$$(3.15) \qquad \begin{cases} y_t = \Delta y + (\alpha_1 - \beta_1 y - \gamma_1 z)y + f, \\ z_t = \Delta z + (\alpha_2 - \beta_2 z - \gamma_2 y)z + g. \end{cases}$$

Roughly speaking, in the above state equation, an increase in the density of one species will result in a decrease in the rate for the density of the other. This shows the competition between two species. Thus, this model represents a quite different situation than that of (3.14). Also, the readers can pose many kinds of control problems.

From the above two models, we see that in general, we have the following type of semilinear parabolic systems as the state equation:

(3.16) $$y_t = \Delta y + F(t, x, y, u),$$

with $y$ taking values in $\mathbb{R}^N$. Also, if we take age into account, then the state equation becomes

(3.17) $$y_t + y_a = \Delta y + F(t, a, x, y, u).$$

Finally, in (3.16), let $F$ be independent of $t$. If this multispecies system has an equilibrium; that is, if all the states are time independent, then we end up with an elliptic system. Control problems then can also be posed for such a system of elliptic equations.

## §4. Fluid Dynamics

Let $\Omega \subset \mathbb{R}^3$ be a domain, which is occupied by a fluid (gas or liquid). We assume that the fluid is *homogeneous* (meaning that it is of uniform chemical composition throughout), *isotropic* (meaning that it behaves in the same way for all directions), and *Newtonian* (meaning that the *stress* and the *strain* are linearly related, which is a fair approximation for many fluids). Let $\rho(x,t)$, $v(x,t)$, and $p(x,t)$ be the density, the velocity, and the pressure of the fluid at the location $x \in \Omega$ and at the time $t$, respectively. By the conservation of mass, we have the following *continuity equation*:

(4.1) $$\rho_t + \nabla \cdot (\rho v) = 0.$$

Next, by Newton's second law of motion, we can derive the well-known Navier–Stokes equation:

(4.2) $$v_t + (v \cdot \nabla)v = \frac{F}{\rho} - \frac{1}{\rho}\nabla p + \frac{\mu}{\rho}\Delta v + \frac{\mu}{3}\nabla(\nabla \cdot v).$$

In the above, $v_t$ is the *local acceleration* of the fluid particle at a fixed point in the space; $(v \cdot \nabla)v$ is the *convective acceleration* of the fluid particle, and it predicts how the flow differs from one space location to the next at the same instant of time; $F$ is the total external force (for example, gravity); $-\frac{1}{\rho}\nabla p$ is the *pressure acceleration* due to the "pumping" action of the flow; $\frac{\mu}{\rho}\Delta v$ is the *viscous deceleration* due to the fluid's frictional resistance to objects moving through it; $\frac{\mu}{3}\nabla(\nabla \cdot v)$ is the acceleration due to the dilatation ($\nabla v$ is called *dilatation*). The parameter $\mu$ is called the *viscosity* of the fluid.

We have seen that (4.1) and (4.2) consist of four equations and five unknown functions ($\rho$, $p$, and $v = (v^1, v^2, v^3)$). Thus, (4.1) and (4.2) are not complete. They are supplemented by the following *equation of state*:

(4.3) $$\rho = \rho(p).$$

## §4. Fluid Dynamics

Then, (4.1)–(4.3) become a complete system of equations. We should note that, in general, it is not easy to solve such a system of equations. Also, we should point out that if the temperature and the internal energy are taken into account, then we will have one more differential equation (called the *energy equation*, which is a little more complicated) and one more equation of state. We shall not give the details here.

Now, let us look at some special cases of the above.

### 1. Incompressible fluid.

For such a fluid, the density $\rho$ is a constant. Thus, (4.3) is dropped out and by (4.1), we see that $v$ is divergence free, i.e.,

$$\nabla \cdot v = 0. \tag{4.4}$$

In this case, (4.2) becomes the following:

$$v_t + (v \cdot \nabla)v = \frac{F}{\rho} - \frac{1}{\rho}\nabla p + \nu \Delta v, \tag{4.5}$$

where $\nu = \mu/\rho$, which is called the *kinematic viscosity*. By normalizing, one may let $\rho = 1$. Then, we have the most familiar form of the Navier–Stokes equation:

$$\begin{cases} v_t + (v \cdot \nabla)v = F - \nabla p + \nu \Delta v, & \text{in } \Omega \times (0, T), \\ \nabla \cdot v = 0, & \text{in } \Omega \times (0, T). \end{cases} \tag{4.6}$$

### 2. Inviscid fluid.

An inviscid fluid is a fluid for which the viscosity vanishes, i.e., $\mu = 0$. Sometimes, such a fluid is called an ideal fluid. Clearly, in this case, (4.2) reads

$$v_t + (v \cdot \nabla)v = \frac{F}{\rho} - \frac{1}{\rho}\nabla p. \tag{4.7}$$

We see that this is a first order partial differential equation. The above equation is also called the *Euler equation*.

### 3. Stokes flow.

Consider a fluid in very slow motion. Then, the total acceleration $v_t + (v \cdot \nabla)v$ of the fluid particle and the acceleration $F/\rho$ due to the external force are negligible. Thus, we may set them to be zero and rewrite (4.2) as follows:

$$\nabla p = \mu \Delta v. \tag{4.8}$$

Any flow satisfying (4.8) is called the *Stokes flow*. Sometimes, such a flow is also called the *creep flow*.

**4. Steady flow.**

If the flow is time invariant, we call it a *steady flow*. In this case, (4.2) is of the following form:

$$(4.9) \qquad (v \cdot \nabla)v = \frac{F}{\rho} - \frac{1}{\rho}\nabla p + \frac{\mu}{\rho}\Delta v + \frac{\mu}{3}\nabla(\nabla \cdot v).$$

In addition, if the fluid is incompressible, then (4.9) can be further reduced to (again let $\rho \equiv 1$)

$$(4.10) \qquad (\nabla \cdot v)v = F - \nabla p + \nu \Delta v.$$

Also, if the flow is inviscid, we have

$$(4.11) \qquad (\nabla \cdot v)v = \frac{F}{\rho} - \frac{1}{\rho}\nabla p.$$

Now, let us concentrate on the viscous incompressible fluid. Thus, we have (4.6). To solve such a system, we need to impose some initial and boundary conditions. Let the initial condition be:

$$(4.12) \qquad v(x, 0) = v_0(x), \qquad x \in \Omega.$$

We impose the boundary condition as follows:

$$(4.13) \qquad v(x,t) = u(x,t), \qquad (x,t) \in \partial\Omega \times (0,T),$$

with

$$(4.14) \qquad \int_{\partial\Omega} u(x,t) \cdot dS = 0, \qquad t \in (0,T).$$

Compatible condition (4.14) follows from the incompressibility of the fluid (see (4.4)) and Green's formula). We may regard (4.13) as the control manner of suction and blowing at the boundary. This kind of control action is very important and physically realizable in many engineering problems involving fluid dynamics.

As before, we can impose many control problems. Consider the cost functional

$$(4.15)$$
$$J(u(\cdot)) = \int_{\Omega \times (0,T)} |v(x,t) - z(x,t)|^2 \, dx \, dt + \int_{\partial\Omega \times (0,T)} |u(x,t)|^2 \, dx \, dt.$$

To minimize the above functional amounts means that we want the velocity field of the flow to be close to some given vector field with some sort of minimal energy. Also, we may consider the cost functional

$$(4.16) \quad J(u(\cdot)) = \int_{\Omega \times (0,T)} |\nabla \times v(x,t)|^2 \, dx \, dt + \int_{\partial\Omega \times (0,T)} |u(x,t)|^2 \, dx \, dt.$$

## §5. Free Boundary Problems

We know that by definition $\nabla \times v$ is the *vorticity* of the flow. Thus, minimizing the above cost functional enables us to control the size of the vorticity of the flow in some sense.

Of course, many other cost functionals are possible. For example, the following is of interest:

$$
(4.17) \quad J(u(\cdot)) = \int_{\Omega \times (0,T)} |\nabla v(x,t) - \nabla z(x,t)|^2 \, dx \, dt \\
+ \int_{\partial\Omega \times (0,T)} g(u(x,t), u_t(x,t)) \, dx \, dt.
$$

We prefer not to get into details. Interested readers are suggested to look at the works by Fattorini–Sritharan [3].

We may also consider some other control problems for fluid dynamics, such as stabilization, controllability, etc.

## §5. Free Boundary Problems

In real applications, there is a class of problems called *free boundary problems*. In this section, we are going to present some of them as well as pose associated control problems.

To begin, let us consider the following situation. Let $\Omega \subset \mathbb{R}^2$ be a bounded domain with a smooth boundary $\partial\Omega$ and $\psi : \Omega \to \mathbb{R}$, $g : \partial\Omega \to \mathbb{R}$ be two given smooth functions with the compatible condition

$$(5.1) \quad \psi \leq g, \quad \text{on } \partial\Omega.$$

The set $\{(x,z) \in \Omega \times \mathbb{R} \mid z \leq \psi(x)\}$ defines a body in $\mathbb{R}^3$. We call this body $\Psi$. Now, consider a membrane that occupies $\Omega$, lies above the body $\Psi$, and is equally stretched in all directions by a uniform tension and loaded by a normal uniformly distributed force $f$. We note that in classical elasticity, a *membrane* is a thin plate offering no resistance to bending, but acting only in tension. We assume that the displacement of the membrane at point $x \in \Omega$ is $y(x)$, and on the boundary $\partial\Omega$, the displacement of the membrane is given by $g(x)$. Namely,

$$(5.2) \quad y(x) = g(x), \quad x \in \partial\Omega.$$

It is assumed that the potential energy of the deformed membrane is proportional to the increase of its surface area. It is not hard to see that the increment of the surface area is given by the following:

$$(5.3) \quad \int_\Omega \{\sqrt{1 + |\nabla y|^2} - 1\} \, dx \approx \frac{1}{2} \int_\Omega |\nabla y|^2 \, dx.$$

Thus, the potential energy of deformation can be represented by

$$(5.4) \quad D(y) = c_0 \int_\Omega |\nabla y|^2 \, dx,$$

for some constant $c_0 > 0$. After normalization, we may let $c_0 = 1/2$. On the other hand, the work done by the external force $f$ during the actual displacement is given by

$$F(y) = \int_\Omega fy \, dx. \tag{5.5}$$

Hence, the total *potential energy* will be

$$E(y) = \frac{1}{2} \int_\Omega |\nabla y|^2 \, dx - \int_\Omega fy \, dx. \tag{5.6}$$

Next, we introduce the following set of admissible displacements:

$$\mathbb{K} \equiv \{v \in V \mid v \geq \psi \text{ in } \Omega, \ v = g \text{ on } \partial\Omega\}. \tag{5.7}$$

Here, $V$ is the space of all functions with finite energy of deformation. (Thus, $V$ actually coincides with the *Sobolev space* $H^1(\Omega)$). Then, by the principle of energy minimization, we know that the true displacement distribution $y(x)$ of the membrane satisfies the following:

$$y \in \mathbb{K}; \quad E(y) \leq E(v), \quad \forall v \in \mathbb{K}. \tag{5.8}$$

The above is called a *variational inequality*. We call the function $\psi$ the *obstacle* for the obvious reason. Consequently, the above variational inequality is also referred to as the *obstacle problem* (we should know that there are many other variational inequalities that are not obstacle problems).

Now, let us derive the equation that $y$ satisfies. To this end, we note that $\mathbb{K}$ is convex. Thus, for any $v \in \mathbb{K}$ and $\lambda \in [0,1]$, $y + \lambda(v-y) \in \mathbb{K}$. Then, by (5.8), we see that

$$\begin{aligned} 0 \geq{} & E(y + \lambda(v-y)) - E(y) \\ ={} & \frac{1}{2} \int_\Omega \{|\nabla y + \lambda \nabla(v-y)|^2 - |\nabla y|^2\} \, dx - \lambda \int_\Omega f\{v-y\} \, dx \\ ={} & \lambda \int_\Omega \{\nabla y \cdot \nabla(v-y) - f\{v-y\}\} \, dx + \frac{\lambda}{2} \int_\Omega |\nabla(v-y)|^2 \, dx. \end{aligned} \tag{5.9}$$

Dividing by $\lambda$ and sending $\lambda \to 0$, we obtain

$$\int_\Omega \nabla y \cdot \nabla(v-y) \, dx \geq \int_\Omega f\{v-y\} \, dx, \quad \forall v \in \mathbb{K}. \tag{5.10}$$

Conversely, if $y \in \mathbb{K}$ satisfies (5.10), then, for any $v \in \mathbb{K}$,

$$\begin{aligned} E(v) ={}& E(y + (v-y)) = E(y) + \int_\Omega \nabla y \cdot \nabla(v-y) \, dx \\ & - \int_\Omega f\{v-y\} \, dx + \frac{1}{2} \int_\Omega |\nabla(v-y)|^2 \, dx \geq E(y). \end{aligned} \tag{5.11}$$

## §5. Free Boundary Problems

This means that $y$ is also a solution of (5.8). Hence, (5.8) and (5.10) are equivalent. Let us assume that (5.10) admits a smooth solution $y$. We want to make some further observations. To this end, we introduce the following:

$$\Omega_0 \equiv \{x \in \Omega \mid y(x) = \psi(x)\}, \tag{5.12}$$

$$\Omega^+ \equiv \{x \in \Omega \mid y(x) > \psi(x)\}. \tag{5.13}$$

The sets $\Omega_0$ and $\Omega^+$ are called the *coincidence set* and the *noncoincidence set*, respectively. Because both $y$ and $\psi$ are continuous (by our assumption), the noncoincidence set $\Omega^+$ is open. Then, for any $\varphi \in C_0^\infty(\Omega^+)$, we can find an $\varepsilon_0 > 0$, such that for all $\varepsilon \in (0, \varepsilon_0]$, $y \pm \varepsilon\varphi \in \mathbb{K}$. Thus, by (5.9),

$$0 \leq \pm\varepsilon \int_\Omega (\nabla y \cdot \nabla \varphi - f\varphi)\, dx = \pm\varepsilon \int_\Omega (-\Delta y - f)\varphi\, dx. \tag{5.14}$$

This implies that

$$-\Delta y = f, \quad \text{in } \Omega^+. \tag{5.15}$$

Finally, for any $\varphi \in C_0^\infty(\Omega)$, $\varphi \geq 0$, we have $y + \varphi \in \mathbb{K}$. Thus, by (5.10),

$$\int_\Omega (-\Delta y - f)\varphi\, dx \geq 0. \tag{5.16}$$

Hence,

$$-\Delta y - f \geq 0, \quad \text{a.e. } x \in \Omega. \tag{5.17}$$

Combining (5.15) and (5.17), we obtain

$$\begin{cases} y - \psi \geq 0, \quad -\Delta y - f \geq 0, \\ (y - \psi)(-\Delta y - f) = 0, \end{cases} \quad \text{a.e. } x \in \Omega. \tag{5.18}$$

Conversely, if $y$ satisfies (5.18), then, for any $v \in \mathbb{K}$,

$$\int_\Omega \nabla y \cdot \nabla(v - y)\, dx = -\int_\Omega (\Delta y)(v - y)\, dx \tag{5.19}$$
$$= \int_{\Omega^+} f\{v - y\}\, dx - \int_{\Omega_0} (\Delta y)(v - y)\, dx \geq \int_\Omega f\{v - y\}\, dx.$$

Hence, in some sense, (5.18) is also equivalent to (5.8). For this reason, people usually identify (5.8), (5.10), and (5.18). We note that (5.18) can also be written as follows:

$$\min\{-\Delta y - f, y - \psi\} = 0, \quad \text{a.e. } x \in \Omega. \tag{5.20}$$

Let us show another way of expressing the above variational inequality. We define a *multifunction* $\beta$ as follows:

(5.21) $$\beta(r) = \begin{cases} 0, & r > 0, \\ (-\infty, 0], & r = 0, \\ \phi, & r < 0. \end{cases}$$

Then, (5.18) can be written as follows

(5.22) $$-\Delta y + \beta(y - \psi) \ni f.$$

Sometimes, such a representation is convenient.

It is seen that the boundary of the coincidence set $\Omega_0$ in $\Omega$ is given by $\partial\Omega_0 \cap \Omega = \partial\Omega^+ \cap \Omega$. This boundary is not a priori known. For this reason, we call it the *free boundary* of the problem. This free boundary is a part of the solution to our original variational inequality problem. Let us take another look at the above problem. Suppose $y$ and $\psi$ are smooth. Because $y - \psi$ attains its minimum on the coincidence set $\Omega_0$, we have $\nabla y = \nabla \psi$. Thus, $y$ can be regarded as a solution of the following problem:

(5.23) $$\begin{cases} -\Delta y = f, & \text{in } \Omega^+, \\ y = g, & \text{on } \partial\Omega, \\ \left. \begin{array}{l} y = \psi, \\ \dfrac{\partial y}{\partial n} = \dfrac{\partial \psi}{\partial n}, \end{array} \right\} & \text{on } \partial\Omega^+ \cap \Omega = \partial\Omega_0 \cap \Omega. \end{cases}$$

It is known that if $\Omega^+$ is a given subdomain of $\Omega$, the above problem will be ill posed, in general, because there are two conditions imposed on the boundary $\partial\Omega^+ \cap \Omega$; the problem is overdetermined. In the present case, however, the boundary $\partial\Omega^+ \cap \Omega$ is "free" to choose so that the problem has a solution. Thus, this boundary is also a part of the solution and the *free boundary* is a natural name for it. Problems with free boundaries are usually referred to as *free boundary problems*.

Now, we consider some interesting optimal control problems. First of all, we may take $f$ and/or $g$ as the control(s). This means that we are designing the shape of the membrane by choosing a suitable external force load $f$ and/or boundary displacement $g$. In this case, we may try to minimize the following cost functional:

(5.24) $$J(f,g) = \int_\Omega |y - z|^2 \, dx,$$

where $z$ is the desired shape. By doing that, we will make the shape of the membrane close to the desired shape $z$ in some average sense. We may also pose the following cost functional:

(5.25) $$J(f,g) = \max_{x \in \Omega} |y(x) - z(x)|.$$

## §5. Free Boundary Problems

Minimizing such a cost functional will enable us to find a shape in which the maximum displacement from the desired shape $z$ is minimized.

There is another very unique optimal control problem for the free boundary problems. Let us explain this now. Suppose that we want the coincidence set $\Omega_0 \equiv \Omega_0(f,g)$ of the designed membrane to be as close to a given set, say, $E \subset \Omega$, as possible. Then, we need to pose the following type cost functional:

$$(5.26) \qquad J(f,g) = \int_\Omega |\chi_{\Omega_0} - \chi_E|^2 \, dx \equiv |\Omega_0 \setminus E| + |E \setminus \Omega_0|,$$

where $|S|$ denotes the Lebesgue measure of the set $S$. But, we should know that the dependence of $\Omega_0$ on $(f,g)$ is not good enough for us to study the cost (5.26) directly. One way to treat the cost is to introduce some approximations of it. The following is one of the approximations:

$$(5.27) \qquad J(f,g) = \int_\Omega \left| \frac{\alpha}{y - \psi + \alpha} - \chi_E \right|^2 dx,$$

where $\alpha > 0$ is a small number. We see that for any $\delta > 0$,

$$(5.28) \qquad 0 \le \frac{\alpha}{y - \psi + \alpha} \begin{cases} = 1, & \text{on } \Omega_0 \equiv \{y = \psi\}, \\ \le \frac{\alpha}{\delta}, & \text{on } \{y - \psi \ge \delta\}. \end{cases}$$

Thus, for $\alpha$ very small (compare with $\delta$), (5.27) is a good approximation of (5.26).

It is possible to take the obstacle $\psi$ as another control variable. Then, we can again formulate all the control problems as above.

Next, let us list some other interesting free boundary problems that can be taken as our state equations of optimal control problems. We will not give the details of the derivations to avoid some lengthy and complicated computations.

### 1. Bending of a plate over an obstacle.

Let $\Omega \subset \mathbb{R}^2$ be occupied by the plate whose thickness is $h > 0$. It lies over the obstacle $\psi$. We still let the displacement of the plate be $y(x)$. Again, the external force is given by $f$. Let the plate be clamped on the boundary:

$$(5.29) \qquad y = 0, \quad \text{on } \partial \Omega.$$

We introduce the set

$$(5.30) \qquad \mathbb{K}_1 \equiv \{v \in H_0^2(\Omega) \mid v = \frac{\partial v}{\partial n} = 0, \text{ on } \partial \Omega \}.$$

Then, the state equation can be written as follows:

$$(5.31) \quad y \in \mathbb{K}_1: \quad \varepsilon \int_\Omega \Delta y \Delta(v - y) \, dx + \int_\Omega \nabla y \cdot \nabla(v - y) \, dx \ge \int_\Omega f\{v - y\} \, dx, \quad \forall v \in \mathbb{K}_1.$$

Here $\varepsilon > 0$ is the normalized rigidity parameter of the plate. We may write the above equation as the following:

$$\varepsilon \Delta^2 y - \Delta y + \beta(y - \psi) \ni f, \qquad \text{in } \Omega, \tag{5.32}$$

with $\beta$ defined by (5.21). For the control problem, we again can take $f$ as the control variable.

*2. Stefan problem.*

Consider a domain $\Omega \subset \mathbb{R}^3$ that is occupied by water and ice. Let $\theta(x,t)$ be the temperature distribution. We assume that at any time $t \geq 0$, the domain $\Omega$ is split into two subdomains $\Omega_t^0$ (occupied by the ice) and $\Omega_t^+$ (occupied by the water). The boundary between the two subdomains is called $\Gamma_t$. Due to heat conduction, the domains $\Omega_t^+$ and $\Omega_t^0$ are changing, or equivalently, the boundary $\Gamma_t$ is *moving*. For this reason, $\Gamma_t$ is sometimes called the *moving boundary*, another name for the free boundary. The fact that the boundary $\Gamma_t$ is moving means that the water and the ice are under the process of exchange. In engineering, the change from water to ice or from ice to water is called a *phase transition*. This happens in many other situations, such as solidification of steel, etc. It is well known that due to the latent heat, when the water becomes ice at $\theta = 0$, a certain amount of heat will be released and when the ice becomes water at $\theta = 0$, a certain amount of heat will be absorbed. Such a phenomenon makes the equation for the temperature a little more complicated than that of the usual heat conduction (see §1). We assume, for simplicity, that the temperature $\theta = 0$ on $\Omega_t^0$. In this case, it turns out that after some normalization, the function $\theta$ satisfies the following:

$$\begin{cases} \theta_t - \Delta\theta \geq 0, \quad \theta \geq 0, \\ (\theta_t - \Delta\theta)\theta = 0, \end{cases} \quad \text{a.e. } (x,t) \in \Omega \times [0,T]. \tag{5.33}$$

This is an evolutionary variational inequality. Actually, it is an *evolutionary obstacle problem* with the obstacle being 0. This problem is usually called the *one-phase Stefan problem*. The name "one-phase" comes from the assumption that $\theta = 0$ in the ice and it is enough just to consider the water part (comparing the two-phase case below). It is also possible for us to write (5.33) as follows:

$$\theta_t - \Delta\theta + \beta(\theta) \ni 0. \tag{5.34}$$

Again, $\beta$ is given by (5.21). Now, let $\Gamma_t$ be represented by $t = \ell(x)$. Then, we can show that $\theta$ satisfies the following:

$$\begin{cases} \theta_t - \Delta\theta = 0, & \text{in } \{(x,t) \in \Omega \times [0,T] \mid \ell(x) < t\}, \\ \theta = 0, & \text{in } \{(x,t) \in \Omega \times [0,T] \mid t \leq \ell(x)\}, \\ \nabla\theta \cdot \nabla\ell = -1, & \text{on } \bigcup_{t \geq 0} \Gamma_t. \end{cases} \tag{5.35}$$

We see again that if $\ell(\cdot)$ is given, the above problem is overdetermined and it is ill posed, in general. For the same reason as before, the boundary $\Gamma_t$ is called the *free boundary*. The boundary conditions on $\partial\Omega$ and the initial condition on $t = 0$ can be imposed as usual. We note that if there is a heat source or sink $f$ in $\Omega$, the right-hand side of (5.34) will become $f$, which can be taken as a control variable. Also, a nonzero Dirichlet or Neumann condition could be imposed. There could be other types of controls. Actually, we see that the situation here is very similar to that in §1, except that we have a "heat equation" with a free boundary.

On the other hand, we know that in the real situation, the temperature of the ice could be below 0. If we take this issue into account, then the situation is more complicated. In this case, after normalization, the temperature $\theta$ satisfies the following:

$$(5.36) \qquad (\gamma(\theta))_t - \Delta\theta = 0, \qquad \mathcal{D}'(\Omega),$$

where $\gamma$ is a multifunction given by the following:

$$(5.37) \qquad \gamma(r) = \begin{cases} r+1, & r > 0, \\ [0,1], & r = 0, \\ r, & r < 0. \end{cases}$$

Again, this is a free boundary problem. But, we should note that this is not a variational inequality. The above problem is called the *two-phase Stefan problem*.

Clearly, various optimal control problems can be posed both for one-phase and two-phase Stefan problems. We will not get into further details.

*3. Other free boundary problems.*

There are very many other free boundary problems that can be taken as our state equations. For the simplicity of our presentation, we just list a very few names without giving the equations.

(i) *Continuous casting problem:* This describes the process of solidification of melted steel. It is basically a variant of the evolutionary or stationary Stefan problem. The optimal control goal can be focused on optimization of the quality of the steel (by controlling the temperature and the location of the free boundary), minimizing the energy loss, etc.

(ii) *Dam problem:* This describes filtration through a porous medium. This a variational inequality. The control goal can be the optimal design of the shape for the dam so that the pressure acting on the dam caused by the fluid is minimized, etc.

(iii) *Electrochemical machining:* This describes the process of metal shaping by electrochemical manners. The governing equations involve Maxwell's equations with free boundary. The control goal is obviously the desired shape and the minimization of time and the cost.

We have left out many other problems (see remarks below).

**Remarks**

The optimal control problem for the heat equation (or more generally the parabolic equations) is probably one of the earliest studied infinite dimensional control problems (see Butkovsky–Lerner [1]). The derivation in §1 is almost standard. Our presentation gives readers an impression of how we approach a real problem: finding a mathematical model according to proper physical laws, deriving the differential equations, and — most importantly from the control theory point of view — identifying the realizable control manners in the models. For related control problems, we refer the readers to the books by Ahmed–Teo [3], Barbu–Precupanu [1], Bensoussan-Da Prato-Delfour-Mitter [1], Butkovsky [2], Curtain–Pritchard [3], Fleming [1], Lasiecka–Triggiani [1], Lions [2–5], Trölzsch [1], and Tsien–Song [1]. These books also cover many control problems with other types of state equations, including some of those discussed below.

In §2, we present another important class of problems. Some of the material is taken from Fung [1] and Russell [4]. We refer the readers to Lagnese [2], and Lagnese–Lions [1] for the problems involving various models of plate and some related control problems. Other related works are G. Chen [1,2], Chen–Russell [1], Kwan–Wang [1], K. S. Liu [1], Liu-Huang-Chen [1], and Z. Liu [1].

In §3, a very general model of population dynamics is presented. Some of the material or ideas are taken from Fife [1] and Murray [1]. There is an enormous amount of literature devoted to the model and analysis of population dynamics. We are only able to mention a few: Langlais [1], and MacCamy [1], Huang–Yong [1], etc. For related control problems, see Song–Yu [3], Lenhart–Yong [1], Lenhart–Bhat [1], Haurie [1], Haurie-Sethi-Hartl [1], Anita [1], Bhat-Huffaker-Lenhart [1,2], Lenhart–Protopopopescu [1], Lenhart-Protopopopescu-Stojanovic [1,2], and Leung–Stojanovic [2].

Flow control has many applications in engineering. The easiest visible example is the control of planes in the air (see Gal-el-Hak [1]) or ships at the sea. However, the mathematical model for such a problem is complicated and there is a lack of complete theory; the results for the corresponding control problem are far from satisfactory. Some of the materials in §4 are taken from Granger [1], Temam [1], and Fattorini–Sritharan [1–3]. For other related works, see Abergel–Temam [2], Abergel–Casas [1], Gunzburger-Hou-Svobodny [1], Lenhart [1], Pironneau [1,2], and Sritharan [1].

In §5, the free boundary problems are briefly presented. Interested readers are suggested to consult the books by Diaz [1], Elliott–Ockendon [1], Friedman [4], and Rodrigues [1] for many details and comprehensive lists of references in that field. The first work on the optimal control of variational inequalities (free boundary problems) was probably that of Mignot [1] in 1976. Later, many authors made contributions to this topic. Among them, we are only able to mention the works we know: Saguez [1], Barbu [1,5], Mignot–Puel [1], Tiba [1,2], Friedman [5,6], Friedman-Huang-Yong [1,2], Friedman–Hoffman [1], Barbu–Friedman [1], Barbu–Tiba [1], Neittaanmäki-Sokolowski-Zolesio [1], and Shi [1].

There is much literature dealing with control problems for other types of state equations. We are only able to mention a very small portion: Kime [1], Lagnese [1], Russell [3] for Maxwell's equations; Yong–Zheng [1,2] for the Cahn–Hilliard equation and the phase field equations; Komornik-Russell-Zhang [1], Russell–Zhang [1] for the KdV equation.

# Chapter 2

# Mathematical Preliminaries

In this chapter we recall some basic concepts and results that are necessary for the presentation of the theories in later chapters. Most proofs for the standard results will be omitted.

## §1. Elements in Functional Analysis

In this section, some basic results of functional analysis are collected. We assume that the readers have some elementary knowledge of linear algebra and real analysis.

### §1.1. Spaces

We begin with the following notion.

**Definition 1.1.** Let $X$ be a linear space over $F$ ($F = \mathbb{R}$ or $\mathbb{C}$).

(i) A map $\varphi : X \to \mathbb{R}$ is called a *norm* on $X$ if it satisfies the following:

(1.1) $\begin{cases} \varphi(x) \geq 0, \quad \forall x \in X; \quad \varphi(x) = 0 \iff x = 0; & \text{(positivity)} \\ \varphi(\alpha x) = |\alpha|\varphi(x), \quad \forall \alpha \in F, x \in X; & \text{(positive-homogeneity)} \\ \varphi(x+y) \leq \varphi(x) + \varphi(y), \quad \forall x, y \in X. & \text{(triangle-inequality)} \end{cases}$

(ii) A map $\psi : X \times X \to F$ is called an *inner product* on $X$ if it satisfies the following:

(1.2) $\begin{cases} \psi(x,x) \geq 0, \quad \forall x \in X; & \text{(positivity)} \\ \psi(x,x) = 0 \iff x = 0; & \\ \psi(x,y) = \overline{\psi(y,x)}, \quad \forall x, y \in X; & \text{(symmetry)} \\ \psi(\alpha x + \beta y, z) = \alpha \psi(x,y) + \beta \psi(y,z), & \\ \quad \forall \alpha, \beta \in F, x, y, z \in X. & \text{(linearity)} \end{cases}$

In (1.2), $\overline{\psi(y,x)}$ is the complex conjugate of $\psi(y,x)$. (If $F = \mathbb{R}$, the bar can be omitted).

Hereafter, we denote a norm on $X$ (if it exists) by $|\cdot|$. Sometimes, $|\cdot|_X$ is used to indicate the norm defined on $X$. Similarly, we denote an inner product on $X$ (if it exists) by $(\cdot,\cdot)$ or $\langle \cdot,\cdot \rangle$, and by $(\cdot,\cdot)_X$ or $\langle \cdot,\cdot \rangle_X$ if the underlying space $X$ needs to be emphasized. If $X$ has a norm (in this case, we also say that $X$ is *endowed* with the norm $|\cdot|$), $(X, |\cdot|)$ is called a *normed linear space*. In a normed linear space $(X, |\cdot|)$, there exists a *topology* induced by the norm $|\cdot|$. Namely, we may talk about the

§1. Elements in Functional Analysis

convergence of a sequence $\{x_n\}$ in $X$. More precisely, let us introduce the following:

**Definition 1.2.** Let $X$ be a normed linear space with the norm $|\cdot|$. We say that a sequence $\{x_n\}$ *strongly converges* to $x \in X$ if

$$(1.3) \qquad \lim_{n\to\infty} |x_n - x| = 0.$$

**Definition 1.3.** A normed linear space $(X, |\cdot|)$ is called a *Banach space* if it is *complete*, i.e., for any sequence $\{x_n\} \subset X$ satisfying

$$(1.4) \qquad \forall \varepsilon > 0, \exists n_0, \text{ such that } |x_n - x_m| < \varepsilon, \quad \forall n, m \geq n_0,$$

there exists an $x \in X$, such that $\{x_n\}$ strongly converges to $x$.

Any sequence $\{x_n\}$ satisfying (1.4) is called a *Cauchy sequence*. The following result gives a possible relationship between the norm and the inner product.

**Proposition 1.4.** (i) *Let $X$ be a linear space over $F$ with an inner product $(\cdot,\cdot)$. Then, the map defined by $|x| = \sqrt{(x,x)}$, $\forall x \in X$, is a norm on $X$.*

(ii) *Let $X$ be a normed linear space over $F$ with the norm $|\cdot|$. Let $|\cdot|$ satisfy the following parallelogram law:*

$$(1.5) \qquad |x+y|^2 + |x-y|^2 = 2(|x|^2 + |y|^2), \qquad \forall x, y \in X.$$

*Then, there exists an inner product $\psi : X \times X \to F$ such that $|x| = \sqrt{\psi(x,x)}$, $\forall x \in X$. More precisely, $\psi(\cdot,\cdot)$ is given by the following: For the case $F = \mathbb{R}$,*

$$(1.6) \qquad \psi(x,y) = \frac{1}{4}(|x+y|^2 - |x-y|^2), \qquad \forall x, y \in X,$$

*and for the case $F = \mathbb{C}$, $(i \stackrel{\Delta}{=} \sqrt{-1})$*

$$(1.7) \qquad \psi(x,y) = \frac{1}{4}(|x+y|^2 - |x-y|^2) + \frac{i}{4}(|x+iy|^2 - |x-iy|^2),$$
$$\forall x, y \in X.$$

From the above, we see that any linear space with an inner product can be regarded as a normed linear space in a natural way. We refer to $|x| = \sqrt{(x,x)}$ as the norm *induced* by $(\cdot,\cdot)$.

**Definition 1.5.** Let $X$ be a linear space with an inner product $(\cdot,\cdot)$ and let $|\cdot|$ be the induced norm. Then, $X$ is called a *Hilbert space* if it is complete under the norm $|\cdot|$.

The most common examples of Banach spaces are the following:

$\ell^p = \{(a_n)_{n\geq 1} \mid |(a_n)|_p \equiv \left(\sum_{n\geq 1}|a_n|^p\right)^{1/p} < \infty\} \quad 1 \leq p < \infty;$

$L^p(\Omega) = \{u : \Omega \to \mathbb{R} \mid |u|_p \equiv \left(\int_\Omega |u(x)|^p dx\right)^{1/p} < \infty\}, \quad 1 \leq p < \infty;$

$L^\infty(\Omega) = \{u : \Omega \to \mathbb{R} \mid |u|_\infty \equiv \operatorname{ess\,sup}_\Omega |u(x)| < \infty\};$

$C(\overline{\Omega}) = \{u : \overline{\Omega} \to \mathbb{R} \mid u(\cdot) \text{ continuous}\}, \quad \Omega \subset \mathbb{R}^n, \text{ bounded}.$

There are some other important spaces like $L^p(0,T;\mathbb{R}^n)$ of $\mathbb{R}^n$-valued $L^p$-functions, $BV([0,T];\mathbb{R}^n)$ of $\mathbb{R}^n$-valued bounded variational functions, and Sobolev spaces $W^{m,p}(\Omega)$ (see §6.1), etc. Also, we know that $L^2(\Omega)$ and $W^{m,2}(\Omega)$ are Hilbert spaces. In what follows, for any Banach space $X$, we define $B_r(x) = \{y \in X \mid |y - x| \leq r\}$ and $\mathcal{O}_r(x) = \{y \in X \mid |y - x| < r\}$ to be the closed and open balls centered at $x$ with radius $r > 0$, respectively. Let us now recall some standard terminology in Banach spaces.

**Definition 1.6.** Let $X$ be a Banach space and $G \subset X$.

(i) $G$ is *open* if for any $x \in G$, $\mathcal{O}_r(x) \subset G$ for some $r > 0$.

(ii) $G$ is *closed* if $X \setminus G \equiv \{x \in X \mid x \notin G\}$ is open.

(iii) The set $\operatorname{Int} G \triangleq \{x \in G \mid \exists r > 0, \ \mathcal{O}_r(x) \subset G\}$ is called the *interior* of $G$; the smallest closed set containing $G$ is called the *closure* of $G$, denoted by $\overline{G}$; and $\partial G \triangleq \overline{G} \setminus \operatorname{Int} G$ is called the *boundary* of $G$.

(iv) $G$ is *compact* if for any family of open sets $\{G_\alpha, \ \alpha \in \Lambda\}$ with $G \subset \bigcup_{\alpha \in \Lambda} G_\alpha$, there exist finitely many $G_\alpha$, called $G_1, \cdots, G_k$, in this family, such that $G \subset \bigcup_{i=1}^k G_i$.

(v) $G$ is *relatively compact* if the closure $\overline{G}$ of $G$ is compact.

(vi) $G$ is *totally bounded* if for any $\varepsilon > 0$, there exists a finite set $\{x_1, \cdots, x_k\} \subset G$, such that $G \subset \bigcup_{i=1}^k B_\varepsilon(x_i)$.

(vii) $G$ is *separable* if it admits a *countable dense subset*, i.e., there exists a countable set $G_0 \equiv \{x_i, i \geq 1\} \subset G$, such that the closure $\overline{G}_0$ of $G_0$ contains $G$. In particular, if $X$ is itself separable, we say that $X$ is a *separable Banach space*.

(viii) $G$ is *nowhere dense* if $\operatorname{Int}(\overline{G}) = \phi$.

(ix) $G$ is of *first category* if there exist at most countably many nowhere dense sets $G_n$, such that $G = \bigcup_{n\geq 1} G_n$; $G$ is of *second category* if it is not of first category.

It is known that the space $L^p(\Omega)$, $W^{m,p}(\Omega)$ $(1 \leq p < \infty)$ and $C(\overline{\Omega})$ are separable, but $L^\infty(\Omega)$ is not. We should point out that all the above concepts, except the total boundedness, are defined for general topological spaces. The following result is usually referred to as the *Baire Category Theorem*.

**Theorem 1.7.** *Any Banach space is of second category.*

§1. Elements in Functional Analysis

This result tells us that if $X$ is a Banach space and if $X = \bigcup_{n\geq 1} G_n$, then, at least one of the $G_n$'s has a nonempty interior. This property is very useful. The following result concerns the compactness.

**Proposition 1.8.** *Let $X$ be a Banach space and $G \subset X$. Then, the following are equivalent:*

(i) *$G$ is relatively compact;*

(ii) *$G$ is totally bounded;*

(iii) *For any sequence $\{x_k\} \subset G$, there exists a (strong) convergence subsequence.*

## §1.2. Linear operators

Let $X$ and $Y$ be two Banach spaces over $F$ and let $\mathcal{D}(A)$ be a subspace of $X$ (not necessarily closed). A map $A : \mathcal{D}(A) \subseteq X \to Y$ is called a *linear operator* if the following holds:

$$(1.8) \qquad A(\alpha x + \beta y) = \alpha A x + \beta A y, \qquad \forall x, y \in \mathcal{D}(A), \ \alpha, \beta \in F.$$

The set $\mathcal{D}(A)$ is called the *domain* of $A$. If $\mathcal{D}(A) = X$ and $A$ maps bounded subsets of $X$ into bounded subsets of $Y$, we say that $A$ is *bounded*. The following result gives several characterizations of linear bounded operators between Banach spaces.

**Proposition 1.9.** *Let $X$ and $Y$ be two Banach spaces and $A : X \to Y$ be a linear map. Then, the following are equivalent:*

(i) *$A$ is a linear bounded operator;*

(ii) *$A$ is bounded on $B_1(0) \equiv \{x \in X \mid |x| \leq 1\}$, i.e., $\sup_{|x|\leq 1} |Ax| < \infty$.*

(iii) *$A$ is continuous on $X$, i.e., $\lim_{x \to x_0} |Ax - Ax_0| = 0$;*

(iv) *$A$ is continuous at $0$, i.e., $\lim_{x \to 0} |Ax| = 0$.*

Because of the above result, linear bounded operators are also called *linear continuous operators*. Now, for any Banach spaces $X$ and $Y$, let $\mathcal{L}(X, Y)$ be the set of all linear bounded operators from $X$ to $Y$. For any $\alpha, \beta \in F$ and $A, B \in \mathcal{L}(X, Y)$, we define $\alpha A + \beta B$ as follows:

$$(1.9) \qquad (\alpha A + \beta B)(x) = \alpha A x + \beta B x, \qquad \forall x \in X.$$

Then, $\mathcal{L}(X, Y)$ is also a linear space. By Proposition 1.9, we may define

$$(1.10) \qquad \|A\| = \sup_{|x|\leq 1} |Ax| = \sup_{x\neq 0} \frac{|Ax|}{|x|}, \qquad \forall x \in X.$$

It is not hard to show that $\|\cdot\|$ defined by (1.10) is a norm on $\mathcal{L}(X, Y)$. Moreover, we can show that $\mathcal{L}(X, Y)$ is a Banach space under this norm. Next, for any linear operator $A : \mathcal{D}(A) \subseteq X \to Y$, we define

$$(1.11) \qquad \begin{cases} \mathcal{N}(A) = \{x \in \mathcal{D}(A) \mid Ax = 0\}, \\ \mathcal{R}(A) = \{Ax \mid x \in \mathcal{D}(A)\}, \\ \mathcal{G}(A) = \{(x, Ax) \mid x \in \mathcal{D}(A)\}. \end{cases}$$

They are called the *kernel*, the *range*, and the *graph* of the operator $A$, respectively. Hereafter, $\mathcal{L}(X,X)$ is simply denoted by $\mathcal{L}(X)$. Let us now state some important results concerning the linear bounded operators.

**Theorem 1.10.** *Let $X$ and $Y$ be Banach spaces and $A : \mathcal{D}(A) \subseteq X \to Y$ be a linear operator. Then, the following hold:*

(i) (Open Mapping Theorem) $A \in \mathcal{L}(X,Y)$ *and* $\mathcal{R}(A) = Y$ *imply that for any open set $G \subset X$, the set $A(G)$ is open in $Y$.*

(ii) (Inverse Mapping Theorem) $A \in \mathcal{L}(X,Y)$ *with* $\mathcal{R}(A) = Y$ *and* $\mathcal{N}(A) = \{0\}$ *imply that* $A^{-1} \in \mathcal{L}(Y,X)$.

(iii) (Closed Graph Theorem) $\mathcal{G}(A)$ *is closed in* $X \times Y$ *implies that* $A \in \mathcal{L}(X,Y)$.

**Theorem 1.11.** (Principle of Uniform Boundedness) *Let $X$ and $Y$ be Banach spaces and $\mathcal{A} \subset \mathcal{L}(X,Y)$. Then,*

$$(1.12) \qquad \sup_{A \in \mathcal{A}} |Ax| < \infty, \quad \forall x \in X \quad \Rightarrow \quad \sup_{A \in \mathcal{A}} \|A\| < \infty.$$

**Corollary 1.12.** (Banach–Steinhaus) *Let $X$ and $Y$ be Banach spaces and $A_n \in \mathcal{L}(X,Y)$. Suppose that for any $x \in X$, $A_n x$ is strongly convergent. Then there exists an $A \in \mathcal{L}(X,Y)$ such that*

$$(1.13) \qquad \lim_{n \to \infty} |A_n x - Ax| = 0,$$

$$(1.14) \qquad \|A\| \leq \varliminf_{n \to \infty} \|A_n\| \leq \sup_{n \geq 1} \|A_n\| < \infty.$$

The first inequality in (1.14) is referred to as the *sequentially lower semicontinuity* of the operator norm $\|\cdot\|$. The proofs of the above results can be found in standard books on functional analysis (see Conway [1], for example).

## §1.3. Linear functionals and dual spaces

Next, let us look at the case $Y = F$ (recall that $F = \mathbb{R}$ or $\mathbb{C}$). Any $f \in \mathcal{L}(X,F)$ is called a *linear bounded functional* (or *linear continuous functional*) on $X$. Hereafter, we denote $X^* = \mathcal{L}(X,F)$ and call it the *dual (space)* of $X$. Sometimes we denote

$$(1.15) \qquad f(x) = \langle f, x \rangle_{X^*,X}, \qquad \forall x \in X.$$

The symbol $\langle \cdot, \cdot \rangle_{X^*,X}$ is referred to as the *duality pairing* between $X^*$ and $X$. It follows from (1.14) that

$$(1.16) \qquad |f|_{X^*} = \sup_{x \in X, |x| \leq 1} |f(x)|, \qquad \forall f \in X^*.$$

## §1. Elements in Functional Analysis

Sometimes, we call the norm $|\cdot|_{X^*}$ defined by the above the *dual norm* of $|\cdot|$. The following types of results are usually referred to as *Riesz Representation Theorems*.

**Theorem 1.13.** (Riesz) *Let $\Omega \subset \mathbb{R}^n$ be a domain. Then,*

(i) $(L^p(\Omega))^* = L^q(\Omega)$, $1 \leq p < \infty$, $1/p + 1/q = 1$.

(ii) *If $\Omega$ is bounded, $C(\overline{\Omega})^* = \mathcal{M}(\overline{\Omega}) \equiv$ the set of all regular signed measures on $\overline{\Omega}$. In particular, $C[0,T]^* = BV_0[0,T] \equiv$ the set of all right continuous bounded variational functions vanishing at 0.*

The following result is very important in later applications.

**Theorem 1.14.** (Hahn–Banach) *Let $X$ be a real linear space and $p : X \to (-\infty, +\infty]$ be a convex function, i.e.,*

(1.17) $\quad p(\lambda x + (1-\lambda)y) \leq \lambda p(x) + (1-\lambda)p(y), \quad \forall \lambda \in [0,1], \ x,y \in X.$

*Let $X_0$ be a subspace of $X$ and $f_0 : X_0 \to \mathbb{R}$ be a linear functional satisfying*

(1.18) $\quad\quad\quad\quad\quad f_0(x) \leq p(x), \quad \forall x \in X_0.$

*Then, there exists a linear functional $f : X \to \mathbb{R}$, such that*

(1.19) $\quad\quad\quad\quad\quad \begin{cases} f(x) = f_0(x), & \forall x \in X_0, \\ f(x) \leq p(x), & \forall x \in X. \end{cases}$

In the above, we refer to $f$ as an *extension* of $f_0$, and $f_0$ as the *restriction* of $f$ on $X_0$. Sometimes, we denote $f\big|_{X_0} = f_0$. Again, the above result is very standard. For a proof, see Barbu–Precupanu [1] or Yosida [1]. The following is the most common form of the Hahn–Banach Theorem.

**Corollary 1.15.** (Hahn–Banach) *Let $X$ be a real Banach space and $X_0$ be a subspace of $X$. Let $f_0 \in X_0^*$. Then there exists an extension $f \in X^*$ of $f_0$ such that $|f|_{X^*} = |f_0|_{X_0^*}$.*

*Proof.* Take $p(x) = |x|$ and apply Theorem 1.14. □

We know that the same result as Corollary 1.15 remains true for *complex Banach spaces* (i.e., $F = \mathbb{C}$).

**Corollary 1.16.** *Let $X$ be a Banach space and $x_0 \in X$. Then, there exists an $f \in X^*$ with $|f|_{X^*} = 1$, such that*

(1.20) $\quad\quad\quad\quad\quad f(x_0) = |x_0|.$

*Proof.* Define $f_0(\alpha x_0) = \alpha |x_0|$, $\alpha \in \mathbb{R}$. Then $f_0$ is a linear bounded functional on $X_0 \equiv \text{span}\{x_0\}$ (the space spanned by $x_0$). By Corollary 1.15, we obtain $f \in X^*$, an extension of $f_0$, which gives (1.20). □

It follows from Corollary 1.16 that for any Banach space we have (compare with (1.16))

$$|x| = \sup_{f \in X^*, |f| \leq 1} |f(x)|, \quad \forall x \in X. \tag{1.21}$$

On the other hand, since $X^* \equiv \mathcal{L}(X, F)$ is also a Banach space, we may talk about the dual of $X^*$, i.e., $(X^*)^* \stackrel{\Delta}{=} X^{**}$. It is important to know that for any $x \in X$, by defining

$$x^{**}(f) = f(x) \equiv \langle f, x \rangle_{X^*, X}, \quad \forall f \in X^*, \tag{1.22}$$

we have $x^{**} \in X^{**}$ and by (1.16) and (1.21), $|x|_X = |x^{**}|_{X^{**}}$. Thus, the map $x \mapsto x^{**}$ makes $X \subset X^{**}$ (topologically and algebraically). An immediate question is when $X = X^{**}$? This leads to the following notion.

**Definition 1.17.** A Banach space $X$ is said to be *reflexive* if $X = X^{**}$, i.e., for any $x^{**} \in X^{**}$, there exists an $x \in X$, such that (1.22) holds.

From Theorem 1.13, we see that $L^p(\Omega)$ with $1 < p < \infty$ is reflexive. Also, the Sobolev spaces $W^{m,p}(\Omega)$ with $1 < p < \infty$ are reflexive (see §6.1). But $C(\overline{\Omega})$, $L^1(\Omega)$, $L^\infty(\Omega)$, and $BV[0,T]$ are not reflexive. On the other hand, we can prove that any Hilbert space is reflexive. Moreover, the following holds:

**Theorem 1.18.** (Riesz) *Let $X$ be a Hilbert space. Then, $X^* = X$. More precisely, for any $f \in X^*$, there exists a $y \in X$, such that*

$$f(x) = (x, y), \quad \forall x \in X, \tag{1.23}$$

*and for any $y \in X$, by defining $f$ as in (1.23), one has $f \in X^*$.*

Next, we introduce some notions of convergence other than the strong convergence (compare Definition 1.3).

**Definition 1.19.** Let $X$ be a Banach space and $X^*$ be its dual.
(i) A sequence $\{x_n\} \subset X$ is said to be *weakly* convergent to $x \in X$ if

$$\lim_{n \to \infty} f(x_n) = f(x), \quad \forall f \in X^*. \tag{1.24}$$

(ii) A sequence $\{x_n^*\} \subset X^*$ is said to be *weakly* * convergent to $x^* \in X^*$ if

$$\lim_{n \to \infty} x_n^*(x) = x^*(x), \quad \forall x \in X. \tag{1.25}$$

We note that a strongly convergent sequence $\{x_n\} \subset X$ is weakly convergent; and for a sequence $\{f_n\} \subset X^*$, we have the implications of the convergence: strong $\Rightarrow$ weak $\Rightarrow$ weak*. But, in general, the converses are not true. In what follows, we sometimes use $x_n \stackrel{w}{\rightharpoonup} x$ and $f_n \stackrel{*}{\rightharpoonup} f$ to denote the weak convergence of $x_n$ to $x$ and the weak* convergence of $f_n$ to $f$.

§1. Elements in Functional Analysis

It is known that any topology can be defined via the so-called *nets* (instead of sequences). Thus, by replacing the sequences $x_n$ and $f_n^*$ by nets $x_\alpha$ and $f_\alpha^*$ above, we have defined two topologies on $X$ and $X^*$, respectively. These topologies are referred to as the *weak topology* on $X$ and the *weak\* topology* on $X^*$, respectively. By the way, the topology induced by the strong convergence is referred to as the *strong topology*. Similar to (1.14), we see that the following hold.

**Proposition 1.20.** *Let $X$ be a Banach space with dual $X^*$. Then the norm $|\cdot|_X$ is weakly sequentially lower semicontinuous and the norm $|\cdot|_{X^*}$ is weakly\* sequentially lower semicontinuous. That is, if $x_n \xrightarrow{w} x$, then $|x| \leq \varliminf_{n\to\infty} |x_n|$ and if $f_n \xrightarrow{*} f$, then $|f| \leq \varliminf_{n\to\infty} |f_n|$.*

In $\mathbb{R}^n$, we know that any bounded closed set $K$ is compact. However, in infinite dimensions, the situation is different. The following result gives a main intrinsic difference between finite and infinite dimensional spaces.

**Theorem 1.21.** *Let $X$ be a Banach space. Then the closed unit ball in $X$ is compact if and only if $X$ is finite dimensional.*

Hence, we see that for an infinite dimensional Banach space $X$, a bounded closed set need not be compact. This is a crucial difference between finite and infinite dimensional spaces. In many cases, this fact will prevent the results in finite dimensions from easily extending to infinite dimensions. However, we have the following interesting results. Recall that the compact sets can be defined in $X$ with the weak topology and in $X^*$ with the weak\* topology.

**Theorem 1.22.** *Let $X$ be a Banach space.*

(i) (Alaoglu) *Any bounded set in $X^*$ is weak\* relatively compact.*

(ii) (Eberlein–Shmul'yan) *The closed unit ball in $X$ is weakly compact if and only if $X$ is reflexive. In particular, if $X$ is reflexive, then any (norm) bounded sequence admits a weak convergence subsequence.*

### §1.4. Adjoint operators

Let $X$ and $Y$ be Banach spaces and $A \in \mathcal{L}(X,Y)$. We define a map $A^* : Y^* \to X^*$ by the following:

$$(1.26) \qquad \langle A^*y^*, x \rangle_{X^*,X} = \langle y^*, Ax \rangle_{Y^*,Y}, \qquad \forall y^* \in Y^*, \, x \in X.$$

Clearly, $A^*$ is linear and bounded. We call $A^*$ the *adjoint operator* of $A$. By definition, it is seen that for any $A, B \in \mathcal{L}(X,Y)$ and $\alpha, \beta \in F$, $(\alpha A + \beta B)^* = \alpha A^* + \beta B^*$. To state further properties of adjoint operators, let us introduce the following notion: For any $X_0 \subset X$,

$$(1.27) \qquad X_0^\perp \equiv \{f \in X^* \mid \langle f, x \rangle = 0, \quad \forall x \in X_0\},$$

and for any $X_1 \subset X^*$,

(1.28) $\quad\quad\quad {}^\perp X_1 \equiv \{x \in X \mid \langle f, x \rangle = 0, \quad \forall f \in X_1\}.$

We call $X_0^\perp$ and ${}^\perp X_1$ the *annihilators* of $X_0$ and $X_1$, respectively. It is clear that $X_0^\perp$ is weakly* closed and ${}^\perp X_1$ is weakly closed. Further, one can show that if $X_0$ and $X_1$ are subspaces, then

(1.29) $\quad\quad\quad {}^\perp(X_0^\perp) = \overline{X_0}, \quad ({}^\perp X_1)^\perp = \overline{X_1}^{w^*}.$

In the case where $X$ is reflexive and $X_1$ is a subspace of $X^*$, ${}^\perp X_1 = X_1^\perp$. Now, let us state some properties of the adjoint operators.

**Proposition 1.23.** *Let $X$ and $Y$ be Banach spaces and $A \in \mathcal{L}(X, Y)$. Then the following hold:*

(i) $(A^*)^*|_X = A$;

(ii) $\|A^*\| = \|A\|$;

(iii) $A^{-1} \in \mathcal{L}(Y, X)$ *if and only if* $(A^*)^{-1} \in \mathcal{L}(Y^*, X^*)$. *In this case,* $(A^*)^{-1} = (A^{-1})^*$;

(iv) $\mathcal{N}(A^*) = \mathcal{R}(A)^\perp$, $\mathcal{N}(A) = {}^\perp\mathcal{R}(A^*)$.

We should point out that when $X$ and $Y$ are Hilbert spaces and $A \in \mathcal{L}(X, Y)$, then, by convention, $(\alpha A)^* = \bar{\alpha} A^*$, where $\bar{\alpha}$ is the conjugate of the number $\alpha$.

Now, we let $X$ be a Hilbert space and $A \in \mathcal{L}(X)$. By the Riesz Theorem (Theorem 1.18), $X^*$ can be identified by itself, $X^* = X$. Thus, $A^* \in \mathcal{L}(X^*) = \mathcal{L}(X)$. If $A^* = A$ holds, this operator is said to be *self-adjoint*.

In a Hilbert space, there is a very important class of adjoint operators that we recall in the following.

**Definition 1.24.** *Let $X$ be a Hilbert space and $P \in \mathcal{L}(X)$. Then $P$ is called an* orthogonal projection *if $\mathcal{R}(P) \equiv X_0$ is closed and*

(1.30) $\quad\quad Px = x, \quad \forall x \in X_0, \quad Py = 0, \quad \forall y \in X_0^\perp.$

We have the following characterization of orthogonal projections on Hilbert spaces.

**Proposition 1.25.** *Let $X$ be a Hilbert space and $P \in \mathcal{L}(X)$. Then, $P$ is an orthogonal projection if and only if*

(1.31) $\quad\quad\quad\quad P^* = P, \quad P^2 = P.$

### §1.5. Spectral theory

Let $X$ be a complex Banach space and $A \in \mathcal{L}(X)$. For any $\lambda \in \mathbb{C}$ and $x \in X$, let $(\lambda - A)x = \lambda x - Ax$. Then $\lambda - A \in \mathcal{L}(X)$. Thus, we may talk

§1. Elements in Functional Analysis 33

about the invertibility of this operator, which leads to the following:

(1.32)
$$\begin{cases} \rho(A) = \{\lambda \in \mathbb{C} \mid (\lambda - A)^{-1} \in \mathcal{L}(X)\}; \quad \sigma(A) = \mathbb{C} \setminus \rho(A); \\ \sigma_p(A) = \{\lambda \in \sigma(A) \mid \mathcal{N}(\lambda - A) \neq \{0\}\}; \\ \sigma_{ap}(A) = \{\lambda \in \sigma(A) \setminus \sigma_p(A) \mid \mathcal{R}(\lambda - A) \neq \overline{\mathcal{R}(\lambda - A)} = X\}; \\ \sigma_r(A) = \{\lambda \in \sigma(A) \setminus \sigma_p(A) \mid \mathcal{R}(\lambda - A) = \overline{\mathcal{R}(\lambda - A)} \neq X\}. \end{cases}$$

We call $\rho(A)$ and $\sigma(A)$ the *resolvent* and the *spectrum* of $A$, respectively; and call $\sigma_p(A)$, $\sigma_{ap}(A)$, and $\sigma_r(A)$ the *point spectrum*, the *approximate point spectrum*, and the *residue spectrum*, respectively. In particular, any $\lambda \in \sigma_p(A)$ is called an *eigenvalue* of $A$ and for such a $\lambda$, let $x \in \mathcal{N}(\lambda - A) \setminus \{0\}$, then $Ax = \lambda x$; we call such an $x$ an *eigenvector* of $A$ corresponding to the eigenvalue $\lambda$. The following result is concerned with the spectrum of $A$.

**Proposition 1.26.** *Let $X$ be a Banach space and $A \in \mathcal{L}(X)$. Then*

(i) *$\sigma(A)$ is a nonempty compact set contained in $B_{r(A)}(0)$, with $r(A)$, called the spectrum radius of $A$, being given by*

(1.33) $$r(A) = \varlimsup_{n \to \infty} \|A^n\|^{1/n}.$$

(ii) *For any analytic function $f$,*

(1.34) $$\sigma(f(A)) = f(\sigma(A)),$$

*where $f(A)$ is defined by*

(1.35) $$f(A) = \int_\Gamma f(\lambda)(\lambda - A)^{-1}\, d\lambda,$$

*with $\Gamma$ being the boundary of any bounded (smooth) domain containing $\sigma(A)$; and the integral is taken counterclockwise along $\Gamma$. In particular, if $f(\lambda) = \sum_{k=0}^n \alpha_k \lambda^k$ is a polynomial, then (1.35) coincides with the following natural definition: $f(A) = \sum_{k=0}^n \alpha_k A^k$.*

(iii) *$\sigma(A^*) = \sigma(A)$ if $X$ is a Banach space; $\sigma(A^*) = \sigma(A)^* \equiv \{\bar{\lambda} \mid \lambda \in \sigma(A)\}$ if $X$ is a Hilbert space; consequently, if $A = A^*$, then $\sigma(A) \subset \mathbb{R}$.*

## §1.6. Compact operators

**Definition 1.27.** Let $X$ and $Y$ be Banach spaces and $A \in \mathcal{L}(X, Y)$. We say that $A$ is *compact* if $A$ maps any bounded set of $X$ into a relatively compact set in $Y$, i.e., if $G$ is bounded in $X$, then the closure of $AG$ is compact in $Y$.

**Definition 1.28.** Let $X$ and $Y$ be Banach spaces and $A \in \mathcal{L}(X, Y)$. We say that $A$ is *completely continuous* if for any sequence $\{x_n\} \subset X$ with $x_n \xrightarrow{w} x$, it holds that $|Ax_n - Ax| \to 0$, i.e., $A$ maps any weakly convergent sequence into a strongly convergent one.

The readers are asked to distinguish the difference between the compact operators and the completely continuous operators. We have the following result.

**Theorem 1.29.** *Let $X$ and $Y$ be Banach spaces.*

(i) *If $A \in \mathcal{L}(X,Y)$ is compact, then $A$ is completely continuous; conversely, if $X$ is reflexive and $A$ is completely continuous, then $A$ is compact.*

(ii) *(Schauder) $A \in \mathcal{L}(X,Y)$ is compact if and only if $A^* \in \mathcal{L}(Y^*, X^*)$ is compact.*

(iii) *If $A_n \in \mathcal{L}(X,Y)$ are compact and $\|A - A_n\| \to 0$, then $A$ is compact.*

(iv) *If $A \in \mathcal{L}(X,Y)$ is compact, $B \in \mathcal{L}(Y,Z)$, and $C \in \mathcal{L}(Z,X)$ with $Z$ being another Banach space, then $BA$ and $AC$ are compact.*

(v) *Let $X$ and $Y$ be Hilbert spaces. Then, $A \in \mathcal{L}(X,Y)$ is compact if and only if there exists a sequence $A_n \in \mathcal{L}(X,Y)$ with $\dim \mathcal{R}(A_n) < \infty$, such that $\|A_n - A\| \to 0$.*

We should point out that if $X$ is *not* reflexive, then $A \in \mathcal{L}(X,Y)$ being completely continuous does not necessarily imply that $A$ is compact. An example is the following: Let $X = \ell^1 \equiv \{x = (x^i) \mid \sum_{i \geq 1} |x^i| < \infty\}$. Then, we can show that any $A \in \mathcal{L}(X)$ is completely continuous; in particular, the identity operator is so, but it is not compact because $\ell^1$ is infinite dimensional (see Theorem 1.21). In some books, compact and completely continuous operators are not distinguished. We adopt the presentation found in Conway [1].

**Theorem 1.30.** *(Riesz) Let $X$ be a Banach space and $A \in \mathcal{L}(X)$ be compact. Then, one and only one of the following holds:*

(i) $\sigma(A) = \{0\}$.

(ii) $\sigma(A) = \{0, \lambda_1, \cdots, \lambda_n\}$ *with*

(1.36) $\quad \lambda_k \neq 0, \quad \lambda_k \in \sigma_p(A), \quad \dim \mathcal{N}(\lambda_k - A) < \infty, \quad 1 \leq k \leq n.$

(iii) $\sigma(A) = \{0, \lambda_1, \lambda_2, \cdots\}$ *with*

(1.37) $\quad \begin{cases} \lambda_k \neq 0, \quad \lambda_k \in \sigma_p(A), \quad \dim \mathcal{N}(\lambda_k - A) < \infty, \quad k \geq 1; \\ \lim_{k \to \infty} \lambda_k = 0. \end{cases}$

**Theorem 1.31.** *(Fredholm Alternative) Let $X$ be a (complex) Banach space, $A \in \mathcal{L}(X)$ be compact, and $\lambda \in \mathbb{C}$, $\lambda \neq 0$. Then, $\mathcal{R}(\lambda - A)$ is closed and*

(1.38) $\quad \dim \mathcal{N}(\lambda - A) = \dim \mathcal{N}(\lambda - A^*) < \infty.$

*Consequently, for any $y \in X$, the equation $(\lambda - A)x = y$ is solvable if and only if $\mathcal{N}(\lambda - A) = \{0\}$. In this case, the solution is unique.*

§1. Elements in Functional Analysis

To state an important corollary of the above theorem, let us first introduce the following notion.

**Definition 1.32.** Let $X$ be a Banach space and $X_0$ be a subspace of $X$. We say that $X_0$ is *finite codimensional* in $X$ if there exist $x_1, \cdots, x_n \in X$, such that

(1.39) $$\text{span}\,\{X_0, x_1, \cdots, x_n\} = X.$$

The smallest number $n$ such that (1.39) holds for some $x_1, \cdots, x_n \in X$ is called the *codimension* of $X_0$, denoted by $\text{codim}\,X_0$.

**Corollary 1.33.** *Let $X$ be a Banach space and $A \in \mathcal{L}(X)$ be compact. Then $\mathcal{R}(I - A)$ is closed and finite codimensional in $X$.*

*Proof.* By Theorem 1.31 with $\lambda = 1$, we know that $\mathcal{R}(I - A)$ is closed. Thus, by (1.29) and Proposition 1.23,

(1.40) $$\mathcal{R}(I - A) = {}^\perp\bigl(\mathcal{R}(I - A)^\perp\bigr) = {}^\perp \mathcal{N}(I - A^*).$$

Next, by (1.38), $\dim \mathcal{N}(I - A^*) < \infty$. Thus, we can find $f_1, \cdots, f_n \in X^*$, linearly independent, such that

(1.41) $$\mathcal{N}(I - A^*) = \text{span}\,\{f_1, \cdots, f_n\}.$$

Now, by Corollary 1.35 below, we can find $x_i \in X$, such that $f_i(x_j) = \delta_{ij}$, $1 \leq i, j \leq n$ (here $\delta_{ii} = 1$ and $\delta_{ij} = 0$ if $i \neq j$). Then, for any $x \in X$,

(1.42) $$y = x - \sum_{i=1}^n f_i(x) x_i \in \bigcap_{i=1}^n \mathcal{N}(f_i) = {}^\perp\bigl(\text{span}\,\{f_1, \cdots, f_n\}\bigr).$$

By (1.41) and (1.40), we see that $y \in \mathcal{R}(I - A)$. Hence, (1.42) implies

(1.43) $$\text{span}\,\{\mathcal{R}(I - A), x_1, \cdots, x_n\} = X.$$

This proves $\text{codim}\,\mathcal{R}(I - A) < \infty$. □

Corollary 1.33 will be useful in Chapter 4. Now, let us prove the following lemma and corollary, which have been used above and will be used later.

**Lemma 1.34.** *Let $X$ be a Banach space and $f_0, \cdots, f_n \in X^*$. Suppose*

(1.44) $$\bigcap_{i=1}^n \mathcal{N}(f_i) \subset \mathcal{N}(f_0).$$

*Then $f_0$ is a linear combination of $f_1, \cdots, f_n$.*

*Proof.* We use induction on $n$. For $n = 1$, (1.44) gives $\mathcal{N}(f_1) \subset \mathcal{N}(f_0)$. If $f_1 = 0$, we must have $f_0 = 0$ and the conclusion is trivial. Let $f_1$ be nonzero. Then, we can find an $x_1 \in X$, such that $f_1(x_1) = 1$ (see Corollary 1.16). Now, for any $x \in X$, we have

$$(1.45) \qquad x - f_1(x)x_1 \in \mathcal{N}(f_1) \subset \mathcal{N}(f_0).$$

This yields $f_0(x) = \alpha f_1(x)$, for all $x \in X$ with $\alpha = f_0(x_1)$. Next, we suppose our claim holds for $n \leq k$ and we prove the claim for $n = k+1$. To this end, let $X_0 = \mathcal{N}(f_{k+1})$ and consider the functionals $f_i|_{X_0}$, $0 \leq i \leq k$. Clearly,

$$(1.46) \qquad \bigcap_{i=1}^{k} \mathcal{N}(f_i|_{X_0}) = \bigcap_{i=1}^{k+1} \mathcal{N}(f_i) \subset \mathcal{N}(f_0) \bigcap \mathcal{N}(f_{k+1}) = \mathcal{N}(f_0|_{X_0}).$$

Thus, by induction, $f_0|_{X_0} = \sum_{i=1}^{k} \alpha_i f_i|_{X_0}$. This implies that

$$(1.47) \qquad \mathcal{N}\Big(f_0 - \sum_{i=1}^{k} \alpha_i f_i\Big) \supset X_0 = \mathcal{N}(f_{k+1}).$$

Then, again by induction, $f_0 - \sum_{i=1}^{k} \alpha_i f_i = \alpha_{k+1} f_{k+1}$. This proves our claim and the lemma is proved. □

**Corollary 1.35.** *Let $X$ be a Banach space and $f_0, \cdots, f_n \in X^*$ be linearly independent. Then, there exist $x_0, \cdots, x_n \in X$, such that $f_i(x_j) = \delta_{ij}$, $0 \leq i, j \leq n$.*

*Proof.* It suffices to show that for each $0 \leq i \leq n$, there exists an $x_i \in X$, such that

$$(1.48) \qquad x_i \in \bigcap_{j \neq i} \mathcal{N}(f_j), \qquad f_i(x_i) \neq 0.$$

Suppose it is not the case. Then, without loss of generality, we may assume (1.44) holds. By Lemma 1.34, we obtain that $f_0$ is a linear combination of $f_1, \cdots, f_n$, contradicting the linear independence of $f_0, \cdots, f_n$. □

## §2. Some Geometric Aspects of Banach Spaces

In this section, we present some results concerning certain geometric properties of Banach spaces and their subsets.

### §2.1. Convex sets

We first introduce the following notion, which is a generalization of Banach spaces. We assume that the readers have the basic knowledge of topological spaces.

**Definition 2.1.** *Let $X$ be a (real) vector space and $\tau$ be a Hausdorff topology on it. We call $(X, \tau)$ a topological vector space if under $\tau$, the maps $(x, y) \mapsto x + y$ and $(\alpha, x) \mapsto \alpha x$ ($x, y \in X, \alpha \in \mathbb{R}$) are continuous.*

§2. Some Geometric Aspects of Banach Spaces

In the above definition, we may allow $\tau$ to be any topology, not necessarily Hausdorff. But, in what follows, we only deal with Hausdorff topology. Thus, for convenience of the presentation, we give a restricted definition.

Clearly, any Banach space $X$ is a topological vector space under its norm topology. Moreover, $X$ is also a topological vector space under its weak topology; and $X^*$ is a topological vector space under its weak* topology. Recall that a subset $\tau_0$ of $\tau$ is called a *basis* of $\tau$ if for any $G \in \tau$, there exists a $G_0 \in \tau_0$, such that $G_0 \subset G$. The following gives a basis of the *neighborhoods* at the origin for the weak topology of $X$:

$$
(2.1) \quad \begin{aligned} N(f_1,\cdots,f_k;\varepsilon) &= \{x \in X \mid |f_i(x)| < \varepsilon, \quad 1 \leq i \leq k\}, \\ & k \geq 1, \quad \varepsilon > 0, \quad f_1,\cdots,f_k \in X^*. \end{aligned}
$$

Because the space is linear, the topology is uniquely determined by a basis of neighborhoods at the origin. Similarly, a basis of the neighborhoods at the origin for the weak* topology of $X^*$ is given by:

$$
(2.2) \quad \begin{aligned} N(x_1,\cdots,x_k;\varepsilon) &= \{f \in X^* \mid |f(x_i)| < \varepsilon, \quad 1 \leq i \leq k\}, \\ & k \geq 1, \quad \varepsilon > 0, \quad x_1,\cdots,x_k \in X. \end{aligned}
$$

We will apply some general results for topological vector spaces to the Banach spaces with weak and/or weak* topology.

Next, for a given topological vector space $X$ (with the topology $\tau$, which is usually suppressed if it can be clearly identified from the context), we can talk about the linear functionals on the space. Again, by a linear functional $f$, we mean a map $f : X \to \mathbb{R}$, which is linear. Similar to the case for Banach spaces (see §1), we have the following result.

**Proposition 2.2.** *Let $X$ be a topological vector space and $f : X \to \mathbb{R}$ be a linear functional. Then, $f$ is continuous if and only if there exists a nonempty open set $G \subset X$, such that $f(G)$ is bounded in $\mathbb{R}$.*

*Proof.* $\Rightarrow$. By the continuity, we see that $G \equiv f^{-1}((-1,1))$ is an open set. For this open set we have $f(G) = (-1,1)$, which is bounded.

$\Leftarrow$. Let $G$ be open such that $|f(x)| \leq M$ for all $x \in G$. Take any $x_0 \in G$. Then, $G - x_0$ is a neighborhood of 0. Since the map $x \mapsto \mu x$ is continuous for any fixed $\mu \in \mathbb{R}$, for any $\varepsilon > 0$, we can find a neighborhood $W$ of 0, such that

$$(2.3) \quad \frac{2M}{\varepsilon} W \subset G - x_0.$$

Thus, for any $x \in W$,

$$(2.4) \quad |f(x)| = \frac{\varepsilon}{2M}|f(\frac{2M}{\varepsilon}x)| \leq \frac{\varepsilon}{2M}\sup_{y \in G}|f(y-x_0)| \leq \varepsilon.$$

Hence, $f$ is continuous at $x = 0$. Therefore it is continuous. □

Next, we recall that a subset $G$ of some vector space $X$ is said to be *convex* if for any $x,y \in G$ and $\lambda \in [0,1]$, one has $\lambda x + (1-\lambda)y \in G$. Let $X$ be a topological vector space and $S$ be a subset of $X$. Denote $\overline{co}\,S$ to be the smallest convex and closed set containing $S$. We call $\overline{co}\,S$ the *convex hull* of the set $S$. Important examples of convex sets are balls and subspaces. It should be pointed out that the intersection of any number of convex sets is convex; but the union of two convex sets is not necessarily convex. Also, if $G_1$ and $G_2$ are convex, then for any $\lambda_1, \lambda_2 \in \mathbb{R}$, the set $\lambda_1 G_1 + \lambda_2 G_2 \equiv \{\lambda_1 x_1 + \lambda_2 x_2 \mid x_1 \in G_1, x_2 \in G_2\}$ is convex.

A topological vector space $(X,\tau)$ is said to be *locally convex* if $\tau$ admits a basis consisting of convex sets. Clearly, any Banach space $X$ with its norm (weak and/or weak*) topology is a locally convex topological space. The following results are very important in convex analysis.

**Theorem 2.3.** (Eidelheit) *Let $X$ be a topological vector space and let $G_1$ and $G_2$ be two convex sets in $X$ with $\text{Int}\,G_1 \neq \phi$ and $(\text{Int}\,G_1) \cap G_2 = \phi$. Then there exists a nonzero continuous linear functional $f$, such that*

$$(2.5) \qquad \sup_{x \in G_1} f(x) \leq \inf_{y \in G_2} f(y).$$

**Theorem 2.4.** *Let $X$ be a locally convex topological vector space and let $G_1$ and $G_2$ be two disjoint closed convex sets in $X$. Suppose that $G_1$ is compact. Then, there exist a continuous linear functional $f$ and some constants $\alpha < \beta$, such that*

$$(2.6) \qquad \sup_{x \in G_1} f(x) \leq \alpha < \beta \leq \inf_{y \in G_2} f(y).$$

The above results are standard. One can find proofs in many topological vector space books (see Barbu–Precupanu [1] or Cristescu [1]). Thus, we omit the proofs here. Let us point out some important consequences.

**Corollary 2.5.** *Let $X$ be a Banach space and $G$ be a convex subset of $X$.*

*(i) If $\text{Int}\,G \neq \phi$ and $x_0 \in X \setminus \text{Int}\,G$, then there exists an $f \in X^*$ with $|f| = 1$, such that*

$$(2.7) \qquad f(x_0) \leq \inf_{y \in G} f(y).$$

*(ii) The strict inequality holds in (2.7) if $G$ is closed and $x_0 \in X \setminus G$.*

Next, we introduce the following notion.

**Definition 2.6.** Let $X$ be a Banach space and $G \subset X$ be convex. A point $x_0 \in \partial G$ is called a *supporting point* of $G$ if there exists an $f \in X^*$ with $|f| = 1$, such that (2.7) holds. In this case, we say that $f$ *supports* the set $G$ at $x_0$.

Clearly, if $G$ is a convex set with $\text{Int}\,G \neq \phi$, then by Corollary 2.5 (i), any $x_0 \in \partial G$ is a supporting point of $G$. On the other hand, if $\text{Int}\,G = \phi$,

§2. Some Geometric Aspects of Banach Spaces

then there may be some $x_0 \in \partial G$ that is not a supporting point of $G$. Here is a simple example.

**Example 2.7.** Let $X = \ell^2 \triangleq \{(a_n)_{n\geq 1} \mid a_n \in \mathbb{R},\ \sum_{n\geq 1} a_n^2 < \infty\}$, and

(2.8) $$G \triangleq \{(a_n)_{n\geq 1} \in \ell^2 \mid a_n \geq 0,\ \forall n \geq 1\}.$$

Then it is easy to show that $G$ is a closed and convex set in $X$. Moreover, we claim $\operatorname{Int} G = \phi$ and $\partial G = G$. In fact, if $\bar{x} = (\bar{a}_n)_{n\geq 1} \in G$, then for any $\varepsilon > 0$, there exists an $n_0 \geq 1$, such that $0 \leq \bar{a}_{n_0} < \varepsilon/3$. Thus, the point

$$\tilde{x} = (\bar{a}_1, \cdots, \bar{a}_{n_0-1}, \bar{a}_{n_0} - \varepsilon/2, \bar{a}_{n_0+1}, \cdots) \in B_\varepsilon(\bar{x}) \setminus G.$$

This shows that $\operatorname{Int} G = \phi$. Now, let $x_0 = (\frac{1}{n})_{n\geq 1} \in G$. We claim that $x_0$ is not a supporting point of $G$. In fact, if there exists an $f = (f_n)_{n\geq 1} \in \ell^2 = (\ell^2)^*$, such that (2.7) holds, then

(2.9) $$\sum_{n\geq 1} \frac{f_n}{n} \leq \sum_{n\geq 1} a_n f_n, \quad \forall (a_n)_{n\geq 1} \in G.$$

For any $k \geq 1$, by taking $(a_n)_{n\geq 1} = \alpha(\delta_{nk})_{n\geq 1} \in G$ ($\delta_{nk} = 0$ if $n \neq k$ and $\delta_{nn} = 1$) with $\alpha \in \mathbb{R}$ and $\alpha \to +\infty$, we see that $f_k \geq 0$ for all $k \geq 1$. Then, by taking $a_n \equiv 0$, we see $f_k = 0$ for all $k \geq 1$, contradicting $f \neq 0$. Thus, $x_0$ is not a supporting point of $G$.

The above example tells us that a closed and convex set in infinite dimensional spaces can have very bad boundary points. The next result gives a nice property of the convex set in Banach spaces.

**Corollary 2.8.** (Mazur) *Let $X$ be a Banach space and $G$ be a convex and (strongly) closed set in $X$. Then $G$ is weakly closed in $X$. Consequently, if $x_n \in X$ converges to $x \in X$ weakly, then there exist $\alpha_{n,k} \geq 0$, $\sum_{k=1}^{K_n} \alpha_{n,k} = 1$, such that*

(2.10) $$\lim_{n\to\infty} \left| \sum_{k=1}^{K_n} \alpha_{n,k} x_{k+n} - x \right| = 0.$$

*Proof.* Let $G$ be convex and closed in $X$. Let $\overline{G}^w$ be the weak closure of $G$. Suppose there exists an $x_0 \in \overline{G}^w \setminus G$. Then, by Corollary 2.5, there exists an $f \in X^*$ with $|f| = 1$, such that for some $\delta > 0$,

(2.11) $$\sup_{y \in G} f(x_0 - y) \leq -\delta.$$

Let $N(x_0, f; \delta) = \{x \in X \mid |f(x - x_0)| < \delta/2\}$. This is a neighborhood of $x_0$ and if $y_0 \in N(x_0, f; \delta) \cap G$, then

(2.12) $$-\delta \geq f(x_0 - y_0) \geq -|f(x_0 - y_0)| > -\frac{\delta}{2},$$

which is impossible. Thus, $N(x_0, f; \delta) \cap G = \phi$, which leads to $x_0 \notin \overline{G}^w$. Hence, we must have $G = \overline{G}^w$.

Next, let $x_n$ converge to $x$ weakly. Then, for any $f \in X^*$, $f(x_n) \to f(x)$. Thus, for any $n \geq 1$,

$$x \in \overline{\{x_k \mid k \geq n\}}^w = \overline{co}\{x_k \mid k \geq n\}. \tag{2.13}$$

The second equality follows from the just proved conclusion. Hence, we can find $\alpha_{n,k} \geq 0$, with $\sum_{k=1}^{K_n} \alpha_{n,k} = 1$, such that

$$\left| \sum_{k=1}^{K_n} \alpha_{n,k} x_{n+k} - x \right| < \frac{1}{n}, \quad \forall n \geq 1. \tag{2.14}$$

Then, (2.10) follows. □

Corollaries 2.5 and 2.8 are commonly seen in many books. The following result is more interesting.

**Corollary 2.9.** *Let $X$ be a Banach space with dual $X^*$. Let $G \subset X^*$ be convex and weak$^*$ closed, and $f_0 \notin G$. Then there exists an $x_0 \in X$, $|x_0| = 1$, such that*

$$f_0(x_0) < \inf_{f \in G} f(x_0). \tag{2.15}$$

The key point in the above result is that the $x_0$ is in $X$ instead of in $X^{**}$. To prove this result, we need the following lemma.

**Lemma 2.10.** *Let $X$ be a Banach space and $X^*$ be its dual endowed with the weak$^*$ topology. Let $g$ be a weak$^*$ continuous linear functional on $X^*$. Then there exists an $x \in X$, such that $g(x^*) = x^*(x)$, for all $x^* \in X^*$.*

*Proof.* Let $g$ be a weak$^*$ continuous linear functional on $X^*$. By Proposition 2.2, there exists a neighborhood $N \equiv N(x_1, \cdots, x_k; \varepsilon)$ of the origin (see (2.2)), such that

$$|g(x^*)| \leq 1, \quad \forall x^* \in N. \tag{2.16}$$

Let $\mathcal{N}(x_i) = \{x^* \in X \mid x^*(x_i) = 0\}$. Then $N \supset \bigcap_{i=1}^k \mathcal{N}(x_i)$, which is a subspace. Thus, (2.16) implies that $g(x^*) = 0$ for all $x^* \in \bigcap_{i=1}^k \mathcal{N}(x_i)$, i.e.,

$$\bigcap_{i=1}^k \mathcal{N}(x_i) \subset \mathcal{N}(g). \tag{2.17}$$

Then, by Lemma 1.34, we have $g = \sum_{i=1}^k \alpha_i x_i \in X$. □

*Proof of Corollary 2.9.* We consider $X^*$ as a topological vector space with the weak$^*$ topology. Because $f_0 \notin G$, both $G$ and $\{f_0\}$ are convex and closed (under weak$^*$ topology), and $\{f_0\}$ is compact, by Theorem 2.4 there exists a linear continuous functional that can be represented by $x_0 \in X$, $|x_0| = 1$ (see Lemma 2.10), such that (2.15) holds. □

**Corollary 2.11.** (Goldstine) *Let $X$ be a Banach space and $B = B_1(0)$ be the unit ball of $X$. Then the weak\* closure $\overline{B}^{w^*}$ of $B$ coincides with the unit ball $B^{**}$ of $X^{**}$. Consequently, $X$ is weak\* dense in $X^{**}$.*

*Proof.* It is clear that $B \subset B^{**}$ and $B^{**}$ is weak\* closed. Thus, $\overline{B}^{w^*} \subset B^{**}$. Now, we let $x^{**} \in B^{**} \setminus \overline{B}^{w^*}$. Then, adopting the weak\* topology in $X^{**}$ and by Theorem 2.4, we can find an $x^* \in X^*$ with $|x^*| = 1$ such that

$$(2.18) \quad 1 = |x^*| = \sup_{y \in B} x^*(y) = \sup_{y^{**} \in \overline{B}^{w^*}} y^{**}(x^*) < x^{**}(x^*) \leq |x^{**}|.$$

This means $x^{**} \notin B^{**}$, a contradiction. □

### §2.2. Convexity of Banach spaces

In this section, we are going to study the strict and uniform convexity of Banach spaces.

**Definition 2.12.** *Let $X$ be a Banach space $X$.*

*(i) $X$ is said to be* strictly convex *if its norm is strictly convex, i.e., $|x| = |y| = 1$ and $|x + y| = 2$ imply $x = y$.*

*(ii) $X$ is said to be* uniformly convex *if for any $\varepsilon > 0$, there exists a $\delta(\varepsilon) > 0$, such that $|x| = |y| = 1$ and $|x - y| \geq \varepsilon$ imply $|x+y| \geq 2(1 - \delta(\varepsilon))$.*

It is immediate that any uniformly convex Banach space is strictly convex. We have the following results, which are useful in applications.

**Proposition 2.13.** *The following are equivalent:*

*(i) $X$ is strictly convex;*

*(ii) Any $z \in X$ with $|z| = 1$ is an extreme point of $B_1(0)$, (i.e., if $z = tx + (1-t)y$ for some $t \in (0,1)$ and $x, y \in B_1(0)$, then $x = y = z$);*

*(iii) For any $f \in X^*$ there exists at most one point $z_0 \in B_1(0)$ such that $f(z_0) = \max_{z \in B_1(0)} f(z)$;*

*(iv) For any $x \neq y$ with $|x| = |y|$, it holds that $|x + y| \neq 2|x|$;*

*(v) For any $x \neq y$ with $|x| = |y|$, it holds that $|x + y| < 2|x|$.*

**Proposition 2.14.** *The following are equivalent:*

*(i) $X$ is uniformly convex;*
*(ii) $|x_n| = |y_n| = 1$ and $|x_n + y_n| \to 2$, then $|x_n - y_n| \to 0$;*
*(iii) $|x_n| \to 1$, $|y_n| \to 1$ and $|x_n + y_n| \to 2$, then $|x_n - y_n| \to 0$.*

The proofs of the above two results are simple. The following gives an interesting property of convex and closed sets in uniformly convex Banach spaces.

**Proposition 2.15.** *Let $X$ be a uniformly convex Banach space. Then, for any convex and closed set $K \subset X$, there exists a unique $x_0 \in K$, such that*

$$|x_0| = \inf\{|x| \mid x \in K\}. \tag{2.19}$$

*Proof.* Let $d = \inf_{x \in K} |x|$. If $d = 0$, by the closeness of $K$, we have $0 \in K$. Thus, $x_0 = 0$. Now, let $d > 0$. Then there exists a sequence $x_n \in K$, such that $|x_n| \to d$. By the convexity of $K$, we have $\frac{1}{2}(x_n + x_m) \in K$. Thus, $2d \le |x_n + x_m|$. Define $y_n = x_n/d$. Then, $|y_n| \to 1$ and $|y_n + y_m| \to 2$. By the above result, $|y_n - y_m| \to 0$. This implies that $\{x_n\}$ is Cauchy. Hence, $x_n \to x_0 \in K$. Clearly, $|x_0| = d$. Now, if we have $y_0 \in K$ also satisfies $|y_0| = d$, then it is necessary that $|x_0 + y_0| = 2d$. By the strict convexity of $X$, we must have $x_0 = y_0$. This proves the uniqueness. □

**Theorem 2.16.** (Milman–Pettis) *Any uniformly convex Banach space is reflexive.*

It is well known that the spaces $L^p(\Omega), W^{m,p}(\Omega)$ $(1 < p < \infty)$ are uniformly convex. But $L^1(\Omega), L^\infty(\Omega), C(\overline{\Omega})$ are not even strictly convex. Thus, we have to go a little further to meet some particular needs in applications.

**Definition 2.17.** *Let $X$ be a Banach space with norm $|\cdot|$. Let $|\cdot|_0$ be another norm on $X$. We say that these two norms are equivalent if there exist constants $\alpha, \beta > 0$, such that*

$$\alpha |x|_0 \le |x| \le \beta |x|_0, \qquad \forall x \in X. \tag{2.20}$$

Clearly, from (2.20), we see that if $(X, |\cdot|)$ is a Banach space and $|\cdot|_0$ is a norm that is equivalent to $|\cdot|$, then, $(X, |\cdot|_0)$ is also a Banach space. Hence, people try to introduce some equivalent norm in the given Banach space hoping that there will be some required convexity of the resulting space. In our applications, which will appear in later chapters, the situation is more difficult. We need the original norm $|\cdot|$ of the underlying Banach space $X$ to be changed to some equivalent one $|\cdot|_0$ so that the dual $(X, |\cdot|_0)^*$ is strictly convex (we do not care whether $(X, |\cdot|_0)$ is strictly convex). The main result of this subsection is the following *renorming theorem*.

**Theorem 2.18.** *Let $X$ be a separable Banach space with norm $|\cdot|$. Then there exists an equivalent norm $|\cdot|_0$, such that the corresponding dual $X^*$ is strictly convex.*

*Proof.* Let $\{x_k\}_{k \ge 1}$ be dense in the unit ball $B_1(0)$ of $X$. Define

$$p(f) = \Big[\sum_{k \ge 1} \frac{|f(x_k)|^2}{2^{2k}}\Big]^{1/2}, \qquad \forall f \in X^*. \tag{2.21}$$

Then, if a net $\{f_\alpha\} \subset X^*$ weakly* converges to $f \in X^*$, by Fatou's Lemma, we have

$$p(f)^2 = \sum_{k \ge 1} \frac{|f(x_k)|^2}{2^{2k}} \le \varliminf_\alpha p(f_\alpha)^2. \tag{2.22}$$

§2. Some Geometric Aspects of Banach Spaces

Thus, $f \mapsto p(f)$ is weak* lower semi continuous. Let

(2.23) $$|f|_1 = |f|_* + p(f), \qquad \forall f \in X^*,$$

where $|\cdot|_*$ is the norm of $X^*$ corresponding to $|\cdot|$. Then we see that $|\cdot|_1$ is weak* lower semi-continuous and it is a norm on $X^*$. On the other hand, for all $f \in X^*$,

(2.24) $$|f|_* \leq |f|_1 \leq |f|_* + \left[\sum_{k \geq 1} \frac{|x_k|}{2^{2k}}\right]^{1/2} |f|_* \leq [1 + \frac{1}{\sqrt{3}}]|f|_*.$$

Thus, norms $|\cdot|_*$ and $|\cdot|_1$ are equivalent. Now, let $f, g \in X^*$ with

(2.25) $$|f|_1 = |g|_1 = 1, \quad |f + g|_1 = 2.$$

Then,

(2.26) $$0 \leq \{|f|_* + |g|_* - |f + g|_*\} + \{p(f) + p(g) - p(f + g)\} = 0.$$

Because the terms in the two sets of brackets on the left-hand side of (2.26) are nonnegative, we obtain

(2.27) $$p(f + g) = p(f) + p(g).$$

Thus, by (2.21), it is necessary that for some constant $\lambda \in \mathbb{R}$,

(2.28) $$f(x_k) = \lambda g(x_k), \qquad \forall k \geq 1,$$

which gives $f = \lambda g$ as $\{x_k\}_{k \geq 1}$ is dense in $B_1(0)$. By (2.25), we obtain $|1 + \lambda| = 2$ and $|\lambda| = 1$. Hence, $\lambda = 1$ and $f = g$. This shows the strict convexity of the space $X^*$ under norm $|\cdot|_1$. Next, we define

(2.29) $$|x|_0 = \sup_{|f|_1 \leq 1} |f(x)|, \qquad x \in X.$$

Clearly, $|\cdot|_0$ is a norm on $X$. On the other hand, by (2.24), for any $x \in X$,

(2.30) $$|x|_0 \leq |x| = \sup_{|f|_* \leq 1} |f(x)| \leq \sup_{|f|_1 \leq 1 + \frac{1}{\sqrt{3}}} |f(x)| \leq (1 + \frac{1}{\sqrt{3}})|x|_0.$$

Thus, norms $|\cdot|$ and $|\cdot|_0$ are equivalent. We claim that $|\cdot|_1$ is the dual norm induced by $|\cdot|_0$. To show this, it suffices to prove that for any $f \in X^*$, $|f|_1 = 1$, we have

(2.31) $$\sup_{|x|_0 \leq 1} f(x) = 1 (= |f|_1).$$

Let us now prove this. First of all, note that

(2.32) $$\{x \in X \mid |x|_0 \leq 1\} = \{x \in X \mid \sup_{|g|_1 \leq 1} |g(x)| \leq 1\}$$
$$\subseteq \{x \in X \mid |f(x)| \leq 1\}.$$

Hence,

(2.33) $$\sup_{|x|_0 \leq 1} f(x) \leq \sup_{\{x \in X \big| |f(x)| \leq 1\}} f(x) \leq 1.$$

On the other hand, we let $B^* = \{g \in X^* \mid |g|_1 \leq 1\}$. Because $|\cdot|_1$ is weak* lower semicontinuous, the set $B^*$ is weak* closed and also convex. Now, for any $r > 1$, the point $rf \notin B^*$. Thus, by Corollary 2.9, there exists an $x_r \in X$ with $|x_r|_0 = 1$, such that

(2.34) $$rf(x_r) > \sup_{g \in B^*} g(x_r) = |x_r|_0 = 1.$$

Then it follows that

(2.35) $$\sup_{|x|_0 \leq 1} f(x) \geq \inf_{r > 1} f(x_r) \geq 1.$$

Hence (2.31) follows from (2.33) and (2.35). □

We have seen that in the above proof, Corollary 2.9 is very crucial. In the literature, there is a more general result concerning the above matter. To state the result, let us first introduce the following notion. A Banach space $X$ is said to be *weakly compactly generated* if there exists a weakly compact subset $X_0$ of $X$ such that $X$ is spanned by $X_0$, i.e., the set of all finite linear combinations of the elements in $X_0$ is (norm) dense in $X$. Clearly, any separable Banach spaces and any reflexive Banach spaces are weakly compactly generated. Now, let us state that result. We will not present a proof here as too many things about the geometry of Banach spaces are involved (see Diestel [1], p.167 for details).

**Theorem 2.19.** *Let $X$ be a weakly compactly generated Banach space. Then there exists an equivalent norm under which the dual space $X^*$ is strictly convex.*

Let $X$ and $Y$ be two Banach spaces and $G \subset X$. Let $F : G \to Y$ be a map (not necessarily linear). We say that $F$ is *continuous* at $x_0 \in G$ if $|F(x) - F(x_0)|_Y \to 0$ whenever $x \in G$, $|x - x_0|_X \to 0$. If $F$ is continuous at each point $x_0 \in G$, we say that $F$ is continuous on $G$. The set of all continuous maps from $G$ to $Y$ is denoted by $C(G; Y)$. In the case $Y = \mathbb{R}$, we simply denote it by $C(G)$. Next, let $G$ be an open set we say that $F$ is *Gâteaux differentiable* at $x_0 \in G$ if there exists an $F_1 \in \mathcal{L}(X, Y)$, such that

(2.36) $$\lim_{\delta \to 0} \left| \frac{F(x_0 + \delta x) - F(x_0)}{\delta} - F_1 x \right|_Y = 0, \quad \forall x \in X.$$

If we replace (2.36) by the following

(2.37) $$\lim_{\delta \to 0} \sup_{|x|_X \leq 1} \left| \frac{F(x_0 + \delta x) - F(x_0)}{\delta} - F_1 x \right|_Y = 0,$$

then $F$ is said to be *Fréchet differentiable* at $x_0$. Moreover, if $F$ is Gâteaux (Fréchet, resp.) differentiable at each point $x_0 \in G$, we say that $F$ is Gâteaux (Fréchet, resp.) differentiable on $G$.

Next, a norm $|\cdot|$ on $X$ is said to be *Gâteaux* (resp. *Fréchet*) *differentiable* if the map $x \mapsto |x|$ is Gâteaux (resp. Fréchet) differentiable on $X \setminus \{0\}$. The following result states a relation between the convexity of the space and the Fréchet differentiability of the norm.

**Proposition 2.20.** *Let $X$ be a Banach space.*

*(i) If $X^*$ is strictly convex, then the norm of $X$ is Gâteaux differentiable;*

*(ii) $X$ is reflexive if and only if there exists an equivalent norm $|\cdot|$ on $X$, which, together with its dual norm $|\cdot|_*$, is Fréchet differentiable.*

The proof can be found in Diestel [1], pp. 23,34, and 167. The point of the above (ii) is that for reflexive Banach spaces, we can assume that both the norm $|\cdot|$ of $X$ and its dual norm $|\cdot|_*$ are Fréchet differentiable. This result has a very interesting consequence. To state it, let us first give the following notion.

**Definition 2.21.** Let $X$ be a Banach space. A continuous function $\psi : X \to [0,1]$ is called a *Fréchet differentiable bump function* if it has the following properties: The set $\{x \in X \mid \psi(x) > 0\}$ is bounded and nonempty; on this set, $\psi$ is Fréchet differentiable.

**Corollary 2.22.** *Let $X$ be a reflexive Banach space. Then $X$ and $X^*$ both admit Fréchet differentiable bump functions.*

*Proof.* We may let the norm $|\cdot|$ on $X$ be Fréchet differentiable. Take any smooth function $\varphi : \mathbb{R} \to [0,1]$ that is not identically zero and with a compact support. Then define $\psi(x) = \varphi(|x|)$. This is a Fréchet differentiable bump function on $X$. We may prove the same thing for $X^*$. □

Corollary 2.22 will be very useful in Chapters 6 and 8.

## §3. Banach Space Valued Functions

In this section, we recall some basic results on Banach space valued functions. Throughout this section, we let $X$ be a Banach space. By a *Banach space valued function*, we mean any map with its image in some Banach space, for example, $f : [0,T] \to X$.

### §3.1. Measurability and integrability

We first consider the measurability of the functions and the Bochner integral.

**Definition 3.1.** (i) Function $f : [0,T] \to X$ is called a *simple function* if there exist finitely many measurable sets $E_i \subset [0,T]$, mutually disjoint, and $x_i \in X$, such that

(3.1) $$f(t) = \sum_i x_i \chi_{E_i}(t), \qquad t \in [0, T].$$

(ii) Function $f : [0, T] \to X$ is said to be *strongly measurable* if there exists a sequence of simple functions $\varphi_k : [0, T] \to X$, such that

(3.2) $$\lim_{k \to \infty} |\varphi_k(t) - f(t)| = 0, \qquad \text{a.e. } t \in [0, T].$$

(iii) Function $f : [0, T] \to X$ is said to be *weakly measurable* if for any $x^* \in X^*$ the scalar function $t \mapsto \langle x^*, f(\cdot) \rangle$ is (Lebesgue) measurable.

**Proposition 3.2.** *Let $f : [0, T] \to X$. If it is strongly measurable, then it is weakly measurable. Conversely, if it is weakly measurable and there exists a set $E \subset [0, T]$ of measure zero, such that the set $\{f(t) \mid t \in [0, T] \setminus E\}$ is separable in $X$, then $f$ is strongly measurable. In particular, if $X$ is separable, then $f$ is strongly measurable if and only if it is weakly measurable.*

Now, for any simple function $f(\cdot) = \sum_i x_i \chi_{E_i}(\cdot)$, we define its *Bochner integral* by

(3.3) $$\int_0^T f(t)\, dt = \sum_i x_i |E_i|,$$

where $|E_i|$ is the Lebesgue measure of the set $E_i$.

**Definition 3.3.** Let $f : [0, T] \to X$ be strongly measurable. We say that $f$ is *Bochner integrable* if there exists a sequence of simple functions $\varphi_k : [0, T] \to X$, such that (3.2) holds and the sequence $\int_0^T \varphi_k(t) dt$ is strongly convergent in $X$. In this case, we define the *Bochner integral* of the function $f$ by

(3.4) $$\int_0^T f(t)\, dt = \lim_{k \to \infty} \int_0^T \varphi_k(t)\, dt.$$

It is not hard to see that the integral defined by (3.4) is independent of the choice of the sequences $\{\varphi_k\}$. Similar to the Lebesgue integral, for any measurable set $E \subset [0, T]$, the Bochner integral of $f$ over the set $E$ is defined by

(3.5) $$\int_E f(t)\, dt = \int_0^T f(t) \chi_E(t)\, dt.$$

The following result is basic.

**Proposition 3.4.** *Let $f : [0, T] \to X$ be strongly measurable. Then $f$ is Bochner integrable if and only if $|f(\cdot)|$ is Lebesgue integrable. Moreover, in this case*

(3.6) $$\left| \int_0^T f(t)\, dt \right| \leq \int_0^T |f(t)|\, dt.$$

§3. Banach Space Valued Functions

The Bochner integral possesses almost the same properties as the Lebesgue integral. We omit the exact statement here. Next, if $f : [0,T] \to X$ is strongly measurable and $|f(\cdot)| \in L^p(0,T)$ for some $p \in [1,\infty)$, then we say that $f$ is $L^p$ Bochner integrable. By Proposition 3.4, we see that any $L^p$ Bochner integrable function (with $p \in [1,\infty)$) $f$ is Bochner integrable. The set of all $L^p$ Bochner integrable functions is denoted by $L^p(0,T;X)$. Then it is seen that the set of Bochner integrable functions is nothing but $L^1(0,T;X)$. For any $f \in L^p(0,T;X)$, we define

$$(3.7) \qquad |f|_{L^p(0,T;X)} = \left\{ \int_0^T |f(t)|^p \, dt \right\}^{1/p}.$$

It is clear that $|\cdot|_{L^p(0,T;X)}$ defined above is a norm under which $L^p(0,T;X)$ is a Banach space. The space $L^\infty(0,T;X)$ is defined to be the set of all $f : [0,T] \to X$ with $|f(\cdot)| \in L^\infty(0,T)$ and the norm can be defined in an obvious way. Similar to finite dimensional space valued functions, we have the following result.

**Proposition 3.5.** *Let $X$ be a reflexive Banach space and $p \in [1,\infty)$, $p' = \frac{p}{p-1}$. Then*

$$(3.8) \qquad L^p(0,T;X)^* = L^{p'}(0,T;X^*).$$

## §3.2. Continuity and differentiability

Next, let us look at the continuity of Banach space valued functions. In infinite dimensional Banach spaces, there are more than one topologies; thus, we have different kinds of continuities for the Banach space valued functions.

**Definition 3.6.** (i) Let $f : [0,T] \to X$. We say that $f$ is *strongly continuous* at $t_0 \in [0,T]$ if for any $\varepsilon > 0$, there exists a $\delta > 0$ such that for any $t \in [0,T]$, with $|t - t_0| < \delta$, one has $|f(t) - f(t_0)| < \varepsilon$. We say that $f : [0,T] \to X$ is *weakly continuous* at $t_0 \in [0,T]$ if for any $x^* \in X^*$, the scalar function $\langle x^*, f(\cdot) \rangle$ is continuous at $t_0$. If $f$ is strongly (weakly, resp.) continuous at each point of $[0,T]$, we say that $f$ is strongly (weakly, resp.) continuous on $[0,T]$.

(ii) Let $X^*$ be the dual space of some Banach space $X$ and $f : [0,T] \to X^*$. We say that $f$ is *weakly\* continuous* at $t_0 \in [0,T]$ if for each $x \in X$ the scalar function $\langle f(\cdot), x \rangle$ is continuous at $t_0 \in [0,T]$. If $f$ is weakly* continuous at each point of $[0,T]$, we say that $f$ is weakly* continuous on $[0,T]$.

(iii) Let $X$ and $Y$ be two Banach spaces and $F : [0,T] \to \mathcal{L}(X,Y)$ be an *operator valued* function. We say that $F$ is *continuous in the operator norm* at $t_0 \in [0,T]$ if for any $\varepsilon > 0$, there exists a $\delta > 0$ such that for any $t \in [0,T]$, with $|t - t_0| < \delta$, one has

$$\|F(t) - F(t_0)\|_{\mathcal{L}(X,Y)} < \varepsilon.$$

We say that $F$ is *strongly continuous* at $t_0 \in [0,T]$ if for any $x \in X$, the $Y$ valued function $F(\cdot)x$ is strongly continuous at $t_0$ as defined in (i). We say that $F$ is *weakly continuous* at $t_0 \in [0,T]$ if for any $x \in X$ and $y^* \in Y^*$, the scalar function $\langle y^*, F(\cdot)x \rangle$ is continuous at $t_0$. The corresponding continuities on the whole interval $[0,T]$ can be defined in an obvious way. In the case where $Y$ is the dual of some Banach space, we can also define the weak* continuity of the function $F$.

We should distinguish the strong continuity for the Banach space valued functions and the operator valued functions. The set of all strong continuous functions from $[0,T]$ to $X$ is denoted by $C([0,T];X)$. For any $f \in C([0,T];X)$, we define

$$(3.9) \qquad |f|_{C([0,T];X)} = \sup_{t \in [0,T]} |f(t)|.$$

Then it is easily seen that (3.9) defines a norm under which $C([0,T];X)$ is a Banach space.

Next, a function $f \in C([0,T];X)$ is said to be (strongly) *differentiable* at $t_0 \in (0,T)$, if there exists a $g \in X$ such that

$$(3.10) \qquad \lim_{s \to t_0} \left| \frac{f(s) - f(t_0)}{s - t_0} - g \right| = 0.$$

We usually denote $g$ by $f'(t_0)$ and call it the (strong) *derivative* of $f$ at $t_0$. If $f$ is differentiable at each point $t_0 \in [0,T]$, we say that $f$ is differentiable on $[0,T]$. In this case, $f' : [0,T] \to X$ is a strongly measurable function. If we further have $f'(\cdot) \in C([0,T];X)$, then $f(\cdot)$ is said to be *continuously differentiable* on $[0,T]$, denoted by $f \in C^1([0,T];X)$. Similarly, we can define a higher order differentiability of $f$. We denote $C^k([0,T];X)$ to be the set of all $k$ times continuously differentiable functions defined on $[0,T]$ taking values in $X$. Next, for $1 \leq p \leq \infty$, we let $W^{k,p}([0,T];X)$ be the completion of $C^k([0,T];X)$ under the norm

$$(3.11) \qquad |f|_{W^{k,p}([0,T];X)} = \sum_{i=1}^{k} |f^{(i)}|_{L^p(0,T;X)}.$$

Recall that in §2.2, we have defined the strong continuity, the Gâteaux and Fréchet differentiabilities of the maps between Banach spaces. The following result is concerned with the compactness of families of continuous Banach space valued functions.

**Theorem 3.7.** (Arzela–Ascoli) *Let $X$ and $Y$ be Banach spaces and $G \subset X$ be a compact set. Let $\mathcal{F} \subset C(G;Y)$ such that for each $x \in G$, the set $\{F(x) \mid F \in \mathcal{F}\}$ is relatively compact in $Y$. Moreover, $\mathcal{F}$ is uniformly bounded and equicontinuous, i.e.,*

$$(3.12) \qquad \sup_{F \in \mathcal{F}, x \in G} |F(x)|_Y < \infty,$$

and for any $\varepsilon > 0$, there exists a $\delta = \delta(\varepsilon) > 0$, such that

(3.13) $\quad |F(x) - F(x')|_Y < \varepsilon, \qquad \forall x, x' \in G, |x - x'|_X < \delta, \quad F \in \mathcal{F}.$

Then there exists a sequence $F_k \in \mathcal{F}$ and $F_0 \in C(G; Y)$, such that

(3.14) $\qquad\qquad \lim_{k \to \infty} \max_{x \in G} |F_k(x) - F_0(x)|_Y = 0.$

In most applications, $X = \mathbb{R}^n$ and $G$ is some bounded closed subset of $\mathbb{R}^n$, for example $X = \mathbb{R}$ and $G = [0, T]$.

## §4. Theory of $C_0$ Semigroups

In this section, we present some basic results on $C_0$ semigroups.

### §4.1. Unbounded operators

We first recall some results on unbounded linear operators. Let $X$ and $Y$ always be Banach spaces unless otherwise stated.

**Definition 4.1.** Let $\mathcal{D}(A)$ be a linear subspace of $X$ (not necessarily closed). Let $A : \mathcal{D}(A) \to Y$ be a linear operator.

(i) We say that $A$ is *densely defined* if $\mathcal{D}(A)$ is dense in $X$.

(ii) We say that $A$ is *closed* if the graph

(4.1) $\qquad\qquad \mathcal{G}(A) \equiv \{(x, y) \in X \times Y \mid x \in \mathcal{D}(A), y = Ax\}$

of $A$ is closed in $X \times Y$.

(iii) We say that $A$ is *closable* if there exists a closed operator $\overline{A} : \mathcal{D}(\overline{A}) \subset X \to Y$, such that

(4.2) $\qquad\qquad \mathcal{D}(A) \subset \mathcal{D}(\overline{A}), \qquad \overline{A}x = Ax, \quad \forall x \in \mathcal{D}(A).$

If operators $A$ and $\overline{A}$ satisfy (4.2), we say that $\overline{A}$ is an *extension* of $A$ or $A$ is a *restriction* of $\overline{A}$, denoted by $A \subseteq \overline{A}$. We know that any linear bounded operator $A \in \mathcal{L}(X)$ is densely defined and closed. However, there are many densely defined closed operators that map some bounded sets to unbounded sets. These operators are called *unbounded linear operators*. We cannot define the norm for unbounded operators. The most typical densely defined (on $X = L^2(0,1)$) and closed *unbounded* operator is the differential operator $D = \frac{d}{dx}$, with the domain $\mathcal{D}(D) = \{y \mid y$ is absolutely continuous with $Dy \in L^2(0,1), y(0) = 0\}$. Clearly, $\mathcal{D}(D)$ is dense in $L^2(0,1)$ and also it can be shown that $D$ is closed. We claim that this operator is unbounded. In fact, $y_k(x) = \sin kx, k \geq 1$ is a bounded sequence in $L^2(0,1)$. But, $Dy_k(x) = k \cos kx$ is unbounded in $L^2(0,1)$.

Next, we note that a linear densely defined operator $A : \mathcal{D}(A) \subset X \to Y$ is not necessarily closed. This can be easily seen from the following extremal example: Let $X_0$ be a dense subspace of $X$, $X_0 \neq X$. Let $\mathcal{D}(A_0) =$

$X_0$ and $A_0 x = 0$ for all $x \in \mathcal{D}(A_0)$. Then $A$ is not closed. In fact, pick any $x_0 \in X \setminus \mathcal{D}(A_0)$ and let $x_n \in \mathcal{D}(A_0)$ with $x_n \xrightarrow{s} x_0$. Because $A_0 x_n = 0$ for all $n \geq 1$, if $A_0$ were closed, we would have $x_0 \in \mathcal{D}(A_0)$, which is a contradiction. Thus, $A_0$ is not closed. However, we see that although $A_0$ is not closed, it has a closed extension, namely the zero operator, which is closed (in fact it is bounded). We will see an example showing that there are densely defined operators that are not closable. The following gives a criterion for operators to be closable.

**Proposition 4.2.** *Linear operator $A : \mathcal{D}(A) \subset X \to Y$ is closable if and only if for all $x_n \in \mathcal{D}(A)$ with $x_n \xrightarrow{s} 0$ and $A x_n \xrightarrow{s} y$, it is necessary that $y = 0$.*

Now, let $A : \mathcal{D}(A) \subset X \to X$ be densely defined. Clearly, the map

(4.3) $$x \mapsto \langle Ax, y \rangle, \qquad x \in \mathcal{D}(A), y \in X^*,$$

is well defined. Suppose $y \in X^*$ is such that

(4.4) $$|\langle Ax, y \rangle| \leq C_y |x|_X, \qquad \forall x \in \mathcal{D}(A).$$

Then the functional $f_y(x) \equiv \langle Ax, y \rangle$ can be extended linearly and boundedly to the whole of $\overline{\mathcal{D}(A)} = X$. Such an extension (still denoted by itself) $f_y$ is in $X^*$ and unique. Hence, we obtain

(4.5) $$\langle Ax, y \rangle = f_y(x) = \langle x, f_y \rangle, \qquad \forall x \in \mathcal{D}(A).$$

Now, we define

(4.6) $$\begin{cases} \mathcal{D}(A^*) = \{ y \in X^* \mid \exists C_y \geq 0, \ |\langle Ax, y \rangle| \leq C_y |x|_X, \ \forall x \in \mathcal{D}(A) \}, \\ A^* y = f_y, \qquad \forall y \in \mathcal{D}(A^*). \end{cases}$$

It is seen that $A^* : \mathcal{D}(A^*) \subset X^* \to X^*$ is a linear operator satisfying

(4.7) $$\langle Ax, y \rangle = \langle x, A^* y \rangle, \qquad \forall x \in \mathcal{D}(A), \ y \in \mathcal{D}(A^*).$$

Operator $A^*$ is called the *adjoint operator* of $A$. It is clear that in the case $A \in \mathcal{L}(X)$, the above definition coincides with the one for bounded operators (see §1.4). We should note that in the case where $A$ is not densely defined, the adjoint operator $A^*$ is not well defined as the extension $f_y$ to the whole of $X$ is not unique. Finally, in the case where $X$ is a Hilbert space, let $X^* = X$ and we say that $A : \mathcal{D}(A) \subset X \to X$ is *symmetric* if $A \subseteq A^*$, namely, $A^*$ is an extension of $A$. Furthermore, we say that $A$ is *self-adjoint* if $A^* = A$ (meaning that $A \subseteq A^*$ and $A^* \subseteq A$). This definition coincides with the one given in §1.4 when the operator is bounded. Now, let us give some basic properties of adjoint operators.

**Proposition 4.3.** (i) *Let $A : \mathcal{D}(A) \subset X \to Y$ be densely defined. Then the graph $\mathcal{G}(A^*) \subset Y^* \times X^*$ of $A^*$ is given by*

(4.8) $$\mathcal{G}(A^*) = \big(\mathcal{J}\mathcal{G}(A)\big)^\perp = (\mathcal{J}^*)^{-1}\big(\mathcal{G}(A)^\perp\big),$$

## §4. Theory of $C_0$ Semigroups

where $\mathcal{J}: X \times Y \to Y \times X$ is defined by $\mathcal{J}(x,y) = (-y,x)$, for all $(x,y) \in X \times Y$, and $\mathcal{J}^*: Y^* \times X^* \to X^* \times Y^*$, $\mathcal{J}^*(y^*, x^*) = (x^*, -y^*)$, for all $(y^*, x^*) \in Y^* \times X^*$. Consequently, $A^*$ is closed.

(ii) Let $A: \mathcal{D}(A) \subset X \to Y$ be densely defined. Then $A$ is closable if and only if $^\perp\mathcal{D}(A^*) = \{0\}$; in particular, if $Y$ is reflexive, then $A$ is closable if and only if $A^*$ is densely defined and $A^{**} = \overline{A}$.

(iii) Let $X$ and $Y$ be Hilbert spaces and let $A: \mathcal{D}(A) \subset X \to Y$ be densely defined and closed. Then, $A^*A: \mathcal{D}(A^*A) \subset X \to X$ is non-negative and self-adjoint. Moreover, there exists a self-adjoint operator $A_0: \mathcal{D}(A_0) \subset X \to X$ and a $K \in \mathcal{L}(X,Y)$, such that

(4.9) $$\begin{cases} A = KA_0, \quad A_0^2 = A^*A, \quad \mathcal{N}(K) = \mathcal{R}(A_0)^\perp, \\ |Kx| = |x|, \quad \forall x \in \mathcal{N}(K)^\perp (\equiv \overline{\mathcal{R}(A_0)}). \end{cases}$$

Any operator $K$ satisfying $|Kx| = |x|$, for all $x \in \mathcal{N}(K)^\perp$ is called a *partial isometry*.

**Proposition 4.4.** *Let $A$ and $B$ be densely defined and closed operators (on some proper spaces). Then the following hold.*

(i) *If $AB$ is densely defined, then*

(4.10) $$(AB)^* \supseteq B^*A^*.$$

*The equality holds if $A$ is bounded or $B$ is boundedly invertible.*

(ii) *If $A + B$ is densely defined, then*

(4.11) $$(A+B)^* \supseteq A^* + B^*.$$

*The equality holds if $A$ or $B$ is bounded.*

The following is an example of a *nonclosable* operator.

*Example 4.5.* Let $\mathbb{N} = \{1, 2, \cdots\}$ be the set of positive integers. It is not hard to give a partition of $\mathbb{N}$ as follows:

(4.12) $$\begin{cases} \bigcup_{k \in \mathbb{N}} \{n_{k,\ell}\}_{\ell \in \mathbb{N}} = \mathbb{N}, \\ \{n_{k,\ell}\}_{\ell \in \mathbb{N}} \cap \{n_{j,\ell}\}_{\ell \in \mathbb{N}} = \phi, \quad k \neq j. \end{cases}$$

Let $X = \ell^2 \equiv \{(x_n)_{n \geq 1} \mid \sum_{n \geq 1} |x_n|^2 < \infty\}$ and define

(4.13) $$\begin{cases} \mathcal{D}(A) = \{x = (x_n)_{n \geq 1} \in \ell^2 \mid \exists n_0 \geq 1, \, x_n = 0, \, n > n_0 \}, \\ Ax = (\sum_{\ell \geq 1} x_{n_{1,\ell}}, \sum_{\ell \geq 1} x_{n_{2,\ell}}, \cdots), \quad \forall x \in \mathcal{D}(A). \end{cases}$$

Note that each term in the definition of $Ax$ is a finite sum as $x$ has only finitely many nonzero components. Now, let $y = (y_n)_{n \geq 1} \in \mathcal{D}(A^*)$ and let

$z = (z_n)_{n\geq 1} = A^*y$. Then, by definition, for any $x \in \mathcal{D}(A)$,

(4.14)
$$\langle y, Ax \rangle = \sum_{k\geq 1}\sum_{\ell\geq 1} y_k x_{n_{k,\ell}} = \langle A^*y, x \rangle$$
$$= \langle z, x \rangle = \sum_{n\geq 1} z_n x_n = \sum_{k\geq 1}\sum_{\ell\geq 1} z_{n_{k,\ell}} x_{n_{k,\ell}}.$$

By setting $x = e_{n_{k,\ell}}$ (the element in $\ell^2$ in which all the components are zero except the $n_{k,\ell}$th component, which is 1), we see that

(4.15) $$z_{n_{k,\ell}} = y_k, \qquad \forall k, \ell \in \mathbb{N}.$$

Note $z \in \ell^2$, which implies $\sum_{\ell\geq 1} |z_{n_{k,\ell}}|^2 < \infty$. Thus, (4.15) yields

(4.16) $$y_k = 0, \qquad \forall k \geq 1.$$

Hence, $y = 0$. This shows that $\mathcal{D}(A^*) = \{0\}$. By Proposition 4.3 (ii), the operator $A$ defined above is not closable.

Similar to bounded operators, for any linear operator (not necessarily bounded) $A : \mathcal{D}(A) \subset X \to X$, we can still define the resolvent $\rho(A) = \{\lambda \in \mathbb{C} \mid (\lambda - A)^{-1} \in \mathcal{L}(X)\}$, the spectrum $\sigma(A) = \mathbb{C} \setminus \rho(A)$, and the point spectrum (or the set of eigenvalues) $\sigma_p(A)$ of $A$. However, we do not define the approximate point spectrum and residue spectrum of unbounded operators.

## §4.2. $C_0$ semigroups

Let us first introduce the following definition.

**Definition 4.6.** Let $\{T(t) \mid t \geq 0\} \subset \mathcal{L}(X)$. We call $T(\cdot)$ a $C_0$ semigroup on $X$ if the following holds:

(4.17)
$$\begin{cases} T(0) = I, \\ T(t+s) = T(t)T(s), & \forall t, s \geq 0, \\ \lim_{s\to 0} |T(s)x - x| = 0, & x \in X. \end{cases}$$

In the case where $T(t)$ is defined for all $t \in \mathbb{R}$ and the second equality in (4.17) holds for all $t, s \in \mathbb{R}$, we call $T(\cdot)$ a $C_0$ group.

The second and third properties in (4.17) are usually referred to as the *semigroup property* and the *strong continuity*, respectively.

**Proposition 4.7.** *Let $T(\cdot)$ be a $C_0$ semigroup on $X$. Then there exist constants $M \geq 1$ and $\omega \in \mathbb{R}$, such that*

(4.18) $$\|T(t)\| \leq Me^{\omega t}, \qquad t \geq 0.$$

## §4. Theory of $C_0$ Semigroups

**Definition 4.8.** Let $T(\cdot)$ be a $C_0$ semigroup on $X$. Let

(4.19)
$$\begin{cases} \mathcal{D}(A) = \{x \in X \mid \text{s-}\lim_{t \to 0} \dfrac{T(t) - I}{t} x \text{ exists }\}, \\ Ax = \text{s-}\lim_{t \to 0} \dfrac{T(t) - I}{t} x, \quad \forall x \in \mathcal{D}(A), \end{cases}$$

where s-lim stands for the strong limit. Operator $A : \mathcal{D}(A) \subset X \to X$ is called the *generator* of the semigroup $T(\cdot)$. In general, the operator $A$ is not bounded. We have the following result.

**Proposition 4.9.** *Let $X$ be a Banach space and $T(t)$ be a $C_0$ semigroup with the generator $A$. Then, $A \in \mathcal{L}(X)$ if and only if $t \mapsto T(t)$ is continuous in the operator norm at $t = 0$, i.e., $\lim_{t \to 0} \|T(t) - I\| = 0$. In this case, $T(t)$ admits the following representation:*

(4.20)
$$T(t) = \sum_{n=0}^{\infty} \frac{A^n t^n}{n!}, \qquad t \geq 0,$$

*where the series converges in the operator norm.*

From the above, we see that if $T(t)$ is *only* strongly continuous at $t = 0$, the generator $A$ must be unbounded. On the other hand, if the generator $A$ of $T(t)$ is unbounded, then, $T(t)$ can not be continuous in the operator norm at $t = 0$.

From (4.19), we see that $\mathcal{D}(A)$ is a linear subspace; in particular, it is nonempty. But *a priori*, it is not clear how "large" the set $\mathcal{D}(A)$ is. Hence, the following result is deep.

**Proposition 4.10.** *Let $T(\cdot)$ be a $C_0$ semigroup on $X$. Then the generator $A$ of $T(\cdot)$ is a densely defined closed operator with $\mathcal{D}(A^\infty) \equiv \bigcap_{n=1}^{\infty} \mathcal{D}(A^n)$ being also dense in $X$. Furthermore, if $S(\cdot)$ is another $C_0$ semigroup on $X$ with the same generator $A$ as $T(\cdot)$, then, $S(\cdot) = T(\cdot)$.*

The above tells us that the $C_0$ semigroup $T(\cdot)$ is uniquely determined by its generator $A$. Because of this, we say that $A$ *generates* the $C_0$ semigroup $T(\cdot)$ if $A$ is the generator of $T(\cdot)$. The following result gives a characterization of the generators of $C_0$ semigroups.

**Theorem 4.11.** (Hille–Yosida) *Let $X$ be a Banach space and $A : \mathcal{D}(A) \subset X \to X$ be a linear operator. Then the following are equivalent:*

*(i) $A$ generates a $C_0$ semigroup $T(\cdot)$ on $X$, such that (4.18) holds with some $\omega \in \mathbb{R}$ and $M \geq 1$.*

*(ii) $A$ is densely defined and closed, such that for the above $\omega \in \mathbb{R}$ and $M \geq 1$, we have $\rho(A) \supset \{\lambda \in \mathbb{C} \mid \operatorname{Re}\lambda > \omega\}$ and*

(4.21)
$$\|(\lambda - A)^{-n}\| \leq \frac{M}{(\operatorname{Re}\lambda - \omega)^n}, \qquad \forall n \geq 0, \ \operatorname{Re}\lambda > \omega.$$

(iii) $A$ is densely defined and closed, such that for the above $\omega \in \mathbb{R}$ and $M \geq 1$, we have some sequence $\lambda_k > 0$, $\lambda_k \to \infty$, with $\lambda_k \in \rho(A)$ and

(4.22) $$\|(\lambda_k - A)^{-n}\| \leq \frac{M}{(\lambda_k - \omega)^n}, \qquad \forall n \geq 0,\ k \geq 1.$$

Because the generator $A$ determines the $C_0$ semigroup $T(\cdot)$ uniquely and in the case $A \in \mathcal{L}(X)$ this $C_0$ semigroup has an explicit expression $e^{At} \equiv \sum_{n \geq 0} \frac{A^n t^n}{n!}$, hereafter, we denote $e^{At}$ the $C_0$ semigroup generated by $A$. The following results collect some other important properties of $C_0$ semigroups and their generators.

**Proposition 4.12.** Let $e^{At}$ be a $C_0$ semigroup on $X$. Let $A_\lambda = \lambda A(\lambda - A)^{-1}$, called the Yosida approximation of $A$. Then

(4.23) $$\lim_{\lambda \to \infty} |A_\lambda x - Ax| = 0, \qquad \forall x \in \mathcal{D}(A),$$

(4.24) $$\lim_{\lambda \to \infty} \sup_{t \in [0,T]} |e^{A_\lambda t} x - e^{At} x| = 0, \qquad \forall x \in X,\ T > 0.$$

**Proposition 4.13.** Let $e^{At}$ be a $C_0$ semigroup on $X$. Then $(e^{At})^*$ is a semigroup on $X^*$ that is weakly* continuous (as an operator valued function). In the case where $X$ is reflexive, $(e^{At})^*$ is also a $C_0$ semigroup on $X^*$ with the generator $A^*$.

Hereafter, we will simply denote $(e^{At})^*$ by $e^{A^* t}$ and keep in mind that this is not necessarily a $C_0$ semigroup unless $X$ is reflexive.

**Proposition 4.14.** Let $X$ be a Banach space and $e^{At}$ be a $C_0$ semigroup on $X$. Then

(i) $\omega_0 \equiv \overline{\lim}_{t \to \infty} \frac{1}{t} \log \|e^{At}\|$ is finite and for any $\lambda$ with $\operatorname{Re} \lambda > \omega_0$, it holds that $\lambda \in \rho(A)$ and

(4.25) $$(\lambda - A)^{-1} x = \int_0^\infty e^{-\lambda t} e^{At} x\, dt, \qquad \forall x \in X.$$

(ii) For any $x \in \mathcal{D}(A)$, it holds that $e^{At} x \in \mathcal{D}(A)$, $t \geq 0$ and

(4.26) $$\begin{cases} \dfrac{d}{dt}(e^{At} x) = A e^{At} x = e^{At} A x, & \forall t > 0, \\ e^{At} x - e^{As} x = \displaystyle\int_s^t e^{Ar} A x\, dr, & \forall 0 \leq s < t < \infty. \end{cases}$$

(iii) Let $g \in L^1_{loc}(0, \infty)$ and $t \geq 0$ be a right-Lebesgue point of $g$, i.e.,

(4.27) $$\lim_{h \to 0} \frac{1}{h} \int_t^{t+h} |g(r) - g(t)|\, dr = 0.$$

§4. Theory of $C_0$ Semigroups

Then

(4.28) $$\lim_{h \to 0} \left| \frac{1}{h} \int_t^{t+h} g(r) e^{Ar} x \, dr - g(t) e^{At} x \right| = 0.$$

(iv) Let $A_h = \frac{1}{h}(e^{Ah} - I)$. Then, for any $x \in X$ and any $T > 0$,

(4.29) $$\lim_{h \to 0} |e^{A_h t} x - e^{At} x| = 0, \qquad \text{uniformly in } t \in [0, T].$$

Next, for any $C_0$ semigroup $e^{At}$, we define

(4.30) $$\mathcal{P}(A) = \{ P \mid P \text{ closed}, \mathcal{D}(P) \supset \mathcal{D}(A); \exists K_t < \infty, \quad t > 0,$$
$$\int_0^1 K_t \, dt < \infty, \quad |P e^{At} x| \leq K_t |x|, \quad \forall x \in \mathcal{D}(A) \}.$$

Clearly, $\mathcal{L}(X) \subset \mathcal{P}(A)$. We have the following result for the perturbation of generators of $C_0$ semigroups.

**Proposition 4.15.** *Let $X$ be a Banach space and $e^{At}$ be a $C_0$ semigroup on $X$. Then, for any $P \in \mathcal{P}(A)$, the operator $A + P$ generates a $C_0$ semigroup $e^{(A+P)t}$ on $X$.*

## §4.3. Special types of $C_0$ semigroups

Let us introduce the following special types of $C_0$ semigroups.

**Definition 4.16.** *Let $e^{At}$ be a $C_0$ semigroup on $X$.*

(i) $e^{At}$ *is a* contraction semigroup *if*

(4.31) $$\| e^{At} \| \leq 1, \qquad \forall t \geq 0.$$

(ii) $e^{At}$ *is* compact *for $t > t_0$ ($t_0 \geq 0$) if for any $t > t_0$, $e^{At}$ is a compact linear operator on $X$. In the case $t_0 = 0$, we simply say that $e^{At}$ is* compact.

(iii) $e^{At}$ *is* differentiable *for $t > t_0$ ($t_0 \geq 0$) if for any $x \in X$, $t \mapsto e^{At} x$ is (strongly) differentiable. If $t_0 = 0$, we simply say that $e^{At}$ is* differentiable.

(iv) $e^{At}$ *is* analytic *if $e^{At}$ has an extension $T(z)$ for $z \in \Delta_\theta \equiv \{ z \in \mathbb{C} \mid |\arg z| < \theta \}$ for some $\theta > 0$, such that $z \mapsto T(z)$ is analytic; and*

(4.32) $$\begin{cases} \lim_{\Delta_\theta \ni z \to 0} |T(z) x - x| = 0, & \forall x \in X; \\ T(z_1 + z_2) = T(z_1) T(z_2), & \forall z_1, z_2 \in \Delta_\theta. \end{cases}$$

Let us give some characterizations.

**Proposition 4.17.** *Let $e^{At}$ be a $C_0$ semigroup on $X$. Then $e^{At}$ is compact for $t > t_0$, if and only if $t \mapsto e^{At}$ is continuous in the operator norm for $t > t_0$ and $(\lambda - A)^{-1}$ is compact for any (or for some) $\lambda \in \rho(A)$.*

**Proposition 4.18.** *Let $e^{At}$ be a $C_0$ semigroup on $X$ satisfying $\| e^{At} \| \leq M e^{\omega t}$.*

(i) If $e^{At}$ is differentiable for $t > t_0$, then for any $b < \frac{1}{t_0}$ ($\frac{1}{0} = \infty$), there exist $a \in \mathbb{R}$ and $C > 0$, such that

(4.33)
$$\begin{cases} \rho(A) \supset \Sigma \equiv \{\lambda \in \mathbb{C} \mid \operatorname{Re} \lambda \geq a - b \log |\operatorname{Im} \lambda|\}, \\ \|(\lambda - A)^{-1}\| \leq C|\operatorname{Im} \lambda|, \quad \forall \lambda \in \Sigma,\ \operatorname{Re} \lambda \leq \omega. \end{cases}$$

(ii) If (4.33) holds for some $a \in \mathbb{R}$ and $b, C > 0$, then $e^{At}$ is differentiable for $t > 3/b$.

**Proposition 4.19.** *Let $e^{At}$ be a $C_0$ semigroup on $X$. Let $0 \in \rho(A)$ and $\|e^{At}\| \leq M$. Then the following are equivalent:*

(i) $e^{At}$ is analytic.

(ii) There exists a $\delta > 0$, such that

(4.34)
$$\begin{cases} \rho(A) \supset \Sigma \equiv \{\lambda \in \mathbb{C} \mid |\arg \lambda| < \frac{\pi}{2} + \delta\} \bigcup \{0\}, \\ \|(\lambda - A)^{-1}\| \leq \frac{M}{|\lambda|}, \quad \forall \lambda \in \Sigma,\ \lambda \neq 0. \end{cases}$$

(iii) $e^{At}$ is differentiable for $t > 0$, such that for some constant $C > 0$,

(4.35)
$$\|Ae^{At}\| \leq \frac{C}{t}, \quad \forall t > 0.$$

(iv) There exists a constant $C > 0$, such that

(4.36)
$$\|(\alpha + i\beta - A)^{-1}\| \leq \frac{C}{|\beta|}, \quad \forall \alpha > 0,\ \beta \neq 0.$$

We point out that the inequality in (4.34) can be replaced by

(4.37)
$$\|(\lambda - A)^{-1}\| \leq \frac{\widetilde{M}}{1 + |\lambda|}, \quad \forall \lambda \in \Sigma.$$

In fact, since $0 \in \rho(A)$, $A^{-1} \in \mathcal{L}(X)$. Thus, for any $|\lambda| < (1 + \|A^{-1}\|)^{-1}$, $(\lambda A^{-1} - I)^{-1} \in \mathcal{L}(X)$. Consequently,

(4.38)
$$(\lambda - A)^{-1} = (\lambda A^{-1} - I)^{-1} A^{-1} = -A^{-1} \sum_{n \geq 0} \lambda^n A^{-n}.$$

Hence, for any $|\lambda| < (1 + \|A^{-1}\|)^{-1}$,

(4.39)
$$\begin{aligned} \|(\lambda - A)^{-1}\| &\leq \frac{\|A^{-1}\|}{1 - |\lambda|\|A^{-1}\|} \leq \|A^{-1}\|(1 + \|A^{-1}\|) \\ &\leq \frac{\|A^{-1}\|(2 + \|A^{-1}\|)}{1 + |\lambda|}. \end{aligned}$$

For $|\lambda| \geq (1 + \|A^{-1}\|)^{-1}$, $\lambda \in \Sigma$, we have

(4.40)
$$\|(\lambda - A)^{-1}\| \leq \frac{M}{\lambda} = \frac{M}{1 + |\lambda|}\left(1 + \frac{1}{|\lambda|}\right) \leq \frac{M(2 + \|A^{-1}\|)}{1 + |\lambda|}.$$

## §4. Theory of $C_0$ Semigroups

Thus, (4.37) holds with $\widetilde{M} = (M + \|A^{-1}\|)(2 + \|A^{-1}\|)$.

Next, we let $A : \mathcal{D}(A) \subset X \to X$ generate an analytic semigroup $e^{At}$ on $X$ and $0 \in \rho(A)$. Then $\rho(A)$ satisfies the first relation in (4.34) for some $\delta > 0$ and (4.37) holds. Thus, for any $\alpha > 0$, we can define

$$(4.41) \qquad (-A)^{-\alpha} x = -\frac{1}{2\pi i} \int_\Gamma z^{-\alpha} (z + A)^{-1} x \, dz, \qquad \forall x \in X,$$

where $\Gamma$ is a path running in $\rho(-A)$ from $\infty e^{-i\theta}$ to $\infty e^{i\theta}$, $\frac{\pi}{2} - \delta < \theta < \pi$, avoiding the origin and the negative real axis, the function $z^{-\alpha}$ taking real positive values on the positive real axis. By (4.37), we see that $(-A)^{-\alpha} \in \mathcal{L}(X)$, for all $\alpha > 0$. We may show that when $\alpha = n$, the above definition of $(-A)^{-n}$ coincides with the usual definition $[(-A)^{-1}]^n$. It can be shown that $(-A)^{-\alpha}$ defined above is injective. Thus, for any $\alpha > 0$, we may define

$$(4.42) \qquad (-A)^\alpha = [(-A)^{-\alpha}]^{-1}.$$

Finally, we define $(-A)^0 = I$. We refer to $(-A)^\alpha$ as the *fractional power* of $-A$. The following result is useful.

**Proposition 4.20.** *Let $A$ generate an analytic semigroup $e^{At}$ on $X$ and $0 \in \rho(A)$.*

*(i) Let $(-A)^\alpha$ be defined as above. Then $(-A)^\alpha \in \mathcal{L}(X)$ for any $\alpha \leq 0$ and $(-A)^\alpha$ is densely defined and closed for any $\alpha > 0$. Moreover,*

$$(4.43) \qquad \begin{cases} \mathcal{D}((-A)^\alpha) \subseteq \mathcal{D}((-A)^\beta), & \alpha \geq \beta, \\ (-A)^{\alpha+\beta} x = (-A)^\alpha (-A)^\beta x, & \forall x \in \mathcal{D}((-A)^\gamma), \ \alpha, \beta \in \mathbb{R}, \\ \gamma = \max\{\alpha, \beta, \alpha + \beta\}. \end{cases}$$

*(ii) The following hold:*

$$(4.44) \qquad \begin{cases} \mathcal{R}(e^{At}) \subset \mathcal{D}((-A)^\alpha), & \forall t > 0, \alpha \in \mathbb{R}, \\ e^{At}(-A)^\alpha x = (-A)^\alpha e^{At} x, & \forall t \geq 0, \ x \in \mathcal{D}((-A)^\alpha), \\ \|(-A)^\alpha e^{At}\| \leq M_\alpha t^{-\alpha}, & \forall t > 0, \ \alpha \in \mathbb{R}, \\ \|e^{At} x - x\| \leq C_\alpha t^\alpha \|(-A)^\alpha x\|, & \forall 0 < \alpha \leq 1, \\ & x \in \mathcal{D}((-A)^\alpha), \ t \geq 0. \end{cases}$$

### §4.4. Examples

In this subsection, let us present some examples of $C_0$ semigroups.

*Example 4.21.* Let $A \in \mathbb{R}^{n \times n}$ be an $(n \times n)$-matrix. Then

$$(4.45) \qquad e^{At} = \sum_{k=0}^\infty \frac{A^k t^k}{k!}, \qquad t \geq 0,$$

is a $C_0$ semigroup on $X = \mathbb{R}^n$. In fact $e^{At}$ is a $C_0$ group. It is well known that $e^{At}$ is the fundamental matrix of the (homogeneous) ordinary differential equation $\dot{y}(t) = Ay(t)$.

*Example 4.22.* Consider an ordinary differential delay equation:

$$(4.46) \quad \begin{cases} \dot{y}(t) = \int_{[-r,0]} [d\eta(\theta)]y(t+\theta), & t > 0, \\ y(0) = y_0, \quad y(t) = \varphi(t), & t \in [-r,0), \end{cases}$$

where $\eta$ is in $BV([-r,0];\mathbb{R}^{n\times n})$ and the integral is understood to be the Lebesgue–Stieltjes integral. Then it is well known that for each $(y_0, \varphi(\cdot)) \in \mathbb{R}^n \times L^2(-r,0;\mathbb{R}^n) \equiv X$, there exists a unique solution $y(\cdot)$ of (4.46), such that $(y(t), y_t(\cdot)) \in X$ for each $t \geq 0$, where $y_t(\cdot) = y(t+\cdot)$. Define $S(t)$ as follows:

$$(4.47) \quad S(t)\begin{pmatrix} y_0 \\ \varphi(\cdot) \end{pmatrix} = \begin{pmatrix} y(t) \\ y_t(\cdot) \end{pmatrix}, \quad t \geq 0.$$

We first claim that $S(t) \in \mathcal{L}(X)$. To prove this (and for later purposes) let us look at the following: For all $0 \leq s < t$, by the Hölder inequality

$$(4.48) \quad \begin{aligned} \int_s^t \Big| \int_{-r}^0 [d\eta(\theta)]y(\tau+\theta) \Big|^2 d\tau &\leq \int_s^t \Big( \int_{-r}^0 |y(\tau+\theta)|\, d|\eta|(\theta) \Big)^2 d\tau \\ &\leq |\eta|([-r,0]) \int_s^t \int_{[-r,0]} |y(\tau+\theta)|^2 \, d|\eta|(\theta)\, d\tau \\ &\leq |\eta|([-r,0]) \int_{[-r,0]} \Big( \int_{s+\theta}^{t+\theta} |y(\tau)|^2 \, d\tau \Big) d|\eta|(\theta) \\ &\leq |\eta|([-r,0])^2 \|y\|_{L^2(s-r,t;\mathbb{R}^n)}^2. \end{aligned}$$

Here $|\eta|$ is the total variation of $\eta$. Thus, for $0 \leq t \leq T$,

$$(4.49) \quad \begin{aligned} |y(t)|^2 &\leq 2|y_0|^2 + 2\Big| \int_0^t \int_{[-r,0]} [d\eta(\theta)]y(\tau+\theta)\, d\tau \Big|^2 \\ &\leq 2|y_0|^2 + 2t \int_0^t \Big| \int_{[-r,0]} [d\eta(\theta)]y(\tau+\theta) \Big|^2 d\tau \\ &\leq 2|y_0|^2 + 2T|\eta|([-r,0])^2 \Big( \|\varphi\|_{L^2(-r,0;\mathbb{R}^n)}^2 + \int_0^t |y(\tau)|^2\, d\tau \Big) \\ &\leq 2\Big(1 + T|\eta|([-r,0])^2\Big)(|y_0|^2 + \|\varphi\|_{L^2(-r,0;\mathbb{R}^n)}^2) \\ &\quad + 2T|\eta|([-r,0])^2 \int_0^t |y(\tau)|^2\, d\tau. \end{aligned}$$

By Gronwall's inequality, we have

$$(4.50) \quad |y(t)|^2 \leq 2\big(1 + T|\eta|([-r,0])^2\big) e^{2T|\eta|([-r,0])^2 t}(|y_0|^2 + \|\varphi\|_{L^2(-r,0;\mathbb{R}^n)}^2).$$

## §4. Theory of $C_0$ Semigroups

Thus, we see that $S(t) \in \mathcal{L}(X)$. One can further check that $S(t)$ is a $C_0$ semigroup on $X$. Moreover, the generator $A$ of $S(t)$ can be identified as follows:

(4.51) $$\begin{cases} \mathcal{D}(A) = \{(y_0, \varphi(\cdot)) \in X \mid \varphi(\cdot) \in W^{1,2}[-r, 0], \varphi(0) = y_0\}, \\ A\begin{pmatrix} \varphi(0) \\ \varphi(\cdot) \end{pmatrix} = \begin{pmatrix} \int_{[-r,0]} [d\eta(\theta)]\varphi(\theta) \\ \dot{\varphi}(\cdot) \end{pmatrix}. \end{cases}$$

Next, we would like to point out that the above $C_0$ semigroup $S(t) = e^{At}$ is compact for $t > r$. To this end, we use (4.48). For any $t > r$,

(4.52) $$\begin{aligned} |\int_{-r}^0 |\dot{y}(t+\bar{\theta})|^2 \, d\bar{\theta} &= \int_{t-r}^t |\dot{y}(\bar{\theta})|^2 \, d\bar{\theta} \\ &= \int_{t-r}^t |\int_{-r}^0 [d\eta(\theta)] y(\bar{\theta}+\theta)|^2 \, d\bar{\theta} \leq |\eta|([-r,0])^2 \|y\|^2_{L^2(t-2r,t;\mathbb{R}^n)} \\ &\leq |\eta|([-r,0])^2 (\|\varphi\|^2_{L^2(-r,0;\mathbb{R}^n)} + \|y\|^2_{L^2(0,t;\mathbb{R}^n)}). \end{aligned}$$

Hence, we see that there exists a constant $C = C(t)$ $(t > r)$, such that

(4.53) $$\|y_t\|_{W^{1,2}(-r,0;\mathbb{R}^n)} \leq C(|y_0| + \|\varphi\|_{L^2(-r,0;\mathbb{R}^n)}).$$

Because the embedding $\mathbb{R}^n \times W^{1,2}(-r, 0; \mathbb{R}^n) \hookrightarrow X$ is compact, we see that $S(t)$ is compact for $t > r$.

*Example 4.23.* Let $\Omega \subset \mathbb{R}^n$ be a bounded domain with a smooth boundary $\partial\Omega$. We consider the Laplacian operator $\Delta \equiv \frac{\partial^2}{\partial x_1^2} + \cdots + \frac{\partial^2}{\partial x_n^2}$. Take $X = L^2(\Omega)$, denote $A_2 = \Delta$, and let $\mathcal{D}(A_2) = W^{2,2}(\Omega) \cap W_0^{1,2}(\Omega)$. Clearly, $\mathcal{D}(A_2)$ is dense in $X$ and we can further show that $A_2$ is closed. In fact, if $(y_m, A_2 y_m) \in \mathcal{G}(A_2)$, the graph of $A_2$, such that $|y_m - y|_X \to 0$ and $|A_2 y_m - z|_X \to 0$, then $\|\Delta y_m\|_{L^2(\Omega)}$ is bounded. Thus, by $L^p$ theory (see §6.2), $\|y_m\|_{W^{2,2}(\Omega)}$ is bounded and there exists a weakly convergent subsequence whose weak limit is denoted by $y$. Then one can show that $y \in \mathcal{D}(A_2)$ and $Ay = z$. This shows the closeness of $A_2$. Next, for any $\lambda > 0$ and $y \in \mathcal{D}(A_2)$, we have

(4.54) $$\langle (\lambda - A_2)y, y \rangle = \lambda \|y\|^2_{L^2(\Omega)} + \int_\Omega |\nabla y|^2 \, dx \geq \lambda \|y\|^2_{L^2(\Omega)}.$$

Thus, it follows that

(4.55) $$\lambda \|y\|_{L^2(\Omega)} \leq \|(\lambda - A_2)y\|_{L^2(\Omega)}, \quad \forall y \in \mathcal{D}(A_2).$$

Hence, $(\lambda - A_2)$ is one-to-one and the range $\mathcal{R}(\lambda - A_2)$ is closed. Furthermore, we can show that $\mathcal{R}(\lambda - A_2) = L^2(\Omega)$. (This amounts to say that the problem $-\Delta y + \lambda y = f$, $y|_{\partial\Omega} = 0$, is solvable for any $f \in L^2(\Omega)$, which is a consequence of the results in §6.2). Hence, (4.55) tells us that $\lambda \in \rho(A_2)$ and

(4.56) $$\|(\lambda - A_2)^{-1}\| \leq \frac{1}{\lambda}, \quad \forall \lambda > 0.$$

Thus, (4.22) holds with $M = 1$ and $\omega = 0$. By Theorem 4.11, $A_2$ generates a $C_0$ semigroup $e^{A_2 t}$ on $X = L^2(\Omega)$. Actually, we can further show that $e^{A_2 t}$ is a contraction, analytic and compact.

This semigroup arises in the study of the (homogeneous) heat equation:

$$(4.57) \quad \begin{cases} y_t - \Delta y = 0, & \text{in } \Omega \times (0, \infty), \\ y|_{\partial\Omega} = 0, & y|_{t=0} = y_0(x), \quad x \in \Omega. \end{cases}$$

We point out that $e^{A_2 t}$ is *not* a $C_0$ group because $e^{A_2 t}$ is *not* defined for $t < 0$. Mathematically, this means that the backward heat equation is *not* well posed. Physically, this means that the process of heat diffusion is *irreversible*.

If we denote $A_p = \Delta$ with $\mathcal{D}(A_p) = W^{2,p}(\Omega) \cap W_0^{1,p}(\Omega)$ and let the underling space be $L^p(\Omega)$ ($1 \leq p < \infty$), with some more careful analysis, one can prove that $A_p$ also generates an analytic and compact semigroup $e^{A_p t}$ on $L^p(\Omega)$. Finally, we can replace $\Delta$ by general second order uniformly elliptic operators (with smooth coefficients) to obtain the same result.

**Example 4.24.** Let $X = W^{1,2}(\Omega) \times L^2(\Omega)$ and

$$(4.58) \quad \begin{cases} \mathcal{D}(A) = \left\{ \begin{pmatrix} y \\ z \end{pmatrix} \mid y \in W^{2,2}(\Omega) \cap W_0^{1,2}(\Omega), \ z \in W^{1,2}(\Omega) \right\}, \\ A \begin{pmatrix} y \\ z \end{pmatrix} = \begin{pmatrix} z \\ \Delta y \end{pmatrix} \equiv \begin{pmatrix} 0 & I \\ \Delta & 0 \end{pmatrix} \begin{pmatrix} y \\ z \end{pmatrix}. \end{cases}$$

Then we can show that $A$ is densely defined and closed. Now, let $\lambda \in \mathbb{R}$ and $|\lambda| > 1$. Consider the following equation

$$(4.59) \quad (\lambda - A) \begin{pmatrix} y \\ z \end{pmatrix} = \begin{pmatrix} f \\ g \end{pmatrix} \in X \equiv W^{1,2}(\Omega) \times L^2(\Omega).$$

This is equivalent to the following:

$$(4.60) \quad \begin{cases} -\Delta y + \lambda^2 y = \lambda f + g, & y|_{\partial\Omega} = 0, \\ z = \lambda y - f. \end{cases}$$

Clearly, for any $f \in W^{1,2}(\Omega)$, $g \in L^2(\Omega)$, there exists a unique solution $(y, z) \in \mathcal{D}(A)$ (see §6.2). Now, let $f, g \in C_0^\infty(\Omega)$. Then, $y, z \in C^\infty(\Omega)$.

§4. Theory of $C_0$ Semigroups

Thus, noting that $|\lambda| > 1$, we have

$$
\begin{aligned}
\|f\|_{W^{1,2}(\Omega)}^2 + \|g\|_{L^2(\Omega)}^2 &= \int_\Omega \left\{ |\nabla f|^2 + f^2 + g^2 \right\} dx \\
&= \int_\Omega \left\{ |\lambda \nabla y - \nabla z|^2 + |\lambda y - z|^2 + |-\Delta y + \lambda z|^2 \right\} dx \\
&= \int_\Omega \left\{ \lambda^2 |\nabla y|^2 - 2\lambda \nabla y \cdot \nabla z + |\nabla z|^2 + \lambda^2 y^2 - 2\lambda y z + z^2 \right. \\
&\qquad \left. + |\Delta y|^2 - 2\lambda z \Delta y + \lambda^2 z^2 \right\} dx \\
&\geq \int_\Omega \left\{ \lambda^2 |\nabla y|^2 + \lambda^2 y^2 + (1 + \lambda^2) z^2 - 2\lambda y z \right\} dx \\
&= (|\lambda| - 1)^2 \left[ \|y\|_{W^{1,2}(\Omega)}^2 + \|z\|_{L^2(\Omega)}^2 \right] + \int_\Omega \left\{ (2|\lambda| - 1) |\nabla y|^2 \right. \\
&\qquad \left. + (|\lambda| - 1) y^2 + |\lambda| z^2 + |\lambda| (y^2 + z^2 - 2\lambda y z) \right\} dx \\
&\geq (|\lambda| - 1)^2 \left[ \|y\|_{W^{1,2}(\Omega)}^2 + \|z\|_{L^2(\Omega)}^2 \right].
\end{aligned}
\tag{4.61}
$$

Thus, it follows that

$$
\|(\lambda - A)^{-1}\| \leq \frac{1}{|\lambda| - 1}, \qquad \forall \lambda \in \mathbb{R},\ |\lambda| > 1.
\tag{4.62}
$$

By Theorem 4.11, we see that $A$ and $-A$ both generate a $C_0$ semigroup on $X$. Thus, $e^{At}$ is actually a $C_0$ group on $X$. This $C_0$ group arises in the study of the following (homogeneous) wave equation:

$$
\begin{cases}
y_{tt} - \Delta y = 0, & \text{in } \Omega \times (0, \infty), \\
y\big|_{\partial \Omega} = 0, \\
y\big|_{t=0} = y_0(x), \quad y_t\big|_{t=0} = y_1(x), & x \in \Omega.
\end{cases}
\tag{4.63}
$$

In fact, if we set $z = y_t$, then (4.63) can be transformed into the following:

$$
w_t = Aw, \qquad w\big|_{t=0} = w_0,
\tag{4.64}
$$

where $w = (y, z_t)$, $w_0 = (y_0, y_1)$ and $A$ is given by (4.58).

We may replace the operator $\Delta$ by more general second order uniformly elliptic operators.

There are many other examples of semigroups. Generically, if we are studying well-posed linear autonomous evolution equations, then in the proper framework, there is always a $C_0$ semigroup associated with it.

To conclude this section, let us point out a fact that has been misunderstood in some literature. We note that if for some $t_0 > 0$ and $x \in X$, $e^{At_0}x \in \mathcal{D}(A)$, then the map $t \mapsto e^{At}x$ is differentiable at $t > t_0$. At $t = t_0$, however, it is only right differentiable and not necessarily (left) differentiable, in general. Here is an example:

*Example 4.25.* Consider the following delay equation in $\mathbb{R}$:

(4.65)
$$\begin{cases} \dot{y}(t) = y(t-r), & t \geq 0, \\ y(0) = x, \quad y(t) = \varphi(t), & t \in [-r, 0). \end{cases}$$

We introduce the following Hilbert space: $X = \mathbb{R} \times L^2(-r, 0)$. Then, by Example 4.22, for any $(x, \varphi(\cdot)) \in X$, (4.65) admits a unique solution $y(\cdot)$ and if we define

(4.66)
$$S(t) \begin{pmatrix} x \\ \varphi(\cdot) \end{pmatrix} = \begin{pmatrix} y(t) \\ y(t+\cdot) \end{pmatrix}, \quad t \geq 0,$$

then $S(t)$ is a $C_0$ semigroup on $X$. Moreover, the generator $A$ is given by the following:

(4.67)
$$\begin{cases} \mathcal{D}(A) = \{(x, \varphi(\cdot)) \in X \mid \varphi(\cdot) \in W^{1,2}[-r, 0], \varphi(0) = x\}, \\ A \begin{pmatrix} x \\ \varphi(\cdot) \end{pmatrix} = \begin{pmatrix} \varphi(-r) \\ \dot{\varphi}(\cdot) \end{pmatrix}. \end{cases}$$

If we denote $z_0 = (x, \varphi(\cdot)) \in X$, and $z(t) = (y(t), y(t+\cdot))$, then

(4.68)
$$z(t) = e^{At} z_0, \quad t \geq 0.$$

Next, we show that

(4.69)
$$e^{Ar} z_0 \in \mathcal{D}(A), \quad A e^{Ar} z_0 = z_0, \quad \forall z_0 \in X,$$

but the function $e^{At} z_0$ could be nondifferentiable at $t = r$. To show (4.69), we take any $z_0 = (x, \varphi(\cdot)) \in X$. Then

(4.70)
$$\frac{1}{\delta}[e^{A(r+\delta)} - e^{Ar}]z_0 = \begin{pmatrix} \frac{1}{\delta} \int_r^{r+\delta} y(\tau - r)\, d\tau \\ \frac{1}{\delta} \int_r^{r+\delta} y(\tau - r + \cdot)\, d\tau \end{pmatrix}$$
$$\to \begin{pmatrix} x \\ \varphi(\cdot) \end{pmatrix} = z_0, \quad \delta \to 0.$$

The convergence is in the $X$ topology. This proves (4.69). On the other hand,

(4.71)
$$\frac{1}{-\delta}[e^{A(r-\delta)} - e^{Ar}]z_0 = \begin{pmatrix} \frac{1}{\delta} \int_{r-\delta}^r y(\tau - r)\, d\tau \\ \frac{1}{\delta} \int_{r-\delta}^r y(\tau - r + \cdot)\, d\tau \end{pmatrix}$$
$$= \begin{pmatrix} \frac{1}{\delta} \int_{-\delta}^0 \varphi(\theta)\, d\theta \\ \frac{1}{\delta} \int_{-\delta}^0 \varphi(\theta + \cdot)\, d\theta \end{pmatrix}.$$

It is clear that the second component approaches $\varphi(\cdot)$ in $L^2(-r, 0)$. But the first component is not necessarily convergent because $\varphi(\cdot)$ is just in $L^2(-r, 0)$. Now, we choose a $\varphi(\cdot) \in L^2(-r, 0)$, such that the limit $\lim_{\delta \to 0} \frac{1}{\delta} \int_{-\delta}^0 \varphi(\theta)\, d\theta$ does not exist. Then, for any $x \in \mathbb{R}$, by setting

§5. Evolution equations

$z_0 = (x, \varphi(\cdot))$, we have that the function $e^{At}z_0$ is nondifferentiable at $t = r$.
□

## §5. Evolution Equations

In this section, we present some results for evolution equations.

### §5.1. Solutions

Let $X$ be a Banach space, $f(\cdot) \in L^1(0, T; X)$ and $x_0 \in X$. Consider

(5.1) $$\begin{cases} \dot{y}(t) = Ax(t) + f(t), & t \in [0, T], \\ y(0) = y_0. \end{cases}$$

Here, $A : \mathcal{D}(A) \subset X \to X$ generates a $C_0$ semigroup $e^{At}$ on $X$. We first introduce some notions of solutions.

**Definition 5.1.** (i) $y \in C([0, T]; X)$ is called a *strong solution* of (5.1) if $y(0) = y_0$, $y$ is strongly differentiable almost everywhere on $[0, T]$, $y(t) \in \mathcal{D}(A)$, a.e. $t \in [0, T]$, and the equation in (5.1) is satisfied almost everywhere.

(ii) $y \in C([0, T]; X)$ is called a *weak solution* of (5.1) if for any $x^* \in \mathcal{D}(A^*)$, $\langle y(\cdot), x^* \rangle$ is absolutely continuous on $[0, T]$ and

(5.2) $$\langle y(t), x^* \rangle = \langle y_0, x^* \rangle + \int_0^t [\langle y(s), A^*x^* \rangle + \langle f(s), x^* \rangle]\, ds,$$
$$\forall t \in [0, T].$$

(iii) $y \in C([0, T]; X)$ is called a *mild solution* of (5.1) if

(5.3) $$y(t) = e^{At}y_0 + \int_0^t e^{A(t-s)}f(s)\, ds, \quad t \in [0, T].$$

The following result is interesting.

**Proposition 5.2.** (Ball) *Suppose $A$ generates a $C_0$ semigroup on some Banach space $X$ and $f \in L^1(0, T; X)$. Then $y \in C([0, T]; X)$ is a mild solution of (5.1) (i.e., (5.3) holds) if and only if $y$ is a weak solution of (5.1) (i.e., (ii) of Definition 5.1 holds).*

*Proof.* $\Rightarrow$ For $x^* \in \mathcal{D}(A^*)$ and $y_0 \in \mathcal{D}(A)$, we have

(5.4) $$\frac{d}{dt}\langle e^{At}y_0, x^* \rangle = \langle e^{At}Ay_0, x^* \rangle = \langle e^{At}y_0, A^*x^* \rangle.$$

Now, for general $y_0 \in X$, let $y_k \in \mathcal{D}(A)$ with $|y_k - y_0|_X \leq \frac{1}{k}$, $k \geq 1$. Denote $g_k(t) = \langle e^{At}y_k, x^* \rangle$ $(k \geq 0)$ and $\tilde{g}_0(t) = \langle e^{At}y_0, A^*x^* \rangle$. Then

(5.5) $$g'_k(t) = \langle e^{At}y_k, A^*x^* \rangle,$$

$$|g_k(t) - g_0(t)| + |g_k'(t) - \widetilde{g}_0(t)| \leq \frac{C}{k}, \quad t \in [0,T], \quad k \geq 1. \tag{5.6}$$

where $C$ is independent of $t$ and $k$. Thus, $g_0'(t) = \widetilde{g}_0(t), t \in [0,T]$. This implies that for any $y_0 \in X$ and $x^* \in \mathcal{D}(A^*)$, $t \mapsto \langle e^{At}y_0, x^* \rangle$ is absolutely continuous and

$$\frac{d}{dt} \langle e^{At}y_0, x^* \rangle = \langle e^{At}y_0, A^*x^* \rangle. \tag{5.7}$$

Now, let $f \in C([0,T];X)$ and $x^* \in \mathcal{D}(A^*)$. Then

$$\frac{d}{dt} \int_0^t \langle e^{A(t-s)}f(s), x^* \rangle \, ds = \langle f(t), x^* \rangle \tag{5.8}$$
$$+ \int_0^t \langle e^{A(t-s)}f(s), A^*x^* \rangle \, ds.$$

Thus, it follows from (5.4) that

$$\frac{d}{dt} \langle y(t), x^* \rangle = \langle e^{At}y_0, A^*x^* \rangle + \langle f(t), x^* \rangle \tag{5.9}$$
$$+ \int_0^t \langle e^{A(t-s)}f(s), A^*x^* \rangle \, ds = \langle y(t), A^*x^* \rangle + \langle f(t), x^* \rangle.$$

Next, for $f \in L^1(0,T;X)$, we find $f_n \in C([0,T];X)$ with $f_n \xrightarrow{s} f$, in $L^1(0,T;X)$. Set

$$y_n(t) = e^{At}y_0 + \int_0^t e^{A(t-s)}f_n(s) \, ds, \quad t \in [0,T]. \tag{5.10}$$

It follows that $|y_n(t) - y(t)|_X \to 0$, uniformly in $t \in [0,T]$. On the other hand, we have proved that

$$\langle y_n(t), x^* \rangle = \langle y_0, v \rangle + \int_0^t [\langle y_n(s), A^*x^* \rangle + \langle f_n(t), x^* \rangle] \, ds. \tag{5.11}$$

Thus, by taking the limits, we see that $y$ is a weak solution of (5.1).

$\Leftarrow$ Let $y$ be a weak solution of (5.1). Then, for any $x^* \in \mathcal{D}(A^*)$, we have

$$\langle y(t) - y_0 - \int_0^t f(s) \, ds, x^* \rangle = \langle \int_0^t y(s) \, ds, A^*x^* \rangle. \tag{5.12}$$

This means (note Proposition 4.3)

$$(\int_0^t y(s) \, ds, y(t) - y_0 - \int_0^t f(s) \, ds) \in {}^\perp(\mathcal{J}^*\mathcal{G}(A^*)) \tag{5.13}$$
$$= {}^\perp(\mathcal{G}(A)^\perp) = \overline{\mathcal{G}(A)} = \mathcal{G}(A), \quad t \in [0,T].$$

## §5. Evolution equations

Thus, $w(t) \stackrel{\Delta}{=} \int_0^t y(s)\,ds \in \mathcal{D}(A)$ for all $t \in [0, T]$ and

(5.14) $$y(t) - y_0 - \int_0^t f(s)\,ds = A \int_0^t y(s)\,ds.$$

Clearly, the above is equivalent to the following:

(5.15) $$\begin{cases} \dot{w}(t) = Aw(t) + y_0 + \int_0^t f(s)\,ds, \\ w(0) = 0. \end{cases}$$

Let $A_\mu = \mu A(\mu I - A)^{-1}$ be the Yosida approximation of the operator $A$. Then, by (5.15), we have

(5.16) $$\begin{aligned} \frac{d}{dt}(e^{-A_\mu t} w(t)) &= e^{-A_\mu t} \dot{w}(t) - A_\mu e^{-A_\mu t} w(t) \\ &= (A - A_\mu) e^{-A_\mu t} w(t) + e^{-A_\mu t}[y_0 + \int_0^t f(s)\,ds]. \end{aligned}$$

Thus, it follows that

(5.17) $$\begin{aligned} w(t) &= e^{A_\mu t} \int_0^t \{(A - A_\mu) e^{-A_\mu s} w(s) + e^{-A_\mu s}[y_0 + \int_0^s f(r)dr]\}\,ds \\ &= \int_0^t (A - A_\mu) e^{A_\mu(t-s)} w(t)\,ds + \int_0^t e^{A_\mu(t-s)}[y_0 + \int_0^s f(r)\,dr]\,ds. \end{aligned}$$

For all $\mu$ large enough, note (5.14), we have

(5.18) $$\begin{aligned} |Aw(s)| &\leq |y(s)| + |y_0| + |\int_0^s f(s)\,ds| \leq C, \\ |A_\mu w(s)| &\leq |\mu(\mu I - A)^{-1}||Aw(s)| \leq C, \end{aligned}$$

and

(5.19) $$\begin{cases} \lim_{\mu \to \infty} (A - A_\mu) w(s) = 0, & \forall s \in [0, T], \\ \lim_{\mu \to \infty} e^{A_\mu t} x = e^{At} x, & \forall x \in X, t \in [0, T], \end{cases}$$

thus, by the Dominated Convergence Theorem, we obtain from (5.17) that

(5.20) $$\begin{aligned} w(t) &= \int_0^t e^{A(t-s)}[y_0 + \int_0^s f(r)\,dr]\,ds \\ &= \int_0^t e^{As} y_0\,ds + \int_0^t e^{As}[\int_0^{t-s} f(r)\,dr]; ds. \end{aligned}$$

Hence,

(5.21) $$y(t) = \dot{w}(t) = e^{At} y_0 + \int_0^t e^{A(t-s)} f(s)\,ds.$$

This shows that $y$ is a mild solution of (5.1). □

## §5.2. Semilinear equations

Next, we consider the following semilinear equation:

$$\text{(5.22)} \quad \begin{cases} \dot{y}(t) = Ay(t) + f(t, y(t)), & t \in [0, T], \\ y(0) = y_0, \end{cases}$$

where $A : \mathcal{D}(A) \subset X \to X$ generates a $C_0$ semigroup $e^{At}$ on $X$ and $f : [0, T] \times X \to X$ satisfies the following:

(i) For each $x \in X$, $f(\cdot, x)$ is strongly measurable.

(ii) There exists a function $L \in L^1(0, T)$, such that

$$\text{(5.23)} \quad \begin{cases} |f(t, x) - f(t, \widehat{x})| \leq L(t)|x - \widehat{x}|, & \forall t \in [0, T], x, \widehat{x} \in X, \\ |f(t, 0)| \leq L(t), & \forall t \in [0, T]. \end{cases}$$

We call $y \in C([0, T]; X)$ a *mild solution* of (5.22) if it is a solution of the following:

$$\text{(5.24)} \qquad y(t) = e^{At} y_0 + \int_0^t e^{A(t-s)} f(s, y(s))\, ds, \qquad t \in [0, T].$$

Hereafter, we will not distinguish the mild solution of (5.22) and the solution of (5.24). The following result is concerned with the existence and uniqueness of the solution to (5.24).

**Proposition 5.3.** *Let the above assumptions concerning $A$ and $f$ hold. Then, for any $y_0 \in X$, (5.24) admits a unique solution $y$. Moreover, if we let $y(\cdot\,; y_0)$ be the solution corresponding to $y_0$, and let the $C_0$ semigroup $e^{At}$ satisfy*

$$\text{(5.25)} \qquad\qquad \|e^{At}\|_{\mathcal{L}(X)} \leq M e^{\omega t}, \qquad t \geq 0,$$

*for some $M \geq 1$ and $\omega \in \mathbb{R}$, then*

$$\text{(5.26)} \quad \begin{cases} |y(t; y_0)| \leq M e^{\omega t + M \int_0^t L(s)\,ds}(1 + |y_0|), \\ |y(t; y_0) - y(t, \widehat{y}_0)| \leq M e^{\omega t + M \int_0^t L(s)\,ds} |y_0 - \widehat{y}_0|, \end{cases} \quad t \geq 0.$$

*Proof.* By Picard iteration, we can obtain the existence and the uniqueness of the mild solution. By Gronwall's inequality, together with assumption (5.23), we can prove (5.26). □

We note that the solution $y$ of (5.24) is not necessarily strongly differentiable because $A$ is unbounded. This causes some inconvenience for many applications. The following convergence result sometimes helps us to overcome such an inconvenience.

## §5. Evolution equations

**Proposition 5.4.** *Let $y$ be the mild solution of (5.24) and let $y^\mu$ be the solution of the following:*

$$(5.27) \quad y^\mu(t) = e^{A_\mu t} y_0 + \int_0^t e^{A_\mu(t-s)} f(s, y^\mu(s))\, ds, \qquad t \in [0,T],$$

*where $A_\mu = \mu A(\mu - A)^{-1}$ is the Yosida approximation of $A$. Then*

$$(5.28) \quad \lim_{\mu \to \infty} \sup_{t \in [0,T]} |y^\mu(t) - y(t)| = 0.$$

*Proof.* Using (5.23) and (5.25), by Gronwall's inequality, we have

$$(5.29) \quad |y^\mu(t) - y(t)| \le C \int_0^T \sup_{r \in [0,T]} |(e^{A_\mu r} - e^{Ar}) f(s, y(s))|\, ds.$$

Then, (5.28) follows from (5.19) and the Dominated Convergence Theorem. $\square$

The following result is an application of the above proposition.

**Proposition 5.5.** *Let $\varphi \in C^1([0,T] \times X)$ with $A^* \varphi_x \in C([0,T] \times X)$. Let $y$ be the solution of (5.24). Then, for $0 \le s < t \le T$,*

$$(5.30) \quad \varphi(t, y(t)) = \varphi(s, y(s)) + \int_s^t \big\{ \varphi_t(r, y(r)) + \langle A^* \varphi_x(r, y(r)), y(r) \rangle + \langle \varphi_x(r, y(r)), f(r, y(r)) \rangle \big\}\, dr.$$

*Proof.* Let $A_\mu = \mu A(\mu I - A)^{-1}$ be the Yosida approximation of $A$ and let $y^\mu$ be the solution of (5.27). Then, because $A_\mu \in \mathcal{L}(X)$, we have

$$(5.31) \quad \begin{aligned} \varphi(t, y^\mu(t)) &= \varphi(s, y^\mu(s)) + \int_s^t \big\{ \varphi_t(r, y^\mu(r)) \\ &\quad + \langle \varphi_x(r, y^\mu(r)), A_\mu y^\mu(r) + f(r, y^\mu(r)) \rangle \big\}\, dr \\ &= \varphi(s, y^\mu(s)) + \int_s^t \big\{ \varphi_t(r, y^\mu(r)) + \langle A_\mu^* \varphi_x(r, y^\mu(r)), y^\mu(r) \rangle \\ &\quad + \langle \varphi_x(r, y^\mu(r)), f(r, y^\mu(r)) \rangle \big\}\, dr. \end{aligned}$$

We note that (5.28) holds. Thus,

$$(5.32) \quad \begin{aligned} |A_\mu^* \varphi_x(r, y^\mu(r)) &- A^* \varphi_x(r, y^\mu(r))| \\ &\le \|\mu(\mu - A^*)^{-1}\|\, |A^* \varphi_x(r, y^\mu(r)) - A^* \varphi_x(r, y(r))| \\ &\quad + |(A_\mu^* - A^*) \varphi(r, y(r))| \to 0, \qquad (\mu \to \infty). \end{aligned}$$

Here, we have used the fact that $\|\mu(\mu - A^*)^{-1}\| = \|\mu(\mu - A)^{-1}\| \le \frac{M\mu}{\mu - \omega} \le C$ and the continuity of $A^* \varphi_x(r, y)$. We then take the limits in (5.31) to get our conclusion. $\square$

The above is a type of *Newton–Leibniz formula* for the function $t \mapsto \varphi(t, y(t))$. The point is that we do not need the differentiability of $y$. This result will be useful in Chapter 7.

## §5.3. Variation of constants formula

This subsection is devoted to the *variation of constants formula* for perturbed evolution equations.

**Lemma 5.6.** *Let $e^{At}$ be a $C_0$ semigroup on $X$ and $B(\cdot) \in L^1(0, T; \mathcal{L}(X))$. Then there exists a unique strongly continuous function $G : \overline{\Delta} \to \mathcal{L}(X)$ with $\Delta = \{(t, s) \in [0, T] \times [0, T] \,|\, 0 \leq s < t \leq T\}$, such that*

(5.33)
$$\begin{cases} G(t, t) = I, & \forall t \in [0, T], \\ G(t, r)G(r, s) = G(t, s), & \forall 0 \leq s \leq r \leq t \leq T, \end{cases}$$

(5.34)
$$\begin{aligned} G(t, s)x &= e^{A(t-s)}x + \int_s^t e^{A(t-r)} B(r) G(r, s) x \, dr \\ &= e^{A(t-s)}x + \int_s^t G(t, r) B(r) e^{A(r-s)} x \, dr, \end{aligned}$$
$$0 \leq s \leq t \leq T, x \in X.$$

*Proof.* Let $s \in [0, T)$ and $x \in X$. Consider the following:

(5.35) $\quad y(t; s, x) = e^{A(t-s)}x + \int_s^t e^{A(t-r)} B(r) y(r; s, x) \, dr, \quad t \in [s, T].$

By Proposition 5.3, (5.35) admits a unique solution. Thus, we can define

(5.36) $\quad G(t, s)x = y(t; s, x), \quad \forall (t, s) \in \overline{\Delta}, \quad x \in X.$

It is clear that $G : \overline{\Delta} \to \mathcal{L}(X)$ is strongly continuous and also, that such a $G(\cdot, \cdot)$ is unique. Hence, we have obtained (5.33) and the first equality in (5.34). Now, we show the second equality in (5.34). To this end, let $s$ be fixed and define

(5.37) $\quad z(t) = e^{A(t-s)}x + \int_s^t G(t, r) B(r) e^{A(r-s)} x \, dr, \quad t \in [s, T].$

## §5. Evolution equations

We claim that $z(t) = y(t; s, x)$ ($t \in [s, T]$). In fact, by (5.35)–(5.37),

$$\int_s^t e^{A(t-r)} B(r) z(r) \, dr = \int_s^t e^{A(t-r)} B(r) e^{A(r-s)} x \, dr$$
$$+ \int_s^t e^{A(t-r)} B(r) \int_s^r G(r, \sigma) B(\sigma) e^{A(\sigma-s)} x \, d\sigma \, dr$$

$$= \int_s^t e^{A(t-r)} B(r) e^{A(r-s)} x \, dr$$

(5.38)
$$+ \int_s^t \int_\sigma^t e^{A(t-r)} B(r) G(r, \sigma) B(\sigma) e^{A(\sigma-s)} x \, dr \, d\sigma$$

$$= \int_s^t e^{A(t-r)} B(r) e^{A(r-s)} x \, ds$$

$$+ \int_s^t [G(t, \sigma) - e^{A(t-\sigma)}] B(\sigma) e^{A(\sigma-s)} x \, d\sigma$$

$$= \int_s^t G(t, \sigma) B(\sigma) e^{A(\sigma-s)} x \, d\sigma = z(t) - e^{A(t-s)} x.$$

Then by uniqueness, we obtain our claim and (5.34) is proved. $\square$

Because of the property (5.33), we refer to the operator valued function $G(\cdot, \cdot)$ as the *evolution operator* generated by $A + B(\cdot)$. Next, we consider the following nonhomogeneous linear equations:

(5.39)
$$\xi(t) = e^{At} \xi_0 + \int_0^t e^{A(t-s)} B(s) \xi(s) \, ds + \int_0^t e^{A(t-s)} g(s) \, ds,$$
$$t \in [0, T],$$

(5.40)
$$\psi(t) = e^{A^*(T-t)} \psi_T + \int_t^T e^{A^*(s-t)} B(s)^* \psi(s) \, ds$$
$$+ \int_t^T e^{A^*(s-t)} h(s) \, ds, \quad t \in [0, T].$$

It is not hard to see that both of the above admit unique solutions. The following result gives the variation of constants formulas.

**Proposition 5.7.** *Let $G(\cdot, \cdot)$ be the evolution operator generated by $A + B(\cdot)$. Then the solution $\xi$ of (5.39) and $\psi$ of (5.40) can be represented by*

(5.41) $$\xi(t) = G(t, 0) \xi_0 + \int_0^t G(t, s) g(s) \, ds, \quad \forall t \in [0, T],$$

(5.42) $$\psi(t) = G^*(T, t) \psi_T + \int_t^T G^*(s, t) h(s) \, ds, \quad \forall t \in [0, T].$$

*Moreover, for any $0 \leq s \leq t \leq T$,*

$$(5.43) \quad \langle \psi(t), \xi(t) \rangle - \langle \psi(s), \xi(s) \rangle = \int_s^t \Big( \langle \psi(r), g(r) \rangle - \langle h(r), \xi(r) \rangle \Big) \, dr.$$

*Proof.* We define $\xi$ by (5.41) and claim that it is the solution of (5.39). In fact, by Lemma 5.6, we have

$$(5.44) \quad \begin{aligned} &\int_0^t e^{A(t-s)} B(s) \xi(s) \, ds \\ &= \int_0^t e^{A(t-s)} B(s) [G(s,0)\xi_0 + \int_0^s G(s,r)g(r) \, dr] \, ds \\ &= \int_0^t e^{A(t-s)} B(s) G(s,0)\xi_0 \, ds \\ &\quad + \int_0^t \int_r^t e^{A(t-s)} B(s) G(s,r)g(r) \, ds \, dr \\ &= (G(t,0) - e^{At})\xi_0 + \int_0^t [G(t,r) - e^{A(t-r)}]g(r) \, dr \\ &= \xi(t) - e^{At}\xi_0 - \int_0^t e^{A(t-r)} g(r) \, dr. \end{aligned}$$

This proves (5.41). Now, by the second equality in (5.34), we have

$$(5.45) \quad G^*(t,s)y = e^{A^*(t-s)}y + \int_s^t e^{A^*(r-s)} B^*(r) G^*(t,r) y \, dr,$$
$$(t,s) \in \overline{\Delta}, \; y \in X^*.$$

Hence, similar to the proof of (5.41), we can prove (5.42). Finally, we prove (5.43). First, it is not hard to see that for any $0 \leq s \leq t \leq T$,

$$(5.46) \quad \begin{cases} \xi(t) = G(t,s)\xi(s) + \int_s^t G(t,r) g(r) \, dr, \\ \psi(s) = G^*(t,s)\psi(t) + \int_s^t G^*(r,s) h(r) \, dr. \end{cases}$$

§6. Elliptic PDEs                                                              71

Thus, it follows that

$$\langle \psi(t), \xi(t)\rangle - \langle \psi(s), \xi(s)\rangle$$
$$= \langle \psi(t), G(t,s)\xi(s) + \int_s^t G(t,r)g(r)\,dr\rangle$$
$$- \langle G^*(t,s)\psi(t) + \int_s^t G^*(r,s)h(r)\,dr, \xi(s)\rangle$$

(5.47)
$$= \int_s^t \langle G^*(t,r)\psi(t), g(r)\rangle\,dr - \int_s^t \langle h(r), G(r,s)\xi(s)\rangle\,dr$$
$$= \int_s^t \langle \psi(r) - \int_r^t G^*(\sigma, r)h(\sigma)\,d\sigma, g(r)\rangle\,dr$$
$$- \int_s^t \langle h(r), \xi(r) - \int_s^r G(r,\sigma)g(\sigma)\,d\sigma\rangle\,dr$$
$$= \int_s^t \Big(\langle \psi(r), g(r)\rangle - \langle h(r), \xi(r)\rangle\Big)\,dr.$$

This proves (5.43).

## §6. Elliptic Partial Differential Equations

In this section, we collect some basic results of elliptic partial differential equations.

### §6.1. Sobolev spaces

We start with some notation. Any $\alpha = (\alpha_1, \alpha_2, \cdots, \alpha_n)$, where $\alpha_j$ are nonnegative integers, is called a *multi-index*. We define $|\alpha| = \sum_{i=1}^n |\alpha_i|$ and

$$D^\alpha = D_1^{\alpha_1} D_2^{\alpha_2} \cdots D_n^{\alpha_n} \equiv \frac{\partial^{\alpha_1}}{\partial x_1^{\alpha_1}} \frac{\partial^{\alpha_2}}{\partial x_2^{\alpha_2}} \cdots \frac{\partial^{\alpha_n}}{\partial x_n^{\alpha_n}}.$$

Next, any open connected set $\Omega$ in $\mathbb{R}^n$ is called a domain, its boundary is denoted by $\partial \Omega$ and its closure is denoted by $\overline{\Omega}$. Hereafter, we let $\Omega$ be a bounded domain. Let $C^m(\Omega)$ and $C^m(\overline{\Omega})$ be the sets of all $m$ times continuously differentiable functions on $\Omega$ and $\overline{\Omega}$, respectively, and let $C_0^m(\Omega)$ be the set of all functions $f \in C^m(\Omega)$, such that the support $\operatorname{supp} f \equiv \{x \in \Omega \mid f(x) \neq 0\}$ is compact in $\Omega$. We simply denote $C^0(\overline{\Omega}) = C(\overline{\Omega})$ and $C_0^0(\Omega) = C_0(\Omega)$. For any $y \in C^m(\overline{\Omega})$, we define

(6.1)
$$\|y\|_{C^m(\overline{\Omega})} = \max_{x \in \overline{\Omega}} \Big\{ \sum_{|\alpha| \leq m} |D^\alpha y(x)| \Big\}.$$

Then $\|\cdot\|_{C^m(\overline{\Omega})}$ is a norm under which $C^m(\overline{\Omega})$ is a Banach space. Next, for any $\sigma \in (0,1)$, let $C^\sigma(\overline{\Omega})$ be the set of all functions $y \in C(\overline{\Omega})$, such that

(6.2)
$$\|y\|_{C^\sigma(\overline{\Omega})} \equiv \sup_{x, \bar{x} \in \overline{\Omega}, x \neq \bar{x}} \frac{|y(x) - y(\bar{x})|}{|x - \bar{x}|^\sigma}$$

is finite. Any function $y \in C^\sigma(\overline{\Omega})$ is said to be *Hölder continuous*. Let $C^{m,\sigma}(\overline{\Omega})$ be the set of all functions $y \in C^m(\overline{\Omega})$ such that $D^\alpha y \in C^\sigma(\overline{\Omega})$ for all multi-index $\alpha$ with $|\alpha| = m$ and define

$$(6.3) \quad \|y\|_{C^{m,\sigma}(\overline{\Omega})} = \|y\|_{C^m(\overline{\Omega})} + \sum_{|\alpha|=m} \|D^\alpha y\|_{C^\sigma(\overline{\Omega})}, \quad \forall y \in C^{m,\sigma}(\overline{\Omega}).$$

It is known that $C^{m,\sigma}(\overline{\Omega})$ is a Banach space under norm (6.3). Let $C_0^{m,\sigma}(\Omega) = C_0^m(\Omega) \bigcap C^{m,\sigma}(\overline{\Omega})$ and let

$$(6.4) \quad C^\infty(\overline{\Omega}) = \bigcap_{m \geq 0} C^m(\overline{\Omega}), \quad C_0^\infty(\Omega) = \bigcap_{m \geq 0} C_0^m(\Omega).$$

Space $C^{m,1}(\overline{\Omega})$ consists of all $C^m(\overline{\Omega})$ functions having Lipschitz continuous $m$th order partial derivatives. In particular, $C^{0,1}(\overline{\Omega})$ is the set of all Lipschitz continuous functions defined on $\overline{\Omega}$. Next, we define

$$(6.5) \quad \|y\|_{W^{m,p}(\Omega)} = \Big( \int_\Omega \sum_{|\alpha| \leq m} |D^\alpha y|^p \, dx \Big)^{1/p}.$$

This is a norm on $C^m(\overline{\Omega})$, under which $C^m(\overline{\Omega})$ is not complete. The completion of $C^m(\overline{\Omega})$ under (6.5) is denoted by $W^{m,p}(\Omega)$. Similarly, the completion of $C_0^m(\Omega)$ under (6.5) is denoted by $W_0^{m,p}(\Omega)$. Thus, $W^{m,p}(\Omega)$ and $W_0^{m,p}(\Omega)$ are Banach spaces. For $p = 2$, we also denote $W^{m,2}(\Omega) = H^m(\Omega)$ and $W_0^{m,p}(\Omega) = H_0^m(\Omega)$. It is known that $H^m(\Omega)$ and $H_0^m(\Omega)$ are Hilbert spaces with the inner product

$$(6.6) \quad (y,z)_m = \int_\Omega \sum_{|\alpha| \leq m} D^\alpha y \overline{D^\alpha z} \, dx.$$

It can be proved that a function $y \in W^{m,p}(\Omega)$, if and only if there exist functions $f_\alpha \in L^p(\Omega)$, $|\alpha| \leq m$, such that

$$(6.7) \quad \int_\Omega y D^\alpha \varphi \, dx = (-1)^{|\alpha|} \int_\Omega f_\alpha \varphi \, dx, \quad \forall \varphi \in C_0^\infty(\Omega), |\alpha| \leq m.$$

Function $f_\alpha$ is referred to as the $\alpha$th *distributional derivative* of $y$. All the spaces $W^{m,p}(\Omega)$ and $W_0^{m,p}(\Omega)$ are called the *Sobolev spaces*.

The smoothness of the boundary for a given domain is important for further discussion. A given domain $\Omega$ is said to be $C^{k,\sigma}$ if for any $x \in \partial\Omega$, there exists a ball $B$, centered at $x$ and some $1 \leq i \leq n$, such that $\partial\Omega \bigcap B$ can be represented by $x_i = \varphi(x_1, \cdots, x_{i-1}, x_{i+1}, \cdots, x_n)$ for some $C^{k,\sigma}$ function $\varphi$. In this case, we also say that the domain $\Omega$ has a $C^{k,\sigma}$ boundary $\partial\Omega$.

The following notion will be useful below.

**Definition 6.1.** Let $X$ and $Y$ be two Banach spaces. We say that $X$ is *continuously embedded* into $Y$, denoted by $X \hookrightarrow Y$, if there exists a continuous injective map $i : X \to Y$ (injective means $\mathcal{N}(i) = \{0\}$). Furthermore,

§6. Elliptic PDEs

if $i$ is compact, then we say that $X$ is *compactly embedded* into $Y$, denoted by $X \hookrightarrow\hookrightarrow Y$.

The following is one of the central results in Sobolev space theory (see Adams [1], Troianiello [1] for details).

**Theorem 6.2.** (Sobolev Embedding Theorem) *Let $\Omega$ be a bounded domain in $\mathbb{R}^n$. Then the following hold:*

(i) *For all $m \geq 0$ and $1 \leq p \leq r \leq \infty$,*

$$(6.8) \qquad W^{m,r}(\Omega) \hookrightarrow W^{m,p}(\Omega).$$

(ii) *Let $\Omega$ be $C^1$, then, for any $0 \leq j < m$, $1 \leq r, p < \infty$, with*

$$(6.9) \qquad \frac{1}{p} > \frac{1}{r} + \frac{j}{n} - \frac{m}{n},$$

*it holds that*

$$(6.10) \qquad W^{m,r}(\Omega) \hookrightarrow\hookrightarrow W^{j,p}(\Omega).$$

(iii) *Let $\Omega$ be $C^1$. Then*

$$(6.11) \quad \begin{cases} W^{k,p}(\Omega) \hookrightarrow L^{\frac{np}{n-kp}}(\Omega), & kp < n, \\ W^{k,p}(\Omega) \hookrightarrow L^q(\Omega), & kp = n, q < \infty, \\ W^{k,p}(\Omega) \hookrightarrow\hookrightarrow L^q(\Omega), & q < \frac{np}{n-kp}, \ (\frac{np}{0} \equiv +\infty) \\ W^{k,p}(\Omega) \hookrightarrow C^{m,\sigma}(\overline{\Omega}) \hookrightarrow\hookrightarrow C^{m,\theta}(\overline{\Omega}), \\ \qquad \frac{n}{p} \notin \mathbb{N}, m = [k - \frac{n}{p}], 0 < \theta < \sigma = k - \frac{n}{p} - m, \\ W^{k,p}(\Omega) \hookrightarrow C^{\frac{k-n}{p-1},\delta}(\overline{\Omega}), & \frac{n}{p} \in \mathbb{N}, \delta \in (0,1). \end{cases}$$

Let us point out some important special cases.

(1) Let $m = 0$ and $1 \leq p \leq r \leq \infty$. Then (6.8) becomes $L^r(\Omega) \hookrightarrow L^p(\Omega)$. This can be easily justified by Hölder's inequality.

(2) Let $m = 1$, $j = 0$, and $1 \leq r = p < \infty$. Then (6.9) is satisfied and (6.10) becomes $W^{1,p}(\Omega) \hookrightarrow\hookrightarrow L^p(\Omega)$. This implies that if $y_k : \Omega \to \mathbb{R}$ is a sequence of functions such that

$$(6.12) \qquad \|y_k\|_{W^{1,p}(\Omega)} \equiv \left\{ \int_\Omega \left[ |y_k|^p + \sum_{i=1}^n \left|\frac{\partial y_k}{\partial x_i}\right|^p \right] dx \right\}^{1/p} \leq C,$$

then there exists a subsequence $y_{k_j}$ such that for some $y \in L^p(\Omega)$, one has

$$(6.13) \qquad \lim_{j \to \infty} \|y_{k_j} - y\|_{L^p(\Omega)} = 0.$$

Such a property is very useful in applications.

(3) Let $k=1$ and $p>n$. Then the fourth relation in (6.11) becomes $W^{1,p}(\Omega) \hookrightarrow C^\sigma(\bar\Omega)$, for all $0 \leq \sigma < 1 - \frac{n}{p}$. This means that if $p > n$, then, any function in $W^{1,p}(\Omega)$ must be Hölder continuous. However, the readers should be careful that when $p \leq n$, functions in $W^{1,p}(\Omega)$ are not necessarily continuous. Likewise, functions in $W^{m,p}(\Omega)$ are not always continuous.

Next, we denote the dual space of $W_0^{m,p}(\Omega)$ by $W^{-m,p'}(\Omega)$, with $p' = \frac{p}{p-1}$. Any $F \in W^{-m,p'}(\Omega)$ has the following form:

$$(6.14) \quad \langle F, \varphi \rangle = \sum_{|\alpha| \leq m} (-1)^{|\alpha|} \int_\Omega f_\alpha D^\alpha \varphi \, dx, \qquad \forall \varphi \in W_0^{m,p}(\Omega),$$

for some $f_\alpha \in L^{p'}(\Omega)$. Thus, $\{f_\alpha, |\alpha| \leq m\} \subset L^{p'}(\Omega)$ gives a representation of the functional $F$. Sometimes, we directly write $F = \sum_{|\alpha| \leq m} D^\alpha f_\alpha$, which is understood as (6.14). Thus, in particular, if $F \in W^{-1,p'}(\Omega)$, then $F = f_0 + \sum_{i=1}^n \frac{\partial}{\partial x_i} f_i$, for some $f_i \in L^{p'}(\Omega)$ $(0 \leq i \leq n)$. Also, if $p=2$, we write $W^{-m,2}(\Omega) = H^{-m}(\Omega)$.

*Remark 6.3.* We should point out the representation (6.14) for linear bounded functionals on $W_0^{m,p}(\Omega)$ is not unique, namely, two different sets of functions $\{f_\alpha, |\alpha| \leq m\}, \{g_\alpha, |\alpha| \leq m\} \in L^{p'}(\Omega)$ can represent the same functional $F \in W^{-m,p'}(\Omega)$. To see this, it suffices to find a nonzero representation for the zero functional. We consider $m=1$ and take any nonzero functions $f_1, \cdots, f_n \in C^\infty(\bar\Omega)$. Set

$$(6.15) \quad f_0(x) = -\sum_{i=1}^n (f_i)_{x_i}(x), \qquad x \in \Omega.$$

Then we can easily check that

$$(6.16) \quad \int_\Omega \Big\{ f_0 \varphi - \sum_{i=1}^n f_i \varphi_{x_i} \Big\} dx = 0, \qquad \forall \varphi \in W_0^{1,p}(\Omega),$$

Thus, $\{f_0, \cdots, f_n\}$ is a nonzero representation of the zero functional.

*Remark 6.4.* We know that $H_0^m(\Omega)$ is a Hilbert space. Thus, by the Riesz Theorem, there exists an isomorphism between $H_0^m(\Omega)$ and $H^{-m}(\Omega) \equiv (H_0^m(\Omega))^*$. But, this does not mean that these two spaces are the same. In the theory of partial differential equations, people do not identify $H_0^m(\Omega)$ with $H^{-m}(\Omega)$. We have seen that the elements in $H_0^m(\Omega)$ are functions with certain regularity; the elements in $H^{-m}(\Omega)$ need not even be functions, and they are usually called the *distributions*.

For a function $y \in C(\bar\Omega)$, sometimes we talk about the value of $y$ on the boundary $\partial\Omega$. However, $y \in W^{m,p}(\Omega)$ may be discontinuous on $\bar\Omega$ for some combinations of $(m,p)$. Thus, the meaning of $y\big|_{\partial\Omega}$ needs to be specified carefully. This leads to the theory of the so-called *trace*, which, by

## §6. Elliptic PDEs

definition, is the "value" of the function (and its distributional derivatives) on the boundary $\partial\Omega$. To make it more precise, we need to introduce the Sobolev spaces on manifolds. Those who are not familiar with a manifold can just think of it as a smooth surface in $\mathbb{R}^n$. Let $\Gamma$ be a compact manifold in $\mathbb{R}^n$. For any $s \in (0,1)$, $p \in [1,\infty)$, and $y \in C^\infty(\Gamma)$, we define

$$(6.17) \quad \|y\|_{W^{s,p}(\Gamma)} = \left\{ \int_\Gamma |y(x)|^p \, dx + \int_{\Gamma \times \Gamma} \frac{|y(x) - y(\widetilde{x})|^p}{|x - \widetilde{x}|^{n-1+sp}} \, dx \, d\widetilde{x} \right\}^{1/p}.$$

This is a norm. The completion of $C^\infty(\Gamma)$ under the above norm is denoted by $W^{s,p}(\Gamma)$ (see Grisvard [1] or Kufner-John-Fučik [1] for details).

*Remark 6.5.* Clearly, the above definition permits $\Gamma$ to be $\Omega$. Thus, the Sobolev space $W^{s,p}(\Omega)$ is well defined for $s \in [0,\infty)$. Here, by convention, $W^{0,p}(\Omega) = L^p(\Omega)$. It can be proved that when $s$ is an integer, the two definitions for $W^{s,p}(\Omega)$ are equivalent, meaning that the two norms (see (6.5) and (6.17)) are equivalent, which leads to the same set of functions used when making the completion of $C^\infty(\overline{\Omega})$ under these norms. Similarly, we may also define $W_0^{s,p}(\Omega)$ for $s \in [0,\infty)$. Furthermore, the dual space of $W_0^{s,p}(\Omega)$ is denoted by $W^{-s,p'}(\Omega)$ ($p' = p/(p-1)$). It can also be proved that the Sobolev Embedding Theorem remains true if we allow $j, k, m$ to be any real numbers.

The following result will be used later, and will be referred to as a *trace theorem*.

**Theorem 6.6.** (Gagliardo) *Let $\Omega$ be a domain in $\mathbb{R}^n$ with a Lipschitz boundary, i.e., $\partial\Omega$ is in $C^{0,1}$. Then there exists a unique linear bounded operator $\gamma : W^{1,p}(\Omega) \to W^{1-\frac{1}{p},p}(\partial\Omega)$, such that*

$$(6.18) \quad \gamma z(x) = z|_{\partial\Omega}(x), \quad \forall x \in \partial\Omega, \ z(\cdot) \in C^{0,1}(\overline{\Omega}).$$

*Moreover, this operator is onto and has a right inverse, i.e., there exists a linear bounded operator $\eta : W^{1-\frac{1}{p},p}(\partial\Omega) \to W^{1,p}(\Omega)$, such that*

$$(6.19) \quad \gamma(\eta(\xi))(\cdot) = \xi(\cdot), \quad \forall \xi(\cdot) \in W^{1-\frac{1}{p},p}(\partial\Omega).$$

### §6.2. Linear elliptic equations

Now, let us consider the following boundary value problem:

$$(6.20) \quad \begin{cases} Ay(x) = f(x), & x \in \Omega, \\ y|_{\partial\Omega} = 0, \end{cases}$$

where $\Omega$ is a bounded domain in $\mathbb{R}^n$ with boundary $\partial\Omega$, $f$ is a given function (or functional), and $A$ is given by

$$(6.21) \quad Ay(x) = -\sum_{i,j=1}^n \left( a_{ij}(x) y_{x_j}(x) \right)_{x_i}.$$

An important example of (6.21) is the case $a_{ij} = \delta_{ij}$. In this case, we have $-Ay(x) = \Delta y(x) \equiv \sum_{i=1}^{n} y_{x_i x_i}(x)$, the so-called Laplacian operator. Problem (6.20) is usually referred to as a *homogeneous Dirichlet problem*. If the right-hand side 0 of the boundary condition in (6.20) is replaced by some suitable function $g$, the problem will be called an *inhomogeneous Dirichlet problem*. We impose the following basic conditions:

(6.22)
$$\begin{cases} a_{ij}(x) = a_{ji}(x), & \forall x \in \Omega, \\ \lambda |\xi|^2 \leq \sum_{ij=1}^{n} a_{ij}(x) \xi_i \xi_j \leq \Lambda |\xi|^2, & \forall x \in \Omega, \xi \in \mathbb{R}^n, \end{cases}$$

for some constants $0 < \lambda \leq \Lambda$. The above conditions are called the *symmetry* and the (uniform) *ellipticity condition*, respectively. It should be pointed out that we may consider more general operators than the $A$ defined in (6.21). For example, we may add first order derivative terms and a zeroth order term. Let us now give the following definition.

**Definition 6.7.** (i) Let $a_{ij} \in C^1(\overline{\Omega})$ and $f \in C(\overline{\Omega})$. A function $y$ is called a *classical solution* of (6.20) if it is $C^2(\overline{\Omega})$, such that (6.20) is satisfied.

(ii) Let $a_{ij} \in W^{1,\infty}(\Omega)$ and $f \in L^p(\Omega)$ for some $p \in (1, \infty)$. A function $y$ is called a *strong solution* of (6.20) if $y \in W^{2,p}(\Omega) \cap W_0^{1,p}(\Omega)$, such that the equation in (6.20) holds almost everywhere in $\Omega$.

(iii) Let $a_{ij} \in L^\infty(\Omega)$ and $f \in W^{-1,p}(\Omega)$ for some $p \in (1, \infty)$. A function $y$ is called a *weak solution* of (6.20) if $y \in W_0^{1,p}(\Omega)$, such that

(6.23)
$$\int_\Omega \sum_{i,j=1}^{n} a_{ij}(x) y_{x_j}(x) \varphi_{x_i}(x) \, dx = \langle f, \varphi \rangle,$$

$$\forall \varphi \in W_0^{1,p'}(\Omega), \quad (p' = \tfrac{p}{p-1}).$$

Now, let us state the following classical result concerning the well-posedness of (6.20).

**Theorem 6.8.** *Let (6.22) hold.*

*(i) Let $\Omega$ be $C^1$, $a_{ij} \in C(\overline{\Omega})$, and $f \in W^{-1,p}(\Omega)$ for some $p \in (1, \infty)$. Then there exists a unique weak solution $y \in W_0^{1,p}(\Omega)$ of (6.20). Moreover, there exists a constant $C > 0$, independent of $f$, such that*

(6.24)
$$\|y\|_{W_0^{1,p}(\Omega)} \leq C \|f\|_{W^{-1,p}(\Omega)}.$$

*(ii) Let $\Omega$ be $C^{1,1}$, $a_{ij} \in C^1(\overline{\Omega})$ and $f \in L^p(\Omega)$ for some $p \in (1, \infty)$. Then, there exists a unique strong solution $y \in W^{2,p}(\Omega) \cap W_0^{1,p}(\Omega)$ of (6.20). Moreover, there exists a constant $C > 0$, independent of $f$, such that*

(6.25)
$$\|y\|_{W^{2,p}(\Omega)} \leq C \|f\|_{L^p(\Omega)}.$$

## §6. Elliptic PDEs

(iii) *Let $\Omega$ be $C^{2,\alpha}$, $a_{ij} \in C^{1,\alpha}(\overline{\Omega})$, and $f \in C^{\alpha}(\overline{\Omega})$ for some $\alpha \in (0,1)$. Then there exists a unique classical solution $y \in C^{2,\alpha}(\overline{\Omega})$ of (6.20). Furthermore, there exists a constant $C > 0$, independent of $f$, such that*

$$(6.26) \qquad \|y\|_{C^{2,\alpha}(\overline{\Omega})} \leq C\|f\|_{C^{\alpha}(\overline{\Omega})}.$$

In the above, (iii) is usually referred to as the *Schauder theory*, and (i) and (ii) are referred to as the $L^p$-*theory*. Also, (6.26) is referred to as the *Schauder estimate* and (6.24) and (6.25) are referred to as the $L^p$-*estimates*. The proof of (i) can be found in Morrey [1, p.156] and (ii) and (iii) can be found in Gilbarg–Trudinger [1]. By Sobolev Embedding Theorem, we know that if $p > n$, the weak solution $y$ is $C^{\alpha}(\overline{\Omega})$ with $0 < \alpha < 1 - n/p$; if $p > n/2$, the strong solution $y$ is $C^{\alpha}(\overline{\Omega})$ with $0 < \alpha < 1 - \frac{n}{2p}$, and, furthermore, if $p > n$, the strong solution is $C^{1,\alpha}(\overline{\Omega})$ with $0 < \alpha < 1 - n/p$. We would like to point out that for $p \geq 2$, the weak solution $y$ of (6.20) is the unique solution of the following variational problem:

**Problem.** Find $y \in W_0^{1,p}(\Omega)$ minimizing the following functional:

$$(6.27) \qquad J(z) = \int_{\Omega} \sum_{i,j=1}^{n} a_{ij}(x) z_{x_j}(x) z_{x_i}(x)\, dx - \langle f, z \rangle.$$

Besides the problem (6.20), there is another very important boundary value problem, which is called the *Neumann problem*. We now explain this. Instead of (6.20), we consider the following:

$$(6.28) \qquad \begin{cases} Ay(x) + a_0(x) y(x) = \sum_{i=1}^{n} (f_i)_{x_i} + f_0(x), & x \in \Omega, \\ \dfrac{\partial y}{\partial \nu_A} + b_0(x) y = g(x), & x \in \partial\Omega. \end{cases}$$

Here, $A$ is the same as before and $a_0, b_0$ satisfy the following conditions:

$$(6.29) \qquad \begin{cases} a_0 \in L^{\infty}(\Omega), & a_0(x) \geq a, \quad \text{a.e. } x \in \Omega; \\ b_0 \in L^{\infty}(\partial\Omega), & b_0(x) \geq b, \quad \text{a.e. } x \in \partial\Omega; \\ a, b \text{ are nonnegative constants}, & a + b > 0. \end{cases}$$

We will allow $f_i \in L^p(\Omega)$. Thus, the right-hand side of the equation in (6.28) is actually some functional. $\frac{\partial y}{\partial \nu_A}$ is called the *conormal derivative* associated with the operator $A$ on the boundary $\partial\Omega$. In the case that $a_{ij}(\cdot) \in C(\overline{\Omega})$ and $\partial\Omega$ is $C^1$, for any $y(\cdot) \in C^1(\overline{\Omega})$, we have

$$(6.30) \qquad \frac{\partial y}{\partial \nu_A}(x) = \sum_{i,j=1}^{n} a_{ij}(x) y_{x_i}(x) \nu_j(x), \qquad x \in \partial\Omega,$$

where $\nu(x) = (\nu_1(x), \cdots, \nu_n(x))$ is the outward normal of $\partial\Omega$ at $x \in \partial\Omega$. In (6.28), this notation is formal. We need to specify the meaning of the boundary value problem (6.28). Unlike the Dirichlet problem, we are only given the definition of the weak solutions (other solutions can be defined accordingly). In what follows, we let $p > n$.

**Definition 6.9.** *A function $y \in W^{1,2}(\Omega)$ is called a weak solution of (6.28) if the following holds:*

$$
(6.31) \quad \int_\Omega \Big\{ \sum_{i,j=1}^n a_{ij} y_{x_j} \varphi_{x_i} + a_0 y \varphi \Big\} \, dx + \int_{\partial\Omega} b_0 y \varphi \, dx
$$
$$
= \int_\Omega \Big\{ \sum_{i=1}^n f_i \varphi_{x_i} + f_0 \varphi \Big\} \, dx + \int_{\partial\Omega} g \varphi \, dx, \quad \forall \varphi \in W^{1,2}(\Omega).
$$

**Theorem 6.10.** *Let (6.22) and (6.29) hold. Let $p > n$ and $q > n - 1$. Then, for any $f_0 \in L^{p/2}(\Omega)$, $f_1, \cdots, f_n \in L^p(\Omega)$ and $g \in L^q(\partial\Omega)$, (6.28) admits a unique weak solution $y \in W^{1,p}(\Omega)$. Moreover, there exist constants $C > 0$ and $\alpha \in (0,1)$, independent of the data $f_i$'s and $g$, such that*

$$
(6.32) \quad \|y\|_{W^{1,2}(\Omega) \cap C^\alpha(\bar\Omega)} \leq C \Big\{ \sum_{i=1}^n \|f_i\|_{L^p(\Omega)} + \|f_0\|_{L^{p/2}(\Omega)} + \|g\|_{L^q(\partial\Omega)} \Big\}.
$$

The above result is quoted from Kinderlehrer–Stampacchia [1]. Sometimes, problem (6.28) is also referred to as the *Robin problem* and only the case in which $b_0(x) \equiv 0$ is referred to as the *Neumann problem*.

### §6.3. Semilinear elliptic equations

Next, we consider the semilinear elliptic equations. By this, we mean the following

$$
(6.33) \quad \begin{cases} Ay(x) = f(x, y(x)), & x \in \Omega, \\ y|_{\partial\Omega} = 0, \end{cases}
$$

where $f : \Omega \times \mathbb{R} \to \mathbb{R}$ is a given map. More precisely, the above is the Dirichlet problem for semilinear elliptic equations. Similar to Definition 6.7, we may define the classical, the strong, and the weak solutions. In what follows, we only concentrate on the weak solutions. Let $\Omega$ be $C^1$. Further, let us make the following assumption on $f$:

(A1) $f$ is measurable in $x$ and $C^1$ in $y$, such that

$$
(6.34) \quad f_y(x, y) \leq 0, \quad \forall (x, y) \in \Omega \times \mathbb{R},
$$

and for any $R > 0$, there exists an $M_R > 0$, such that

$$
(6.35) \quad |f(x,y)|, |f_y(x,y)| \leq M_R, \quad \forall x \in \Omega, \ |y| \leq R.
$$

**Theorem 6.11.** *Let (6.22) and (A1) hold. Let $a_{ij} \in C(\bar\Omega)$. Then, for any $p \in [1, \infty)$, (6.33) admits a unique weak solution $y \in W_0^{1,p}(\Omega) \cap L^\infty(\Omega)$.*

## §6. Elliptic PDEs

*Sketch of the proof.* Let $1 \leq p < \infty$ and $m > 0$. We define

(6.36) $$f_m(x,y) = \begin{cases} f(x,-m), & y < -m, \\ f(x,y), & |y| \leq m, \\ f(x,m), & y > m. \end{cases}$$

Consider the following truncated problem:

(6.37) $$\begin{cases} Ay^m = f_m(x, y^m), & \text{in } \Omega, \\ y^m|_{\partial\Omega} = 0. \end{cases}$$

Now, for any $z(\cdot) \in L^p(\Omega)$, by (6.36), we see that $|f_m(x, z(x))| \leq M_m$. Thus, by Theorem 6.8 (i), there exists a unique weak solution $z^m \in W_0^{1,p}(\Omega)$ of the following: ($z(\cdot)$ is fixed)

(6.38) $$\begin{cases} Az^m = f_m(x, z), & \text{in } \Omega, \\ z^m|_{\partial\Omega} = 0. \end{cases}$$

Moreover, the following holds:

(6.39) $$\|z^m\|_{W_0^{1,p}(\Omega)} \leq C\|f_m(\cdot, z(\cdot))\|_{W^{-1,p}(\Omega)} \leq \widetilde{C} M_m |\Omega|^{1/p}.$$

Here $\widetilde{C} > 0$ is an absolute constant that is independent of $z$, in particular. Thus, we see that the map $z \mapsto z^m$ defined through (6.38) is continuous and compact (i.e., it maps bounded sets into relatively compact sets) from some fixed ball in $L^p(\Omega)$ into itself. Hence, by the Schauder fixed point theorem, there exists a fixed point $y^m$ of this map. Clearly, $y^m \in W_0^{1,p}(\Omega)$ is a solution of (6.38). Next, by De Giorgi–Moser method, we can show that

(6.40) $$\|y^m\|_{L^\infty(\Omega)} \leq C, \qquad \forall m > 0,$$

where the constant $C$ is independent of $m > 0$. Here, the monotonicity condition (6.34) plays an essential role. Then, if we take $m > C$, $y^m = y$ is a weak solution of (6.33). Finally, using (6.34) again, we obtain the uniqueness. □

To conclude this section, let us look at the semilinear equation with a semilinear boundary condition. More precisely, we consider the following problem:

(6.41) $$\begin{cases} Ay(x) = f(x, y(x)), & x \in \Omega, \\ \dfrac{\partial y}{\partial \nu_A} = g(x, y(x)), & x \in \partial\Omega. \end{cases}$$

Here, the operator $A$ is the same as before, and $f : \Omega \times \mathbb{R} \to \mathbb{R}$ and $g : \partial\Omega \times \mathbb{R} \to \mathbb{R}$ are two given functions. We assume the following: (compare with (A1))

(A2) $f$ satisfies (A1) with (6.34) replaced by the following:

(6.42) $$f_y(x,y) \leq -a, \quad \forall (x,y) \in \Omega \times \mathbb{R},$$

for some constant $a \geq 0$, and $g$ satisfies (A1) with $\Omega$ replaced by $\partial\Omega$ and (6.34) replaced by the following:

(6.43) $$g_y(x,y) \leq -b, \quad \forall (x,y) \in \partial\Omega \times \mathbb{R},$$

for some constant $b \geq 0$. Moreover,

(6.44) $$a + b > 0.$$

Similar to Definition 6.9, we can define the weak solution of (6.41) (just replace $-a_0(x)y$ and $-b_0(x)y+g(x)$ by $f(x,y)$ and $g(x,y)$, respectively, and set $f_i(x)$'s to be zero). Our result is the following:

**Theorem 6.12.** *Let (6.22) and (A2) hold. Then (6.41) admits a unique weak solution $y(\cdot) \in W^{1,2}(\Omega) \bigcap C^\alpha(\bar{\Omega})$, for some $\alpha \in (0,1)$.*

The proof is very similar to that of Theorem 6.11; therefore, we omit it here. Later, we will discuss elliptic variational inequalities and quasilinear equations.

**Remarks**

The material of §1 is almost standard. Good references are Conway [1], Dunford–Schwartz [1], and Yosida [1]. The material in §2.1 is selected from Barbu–Precupanu [1], Cristescu [1], and Conway [1], with some modifications. Very nice references about the convexity of the Banach spaces are Goebel-Reich [1] and Istratescu [1]. Concerning the renorming theorem in §2.2, we have taken some results from Diestel [1]. However, our proof of Theorem 2.18 is elementary and different from that given in Diestel [1]. Some other relevant references in this direction are the following: Bourgin [1], Diestel [2], Diestel–Uhl [1]. §3 is mainly selected from Deimling [1], Dunford–Schwartz [1], Hille–Phillips [1], and Zeidler [1].

Most of the material in §4 is taken from Pazy [1], Hille–Phillips [1] and Yosida [1]. Example 4.5 is taken from Weidmann [1], pp.70–71. Also, the books by Bensoussan-Da Prato-Delfour-Mitter [1] and Hale [1] were consulted. Example 4.25 is given here for clarifying some ambiguity in the literature.

Some of the material in §5 is taken from Balakrishnan [1], Barbu [1], and Pazy [1]. Proposition 5.2 is from Ball [1]. Lemma 5.6 is from Curtain–Pritchard [1].

Most material in §6 is selected from the following: Adams [1], Gilbarg–Trudinger [1], Grisvard [1], Kinderlehrer–Stampacchia [1], Kufner-John-Fučík [1], Ladyzhenskaya–Ural'tseva [1], Morrey [1], Stampacchia [1], and Troianiello [1].

# Chapter 3
# Existence Theory of Optimal Controls

In this chapter we will present an existence theory for optimal controls. The first two sections contain some necessary preliminaries for the rest of the chapter. We have tried to make them self-contained. The readers who are only interested in the results of existence of optimal controls can skip these first two sections.

## §1. Souslin Space

This section is concerned with the theory of Souslin space, which will play an important role in sequel. We assume that the readers have a basic knowledge of topological spaces.

### §1.1. Polish space

**Definition 1.1.** A topological space $X$ is called a *Polish space* if it has a countable base, is metrizable, and the space is complete under a metric compatible with the topology of the space.

Hereafter, if $X$ is a Polish space, we always fix a metric $d$ that is compatible with the topology of $X$ and $X$ is complete under $d$. We will refer to such a $d$ as a *compatible metric* of $X$. We note that the properties having a countable base, being metrizable, and being complete under some metric are topological properties. Thus, if a space is homeomorphic to a Polish space, then it is a Polish space. This observation will be useful in sequel. In the case where $X$ is a Polish space, we also say that $X$ is *Polish*.

**Proposition 1.2.** (i) *Let $X_n$ be a sequence of Polish spaces. Then the product space $\prod_n X_n$ and the sum space $\sum_n X_n$ are all Polish spaces.*

(ii) *Let $X$ be a Polish space. Then any closed subset $F \subset X$ and any open subset $U \subset X$ are all Polish spaces.*

*Proof.* (i) Let the corresponding metric of $X_n$ be $d_n$. Without loss of generality, we may assume that $\operatorname{diam} X_n \equiv \sup_{x,y \in X} d_n(x,y) < 1$; otherwise, we replace $d_n(x,y)$ by $\frac{d_n(x,y)}{1+d_n(x,y)}$. Let

$$(1.1) \qquad \rho(x,y) = \sup_n \{\frac{1}{n} d_n(x_n, y_n)\}, \quad \forall x = (x_n),\ y = (y_n) \in \prod_n X_n.$$

Then it is easy to see that $\rho(\cdot,\cdot)$ is a metric on $X \equiv \prod_n X_n$, under which $X$ is complete. Also, it is clear that $X$ has a countable base. Thus, $X$ is a Polish space.

Now, let $\widehat{X} = \sum_n X_n$ and define

(1.2) $$d(x,y) = \begin{cases} d_n(x,y), & x,y \in X_n, \\ 1, & \text{otherwise.} \end{cases}$$

Then $d$ is a metric and $\widehat{X}$ is complete under this metric. Also, we still have a countable base for $\widehat{X}$. Thus, $\widehat{X}$ is a Polish space.

(ii) Let $d$ be a compatible metric of $X$. Let $F \subset X$ be a closed set. Then, as a subspace, it also has a countable base and is metrizable. By the closeness of $F$, it is complete under $d$. Thus, $F$ is a Polish space.

Now, let $U$ be an open set in $X$ and $U \neq X$ (otherwise, we are done). Let

(1.3) $$V = \{(t,x) \in \mathbb{R} \times X \mid td(x, X \setminus U) = 1\}.$$

It is clear that $V$ is closed in $\mathbb{R} \times X$. By the above proof, we know that $V$ is a Polish space. Let $P_X : \mathbb{R} \times X \to X$ be the canonical projection: $P_X(t,x) = x$, $\forall x \in X$. Then, $P_X : V \to U$ is continuous and onto. Moreover, for any $x \in U$, we may define

(1.4) $$P_X^{-1}(x) = \left(\frac{1}{d(x, X \setminus U)}, x\right) \in V.$$

It is easy to see that $P_X^{-1} : U \to V$ is continuous and onto. Thus, $P_X$ is a homeomorphism between $V$ and $U$. Consequently, $U$ must also be a Polish space. $\square$

**Proposition 1.3.** *Let $X$ be a topological space with a countable base and $\{A_n\}$ be a sequence of Polish subspaces of $X$. Then $\bigcap_n A_n$ is a Polish subspace of $X$.*

*Proof.* Let $A = \bigcap_n A_n$ and define $f : X \to X^{\mathbb{N}}$ as the diagonal map, i.e.,

(1.5) $$f(x) = (x, x, , \cdots) \in X^{\mathbb{N}}, \qquad \forall x \in X.$$

We see that $f$ is a homeomorphism between $X$ and $f(X)$. Also,

(1.6) $$f(A) = \left(\prod_n A_n\right) \bigcap f(X).$$

It can be seen that $f(X)$ is closed in $X^{\mathbb{N}}$. Because $X$ has a countable base, then so has $f(X)$. By Proposition 1.2, $\prod_n A_n$ is Polish. Thus, $f(A)$, as a closed subset of $\prod_n A_n$, is Polish. Because $f$ is a homeomorphism, we obtain that $A \equiv \bigcap_n A_n$ is also Polish. $\square$

**Theorem 1.4.** *Let $Y$ be a subset of some Polish space $X$. Then $Y$ is Polish if and only if $Y$ is the intersection of a sequence of open sets in $X$.*

*Proof.* Sufficiency. By Propositions 1.2 and 1.3, we see that $Y$ is Polish.

## §1. Souslin Space

Necessity. Let $d$ and $\bar{d}$ be compatible metrics of $Y$ and $X$, respectively. Let $\overline{Y}$ be the closure of $Y$ in $(X,\bar{d})$. Then, for any $x \in \overline{Y}$ and any open set $U$ (of $X$) containing $x$, $U \cap Y \neq \phi$. Next, for any $n$, let

(1.7)
$$Y_n = \{x \in \overline{Y} \mid \exists \text{ open set } U \text{ of } X, \, x \in U$$
$$\text{diam}_d(U \cap Y) \leq \frac{1}{n}, \, \text{diam}_{\bar{d}} U < \frac{1}{n}\},$$

where $\text{diam}_d$ and $\text{diam}_{\bar{d}}$ have an obvious meaning. Then $Y_n$ is open in $\overline{Y}$ and contains $Y$, for all $n$. Thus, $Y \subseteq \bigcap_n Y_n$. Now, for any $x \in \bigcap_n Y_n$ (noting that $Y_n$ is monotone decreasing), we can find a decreasing sequence of open sets $U_n$ in $(X, \bar{d})$, such that

(1.8)
$$\text{diam}_d(U_n \cap Y) \leq \frac{1}{n},$$

(1.9)
$$x \in U_n, \quad \text{diam}_{\bar{d}} U_n < \frac{1}{n}.$$

Then, pick any $x_n \in U_n \cap Y$. By (1.8) and the monotonicity of $U_n \cap Y$, we see that $\{x_n\}$ is Cauchy in $(Y, d)$. Thus, by the completeness of $(Y, d)$, we have $x_n \to \hat{x}$, in $(Y, d)$. On the other hand, by (1.9), we have $x_n \to x$, in $(X, \bar{d})$. Hence, $x = \hat{x} \in Y$. This shows that

(1.10)
$$Y = \bigcap_n Y_n.$$

Because $Y_n$ is open in $\overline{Y}$, we can find an open set $H_n$ in $(X, \bar{d})$, such that

(1.11)
$$Y_n = H_n \cap \overline{Y}, \quad n \geq 1.$$

By the metrizability of $X$, we see that $\overline{Y}$ is the intersection of a sequence of open sets in $X$. Let $U_m$ be open in $X$ such that $\overline{Y} = \bigcap_m U_m$. Then

(1.12)
$$Y = \bigcap_n Y_n = \bigcap_n H_n \cap \left(\bigcap_m U_m\right) = \bigcap_{m,n}(U_m \cap H_n).$$

$U_m \cap H_n$ is open in $X$; thus, we are done. □

The following corollaries will be useful.

**Corollary 1.5.** *Space $X$ is Polish if and only if $X$ is homeomorphic to the intersection of a sequence of open sets in $[0,1]^{\mathbb{N}}$.*

*Proof.* Sufficiency. By Proposition 1.2, $[0,1]^{\mathbb{N}}$ is Polish. Thus, by Theorem 1.4, we know that $X$ is Polish.

Necessity. Let $(X, d)$ be complete and separable. Let $\{a_n\}_{n \geq 1}$ be dense in $X$ and define

(1.13)
$$\varphi(x) = \left(\frac{d(x, a_n)}{1 + d_n(x, a_n)}\right)_{n \geq 1}, \quad \forall x \in X.$$

Then it is seen that $\varphi: X \to \varphi(X) \subset [0,1]^{\mathbb{N}}$ is a homeomorphism. Thus, $\varphi(X)$ is a Polish space and hence, by Theorem 1.4, it is the intersection of a sequence of open sets in $[0,1]^{\mathbb{N}}$. Thus, our conclusion follows. □

By noting the fact that the space $[0,1]^{\mathbb{N}}$ is compact, we immediately obtain the following:

**Corollary 1.6.** *Space $X$ is Polish if and only if there exists a compact metric space $M$, such that $X$ is homeomorphic to the intersection of a sequence of open sets in $M$.*

**Corollary 1.7.** *Let $X$ and $Y$ be Polish, $f: X \to Y$ be continuous. Then, for any Polish subspace $Z$ of $Y$, $f^{-1}(Z)$ is Polish in $X$.*

*Proof.* By Theorem 1.4 we know that there exists a sequence of open sets $Y_n \subset Y$, such that $Z = \bigcap_n Y_n$. By the continuity of $f$, we see that each $f^{-1}(Y_n)$ is open in $X$. On the other hand, we have

$$(1.14) \qquad f^{-1}(Z) = \bigcap_n f^{-1}(Y_n).$$

Hence, our conclusion follows from Theorem 1.4. □

The above result says that the pre-image of a Polish space under a continuous map is Polish. However, we should note that, in general, the image of a Polish space under a continuous map is not necessarily Polish.

### §1.2. Souslin space

**Definition 1.8.** A topological space $X$ is called a *Souslin space* if it is a Hausdorff space and there exist a Polish space $P$ and a continuous map $g: P \to X$, such that

$$(1.15) \qquad g(P) = X.$$

A subset $A \subset X$ is said to be *Souslinian* if as a subspace, $A$ is a Souslin space. Sometimes, we also call such an $A$ a *Souslin set*.

**Proposition 1.9.** (i) *Any Polish space is Souslinian.*

(ii) *Any Souslin space is separable.*

(iii) *Any closed and open subsets of a Souslin space are Souslinian.*

*Proof.* (i) is obvious.

(ii) and (iii). Let $X$ be a Souslin space, $P$ be a Polish space, and $g: P \to X$ be a continuous map with property (1.15). Then, by the separability of $P$, we obtain (ii). Now for any closed (resp. open) subset $A$ of $X$, by the continuity of $g$, $g^{-1}(A)$ is closed (resp. open) in $P$. By Proposition 1.2, $g^{-1}(A)$ is Polish and thus (iii) follows. □

**Proposition 1.10.** *Let $X$ be a Souslin space, $Y$ be a Hausdorff space and $f: X \to Y$ be continuous. Then*

## §1. Souslin Space

(i) $f(X)$ is Souslinian.

(ii) If $A \subset Y$ is Souslinian, then $f^{-1}(A)$ is Souslinian.

*Proof.* (i) Let $P$ be Polish and $g : P \to X$ be continuous and onto. Then $f \circ g : P \to f(X)$ is continuous and onto. Thus, $f(X)$ is Souslinian.

(ii) Let $P$ and $Q$ be two Polish spaces, $g : P \to X$ and $h : Q \to A$ be continuous and onto. Let

(1.16) $$R = \{(x, y) \in P \times Q \mid f(g(x)) = h(y)\}.$$

Then we see that $R$ is nonempty and closed in $P \times Q$. Thus, it is Polish. Let $p_1 : P \times Q \to P$ be the canonical projection and $\varphi = p_1 \big|_R : R \to P$ be the restriction of $p_1$. Then, by some simple calculation,

(1.17) $$f^{-1}(A) = g(\varphi(R)).$$

Hence, our conclusion follows from the continuity of $g \circ \varphi$. □

**Proposition 1.11.** *The product and the sum of countably many Souslin spaces are Souslinian.*

*Proof.* Let $X_n$ be Souslinian, $P_n$ be Polish, and $g_n : P_n \to X_n$ be continuous and onto. By Proposition 1.2, we know that $\prod_n P_n$ and $\sum_n P_n$ are Polish spaces. Let $g : \prod_n P_n \to \prod_n X_n$ be defined by

(1.18) $$g(\{p_n\}) = \{g_n(p_n)\}, \qquad \forall \{p_n\} \in \prod_n P_n.$$

Then $g$ is continuous and onto. Because each $X_n$ is Hausdorff, then so is $\prod_n X_n$. Thus, $\prod_n X_n$ is Souslin. Now, if we set $g : \sum_n P_n \to \sum_n X_n$ as

(1.19) $$g(p) = g_n(p), \qquad \forall p \in X_n, \quad n \geq 1,$$

then we see that $\sum_n X_n$ is Souslinian. □

**Proposition 1.12.** *Let $X$ be Hausdorff and $A_n$ be Souslinian in $X$. Then $\bigcup_n A_n$ and $\bigcap_n A_n$ are Souslinian.*

*Proof.* It is clear that there exists a continuous onto map $g : \sum_n A_n \to \bigcup_n A_n$. Thus, by Proposition 1.11, $\bigcup_n A_n$ is Souslinian. Now, let $f : X \to X^{\mathbb{N}}$ be the diagonal map (see (1.5)). Then, we know that $\bigcap_n A_n$ is homeomorphic to a closed subset of $\prod_n A_n$. Hence, by Propositions 1.9 and 1.11, we see that $\bigcap_n A_n$ is Souslinian. □

From Proposition 1.12 we see that if $X$ is a Hausdorff space, then the set of all Souslin sets in $X$ is closed under countable unions and intersections. However, we should point out that, in general, the set of all Souslin sets in $X$ is *not* necessarily a $\sigma$-field. The reason is that if $A \subset X$ is a Souslin set, the set $X \setminus A$ need not be a Souslin set. But, it is very interesting and important that we have the following result.

**Proposition 1.13.** *Let $X$ be a Souslin space. Then, any Borel set $A$ in $X$ is Souslinian.*

*Proof.* Let $\mathcal{B}(X)$ be the Borel $\sigma$-field on $X$. Define

$$(1.20) \qquad \mathcal{F} = \{A \subset X \mid A \text{ and } A^c \text{ are all Souslinian }\}.$$

Then $\mathcal{F}$ is closed under countable unions and intersections. Moreover, by the definition, $A \in \mathcal{F}$ implies $A^c \in \mathcal{F}$ and $X \in \mathcal{F}$. Thus, $\mathcal{F}$ is a $\sigma$-field. On the other hand, all the closed sets are in $\mathcal{F}$. Thus, $\mathcal{B}(X) \subset \mathcal{F}$. □

We should note that the set $\mathcal{F}$ defined by (1.20) is not the set of all Souslin sets in $X$.

**Corollary 1.14.** *Let $X$ be a Souslin space, $Y$ be a Hausdorff space, and $f : X \to Y$ be a continuous map. Then, for any $B \in \mathcal{B}(X)$, $f(B)$ is Souslinian in $Y$.*

*Proof.* Our conclusion follows from Propositions 1.10 and 1.13. □

## §1.3. Capacity and capacitability

Let us introduce the following important notion.

**Definition 1.15.** Let $X$ be a Hausdorff space, and $2^X$ be the set of all subsets of $X$. A map $f : 2^X \to [-\infty, \infty]$ is called a *capacity* on $X$ if the following conditions are satisfied:

(i) Monotonicity: $A \subset B$, implies $f(A) \leq f(B)$.

(ii) Left continuity: $\forall A_n \subset X$, with $A_n \subset A_{n+1}$, $f(A_n) \to f(\bigcup_n A_n)$.

Moreover, $f$ is called a *right continuous capacity* if we additionally have

(iii) Right continuity: For any compact set $K \subset X$,

$$(1.21) \qquad f(K) = \inf\{f(U) \mid K \subset U \subset X, \ U \text{ open }\}.$$

**Definition 1.16.** *Let $f$ be a capacity on $X$. A set $A \subset X$ is said to be $f$-capacitable if*

$$(1.22) \qquad f(A) = \sup\{f(K) \mid K \subset A, \ K \text{ compact }\}.$$

The following result is very useful.

**Theorem 1.17.** *Let $X$ be a Hausdorff space and $f$ be a right continuous capacity on $X$. Then any Souslin subspace $Y \subset X$ is $f$-capacitable.*

*Proof.* Let $P$ be a Polish space and $g : P \to Y$ be a continuous onto map. By Corollary 1.6, we may assume that $M$ is a compact metric space and $G_n$ are open sets in $M$, such that

$$(1.23) \qquad P = \bigcap_{n \geq 0} G_n, \qquad G_0 = M.$$

§1. Souslin Space

Then there exist closed sets (and hence they are compact) $K_{nm} \subset M$, such that for any given $n \geq 0$, the sequence $\{K_{nm}\}_{m\geq 1}$ is nondecreasing and

$$(1.24) \qquad G_n = \bigcup_{m\geq 1} K_{nm}, \qquad n \geq 0.$$

If $f(Y) = -\infty$, then by the monotonicity of $f$, we see that $Y$ is $f$-capacitable. Now, let us assume $f(Y) > -\infty$. To prove our conclusion, it suffices to prove the following: For any $a \in \mathbb{R}$, with $f(Y) > a$, there exists a compact set $K \subset Y$, such that

$$(1.25) \qquad f(K) \geq a.$$

To prove this, we inductively define monotonically decreasing compact subsets $T_n \subset M$, such that for all $n \geq 0$,

$$(1.26) \qquad T_n \subset G_n, \quad f(g(P \cap T_n)) > a.$$

Setting $T_0 = M$, we see that (1.26) holds for $n = 0$. If $T_n$ is defined, then (1.26) holds for $n$. Then, noting the fact that

$$(1.27) \qquad G_{n+1} = \bigcup_{m\geq 1} K_{n+1,m} \supset P \supset P \cap T_n,$$

we have

$$(1.28) \qquad P \cap T_n = \bigcup_{m\geq 1} [P \cap T_n \cap K_{n+1,m}].$$

By the left continuity of the capacity $f$, we see that

$$(1.29) \qquad \begin{aligned} a < f(g(P \cap T_n)) &= f(\bigcup_{m\geq 1} g(P \cap T_n \cap K_{n+1,m})) \\ &= \lim_{m\to\infty} f(g(P \cap T_n \cap K_{n+1,m})). \end{aligned}$$

Hence, there exists a $k \geq 1$, such that

$$(1.30) \qquad f(g(P \cap T_n \cap K_{n+1,k})) > a.$$

Then, by letting $T_{n+1} = T_n \cap K_{n+1,k}$, we can easily see that (1.26) holds for $n+1$. This gives the required sequence $\{T_n\}$. Now, set

$$(1.31) \qquad T \equiv \bigcap_{n\geq 0} T_n \subset \bigcap_{n\geq 0} G_n = P.$$

Clearly, $T$ is nonempty and compact. Thus, $K \equiv g(T)$ is compact in $Y$. Now, for any open set $U \supset K$, the set $V \equiv g^{-1}(U)$ is open in $P$ containing

$T$. We can find an open set $V'$ in $M$, such that $V = P \bigcap V'$. Then, in $M$, we have

$$(1.32) \qquad V' \supset V \supset T = \bigcap_n T_n.$$

By the compactness and the monotonicity of $T_n$'s, we see that there exists an $n_0 \geq 0$, such that

$$(1.33) \qquad V' \supset T_n, \qquad \forall n \geq n_0.$$

Hence,

$$(1.34) \qquad U = g(V) = g(P \bigcap V') \supset g(P \bigcap T_n), \qquad \forall n \geq n_0.$$

By the monotonicity of $f$, it follows that

$$(1.35) \qquad f(U) \geq f(g(P \bigcap T_n)) > a.$$

Because $U$ is an arbitrary open set containing the compact set $K$, from the right continuity of $f$, we see that (1.25) holds and thus $Y$ is $f$-capacitable. □

The following result is the aim of our studying the Souslin space theory.

**Theorem 1.18.** *Let $Y$ be a Souslin set in $\mathbb{R}^n$. Then it is Lebesgue measurable.*

*Proof.* For any $A \subset \mathbb{R}^n$, we define the *outer measure* of $A$ to be

$$(1.36) \qquad m^*(A) = \inf\{|G| \mid G \supset A, \ G \text{ open}\},$$

where $|G|$ is the Lebesgue measure of the set $G \subset \mathbb{R}^n$. It can now be shown that $m^*$ is a right continuous capacity on $\mathbb{R}^n$. Thus, by Theorem 1.17, we know that any Souslin set $Y$ is $m^*$-capacitable. That means

$$(1.37) \qquad m^*(Y) = \sup\{|K| \mid K \subset Y, \ K \text{ compact}\}.$$

Hence, by a well-known criterion of the Lebesgue measurability, we see that $Y$ is Lebesgue measurable. □

Let us now take a quick look at some interesting consequences of the above theorem.

**Corollary 1.19.** *Let $X$ be a Souslin space, and let $f : X \to \mathbb{R}^n$ be continuous. Then, for any Borel set $A \subset X$, $f(A)$ is Lebesgue measurable. In particular, if $A$ is an open or closed set in $X$, then $f(A)$ is Lebesgue measurable.*

*Proof.* From Corollary 1.14, we know that $f(A)$ is Souslinian in $\mathbb{R}^n$. Then our result follows from Theorem 1.18. □

## §2. Multifunctions and Selection Theorems

**Corollary 1.20.** *Let $Z$ be a Lebesgue nonmeasurable set in $\mathbb{R}$. Let*

(1.38) $$A = \{(z,z) \mid z \in Z\} \subset \mathbb{R}^2.$$

*Then $A$ is a set of Lebesgue measure 0 in $\mathbb{R}^2$, but it is not a Souslin set (thus, it is not a Borel set).*

The proof is immediate (by contradiction).

It is possible to come up with some other interesting corollaries like the above. One can imagine that it is very difficult to prove results like Corollaries 1.19 and 1.20 without the theory of Souslin space.

## §2. Multifunctions and Selection Theorems

Let $T$ and $X$ be two topological spaces. Let $2^X$ be the set of all nonempty subsets of $X$. We call any map $\Gamma : T \to 2^X$ a *multifunction*.

### §2.1. Continuity

In this subsection, we discuss various continuities of multifunctions.

**Definition 2.1.** Multifunction $\Gamma : T \to 2^X$ is said to be *upper semicontinuous* if for any closed subset $F \subset X$, the set

(2.1) $$\Gamma^{-1}(F) \equiv \{t \in T \mid \Gamma(t) \bigcap F \neq \phi\}$$

is closed in $T$; $\Gamma$ is said to be *lower semicontinuous* if for any open subset $U \subset X$, $\Gamma^{-1}(U)$ is open in $T$; $\Gamma$ is said to be *continuous* if it is both upper and lower semicontinuous.

We note that when $\Gamma$ is single valued, the above three kinds of continuities are equivalent.

**Proposition 2.2.** *Let $\Gamma : T \to 2^X$. Then*

(i) *$\Gamma$ is upper semicontinuous if and only if for each $t \in T$ and any open set $V \supset \Gamma(t)$, there exists an open set $U \ni t$, such that*

(2.2) $$\Gamma(U) \subset V.$$

(ii) *$\Gamma$ is lower semicontinuous if and only if for each $t \in T$ and open set $V \subset X$, with*

(2.3) $$\Gamma(t) \bigcap V \neq \phi,$$

*there exists an open set $U \ni t$, such that*

(2.4) $$\Gamma(s) \bigcap V \neq \phi, \qquad \forall s \in U.$$

*Proof.* The statement of (i) can be restated as follows: $\Gamma$ is upper semicontinuous if and only if for any open set $V \subset X$, the set $\{t \in T \mid \Gamma(t) \subseteq V\}$ is open. We note that for any set $G \subseteq X$ (recall $G^c = X \setminus G$)

(2.5) $$t \notin \Gamma^{-1}(G^c) \iff \Gamma(t) \cap G^c = \phi \iff \Gamma(t) \subseteq G.$$

Thus,

(2.6) $\quad \Gamma^{-1}(G^c)^c = \{t \in T \mid \Gamma(t) \subseteq G\}, \qquad \forall G \subseteq X.$

Now, by Definition 2.1, $\Gamma$ is upper semicontinuous if and only if for any open set $G$, $\Gamma^{-1}(G^c)^c$ is open. Thus, by (2.6), we obtain (i).

(ii) Note that (2.3) is the same as $t \in \Gamma^{-1}(V)$ and (2.4) is equivalent to $U \subseteq \Gamma^{-1}(V)$. Thus, (ii) is nothing but a restatement of the definition of the lower semicontinuity for $\Gamma$. □

We see that Definition 2.1 describes a global nature of lower and upper semicontinuity for $\Gamma$ (topologically) whereas Proposition 2.2 describes a local nature, which allows us to define the upper and lower semicontinuity of $\Gamma$ at an individual point.

Now, let us look at the case where both $T$ and $X$ are metric spaces. Let $d$ be the metric in $T$ or $X$, which can be identified from the context. We introduce the following:

(2.7) $\quad \rho_H(A, B) = \max\left\{ \sup_{a \in A} d(a, B), \sup_{b \in B} d(b, A) \right\}, \qquad \forall A, B \in 2^X.$

Let $\mathcal{K}(X)$ be the set of all nonempty compact subsets in $X$. It is easy to check whether $\rho_H$ is a metric on $\mathcal{K}(X)$, called the *Hausdorff metric*, under which $\mathcal{K}(X)$ is a complete metric space. Thus, for a multifunction $\Gamma : T \to 2^X$, if for each $t \in T$, we have $\Gamma(t) \in \mathcal{K}(X)$, then we may regard $\Gamma : T \to \mathcal{K}(X)$ as a single valued map and talk about the continuity in the usual sense. It is natural to ask whether the continuity defined in Definition 2.1 is compatible with the usual one. The following result gives a positive answer.

**Proposition 2.3.** *The map $\Gamma : T \to \mathcal{K}(X)$ is continuous if and only if $\Gamma$ is continuous in the sense of Definition 2.1.*

*Proof.* Necessity. For any $t_0 \in T$ and open set $V \supset \Gamma(t_0)$, there exists a $\delta > 0$, such that

(2.8) $\quad \Gamma(t) \subset \mathcal{O}_{\rho_H(\Gamma(t),\Gamma(t_0))}(\Gamma(t_0)) \subset V, \qquad \forall d(t, t_0) < \delta,$

where $\mathcal{O}_\delta(A) \equiv \{x \in X \mid d(x, A) < \delta\}$ is an open set. Thus, by Proposition 2.2, $\Gamma$ is upper semicontinuous. Next, let $V \subset X$ be an open set and $t_0 \in \Gamma^{-1}(V)$. Then, there exist $\varepsilon > 0$ and $x_0 \in \Gamma(t_0) \cap V$, such that $\mathcal{O}_\varepsilon(x_0) \subset V$. On the other hand, there exists a $\delta > 0$, such that

(2.9) $\quad d(x_0, \Gamma(t)) \leq \rho_H(\Gamma(t_0), \Gamma(t)) < \varepsilon/2, \qquad \forall t \in \mathcal{O}_\delta(t_0).$

This implies

(2.10) $\quad \phi \neq \Gamma(t) \cap \mathcal{O}_\varepsilon(x_0) \subset \Gamma(t) \cap V, \qquad \forall t \in \mathcal{O}_\delta(t_0).$

Hence, $\mathcal{O}_\delta(t_0) \subset \Gamma^{-1}(V)$, which yields the openness of $\Gamma^{-1}(V)$. Thus, $\Gamma$ is lower semicontinuous.

## §2. Multifunctions and Selection Theorems

Sufficiency. Let $t_0 \in T$ be fixed. For any $\varepsilon > 0$, by the upper semicontinuity of $\Gamma$, we can find a $\delta > 0$, such that

$$\Gamma(\mathcal{O}_\delta(t_0)) \subset \mathcal{O}_\varepsilon(\Gamma(t_0)). \tag{2.11}$$

This implies

$$\sup_{x \in \Gamma(t)} d(x, \Gamma(t_0)) \leq \varepsilon, \qquad \forall t \in \mathcal{O}_\delta(t_0). \tag{2.12}$$

On the other hand, by the compactness of $\Gamma(t_0)$, we can find $x^1, x^2, \cdots, x^k \in \Gamma(t_0)$, such that

$$\bigcup_{i=1}^{k} \mathcal{O}_\varepsilon(x^i) \supset \Gamma(t_0). \tag{2.13}$$

By the lower semicontinuity of $\Gamma$, for any $\varepsilon > 0$, $\Gamma^{-1}(\mathcal{O}_\varepsilon(x^i))$ is open and contains $t_0$ (because $x^i \in \Gamma(t_0)$). Then we can find a $\delta > 0$, such that

$$\mathcal{O}_\delta(t_0) \subset \bigcap_{i=1}^{k} \Gamma^{-1}(\mathcal{O}_\varepsilon(x^i)), \tag{2.14}$$

i.e.,

$$\Gamma(t) \bigcap \mathcal{O}_\varepsilon(x^i) \neq \phi, \qquad \forall t \in \mathcal{O}_\delta(t_0),\ 1 \leq i \leq k. \tag{2.15}$$

Then, for any $x \in \Gamma(t_0)$, by taking $i$ with $x \in \mathcal{O}_\varepsilon(x^i)$, we have

$$d(x, \Gamma(t)) \leq d(x, x^i) + d(x^i, \Gamma(t)) < 2\varepsilon, \qquad \forall t \in \mathcal{O}_\delta(t_0). \tag{2.16}$$

This gives

$$\sup_{x \in \Gamma(t_0)} d(x, \Gamma(t)) \leq 2\varepsilon, \qquad \forall t \in \mathcal{O}_\delta(t_0). \tag{2.17}$$

Combining the above with (2.12), we obtain the continuity of $\Gamma$ with respect to $\rho_H$. □

From the above result, we see that the notion given in Definition 2.1 is a reasonable generalization of the continuity for multifunctions between metric spaces to those between topological spaces. In applications, some weaker notions are also often used.

**Definition 2.4.** A multifunction $\Gamma : T \to 2^X$ is said to be *pseudo-continuous* at $t \in T$ if

$$\bigcap_{\varepsilon > 0} \overline{\Gamma(\mathcal{O}_\varepsilon(t))} = \Gamma(t). \tag{2.18}$$

We say that $\Gamma$ is pseudo-continuous on $T$ if it is pseudo-continuous at each point $t \in T$.

We note that in (2.18), the relation "$\supseteq$" is always true. Thus, the equality in (2.18) can be replaced by "$\subseteq$". Also, we should note that when $\Gamma$ is pseudo-continuous, then it takes closed set values for each $t \in T$.

**Proposition 2.5.** *Let $\Gamma : T \to 2^X$ be a multifunction taking closed set values. Then $\Gamma$ is pseudo-continuous if and only if the graph*

$$(2.19) \qquad \mathcal{G}(\Gamma) \equiv \{(t,x) \in T \times X \mid x \in \Gamma(t)\}$$

*is closed in $T \times X$.*

*Proof.* Necessity. Let $(t_n, x_n) \in \mathcal{G}(\Gamma)$, with $(t_n, x_n) \to (t, x)$ in $T \times X$. Then, for any $\varepsilon > 0$, there exists an $n_\varepsilon$, such that $n \geq n_\varepsilon$ implies $t_n \in \mathcal{O}_\varepsilon(t)$. Thus,

$$(2.20) \qquad x_n \in \Gamma(t_n) \subset \Gamma(\mathcal{O}_\varepsilon(t)), \qquad \forall n \geq n_\varepsilon.$$

Let $n \to \infty$; we obtain $x \in \overline{\Gamma(\mathcal{O}_\varepsilon(t))}$, $\forall \varepsilon > 0$. It then follows from (2.18) that $x \in \Gamma(t)$, which implies $\mathcal{G}(\Gamma)$ is closed.

Sufficiency. Let $x \in \bigcap_{\varepsilon>0} \overline{\Gamma(\mathcal{O}_\varepsilon(t))}$. Then there exist sequences $\varepsilon_n \downarrow 0$, $t_n \in \mathcal{O}_{\varepsilon_n}(t)$, and $x_n \in \Gamma(t_n)$, with $x_n \to x$. Thus,

$$(2.21) \qquad (t_n, x_n) \to (t, x), \qquad (t_n, x_n) \in \mathcal{G}(\Gamma).$$

Hence, by the closeness of $\mathcal{G}(\Gamma)$, we have $(t, x) \in \mathcal{G}(\Gamma)$, which yields $x \in \Gamma(t)$. We have proved the relation "$\subseteq$" in (2.18) and our conclusion follows because the other relation is trivially true. $\square$

**Proposition 2.6.** *Let $\Gamma : T \to 2^X$ be a multifunction taking closed set values and upper semicontinuous. Then it is pseudo-continuous.*

*Proof.* By the upper semicontinuity of $\Gamma$, we see that for any given $t \in T$, any $\varepsilon > 0$, there exists a $\delta > 0$, such that $\Gamma(\mathcal{O}_\delta(t)) \subseteq \mathcal{O}_\varepsilon(\Gamma(t))$. Thus,

$$(2.22) \qquad \bigcap_{\delta>0} \overline{\Gamma(\mathcal{O}_\delta(t))} \subseteq \bigcap_{\varepsilon>0} \overline{\mathcal{O}_\varepsilon(\Gamma(t))} = \overline{\Gamma(t)} = \Gamma(t).$$

Hence, $\Gamma$ is pseudo-continuous. $\square$

We should note that, in general, the converse of the above proposition is not true. The following is a simple example.

*Example 2.7.* Let $X = \mathbb{R}$, $T = [0,1]$. Define

$$(2.23) \qquad \Gamma(t) = \begin{cases} [0,1] \cup \{\frac{1}{t}\}, & t \neq 0, \\ [0,1], & t = 0. \end{cases}$$

Then $\Gamma$ takes closed set values and at $t = 0$, we have

$$(2.24) \qquad \bigcap_{\varepsilon>0} \overline{\Gamma(\mathcal{O}_\varepsilon(0))} = [0,1] = \Gamma(0).$$

Thus, $\Gamma$ is pseudo-continuous at $t = 0$. On the other hand, the set $F = [2, \infty) \subset X$ is closed, but

$$\Gamma^{-1}(F) = \{t \in T \mid \Gamma(t) \cap F \neq \phi\} = (0, \tfrac{1}{2}] \tag{2.25}$$

is not closed in $T$. Thus, $\Gamma$ is not upper semicontinuous (at $t = 0$).

However, we have the following result.

**Proposition 2.8.** *Let $X$ be a compact metric space and $\Gamma : T \to 2^X$ be a multifunction taking closed set values. Then $\Gamma$ is upper semicontinuous if and only if it is pseudo-continuous.*

*Proof.* Necessity follows from Proposition 2.6.

Sufficiency. Suppose $\Gamma$ is not upper semicontinuous. Then by Proposition 2.2, there exists a $t \in T$ and $\varepsilon > 0$, such that for some sequence $t_n \to t$ and $x_n \in \Gamma(t_n)$, one has

$$d(x_n, \Gamma(t)) \geq \varepsilon, \qquad n \geq 1. \tag{2.26}$$

As $X$ is sequentially compact, we may assume that $x_n \to x$. Then by the pseudo-continuity of $\Gamma$ and Proposition 2.5, we have $x \in \Gamma(t)$. But this contradicts (2.26). Hence, $\Gamma$ is upper semicontinuous. □

Note that in Example 2.7, $X$ is not compact. In the existence theory of optimal controls for distributed parameter systems, sometimes we encounter multifunctions valued in some infinite dimensional spaces that, of course, lack compactness. Thus, in those cases, pseudo-continuity will be more useful than upper semicontinuity.

Next, for any multifunction $\Gamma : T \to 2^X$, we introduce the following function:

$$\varphi_\Gamma(t, x) = d(x, \Gamma(t)) \equiv \inf_{y \in \Gamma(t)} d(x, y), \qquad \forall (t, x) \in T \times X. \tag{2.27}$$

Then it is clear that

$$|\varphi_\Gamma(t, x) - \varphi_\Gamma(t, \bar{x})| \leq d(x, \bar{x}), \qquad \forall t \in T, \quad x, \bar{x} \in X, \tag{2.28}$$

$$|\varphi_\Gamma(t, x) - \varphi_\Gamma(\bar{t}, x)| \leq \rho_H(\Gamma(t), \Gamma(\bar{t})), \qquad \forall t, \bar{t} \in T, \quad x \in X. \tag{2.29}$$

Thus, in the case where $\Gamma : T \to \mathcal{K}(X)$ is continuous in $\rho_H$, the function $(t, x) \mapsto \varphi_\Gamma(t, x)$ is continuous. As a matter of fact, we have the following more delicate result.

**Proposition 2.9.** *Let $\Gamma : T \to 2^X$ be a multifunction taking closed set values. Then among the following statements, (i) $\Rightarrow$ (ii) $\Rightarrow$ (iii):*

  (i) *$\Gamma$ is upper semicontinuous.*

  (ii) *For all $x \in X$, the function $t \mapsto \varphi_\Gamma(t, x)$ is lower semicontinuous.*

(iii) $\Gamma$ is pseudo-continuous.

*Proof.* (i) $\Rightarrow$ (ii). Let $t_0 \in T$ be fixed. For any $\varepsilon > 0$, there exists a $\delta > 0$, such that $\Gamma(\mathcal{O}_\delta(t_0)) \subset \mathcal{O}_\varepsilon(\Gamma(t_0))$. Hence, for any $t \in \mathcal{O}_\delta(t_0)$, there exist an $x_t \in \Gamma(t)$ and $x_0 \in \Gamma(t_0)$, such that

$$\begin{aligned}d(x,\Gamma(t)) &\geq d(x,x_t) - \varepsilon \geq d(x,x_0) - d(x_0,x_t) - \varepsilon \\ &\geq d(x,\Gamma(t_0)) - 2\varepsilon.\end{aligned} \quad (2.30)$$

Thus, (ii) holds.

(ii) $\Rightarrow$ (iii). By (2.28), we see that $(t,x) \mapsto \varphi_\Gamma(t,x)$ is also lower semicontinuous. Hence, by the assumption that $\Gamma$ takes closed set values and the nonnegativity of the function $\varphi_\Gamma$, we have that the set

$$\mathcal{G}(\Gamma) = \varphi_\Gamma^{-1}(\{0\}) = \{(t,x) \in T \times X \mid \varphi_\Gamma(t,x) \leq 0\} \quad (2.31)$$

is closed. Hence, (iii) follows from Proposition 2.5. $\square$

By Proposition 2.8, we see that in the case where $X$ is compact, the above three statements are equivalent.

## §2.2. Measurability

We now consider the measurability of multifunctions. To this end, let $T$ be a Lebesgue measurable set in $\mathbb{R}^n$ and $X$ be a Polish space. From Proposition 1.9, we know that $X$ is also a Souslin space. Recall that $|S|$ is the Lebesgue measure of the Lebesgue measurable set $S \subset \mathbb{R}^n$.

**Definition 2.10.** A multifunction $\Gamma : T \to 2^X$ is said to be *Lebesgue* (resp. *Borel, Souslin*) *measurable* if for any closed set $F \subset X$, the set $\Gamma^{-1}(F)$ is Lebesgue (resp. Borel, Souslin) set in $T$.

Note that in the case where $\Gamma$ is single valued, the Lebesgue (resp. Borel) measurability coincides with the usual one, whereas, the notion of Souslin measurability is introduced just for simplicity of the terminology. Hereafter, by just saying that $\Gamma$ is measurable, we always refer to Lebesgue measurability. The following result collects many useful criteria for the measurability of multifunctions.

**Theorem 2.11.** (Himmelberg-Jacobs-Van Vleck) *Let $\Gamma : T \to 2^X$ be a multifunction taking closed set values. Then, the following are equivalent:*

(i) *$\Gamma$ is measurable.*

(ii) *For any open set $U \subset X$, the set $\Gamma^{-1}(U)$ is measurable.*

(iii) *For any Souslin set $S \subset X$, the set $\Gamma^{-1}(S)$ is measurable.*

(iv) *For any $x \in X$, the function $\varphi_\Gamma(\cdot,x)$ is measurable.*

(v) *For any $\varepsilon > 0$, there exists a closed set $T_\varepsilon \subset T$, with $|T \setminus T_\varepsilon| < \varepsilon$ such that $\varphi_\Gamma|_{T_\varepsilon \times X}$ is continuous.*

§2. Multifunctions and Selection Theorems

(vi) *For any $\varepsilon > 0$, there exists a Souslin set $T_\varepsilon \subset T$, with $|T \setminus T_\varepsilon| < \varepsilon$ such that $\mathcal{G}(\Gamma|_{T_\varepsilon})$ is Souslinian in $T \times X$.*

(vii) *There exists a Souslin set $T' \subset T$, with $|T \setminus T'| = 0$, such that $\mathcal{G}(\Gamma|_{T'})$ is Souslinian in $T \times X$.*

(viii) *For any $\varepsilon > 0$, there exists a closed set $T_\varepsilon \subset T$, with $|T \setminus T_\varepsilon| < \varepsilon$, such that $\Gamma|_{T_\varepsilon}$ is pseudo-continuous.*

(ix) *For any $\varepsilon > 0$, there exists a closed set $T_\varepsilon \subset T$, with $|T \setminus T_\varepsilon| < \varepsilon$, such that $\mathcal{G}(\Gamma|_{T_\varepsilon})$ is closed.*

*Proof.* (i)⇒(ii). Because $X$ is a Polish space, we know that for any open set $U \subset X$, there exists a sequence of closed sets $\{F_k\}$, such that $U = \bigcup_k F_k$. On the other hand, we have

$$\Gamma^{-1}(U) = \bigcup_k \Gamma^{-1}(F_k), \tag{2.32}$$

and each $\Gamma^{-1}(F_k)$ is measurable. Thus, (ii)

(ii)⇒(iv). For any $x \in X$ and $\lambda > 0$, we have

$$\{t \in T \mid \varphi_\Gamma(t,x) < \lambda\} = \{t \in T \mid \Gamma(t) \bigcap \mathcal{O}_\lambda(x) \neq \phi\} \tag{2.33}$$
$$= \Gamma^{-1}(\mathcal{O}_\lambda(x)),$$

which is measurable and thus, (iv) follows.

(iv)⇒(v). Because $X$ is separable, we may let $\{x_k \mid k \geq 1\}$ be dense in $X$. By Luzin's theorem, there exists a closed set $T^k \subset T$ with $|T \setminus T^k| < \varepsilon/2^k$, such that $\varphi_\Gamma(\cdot, x_k)$ is continuous on $T^k$. Set $T_\varepsilon = \bigcap_k T^k$. We see that $T_\varepsilon$ is closed and $|T \setminus T_\varepsilon| < \varepsilon$. We can easily show that $\varphi_\Gamma$ is continuous on $T_\varepsilon \times X$. This proves (v).

(v)⇒(vi). Take $T_\varepsilon$ as in (v). Then it is Souslinian. By the continuity of $\varphi_\Gamma$ on $T_\varepsilon \times X$, we have from Proposition 2.9 that $\Gamma(\cdot)$ is pseudo-continuous. Thus, it follows from Proposition 2.5 that $\mathcal{G}(\Gamma|_{T_\varepsilon})$ is closed and thus it is Souslinian.

(vi)⇒(vii). Let $T_k$ be Souslinian with $|T \setminus T_k| < 1/k$ and let $\mathcal{G}(\Gamma|_{T_k})$ be Souslinian. Set $T' = \bigcup_k T_k$. Then it is a Souslin set and $|T \setminus T'| = 0$. On the other hand, $\mathcal{G}(\Gamma|_{T'}) = \bigcup_k \mathcal{G}(\Gamma|_{T_k})$, which is the union of a sequence of Souslin sets. Thus, $\mathcal{G}(\Gamma|_{T'})$ is Souslinian.

(vii)⇒(iii). Let $T'$ be as in (vii). Let $P_T : T \times X \to T$ be the canonical projection: $P_T(t,x) = t$. Then, for any Souslin set $S \subset X$,

$$\Gamma^{-1}(S) = P_T\big(\mathcal{G}(\Gamma) \bigcap (T \times S)\big) \tag{2.34}$$
$$= P_T[\mathcal{G}(\Gamma) \bigcap (T' \times S)] \bigcup P_T[\mathcal{G}(\Gamma) \bigcap ((T \setminus T') \times S)].$$

Because $\mathcal{G}(\Gamma)$, $T' \times S$ are Souslin sets, then so is their intersection. By Proposition 1.10, we know that the first term on the right-hand side of

(2.34) is Souslinian and thus by Theorem 1.18, it is measurable, whereas, the second term on the right-hand side of (2.34) is of measure zero because $|T \setminus T'| = 0$. Hence, $\Gamma^{-1}(S)$ is measurable and (iii) follows.

(iii)$\Rightarrow$(i) is obvious.
(v)$\Rightarrow$(viii) follows from Proposition 2.9.
(viii)$\Rightarrow$(ix) follows from Proposition 2.5.
(ix)$\Rightarrow$(vi) is obvious.

Hence, we have proved the following implications:
(i)$\Rightarrow$(ii)$\Rightarrow$(iv)$\Rightarrow$(v)$\Rightarrow$(vi)$\Rightarrow$(vii)$\Rightarrow$(iii)$\Rightarrow$(i); (v)$\Rightarrow$(viii)$\Rightarrow$(ix)$\Rightarrow$(vi).

This proves Theorem 2.11. □

We let $\mathcal{L}(T)$, $\mathcal{S}(T)$, and $\mathcal{B}(T)$ be the sets of all Lebesgue, Souslin, and Borel sets in $T$, respectively. From Definition 2.10, we have the following result.

**Corollary 2.12.** *Let $\Gamma : T \to 2^X$. Then $\Gamma$ is Borel measurable $\Rightarrow$ $\Gamma$ is Souslin measurable $\Rightarrow$ $\Gamma$ is Lebesgue measurable.*

Let us give some relationships between the continuity and the measurability of multifunctions.

**Proposition 2.13.** *Let $T$ be Souslinian and $\Gamma : T \to 2^X$ be pseudo-continuous. Then, for any Souslin set $S \subset X$, $\Gamma^{-1}(S) \in \mathcal{S}(T)$. In particular, $\Gamma$ is Souslin measurable.*

*Proof.* Let $S \in \mathcal{S}(T)$. Because $\Gamma$ is pseudo-continuous, $\mathcal{G}(\Gamma)$ is closed in $T \times X$. On the other hand, $T \times S$ is Souslinian. Thus, from

$$(2.35) \qquad \Gamma^{-1}(S) = P_T(\mathcal{G}(\Gamma) \bigcap (T \times S)),$$

we see that $\Gamma^{-1}(S)$ is Souslinian. □

We note that the projection of a closed set is not necessarily a Borel set. Thus, it is seen from (2.35) that a pseudo-continuous multifunction $\Gamma$ (defined on some Borel set $T$) is not necessarily Borel measurable. From Definitions 2.1 and 2.10, we have the following.

**Proposition 2.14.** *Let $T$ be a Borel set and $\Gamma : T \to 2^X$ be upper or lower semicontinuous. Then $\Gamma$ is Borel measurable.*

**Proposition 2.15.** *Let $\Gamma_1, \Gamma_2 : T \to 2^X$ be measurable taking closed set values. Then*

(i) $\Gamma_1(\cdot) \bigcup \Gamma_2(\cdot)$ *is measurable.*

(ii) *The set $T_0 = \{t \in T \mid \Gamma_1(t) \bigcap \Gamma_2(t) \neq \phi\}$ is measurable and $\Gamma_1(\cdot) \bigcap \Gamma_2(\cdot)$ is measurable on $T_0$.*

*Proof.* (i) For all $(t, x) \in T \times X$, we have

$$(2.36) \qquad \varphi_{\Gamma_1 \cup \Gamma_2}(t, x) = \min\{\varphi_{\Gamma_1}(t, x), \varphi_{\Gamma_2}(t, x)\},$$

§2. Multifunctions and Selection Theorems

By Theorem 2.11, we know that for each $x \in X$, $\varphi_{\Gamma_1}(\cdot, x)$ and $\varphi_{\Gamma_2}(\cdot, x)$ are measurable, and therefore so is $\varphi_{\Gamma_1 \cup \Gamma_2}(\cdot, x)$. Again by Theorem 2.11, we obtain the measurability of $\Gamma_1(\cdot) \bigcup \Gamma_2(\cdot)$.

(ii) By Theorem 2.11, there exist Souslin sets $T_i' \subset T$ with $|T \setminus T_i'| = 0$, such that $\mathcal{G}(\Gamma_i|_{T_i'})$ are Souslin sets ($i = 1, 2$). Then

$$(2.37) \qquad T_0 = P_T\big(\mathcal{G}(\Gamma_1 \bigcap \Gamma_2)\big) = P_T\Big(\mathcal{G}(\Gamma_1|_{T_1'}) \bigcap N\Big),$$

with some $N \subset T \setminus (T_1' \cap T_2')$. Hence, noting that $|N| = 0$, we see that $T_0$ is measurable. Now, let $T_0' \subset T_0$ be a Souslin set such that $|T_0 \setminus T_0'| = 0$. For any closed set $F \subset X$, we have

$$(2.38) \qquad \begin{aligned} (\Gamma_1 \bigcap \Gamma_2)^{-1}(F) &= P_T\Big(\mathcal{G}(\Gamma_1 \bigcap \Gamma_2) \bigcap (T_0 \times F)\Big) \\ &= P_T\Big(\mathcal{G}(\Gamma_1|_{T_1'}) \bigcap \mathcal{G}(\Gamma_2|_{T_2'}) \bigcap (T_0' \times F)\Big) \bigcup \widetilde{N}, \end{aligned}$$

with some $\widetilde{N} \subset T_0 \setminus (T_0' \cap T_1' \cap T_2')$. Because $T_0' \times F$ is a Souslin set and $|\widetilde{N}| = 0$, we obtain the measurability of $(\Gamma_1 \bigcap \Gamma_2)^{-1}(F)$, which proves our conclusion. □

We should note that in order to define the Borel measurability of multifunction $\Gamma : T \to 2^X$, we only need $T$ to be a topological space; to define the Souslin measurability of $\Gamma$, we need $T$ to be a Hausdorff space; and to define the Lebesgue measurability of $\Gamma$, we need $T$ to be a Lebesgue measurable set in $\mathbb{R}^n$.

The next result gives the measurability for composition of multifunctions.

**Proposition 2.16.** *Let $X_1$, $X_2$, and $X_3$ be Polish spaces. Let $\Gamma_1 : X_1 \to 2^{X_2}$, $\Gamma_2 : X_2 \to 2^{X_3}$. Then*

(i) $\Gamma_2 \circ \Gamma_1$ *is Borel measurable provided both $\Gamma_1$ and $\Gamma_2$ are Borel measurable.*

(ii) $\Gamma_2 \circ \Gamma_1$ *is measurable provided $X_1$ is a Lebesgue measurable set in $\mathbb{R}^n$, $\Gamma_1$ is measurable, and $\Gamma_2$ is Souslin measurable.*

*Proof.* We note that for any $F \subset X_3$,

$$(2.39) \qquad (\Gamma_2 \circ \Gamma_1)^{-1}(F) = \Gamma_1^{-1} \circ \Gamma_2^{-1}(F).$$

Thus, our conclusion follows from Definition 2.10 and Theorem 2.11. □

**Corollary 2.17.** *Let $X_1$ and $X_2$ be Polish spaces and $T \subset \mathbb{R}^n$ be a Lebesgue measurable set. Let $\Gamma_1 : T \to 2^{X_1}$ and $\Gamma_2 : T \to 2^{X_2}$ be measurable and taking closed set values. Then $\Gamma_1 \times \Gamma_2 : T \to 2^{X_1 \times X_2}$ is measurable.*

*Proof.* Let $P_1 : X_1 \times X_2 \to X_1$ and $P_2 : X_1 \times X_2 \to X_2$ be the canonical projections. Then

$$\tag{2.40} \begin{cases} P_1^{-1}(x_1) = \{x_1\} \times X_2, & \forall x_1 \in X_1, \\ P_2^{-1}(x_2) = X_2 \times \{x_2\}, & \forall x_2 \in X_2. \end{cases}$$

It is clear that $P_1^{-1} : X_1 \to 2^{X_1 \times X_2}$ and $P_2^{-1} : X_2 \to 2^{X_1 \times X_2}$ are continuous multifunctions. On the other hand,

$$\tag{2.41} \Gamma_1(t) \times \Gamma_2(t) = P_1^{-1}(\Gamma_1(t)) \bigcap P_2^{-1}(\Gamma_2(t)), \qquad t \in T.$$

By Proposition 2.16, $P_1^{-1} \circ \Gamma_1$ and $P_2^{-1} \circ \Gamma_2$ are measurable. Then, by Proposition 2.15, we complete the proof. □

**Corollary 2.18.** *Let $T$ be a Lebesgue measurable set in $\mathbb{R}^n$ and $X$ and $Y$ be Polish spaces. Let $U : T \times X \to 2^Y$ be Souslin measurable and $\xi : T \to X$ be measurable. Then, $\Gamma(\cdot) \equiv U(\cdot, \xi(\cdot)) : T \to 2^Y$ is measurable. In addition, if $U$ is pseudo-continuous and $\xi$ is continuous, then $\Gamma$ is Souslin measurable.*

*Proof.* Let $s(t) = (t, \xi(t))$. Then $s : T \to T \times X$ is measurable and

$$\tag{2.42} \Gamma(t) = U(s(t)), \qquad t \in T.$$

By Proposition 2.16, we obtain the measurability of $\Gamma$. The proof of the second conclusion is similar (note Proposition 2.13). □

Next, let $T \subset \mathbb{R}^n$ be a Lebesgue measurable set and $X$ and $Y$ be Polish spaces. Let $Q \subset Y$ and $f : T \times X \to Y$. We ask the following question: Under what conditions is the following set

$$\tag{2.43} \Gamma(t) = \{x \in X \mid f(t, x) \in Q\}$$

measurable? Such a situation will be frequently encountered in studying the existence theory of optimal controls for distributed parameter systems. Let us give the following result first.

**Theorem 2.19.** *Let $Q \subset Y$ be a closed set. Let $f : T \times X \to Y$ satisfy the following:*

(i) *For any $x \in X$, the map $f(\cdot, x)$ is measurable,*

(ii) *$f(\cdot, \cdot)$ is locally uniformly continuous in $x$, i.e., for any $(t_0, x_0) \in T \times X$ and $\delta > 0$, there exists a modulus of continuity $\omega(\cdot) \equiv \omega(\cdot, t_0, x_0, \delta)$, such that*

$$\tag{2.44} d(f(t, x), f(t, \widehat{x})) \leq \omega(d(x, \widehat{x})), \quad \forall x, \widehat{x} \in \mathcal{O}_\delta(x_0), \ t \in \mathcal{O}_\delta(t_0).$$

*Then the multifunction $\Gamma(\cdot)$ defined by (2.43) is measurable.*

*Proof.* Let $X_0 = \{x_k \mid k \geq 1\}$ be dense in $X$. For any $\varepsilon > 0$, there exists a closed set $T_\varepsilon^k \subset T$, with

$$\tag{2.45} |T \setminus T_\varepsilon^k| < \frac{\varepsilon}{2^k}, \qquad k \geq 1,$$

§2. Multifunctions and Selection Theorems       99

such that $f(\cdot, x_k)$ is continuous on $T_\varepsilon^k$. Set $T_\varepsilon = \bigcap_{k\geq 1} T_\varepsilon^k$. Then we see that $|T \setminus T_\varepsilon| < \varepsilon$. As in the proof of (iv)$\Rightarrow$(v) of Theorem 2.11, we can prove that $f$ is continuous on $T_\varepsilon \times X$ and $T_\varepsilon$ is closed. Thus,

$$\{(t,x) \in T_\varepsilon \times X \mid f(t,x) \in Q\} = \mathcal{G}(\Gamma|_{T_\varepsilon}) \tag{2.46}$$

is closed. On the other hand, by the continuity of $f(t,\cdot)$, we see that $\Gamma(t)$ is closed for each $t \in T$. Hence, by (ix) of Theorem 2.11, we obtain the measurability of $\Gamma(\cdot)$. □

In the following result, to ensure the same conclusion, different conditions are imposed on $f(\cdot,\cdot)$.

**Theorem 2.20.** *Let $f : T \times X \to Y$ be Souslin measurable and $Q \subset Y$ be closed. Then the multifunction $\Gamma(\cdot)$ defined by (2.43) is measurable. In particular, this is the case if $f$ is Borel measurable.*

*Proof.* We first let $T' \subset T$ be a Souslin set with the property that $|T \setminus T'| = 0$. For any closed set $F \subset X$,

$$\begin{aligned}\Gamma^{-1}(F) &= P_T\Big(f^{-1}(Q) \bigcap (T \times F)\Big) \\ &= P_T\Big(f^{-1}(Q) \bigcap (T' \times F)\Big) \bigcup P_T\Big(f^{-1}(Q) \bigcap ((T \setminus T') \times F)\Big).\end{aligned} \tag{2.47}$$

Hence, we see that $\Gamma(\cdot)$ is measurable. □

Note that in general, the Lebesgue measurability of $f(\cdot,\cdot)$ is not enough to ensure the measurability of $\Gamma(\cdot)$.

**Corollary 2.21.** *Let $y : T \to Y$ be Lebesgue measurable and $g : T \times X \to Y$. Let*

$$\Gamma(t) \equiv \{x \in X \mid g(t,x) = y(t)\} \neq \phi, \qquad \forall t \in T. \tag{2.48}$$

*Then $\Gamma(\cdot)$ is measurable if one of the following conditions holds:*

(i) *$g$ satisfies conditions (i) and (ii) of Theorem 2.19.*

(ii) *$g$ is Souslin measurable.*

*Proof.* (i) Let $f(t,x) = g(t,x) - y(t)$ and $Q = \{0\}$. Then Theorem 2.19 applies.

(ii) By Theorem 2.11, we may find a Souslin set $T' \subset T$ with $|T \setminus T'| = 0$, such that $y : T' \to Y$ is Souslin measurable. Then, by setting $f(t,x) = g(t,x) - y(t)$, we know that $f : T' \times X \to Y$ is Souslin measurable. Setting $Q = \{0\}$ and applying Theorem 2.20, we see that $\Gamma|_{T'} : T' \to X$ is measurable. Thus, for any closed set $F \subset X$, from

$$\Gamma^{-1}(F) = \Gamma|_{T'}^{-1}(F) \bigcup \Big(\Gamma^{-1}(F) \bigcap (T \setminus T')\Big), \tag{2.49}$$

we see that $\Gamma(\cdot)$ is measurable. □

## §2.3. Measurable selection theorems

Next, we are going to present some results on measurable selections for multifunctions. Again, in this subsection, we let $T \subset \mathbb{R}^n$ be some Lebesgue measurable set and $X$ be a Polish space.

**Definition 2.22.** Let $\Gamma : T \to 2^X$ be a multifunction. Function $f : T \to X$ is called a *selection* of $\Gamma(\cdot)$ if

$$f(t) \in \Gamma(t), \quad \text{a.e. } t \in T. \tag{2.50}$$

If such an $f$ is measurable, then $f$ is called a *measurable selection* of $\Gamma(\cdot)$.

The following is the main result concerning the existence of measurable selections.

**Theorem 2.23.** *Let $\Gamma : T \to 2^X$ be measurable taking closed set values. Then $\Gamma(\cdot)$ admits a measurable selection.*

*Proof.* Without loss of generality, we assume that diam $X < 1$. By the separability of $X$, we may let $X_0 = \{x_k \mid k \geq 1\}$ be dense in $X$. Define

$$f_0(t) = x_1, \quad t \in T. \tag{2.51}$$

Then, $f_0(\cdot)$ is measurable and

$$d(f_0(t), \Gamma(t)) < 1, \quad t \in T. \tag{2.52}$$

Suppose we have defined $f_{n-1}(\cdot)$, as measurable, satisfying

$$d(f_{n-1}(t), \Gamma(t)) < \frac{1}{2^{n-1}}, \quad t \in T. \tag{2.53}$$

$$d(f_{n-1}(t), f_{n-2}(t)) < \frac{1}{2^{n-2}}, \quad t \in T. \tag{2.54}$$

Then, define

$$\begin{cases} C_i^n = \{t \in T \mid d(x_i, \Gamma(t)) < \frac{1}{2^n}\}, \\ D_i^n = \{t \in T \mid d(x_i, f_{n-1}(t)) < \frac{1}{2^{n-1}}\}, \\ A_i^n = C_i^n \cap D_i^n, \end{cases} \quad i \geq 1. \tag{2.55}$$

We claim that

$$T = \bigcup_{i=1}^{\infty} A_i^n. \tag{2.56}$$

In fact, for any $t \in T$, by (2.53), there exists an $x \in \Gamma(t)$, such that $d(f_{n-1}(t), x) < \frac{1}{2^{n-1}}$. By the density of $X_0$ in $X$, there is an $i \geq 1$, such that

$$d(x_i, x) < \frac{1}{2^n}, \quad d(f_{n-1}(t), x_i) < \frac{1}{2^{n-1}}. \tag{2.57}$$

## §2. Multifunctions and Selection Theorems

The above implies $t \in C_i^n$ and $t \in D_i^n$ (see (2.55)). Hence, (2.56) holds. Now, set

$$\tag{2.58} B_i^n = \{x \in X \mid d(x, x_i) < \frac{1}{2^n}\} \equiv \mathcal{O}_{1/2^n}(x_i).$$

It follows that

$$\tag{2.59} C_i^n = \Gamma^{-1}(B_i^n), \qquad D_i^n = f_{n-1}^{-1}(B_i^{n-1}).$$

Hence, by the measurability of $\Gamma$ and $f_{n-1}$, we see that $C_i^n$ and $D_i^n$ are measurable and so is $A_i^n$. Next, we define

$$\tag{2.60} f_n(t) = x_i, \qquad \forall t \in A_i^n \setminus \bigcup_{j=1}^{i-1} A_j^n.$$

It is clear that $f_n(\cdot)$ is measurable and for any $t \in T$, there exists an $i$, such that

$$\tag{2.61} t \in A_i^n \setminus \bigcup_{j=1}^{i-1} A_j^n \subset C_i^n \bigcap D_i^n.$$

Thus, by $t \in C_i^n$, we have

$$\tag{2.62} d(f_n(t), \Gamma(t)) < \frac{1}{2^n},$$

and by $t \in D_i^n$, we have

$$\tag{2.63} d(f_n(t), f_{n-1}(t)) < \frac{1}{2^{n-1}}.$$

That means we can find a sequence of measurable functions $f_n : T \to X$, such that (2.62) and (2.63) hold for all $n \geq 1$. By letting $n \to \infty$, noting the closeness of $\Gamma(t)$, we obtain a measurable function $f : T \to X$ satisfying

$$\tag{2.64} f(t) \in \Gamma(t), \qquad \forall t \in T.$$

Thus, our conclusion follows. $\square$

It is clear that the condition of Theorem 2.23 can be slightly relaxed, namely, we require only that $\Gamma(t)$ take closed set values for almost all $t \in T$.

**Theorem 2.24.** *Let* $\Gamma : T \to 2^X$ *be measurable taking closed set values. Let* $f : T \times X \to Y$ *satisfy one of the following:*

(i) $f(\cdot, x)$ *is measurable for each* $x \in X$ *and* $f(t, x)$ *is locally uniformly continuous in* $x$ *(see Theorem 2.19).*

(ii) $f(\cdot, \cdot)$ *is Souslin measurable.*

*Moreover, $Q \subset Y$ is a Souslin set such that for almost all $t \in T$,*

$$\tag{2.65} \Lambda(t) \equiv \{x \in X \mid f(t, x) \in Q\}$$

*is closed and*

(2.66) $$\Lambda(t) \bigcap \Gamma(t) \neq \phi.$$

*Then there exists a measurable function $h : T \to X$, such that*

(2.67) $$\begin{cases} h(t) \in \Gamma(t), & \text{a.e. } t \in T, \\ f(t, h(t)) \in Q, & \text{a.e. } t \in T. \end{cases}$$

*Proof.* By Theorems 2.19 and 2.20, we know that $\Lambda : T \to 2^X$ is measurable. Thus, by Proposition 2.15, $\Lambda \bigcap \Gamma : T \to 2^X$ is measurable and taking closed set values for almost all $t \in T$. Hence, by Theorem 2.23, there exists a measurable function $h : T \to X$ satisfying

(2.68) $$h(t) \in \Lambda(t) \bigcap \Gamma(t), \qquad \forall t \in T.$$

Thus, our conclusion follows. □

**Corollary 2.25.** *Let $D$ be a Polish space, $Z$ be a complete metric space, and $T \subset \mathbb{R}^n$ be a measurable set. Let $g : D \to Z$ be locally uniformly continuous and $z : T \to Z$ be measurable satisfying*

(2.69) $$z(t) \subseteq g(D), \qquad \text{a.e. } t \in T.$$

*Then there exists a measurable function $h : T \to D$, such that*

(2.70) $$z(t) = g(h(t)), \qquad \text{a.e. } t \in T.$$

*Proof.* Let $\Gamma(t) \equiv D$, $Q = \{0\}$, and

(2.71) $$f(t, x) = g(x) - z(t), \qquad \forall (t, x) \in T \times D.$$

Then Theorem 2.24 applies. □

In Berkovitz [1] a similar result was presented, with an additional condition that $D$ is the union of a sequence of compact metric spaces. This is not the case if $D$ is, say, an infinite dimensional Banach space.

**Corollary 2.26.** *(Filippov) Let $\Gamma : T \to 2^X$ be measurable taking closed set values. Let $f : T \times X \to Y$ be Souslin measurable and for each $x \in X$, $f(\cdot, x)$ is measurable; for almost all $t \in T$, $f(t, \cdot)$ is continuous. Let $y : T \to Y$ be Lebesgue measurable satisfying*

(2.72) $$y(t) \in f(t, \Gamma(t)), \qquad \text{a.e. } t \in T.$$

*Then there exists a measurable function $h : T \to X$, such that*

(2.73) $$\begin{cases} h(t) \in \Gamma(t), & \text{a.e. } t \in T, \\ y(t) = f(t, h(t)), & \text{a.e. } t \in T. \end{cases}$$

*Proof.* By (ii) of Corollary 2.21, we know that if we set

(2.74) $$\Lambda(t) \equiv \{x \in X \mid f(t,x) = y(t)\}, \qquad t \in T.$$

then $\Lambda : T \to 2^X$ is measurable. By the continuity of $f(t,\cdot)$, $\Lambda(t)$ is closed for almost all $t \in T$. Then our conclusion follows from an argument similar to that used in proving Theorem 2.24. □

Corollary 2.26 is usually referred to as *Filippov's Lemma*. But, we should note that our $X$ and $Y$ are only a Polish space and a Hausdorff space, respectively.

## §3. Evolution Systems with Compact Semigroups

We first consider the existence theory of the optimal controls for semilinear evolutionary distributed parameter systems in which the semigroup involved is compact. This is the case for systems governed by semilinear parabolic partial differential equations with proper boundary conditions.

Let us start with the following assumptions.

(P1) $X$ is a reflexive Banach space, $U$ is a Polish space, and $T > 0$ is a constant.

(P2) $A : \mathcal{D}(A) \subset X \to X$ generates a compact semigroup $e^{At}$ on $X$.

(P3) $f : [0,T] \times X \times U \to X$ is Borel measurable in $(t,x,u)$ and continuous in $(x,u)$ for almost all $t \in [0,T]$; $f^0 : [0,T] \times X \times U \to \mathbb{R}$ is Borel measurable in $(t,x,u)$, lower semicontinuous in $(x,u)$, and there exists a constant $K \geq 0$, such that

(3.1) $$f^0(t,x,u) \geq -K, \qquad \forall (t,x,u) \in [0,T] \times X \times U.$$

(P4) $\Gamma : [0,T] \times X \to 2^U$ is pseudo-continuous.

(P5) $Q \subset X$ is closed and $S \subset X \times X$ is bounded and weakly closed.

Next, let $\mathcal{U}[0,T] = \{u : [0,T] \to U \mid u(\cdot) \text{ is measurable}\}$. Any element in $\mathcal{U}[0,T]$ is called a control (on $[0,T]$). The evolution system we are considering is the following

(3.2) $$\dot{y}(t) = Ay(t) + f(t,y(t),u(t)), \qquad t \in [0,T].$$

As in Chapter 2, §5, a (mild) solution $y(\cdot)$ of (1.2) is defined as a solution of the following integral equation:

(3.3) $$y(t) = e^{At}y(0) + \int_0^t e^{A(t-s)} f(s,y(s),u(s))\, ds, \qquad t \in [0,T].$$

Any solution $y(\cdot)$ of (3.3) is referred to as a state trajectory of the evolution system corresponding to the initial state $y(0)$ and the control $u(\cdot)$. Note that in the above, we do not assume the uniqueness and/or existence of solutions to (3.3). This allows that for any given control $u(\cdot)$ and initial state $y(0)$, there may be more than one or no "response" from the system. Thus, the following notion is necessary.

**Definition 3.1.** A pair $(y(\cdot), u(\cdot))$ is said to be *feasible* if (3.3) is satisfied and

$$(3.4) \qquad u(t) \in \Gamma(t, y(t)), \qquad \text{a.e. } t \in [0, T].$$

The pair is said to be *admissible* if it is feasible and

$$(3.5) \qquad \begin{cases} (y(0), y(T)) \in S, \quad y(t) \in Q, \quad t \in [0, T], \\ f^0(\cdot, y(\cdot), u(\cdot)) \in L^1(0, T). \end{cases}$$

In the case where $(y(\cdot), u(\cdot))$ is feasible (admissible, resp.), we refer to $y(\cdot)$, $u(\cdot)$, and $(y(\cdot), u(\cdot))$ as *feasible (admissible, resp.) trajectory, control*, and *pair*, respectively.

Hereafter, we let

$$\mathcal{A} = \{(y(\cdot), u(\cdot)) \in C([0,T]; X) \times \mathcal{U}[0,T] \mid (y(\cdot), u(\cdot)) \text{ is feasible }\},$$
$$\mathcal{Y} = \{y(\cdot) \in C([0,T]; X) \mid \exists u(\cdot) \in \mathcal{U}[0,T], \text{ such that } (y(\cdot), u(\cdot)) \in \mathcal{A}\},$$
$$\mathcal{U} = \{u(\cdot) \in \mathcal{U}[0,T] \mid \exists y(\cdot) \in C([0,T]; X), \text{ such that } (y(\cdot), u(\cdot)) \in \mathcal{A}\},$$

and define $\mathcal{A}_{ad}$, $\mathcal{Y}_{ad}$, and $\mathcal{U}_{ad}$ similar to $\mathcal{A}$, $\mathcal{Y}$, and $\mathcal{U}$, replacing feasibility by admissibility. Next, let us introduce the cost functional:

$$(3.6) \qquad J(y(\cdot), u(\cdot)) = \int_0^T f^0(t, y(t), u(t))\, dt, \qquad \forall (y(\cdot), u(\cdot)) \in \mathcal{A}_{ad}.$$

Then our optimal control problem can be stated as follows.

**Problem (P).** Find $(\bar{y}(\cdot), \bar{u}(\cdot)) \in \mathcal{A}_{ad}$, such that

$$(3.7) \qquad J(\bar{y}(\cdot), \bar{u}(\cdot)) = \min_{(y(\cdot), u(\cdot)) \in \mathcal{A}_{ad}} J(y(\cdot), u(\cdot)).$$

If such a pair $(\bar{y}(\cdot), \bar{u}(\cdot))$ exists, we call $\bar{y}(\cdot)$, $\bar{u}(\cdot)$ and $(\bar{y}(\cdot), \bar{u}(\cdot))$ an *optimal trajectory, control*, and *pair*, respectively.

The following result will play an important role in sequel.

**Lemma 3.2.** *Let $e^{At}$ be a compact $C_0$ semigroup on some Banach space $X$. Let $p > 1$ and define*

$$(3.8) \qquad \mathcal{S}(g(\cdot)) = \int_0^{\cdot} e^{A(\cdot - s)} g(s)\, ds, \qquad \forall g(\cdot) \in L^p(0, T; X).$$

*Then $\mathcal{S} : L^p(0, T; X) \to C([0, T]; X)$ is compact.*

*Proof.* Let $g_k(\cdot) \in L^p(0, T; X)$ with $\|g_k(\cdot)\|_{L^p(0,T;X)} \leq 1$, $\forall k \geq 1$. We need to prove that $\{\mathcal{S}(g_k(\cdot))\}_{k \geq 1}$ is relatively compact in $C([0, T]; X)$. To this end, we first prove that for each $t \in [0, T]$, the set $\{\mathcal{S}(g_k(\cdot))(t)\}_{k \geq 1}$ is

## §3. Evolution Systems with Compact Semigroups

relatively compact in $X$. In fact, the case where $t = 0$ is trivial. We let $t \in (0, T]$. Then, for any $\varepsilon > 0$, there exists a $\delta \in (0, t]$, such that (note that $p > 1$)

$$\text{(3.9)} \qquad |\int_{t-\delta}^{t} e^{A(t-s)} g_k(s)\, ds| < \frac{\varepsilon}{2}, \qquad \forall k \geq 1.$$

Next, let

$$\text{(3.10)} \qquad y_k \equiv \int_0^{t-\delta} e^{A(t-\delta-s)} g_k(s)\, ds, \qquad \forall k \geq 1.$$

Then it is seen that the set $\{y_k\}_{k \geq 1}$ is bounded in $X$. Thus, by the compactness of $e^{A\delta}$, we can find a finite set $\{z_i,\, 1 \leq i \leq m\}$ in $X$, such that

$$\text{(3.11)} \qquad \{e^{A\delta} y_k\}_{k \geq 1} \subset \bigcup_{i=1}^{m} \mathcal{O}_{\varepsilon/2}(z_i).$$

Consequently,

$$\text{(3.12)} \qquad \{\mathcal{S}(g_k(\cdot))(t)\}_{k \geq 1} \subset \bigcup_{i=1}^{m} \mathcal{O}_{\varepsilon}(z_i).$$

Hence, for each $t \in [0, T]$, $\{\mathcal{S}(g_k(\cdot))(t)\}_{k \geq 1}$ is relatively compact in $X$. Next, we show that $\{\mathcal{S}(g_k(\cdot))\}_{k \geq 1}$ is equicontinuous on $[0, T]$. In fact, for $t' > t > 0$ and $0 < \delta \leq t$,

$$\text{(3.13)} \qquad \begin{aligned} \mathcal{S}(g_k(\cdot))(t') - \mathcal{S}(g_k(\cdot))(t) &= \int_t^{t'} e^{A(t'-s)} g_k(s)\, ds \\ &+ \int_0^{t-\delta} (e^{A(t'-s)} - e^{A(t-s)}) g_k(s)\, ds \\ &+ \int_{t-\delta}^{t} (e^{A(t'-s)} - e^{A(t-s)}) g_k(s)\, ds \\ &\equiv I_1 + I_2 + I_3. \end{aligned}$$

By some direct estimation, we can find some constant $C$, independent of $k$, such that

$$\text{(3.14)} \qquad \begin{aligned} &|I_1| \leq C|t'-t|^{1/p'}, \qquad |I_3| \leq C\delta^{1/p'}, \\ &|I_2| \leq C\Big(\int_\delta^t \|e^{A(t'-t+s)} - e^{As}\|^{p'}\, ds\Big)^{1/p'}, \end{aligned}$$

with $p' = p/(p-1)$. From Chapter 2, Proposition 4.17, we know that $t \mapsto e^{At}$ is continuous in the operator norm on $(0, \infty)$. Thus, we obtain the equicontinuity of the set $\{\mathcal{S}(g_k(\cdot))\}$ on $[0, T]$. Then by the Arzelà–Ascoli Theorem (see Chapter 2, Theorem 3.7), we obtain the compactness of the operator $\mathcal{S}$. □

**Corollary 3.3.** Let $e^{At}$ be a compact $C_0$ semigroup on some Banach space $X$ and $p > 1$. Let $g_k(\cdot) \in L^p(0,T;X)$ satisfy

(3.15) $$g_k(\cdot) \xrightarrow{w} g(\cdot), \qquad \text{in } L^p(0,T;X).$$

Then

(3.16) $$\lim_{k\to\infty} \sup_{0\leq t\leq T} \left| \int_0^t e^{A(t-s)}(g_k(s) - g(s))\,ds \right| = 0.$$

It is important to know that the conclusion of Lemma 3.2 is not true in general for $p = 1$. Here is a simple example.

*Example 3.4.* Let $X$ be a separable infinite dimensional Hilbert space with an orthonormal basis $\{\varphi_n, n \geq 1\}$. Let $A$ be a diagonal operator:

(3.17) $$Ax = -\sum_{n\geq 1} n \langle x, \varphi_n \rangle \varphi_n, \qquad \forall x \in X.$$

Then we know that the corresponding $C_0$ semigroup $e^{At}$ is given by

(3.18) $$e^{At}x = \sum_{n\geq 1} e^{-nt} \langle x, \varphi_n \rangle \varphi_n, \qquad \forall x \in X.$$

Now, we take

(3.19) $$g_n(t) = n\chi_{[T-1/n,T]}(t)\varphi_n, \qquad n \geq 1.$$

Clearly, for all $n \geq 1$,

(3.20) $$\int_0^T e^{A(T-s)} g_n(s)\,ds = n \int_{T-1/n}^T e^{-n(T-s)}\varphi_n\,ds = (1 - e^{-1})\varphi_n.$$

This implies that for $p = 1$, the map $g(\cdot) \mapsto \mathcal{S}(g(\cdot))(T)$ is not compact from $L^1(0,T;X)$ to $X$. Hence, the map $\mathcal{S}: L^1(0,T;X) \to C([0,T];X)$ must not be compact.

## §4. Existence of Feasible Pairs and Optimal Pairs

### §4.1. Cesari property

First, let us introduce the following notion.

**Definition 4.1.** Let $Y$ be a Banach space and $Z$ be a metric space. Let $\Lambda: Z \to 2^Y$ be a multifunction. We say $\Lambda$ possesses the *Cesari property* at $z_0 \in Z$, if

(4.1) $$\bigcap_{\delta > 0} \overline{co}\,\Lambda(\mathcal{O}_\delta(z_0)) = \Lambda(z_0),$$

where $\overline{co}\,D$ is the *closed convex hull* of $D$ (see Chapter 2, §2.1). If $\Lambda$ has the Cesari property at every point $z \in Q \subseteq Z$, we simply say that $\Lambda$ has the Cesari property on $Q$.

## §4. Existence of Feasible Pairs and Optimal Pairs

**Proposition 4.2.** *Let $Y$ be a Banach space and $Z$ be a metric space. Let $\Lambda : Z \to 2^Y$ be upper semicontinuous, convex, and closed valued. Then $\Lambda$ has the Cesari property on $Z$.*

*Proof.* Let $z_0 \in Z$ be fixed. For any $\varepsilon > 0$, there exists a $\delta > 0$ such that $\Lambda(\mathcal{O}_\delta(z_0)) \subset \mathcal{O}_\varepsilon(\Lambda(z_0))$. Because $\Lambda(z_0)$ is convex, so is $\mathcal{O}_\varepsilon(\Lambda(z_0))$. Consequently, $\overline{\text{co}}\,\Lambda(\mathcal{O}_\delta(z_0)) \subset \overline{\mathcal{O}_\varepsilon(\Lambda(z_0))}$. Then it follows that

$$(4.2) \qquad \bigcap_{\delta > 0} \overline{\text{co}}\,\Lambda(\mathcal{O}_\delta(z_0)) \subset \bigcap_{\varepsilon > 0} \overline{\mathcal{O}_\varepsilon(\Lambda(z_0))} = \overline{\Lambda(z_0)} = \Lambda(z_0),$$

which proves our proposition. $\square$

Now, for any $(t, x) \in [0, T] \times X$, let us introduce the following set:

$$(4.3) \qquad \mathcal{E}(t, x) = \{(z^0, z) \in \mathbb{R} \times X \mid z^0 \geq f^0(t, x, u), \; z = f(t, x, u), \; u \in \Gamma(t, x) \}.$$

In the proof of the existence of optimal pairs for Problem (P), the following hypotheses will play a crucial role.

(P6) For almost all $t \in [0, T]$, the map $\mathcal{E}(t, \cdot) : X \to 2^{\mathbb{R} \times X}$ has the Cesari property on $Q$ ($Q \subset X$ is the constraint for $y(t)$, see (P5) and (3.5)).

The following result gives a sufficient condition ensuring (P6).

**Proposition 4.3.** *Let the following hold:*

*(P3)' For almost all $t \in [0, T]$, the map $f(t, \cdot, \cdot)$ is continuous uniformly in $u \in U$ and $f^0(t, \cdot, \cdot)$ is lower semicontinuous uniformly in $u \in U$, i.e., for any given $x \in X$ and any $\varepsilon > 0$, there exists a $\sigma = \sigma(x) > 0$, such that whenever $x' \in \mathcal{O}_\sigma(x)$ and $d(u, u') < \sigma$, one has*

$$(4.4) \qquad \begin{cases} |f(t, x', u') - f(t, x, u)| < \varepsilon, \\ f^0(t, x', u') > f^0(t, x, u) - \varepsilon. \end{cases}$$

*(P4)' For almost all $t \in [0, T]$, the map $\Gamma(t, \cdot) : X \to 2^U$ is upper semicontinuous on $X$; in particular, this is the case if $\Gamma(t, x) \equiv \Gamma(t)$, for all $x \in X$.*

*Then, for given $t \in [0, T]$, $\mathcal{E}(t, \cdot)$ has the Cesari property at $x$ if and only if the set $\mathcal{E}(t, x)$ is convex and closed.*

*Proof.* It suffices to prove the sufficiency. Let $t \in [0, T]$ be given such that (P3)' and (P4)' hold at this $t$ and let $x \in X$ be fixed such that $\mathcal{E}(t, x)$ is convex and closed. By (P3)', for any $\varepsilon > 0$, there exists a $\sigma = \sigma(x) > 0$, such that whenever $x' \in \mathcal{O}_\sigma(x)$ and $d(u, u') < \sigma$, (4.4) holds. Next, by (P4)', we can find $0 < \delta = \delta(x) < \sigma$, such that

$$(4.5) \qquad \Gamma(t, \mathcal{O}_\delta(x)) \subset \mathcal{O}_\sigma(\Gamma(t, x)).$$

Thus, for any $(y_0^\delta, y^\delta) \in \mathcal{E}(t, \mathcal{O}_\delta(x))$, there exist an $x^\delta \in \mathcal{O}_\delta(x)$ and a $u^\delta \in \Gamma(t, x^\delta) \subset \mathcal{O}_\sigma(\Gamma(t, x))$, such that

(4.6) $\qquad y_0^\delta \geq f^0(t, x^\delta, u^\delta), \qquad y^\delta = f(t, x^\delta, u^\delta).$

Then it is clear that there exists a $u \in \Gamma(t, x)$ with $d(u^\delta, u) < \sigma$. Hence, noting the fact that $\delta \leq \sigma$, we see from (4.4) and (4.6),

(4.7) $\qquad \begin{cases} y_0^\delta \geq f^0(t, x^\delta, u^\delta) > f^0(t, x, u) - \varepsilon, \\ |y^\delta - f(t, x, u)| = |f(t, x^\delta, u^\delta) - f(t, x, u)| < \varepsilon. \end{cases}$

This means $(y_0^\delta, y^\delta) \in \mathcal{O}_\varepsilon(\mathcal{E}(t, x))$. Hence, the set $\mathcal{E}(t, \cdot)$ is upper semicontinuous. Then our conclusion follows from Proposition 4.2. □

The above proposition looks very similar to the result of Berkovitz [1, pp.72–74] for a finite dimensional situation. However, here, we are in infinite dimensional space and we have not assumed any compactness! Hence, our result is an improvement of that found in Berkovitz [1].

In the above result, we see that (P3)' and (P4)' are not very restrictive. Thus, (P6) essentially says that $\mathcal{E}(t, x)$ is convex and closed. The following result gives a sufficient condition for $\mathcal{E}(t, x)$ being convex.

**Proposition 4.4.** Let $(t, x) \in [0, T] \times X$ be fixed. Let $f(t, x, \Gamma(t, x))$ be convex and let there exist a convex function $\varphi(\,\cdot\,; t, x) : X \to \mathbb{R}$, such that

(4.8) $\qquad f^0(t, x, u) = \varphi(f(t, x, u); t, x), \qquad \forall u \in \Gamma(t, x).$

Then $\mathcal{E}(t, x)$ is convex.

*Proof.* Let $(y_i^0, y_i) \in \mathcal{E}(t, x)$, $i = 1, 2$. We can find $u_1, u_2 \in \Gamma(t, x)$, such that

(4.9) $\qquad \begin{cases} y_i^0 \geq f^0(t, x, u_i), \\ y_i = f(t, x, u_i), \end{cases} \quad i = 1, 2.$

By the convexity of $f(t, x, \Gamma(t, x))$, for $\lambda \in (0, 1)$, there exists a $u_3 \in \Gamma(t, x)$, such that

(4.10) $\qquad \lambda y_1 + (1 - \lambda) y_2 \equiv \lambda f(t, x, u_1) + (1 - \lambda) f(t, x, u_2) = f(t, x, u_3).$

By (4.8), we have

(4.11) $\begin{aligned} \lambda y_1^0 + (1-\lambda) y_2^0 &\geq \lambda f^0(t, x, u_1) + (1 - \lambda) f^0(t, x, u_2) \\ &= \lambda \varphi(f(t, x, u_1); t, x) + (1 - \lambda) \varphi(f(t, x, u_2); t, x) \\ &\geq \varphi\bigl(\lambda f(t, x, u_1) + (1 - \lambda) f(t, x, u_2); t, x\bigr) \\ &= \varphi(f(t, x, u_3); t, x) = f^0(t, x, u_3). \end{aligned}$

Hence, $\mathcal{E}(t, x)$ is convex. □

§4. Existence of Feasible Pairs and Optimal Pairs

It is easy to check that if $f(t,x,u)$ is linear in $u$, $f^0(t,x,u)$ is convex in $u$, and $\Gamma(t,x)$ is convex, then $\mathcal{E}(t,x)$ is convex. The above result gives some other situations that guarantee the convexity of $\mathcal{E}(t,x)$. By some simple observation, we can see that in order for the set $\mathcal{E}(t,x)$ to be convex, it is necessary that $f(t,x,\Gamma(t,x))$ be convex and the functions $f(t,x,\cdot)$, $f^0(t,x,\cdot)$ and the set $\Gamma(t,x)$ be compatible in some sense. Condition (4.8) is one such compatibility condition.

Next, we introduce the following assumptions, which are comparable with (P3) and (P6):

(P3)″ The map $f:[0,T]\times X\times U\to X$ is Borel measurable in $(t,x,u)$, continuous in $(x,u)$, and for some constant $L>0$,

(4.12) $$\begin{cases} |f(t,x,u)-f(t,\hat{x},u)|\leq L|x-\hat{x}|, \\ \qquad\qquad\qquad\qquad \forall t\in[0,\infty),\ x,\hat{x}\in X,\ u\in U, \\ |f(t,0,u)|\leq L, \qquad \forall t\in[0,\infty),\ u\in U. \end{cases}$$

(P6)′ For almost all $t\in[0,\infty)$, the set $f(t,x,\Gamma(t,x))$ satisfies the following:

(4.13) $$\bigcap_{\delta>0}\overline{co}\,f(t,\mathcal{O}_\delta(x),\Gamma(\mathcal{O}_\delta(t,x)))=f(t,x,\Gamma(t,x)).$$

(P6)″ Let $\Lambda(t,x)=f(t,x,\Gamma(t,x))$. For almost all $t\in[0,T]$, the map $\Lambda(t,\cdot):X\to 2^X$ has the Cesari property.

It is clear that (P6)′ implies (P6)″. In fact, one has

(4.14) $$\Lambda(t,\mathcal{O}_\delta(x))\subseteq f(t,\mathcal{O}_\delta(x),\Gamma(\mathcal{O}_\delta(t,x))).$$

Furthermore, we have the following result.

**Proposition 4.5.** *Let* $\Gamma:[0,T]\times X\to 2^U$ *be upper semicontinuous taking closed set values and* $f(t,x,u)$ *be uniformly continuous in* $(x,u)$ *for any* $t\in[0,T]$. *Then the following are equivalent:*

(i) (P6)′ *holds,*

(ii) (P6)″ *holds,*

(iii) $f(t,x,\Gamma(t,x))$ *is closed and convex.*

*Proof.* (i)⇒(ii)⇒(iii) are immediate.

(iii)⇒(i). By the uniform continuity of $f(t,x,u)$ in $(x,u)$, for any $\varepsilon>0$, there exists a $\sigma>0$, such that

(4.15) $$f(t,\mathcal{O}_\sigma(x),\mathcal{O}_\sigma(\Gamma(t,x)))\subset\mathcal{O}_\varepsilon(f(t,x,\Gamma(t,x))).$$

On the other hand, by the upper semicontinuity of $\Gamma$, we can find a $\delta\in(0,\sigma]$, such that

(4.16) $$\Gamma(\mathcal{O}_\delta(t,x))\subset\mathcal{O}_\sigma(\Gamma(t,x)).$$

Thus, (i) follows from the convexity and closeness of $f(t, x, \Gamma(t, x))$. □

## §4.2. Existence theorems

In this subsection, we present some existence theorems. The following result asserts the existence of feasible pairs.

**Theorem 4.6.** *Let (P1), (P2), (P3)″, (P4), and (P6)′ hold. Then, for any $x \in X$, the set*

$$(4.17) \qquad \mathcal{A}^x[0,T] \triangleq \{(y(\cdot), u(\cdot)) \in \mathcal{A} \mid y(0) = x\} \neq \phi.$$

*Moreover, $\mathcal{Y}^x[0,T] \triangleq \{y(\cdot) \in \mathcal{Y} \mid y(0) = x\}$ is compact in $C([0,T];X)$.*

*Proof.* For any $k \geq 1$, let $t_j = \frac{j}{k}T$, $0 \leq j \leq k-1$. We set

$$(4.18) \qquad u_k(t) = \sum_{j=0}^{k-1} u^j \chi_{[t_j, t_{j+1})}(t), \qquad t \in [0,T].$$

Here, the sequence $\{u^j\}$ is constructed as follows: First, we take $u^0 \in \Gamma(0, x)$. By (4.12), we know that there exists a unique $y_k(\cdot)$ satisfying

$$(4.19) \qquad y_k(t) = e^{At}x + \int_0^t e^{A(t-s)} f(s, y_k(s), u^0)\, ds, \qquad \forall t \in [t_0, t_1].$$

Then, take $u^1 \in \Gamma(t_1, y_k(t_1))$. We can continue this procedure to obtain $y_k(\cdot)$ on $[t_1, t_2]$, etc. By induction, we end up with the following:

$$(4.20) \qquad \begin{cases} y_k(t) = e^{At}x + \int_0^t e^{A(t-s)} f(s, y_k(s), u_k(s))\, ds, & \forall t \in [0,T], \\ u_k(t) \in \Gamma(t_j, y_k(t_j)), & t \in [t_j, t_{j+1}),\ 0 \leq j \leq k-1. \end{cases}$$

By Gronwall's inequality and (4.12), we see that for some absolute constant $C > 0$,

$$(4.21) \qquad \begin{cases} |y_k(t)| \leq C, & \forall t \in [0,T], \\ |f(t, y_k(t), u_k(t))| \leq C, & \text{a.e. } t \in [0,T], \end{cases} \qquad k \geq 1.$$

It follows from Lemma 3.2 that the sequence $\{y_k(\cdot)\}_{k \geq 1}$ is relatively compact in $C([0,T];X)$. Thus, we may assume

$$(4.22) \qquad y_k(\cdot) \xrightarrow{s} \bar{y}(\cdot), \qquad \text{in } C([0,T];X),$$

for some $\bar{y}(\cdot) \in C([0,T];X)$. Also, we may let

$$(4.23) \qquad f(\cdot, y_k(\cdot), u_k(\cdot)) \xrightharpoonup{*} \bar{f}(\cdot), \qquad \text{in } L^\infty(0,T;X),$$

## §4. Existence of Feasible Pairs and Optimal Pairs

for some $\bar{f}(\cdot) \in L^\infty(0,T;X)$. By the compactness of the operator $\mathcal{S}$ (see (3.8)), we have

$$(4.24) \qquad \bar{y}(t) = e^{At}x + \int_0^t e^{A(t-s)}\bar{f}(s)\,ds, \qquad t \in [0,T].$$

By (4.22), for any $\delta > 0$, there exists a $k_0$, such that

$$(4.25) \qquad y_k(t) \in \mathcal{O}_\delta(\bar{y}(t)), \qquad \forall t \in [0,T], \quad k \geq k_0.$$

On the other hand, by the definition of $u_k(\cdot)$, for $k$ large, one has

$$(4.26) \qquad \begin{aligned} u_k(t) &\in \Gamma(t_j, y_k(t_j)) \subset \Gamma(\mathcal{O}_\delta(t, \bar{y}(t))), \\ &\forall t \in [t_j, t_{j+1}), \quad 0 \leq j \leq k-1. \end{aligned}$$

Next, by (4.23) and the Mazur Theorem (Chapter 2, Corollary 2.8), we may let $\alpha_{ij} \geq 0$, $\sum_{i \geq 1} \alpha_{ij} = 1$, such that for some $p > 1$,

$$(4.27) \qquad \psi_j(\cdot) \equiv \sum_{i \geq 1} \alpha_{ij} f(\cdot, y_{i+j}(\cdot), u_{i+j}(\cdot)) \xrightarrow{s} \bar{f}(\cdot), \quad \text{in } L^p(0,T;X).$$

Then, we can assume

$$(4.28) \qquad \psi_j(t) \xrightarrow{s} \bar{f}(t), \qquad \text{in } X, \quad \text{a.e. } t \in [0,T].$$

On the other hand, by (4.25) and (4.26), we see that for $j$ large, one has

$$(4.29) \qquad \psi_j(t) \in \text{co}f(t, \mathcal{O}_\delta(\bar{y}(t)), \Gamma(\mathcal{O}_\delta(t, \bar{y}(t)))), \qquad \text{a.e. } t \in [0,T].$$

Thus, for any $\delta > 0$,

$$(4.30) \qquad \bar{f}(t) \in \overline{\text{co}}f(t, \mathcal{O}_\delta(\bar{y}(t)), \Gamma(\mathcal{O}_\delta(t, \bar{y}(t)))), \qquad \text{a.e. } t \in [0,T].$$

By (P6)',

$$(4.31) \qquad \bar{f}(t) \in f(t, \bar{y}(t), \Gamma(t, \bar{y}(t))), \qquad \text{a.e. } t \in [0,T].$$

By Corollary 2.18, we know that $\Gamma(\cdot, \bar{y}(\cdot))$ is Souslin measurable. Thus, by Corollary 2.26, there exists a $\bar{u}(\cdot) \in \mathcal{U}[0,T]$, such that

$$(4.32) \qquad \begin{cases} \bar{u}(t) \in \Gamma(t, \bar{y}(t)), & \text{a.e. } t \in [0,T], \\ \bar{f}(t) = f(t, \bar{y}(t), \bar{u}(t)), & \text{a.e. } t \in [0,T]. \end{cases}$$

Combining (4.24) and (4.32), we see that $(\bar{y}(\cdot), \bar{u}(\cdot)) \in \mathcal{A}^x[0,T]$. Thus, (4.17) follows. Finally, let $\{y^k(\cdot)\}_{k \geq 1} \subset \mathcal{Y}^x[0,T]$ with $|y^k(\cdot)|_{C([0,T];X)} \leq C$, $\forall k \geq 1$. Then, by the above proof, we see that $\{y^k(\cdot)\}_{k \geq 1}$ is relatively compact in $C([0,T];X)$. Moreover, if for some subsequence (still denoted by itself), one has $y^k(\cdot) \xrightarrow{s} \hat{y}(\cdot)$, in $C([0,T];X)$, then, by (P6)', we must have $\hat{y}(\cdot) \in \mathcal{Y}^x[0,T]$. Thus, $\mathcal{Y}^x[0,T]$ is compact in $C([0,T];X)$. □

We should note that in the case $\Gamma(t,x) \equiv \Gamma(t), \forall x \in X$, Theorem 4.6 holds under much simpler conditions and the proof becomes much simpler.

Next, we present the existence of optimal pairs for Problem (P).

**Theorem 4.7.** *Let (P1)–(P6) hold. Let there exist a minimizing sequence $\{y_k(\cdot), u_k(\cdot)\} \in \mathcal{A}_{ad}$, such that for some $p > 1$ and $C > 0$,*

$$(4.33) \qquad \|f(\cdot, y_k(\cdot), u_k(\cdot))\|_{L^p(0,T;X)} \leq C, \qquad \forall k \geq 1.$$

*Then, Problem (P) admits at least one optimal pair.*

*Proof.* Without loss of generality, we may assume

$$(4.34) \qquad f(\cdot, y_k(\cdot), u_k(\cdot)) \xrightarrow{w} \bar{f}(\cdot), \qquad \text{in } L^p(0,T;X),$$

for some $\bar{f}(\cdot) \in L^p(0,T;X)$. Then, by Corollary 3.3, we obtain

$$(4.35) \qquad \int_0^t e^{A(t-s)} f(s, y_k(s), u_k(s)) ds \xrightarrow{s} \int_0^t e^{A(t-s)} \bar{f}(s)\, ds,$$

uniformly in $t \in [0,T]$. Also, by the boundedness of $S$ (see (P5)), the reflexivity of $X$, and the compactness of $e^{At}(t > 0)$, we may assume

$$(4.36) \qquad \begin{cases} y_k(0) \xrightarrow{w} x_0, \\ e^{At} y_k(0) \xrightarrow{s} e^{At} x_0, \qquad \forall t \in (0,T]. \end{cases}$$

Let

$$(4.37) \qquad \bar{y}(t) = e^{At} x_0 + \int_0^t e^{A(t-s)} \bar{f}(s) ds, \qquad t \in [0,T].$$

Then it follows that

$$(4.38) \qquad \lim_{k \to \infty} |y_k(t) - \bar{y}(t)| = 0, \qquad t \in (0,T].$$

By (4.34) and the Mazur Theorem (Chapter 2, Corollary 2.8), we may let $\alpha_{ij} \geq 0, \sum_{i \geq 1} \alpha_{ij} = 1$, such that

$$(4.39) \qquad \psi_j(\cdot) \equiv \sum_{i \geq 1} \alpha_{ij} f(\cdot, y_{i+j}(\cdot), u_{i+j}(\cdot)) \xrightarrow{s} \bar{f}(\cdot), \qquad \text{in } L^p(0,T;X).$$

We set

$$(4.40) \qquad \psi_j^0(\cdot) \equiv \sum_{i \geq 1} \alpha_{ij} f^0(\cdot, y_{i+j}(\cdot), u_{i+j}(\cdot)),$$

and set

$$(4.41) \qquad \bar{f}^0(t) = \lim_{j \to \infty} \psi_j^0(t) \geq -K, \qquad \text{a.e. } t \in [0,T].$$

Then, by (P6),

(4.42) $\quad (\bar{f}^0(t), \bar{f}(t)) \in \mathcal{E}(t, \bar{y}(t)), \qquad$ a.e. $t \in [0, T],$

and (by Fatou's Lemma)

(4.43) $\quad \displaystyle\int_0^T \bar{f}^0(t)\, dt \leq \lim_{k \to \infty} J(y_k(\cdot), u_k(\cdot)) = \inf_{(y(\cdot), u(\cdot)) \in \mathcal{A}_{ad}} J(y(\cdot), u(\cdot)).$

By Corollaries 2.18 and 2.26 (see the proof of Theorem 4.6), we can find a measurable selection $\bar{u}(\cdot)$ of $\Gamma(\cdot, \bar{y}(\cdot))$, such that

(4.44) $\quad \begin{cases} \bar{f}^0(t) \geq f^0(t, \bar{y}(t), \bar{u}(t)), \\ \bar{f}(t) = f(t, \bar{y}(t), \bar{u}(t)), \end{cases}$ a.e. $t \in [0, T].$

Thus, noting (P5), we can easily see that $(\bar{y}(\cdot), \bar{u}(\cdot)) \in \mathcal{A}_{ad}$, and

(4.45) $\quad J(\bar{y}(\cdot), \bar{u}(\cdot)) \leq \displaystyle\int_0^T \bar{f}^0(t)\, dt \leq \inf_{(y(\cdot), u(\cdot)) \in \mathcal{A}_{ad}} J(y(\cdot), u(\cdot)).$

This means that $(\bar{y}(\cdot), \bar{u}(\cdot))$ is an optimal pair. $\quad\square$

We should note that (4.33) is a very weak condition. This condition holds if (P3)″ holds.

*Remark 4.8.* If the semigroup $e^{At}$ is compact and analytic, then the map $f$ can be more general. For example, in the case where $A$ is the Laplacian in some bounded domain in $\mathbb{R}^n$ with suitable boundary conditions, the nonlinear term $f$ is allowed to contain the first order spatial derivatives of the state. Of course, the assumptions ensuring the above results should be changed properly.

## §5. Second Order Evolution Systems

### §5.1. Formulation of the problem

In this section we will discuss optimal control problems with the systems governed by the following evolution equation:

(5.1) $\quad \ddot{y}(t) + Ay(t) = f(t, y(t), u(t)), \qquad t \in [0, T],$

with some symmetric operator $A$. The motivation is the controlled wave or beam equation. We will see that the theory developed in the previous section is not applicable here. To make the presentation shorter and catch the essence of the results of this section, we will omit some standard and lengthy details. Let us start with some basic assumptions.

(W1) $V$ and $H$ are separable Hilbert spaces with duals $V'$ and $H' \equiv H$, respectively. The embedding $V \hookrightarrow H$ is dense and compact. The duality pairing between $V$ and $V'$ is $\langle \cdot, \cdot \rangle$ and the inner product of $H$ is $(\cdot, \cdot)$. $U$ is a Polish space and $T$ is a positive constant.

(W2) $A \in \mathcal{L}(V, V')$ is symmetric and coercive, i.e.,

(5.2) $$\begin{cases} \langle Ax, y \rangle = \langle x, Ay \rangle, & \forall x, y \in V, \\ \langle Ax, x \rangle \geq \alpha |x|_V^2, & \forall x \in V, \end{cases}$$

for some constant $\alpha > 0$.

From (W1), we know that $V \hookrightarrow H = H' \hookrightarrow V'$. The space $H$ is usually referred to as the *pivot space*. Now, for $g(\cdot) \in L^2(0, T; V')$ and $(y_0, y_1) \in V \times H$, we consider the following evolution equation

(5.3) $$\ddot{y}(t) + Ay(t) = g(t), \quad \text{a.e. } t \in [0, T], \text{ in } V',$$

with the initial conditions

(5.4) $$y(0) = y_0, \quad \text{in } H,$$

(5.5) $$\dot{y}(0) = y_1, \quad \text{in } V'.$$

We note that (5.4) and (5.5) stand for the following, respectively:

(5.6) $$\lim_{t \to 0} |y(t) - y_0|_H = 0, \quad \lim_{t \to 0} |\dot{y}(t) - y_1|_{V'} = 0.$$

Before introducing the definition of a solution to (5.3)–(5.5), we first give the following result. Its proof is lengthy and straightforward; we leave it to the readers.

**Proposition 5.1.** *The following are equivalent:*

(i) *Function* $y(\cdot) \in L^2(0, T; V) \cap W^{1,2}([0, T]; H) \cap W^{2,2}([0, T]; V')$ *satisfies (5.3)–(5.5).*

(ii) *Function* $y(\cdot) \in L^2(0, T; V) \cap W^{1,2}([0, T]; H)$ *satisfies (5.4) and*
(5.7)
$$\int_0^T \langle g(t), v(t) \rangle \, dt = -(y_1, v(0)) - \int_0^T (\dot{y}(t), \dot{v}(t)) \, dt$$
$$+ \int_0^T \langle Ay(t), v(t) \rangle \, dt,$$
$$\forall v(\cdot) \in L^2(0, T; V) \cap W^{1,2}([0, T]; H), \quad v(T) = 0 \quad \text{in } H.$$

(iii) *Function* $y(\cdot) \in L^2(0, T; V) \cap W^{1,2}([0, T]; H)$ *satisfies (5.4) and (5.5) and for any* $v \in V$, $(\dot{y}(\cdot), v)$ *is absolutely continuous in* $[0, T]$ *and*

(5.8) $$\frac{d}{dt}(\dot{y}(t), v) + \langle Ay(t), v \rangle = \langle g(t), v \rangle, \quad \text{a.e. } t \in [0, T].$$

Based on the above result, we introduce the following definition.

**Definition 5.2.** A function $y(\cdot) \in L^2(0, T; V) \cap W^{1,2}([0, T]; H)$ is called a solution of (5.3)–(5.5) if one of (i)–(iii) in Proposition 5.1 holds.

## §5. Second Order Evolution Systems

By the standard Galerkin type method, we can prove the following result.

**Proposition 5.3.** *Let $A$ satisfy (W2) and $(y_0, y_1) \in V \times H$, $g(\cdot) \in L^1(0,T;H)$. Then (5.3)–(5.5) admits a unique solution*

(5.9)
$$y(\cdot) \in L^\infty(0,T;V) \bigcap W^{1,\infty}([0,T];H) \bigcap W^{2,2}([0,T];V')$$
$$\hookrightarrow C([0,T];H) \bigcap C^1([0,T];V').$$

*Moreover, it holds that*

(5.10)
$$|\dot{y}(t)|^2 + \langle Ay(t), y(t) \rangle \le \left\{ [|y_1|^2 + \langle Ay_0, y_0 \rangle]^{1/2} + \int_0^t |g(s)|\, ds \right\}^2,$$
$$t \in [0,T].$$

Next, we introduce another kind of solution, the mild solution. To this end, we let

(5.11)
$$\widetilde{A} = \begin{pmatrix} 0 & I \\ -A & 0 \end{pmatrix} : V \times V' \to V' \times V',$$

and let

(5.12)
$$\begin{cases} \mathcal{D}(\widehat{A}) = \{(x,y) \in V \times H \mid \widetilde{A}(x,y) \in V \times H\} \\ \qquad\qquad = \{x \in V \mid Ax \in H\} \times V, \\ \widehat{A} = \widetilde{A}\big|_{\mathcal{D}(\widehat{A})} : \mathcal{D}(\widehat{A}) \subset V \times H \to V \times H. \end{cases}$$

We have the following result (similar to Chapter 2, Example 4.24).

**Proposition 5.4.** *The operator $\widehat{A}$ generates a $C_0$ group on the space $V \times H$.*

Let us set (formally) $z = (y, \dot{y})$, $z_0 = (y_0, y_1)$, $h = (0, g)$. Then (5.3)–(5.5) read

(5.13)
$$\begin{cases} \dot{z}(t) = \widehat{A} z(t) + h(t), \\ z(0) = z_0. \end{cases}$$

As usual, a function $z(\cdot) \in C([0,T];Z)$ is called a *mild solution* of (5.13) if it satisfies the following integral equation:

(5.14)
$$z(t) = e^{\widehat{A}t} z_0 + \int_0^t e^{\widehat{A}(t-s)} h(s)\, ds, \qquad t \subset [0,T].$$

On the other hand, we have the notion of a weak solution to (5.13) as follows: (see Chapter 2, §5.1)

**Definition 5.5.** A function $z(\cdot) \in C([0,T];Z)$ is called a *weak solution* of (5.13), if for any $z^* \in \mathcal{D}(\widehat{A}^*)$, the map $\langle z(\cdot), z^* \rangle$ is absolutely continuous on $[0, T]$, and

(5.15) $\quad \begin{cases} \dfrac{d}{dt} \langle z(t), z^* \rangle = \langle z(t), \widehat{A}^* z^* \rangle + \langle h(t), z^* \rangle, & \text{a.e. } t \in [0, T], \\ \langle z(t), z^* \rangle \big|_{t=0} = \langle z_0, z^* \rangle. \end{cases}$

By Proposition 5.2 of Chapter 2, we know that $z(\cdot) \in C([0,T]; Z)$ is a mild solution of (5.13) (i.e., (5.14) holds) if and only if $z(\cdot)$ is a weak solution of (5.13) (in the sense of Definition 5.5). From this result, we end up with the following theorem.

**Theorem 5.6.** *Let (W2) hold and $g(\cdot) \in L^1(0, T; H)$.*

*(i) If $y(\cdot)$ is a solution of (5.3)–(5.5) in the sense of Definition 5.2, then $(y(\cdot), \dot{y}(\cdot))$ is a mild solution of (5.13).*

*(ii) If $(y(\cdot), y_1(\cdot))$ is a mild solution of (5.13), then $y_1(\cdot) = \dot{y}(\cdot)$ and $y(\cdot)$ is a solution of (5.3)–(5.5) in the sense of Definition 5.2.*

*Proof.* First of all, we can easily check that

(5.16) $\quad \begin{cases} \mathcal{D}(\widehat{A}^*) = H \times V, \\ \widehat{A}^* \begin{pmatrix} \varphi \\ \psi \end{pmatrix} = \begin{pmatrix} -A\psi \\ \varphi \end{pmatrix} \equiv \begin{pmatrix} 0 & -A \\ I & 0 \end{pmatrix} \begin{pmatrix} \varphi \\ \psi \end{pmatrix}, \quad \forall \begin{pmatrix} \varphi \\ \psi \end{pmatrix} \in \mathcal{D}(\widehat{A}^*). \end{cases}$

(i) Let $y(\cdot)$ be a solution of (5.3)–(5.5) in the sense of Definition 5.2. Then, by setting $y_1(\cdot) = \dot{y}(\cdot)$, we have

(5.17) $\quad \dfrac{d}{dt}(y_1(t), \psi) = \langle -Ay(t), \psi \rangle + (g(t), \psi), \quad \text{a.e. } t \in [0, T], \; \forall \psi \in V.$

It is easy to see that (since $y(\cdot) \in W^{1,2}([0,T]; H)$)

(5.18) $\quad \dfrac{d}{dt}(y(t), \varphi) = (y_1(t), \varphi), \quad \text{a.e. } t \in [0, T], \; \forall \varphi \in H.$

Thus, we have

(5.19) $\quad \begin{aligned} \dfrac{d}{dt} \langle \begin{pmatrix} y(t) \\ y_1(t) \end{pmatrix}, \begin{pmatrix} \varphi \\ \psi \end{pmatrix} \rangle &= (y_1(t), \varphi) + \langle -Ay(t), \psi \rangle + (g(t), \psi) \\ &= \langle \begin{pmatrix} y(t) \\ y_1(t) \end{pmatrix}, \widehat{A}^* \begin{pmatrix} \varphi \\ \psi \end{pmatrix} \rangle + \langle \begin{pmatrix} 0 \\ g(t) \end{pmatrix}, \begin{pmatrix} \varphi \\ \psi \end{pmatrix} \rangle, \\ &\forall \begin{pmatrix} \varphi \\ \psi \end{pmatrix} \in H \times V \equiv \mathcal{D}(\widehat{A}^*). \end{aligned}$

Thus, $(y(\cdot), y_1(\cdot))$ is a weak solution of (5.13). By Proposition 5.2 of Chapter 2, it is a mild solution of (5.13).

(ii) Let $(y(\cdot), y_1(\cdot))$ be a mild solution of (5.13). Then it is a weak solution of (5.13) by Proposition 5.2, Chapter 2, again. That means that (5.17) and (5.18) hold. Then it can be shown that $y(\cdot) \in W^{1,2}([0,T]; H)$

## §5. Second Order Evolution Systems

and $y_1(\cdot) = \dot{y}(\cdot)$. Then (5.17) is exactly the same as (5.8) and thus our conclusion follows. □

Hereafter, $y(\cdot)$ is referred to as a mild solution of (5.3)–(5.5) if $(y(\cdot), \dot{y}(\cdot))$ is a mild solution of (5.13). We have seen that the solutions defined in Definition 5.2 are mild solutions. We will simply refer to them as the solutions of (5.3)–(5.5). Because of Theorem 5.6, we regard (5.13) as an equivalent form of (5.3)–(5.5).

Next, we state our optimal control problem. To this end, let Assumptions (W3), (W4), and (W6) be the same as (P3), (P4), and (P6), with $X$ being replaced by $H$ and we also assume

(W5) The set $Q \subset H \times V'$ is closed and $S \subset H \times H \times V' \times V'$ is bounded and weakly closed.

Our controlled evolution system is given by (5.1). For any given $y_0 \in V$, $y_1 \in H$ and $u(\cdot) \in \mathcal{U}[0,T]$, we may talk about the mild (or weak) solution of the following state equation:

$$(5.20) \quad \begin{cases} \ddot{y}(t) + Ay(t) = f(t, y(t), u(t)), & \text{a.e. } t \in [0,T], \\ y(0) = y_0, \quad \dot{y}(0) = y_1. \end{cases}$$

More precisely, we have the following:

**Definition 5.7.** A function $y(\cdot) \in L^2(0,T;V) \cap W^{1,2}([0,T];H)$ is called a mild solution of (5.20) (corresponding to $y_0 \in V$, $y_1 \in H$, and $u(\cdot) \in \mathcal{U}[0,T]$) if $f(\cdot, y(\cdot), u(\cdot)) \in L^2(0,T;H)$, and $y(\cdot)$ is a mild solution of (5.5) with $g(\cdot)$ being $f(\cdot, y(\cdot), u(\cdot))$.

**Definition 5.8.** A pair $(y(\cdot), u(\cdot)) \in (L^2(0,T;V) \cap W^{1,2}([0,T];H)) \times \mathcal{U}[0,T]$ is said to be *feasible* if $y(\cdot)$ is a mild solution of (5.20) corresponding to $u(\cdot)$ and

$$(5.21) \quad u(t) \in \Gamma(t, y(t)), \quad \text{a.e. } t \in [0,T].$$

Moreover, if the following also hold:

$$(5.22) \quad \begin{cases} (y(0), y(T), \dot{y}(0), \dot{y}(T)) \in S \subset H \times H \times V' \times V', \\ (y(t), \dot{y}(t)) \in Q \subset H \times V', \quad \text{a.e. } t \in [0,T], \\ f^0(\cdot, y(\cdot), u(\cdot)) \in L^1(0,T), \end{cases}$$

we call $(y(\cdot), u(\cdot))$ an *admissible pair*.

Set

$\mathcal{A} = \{(y(\cdot), u(\cdot)) \in C([0,T];H) \times \mathcal{U}[0,T] \mid (y(\cdot), u(\cdot)) \text{ is feasible }\},$

$\mathcal{A}_{ad} = \{(y(\cdot), u(\cdot)) \in C([0,T];H) \times \mathcal{U}[0,T] \mid (y(\cdot), u(\cdot)) \text{ is admissible }\}.$

Next, we introduce the cost functional.

$$(5.23) \quad J(y(\cdot), u(\cdot)) = \int_0^T f^0(t, y(t), u(t))\, dt, \quad \forall (y(\cdot), u(\cdot)) \in \mathcal{A}_{ad}.$$

Our optimal control problem can be stated as follows:

**Problem (W).** Find $(\bar{y}(\cdot), \bar{u}(\cdot)) \in \mathcal{A}_{ad}$, such that

(5.24) $$J(\bar{y}(\cdot), \bar{u}(\cdot)) = \min_{(y(\cdot), u(\cdot)) \in \mathcal{A}_{ad}} J(y(\cdot), u(\cdot)).$$

If such a pair exists, we refer to $\bar{y}(\cdot)$, $\bar{u}(\cdot)$, and $(\bar{y}(\cdot), \bar{u}(\cdot))$ as an optimal trajectory, control, and pair, respectively.

Problem (W) seems the same as Problem (P). However, we claim that there is an essential difference between them. To see this, let us take a look at the equivalent form of the state equation (5.20): Setting (formally) $z = (y, \dot{y})$, $z_0 = (y_0, y_1)$, $F(t, z, u) = (0, f(t, y, u))$,

(5.25) $$\begin{cases} \dot{z}(t) = \widehat{A} z(t) + F(t, z(t), u(t)), \\ z(0) = z_0, \end{cases}$$

where $\widehat{A}$ is given by (5.12). Then (5.25) looks similar to the state equation for Problem (P). But, by Proposition 5.4, $e^{\widehat{A}t}$ is a $C_0$ group on the underlying space $X \equiv V \times H$, instead of a $C_0$ semigroup! Thus, it cannot be compact! Consequently, the theory presented in the previous section does not apply here.

## §5.2. Existence of optimal controls

Before we state and prove our main theorem of this section, let us first give the following preliminary result.

**Lemma 5.9.** (Aubin–Lions) *Let* $X_0 \hookrightarrow\hookrightarrow X_1 \hookrightarrow X_2$ *be Banach spaces with* $X_0$ *and* $X_2$ *being reflexive. Then, for any* $1 < p, q < \infty$,

(5.26) $$L^p(0, T; X_0) \bigcap W^{1,q}([0, T]; X_2) \hookrightarrow\hookrightarrow L^p(0, T; X_1).$$

*Proof.* First of all, we claim that for any $\delta > 0$, there exists a $C_\delta > 0$, such that

(5.27) $$|x|_{X_1} \leq \delta |x|_{X_0} + C_\delta |x|_{X_2}, \qquad \forall x \in X_0.$$

If this is not the case, then there exists a $\delta > 0$ and a weakly convergent sequence $\{x_n\} \subset X_0$, $|x_n|_{X_0} = 1$, such that

(5.28) $$|x_n|_{X_1} \geq \delta + n |x_n|_{X_2}, \qquad n \geq 1.$$

By the compact embedding $X_0 \hookrightarrow\hookrightarrow X_1$, we may assume that $x_n \xrightarrow{s} \bar{x}$ in $X_1$. Then (5.28) implies that $x_n \xrightarrow{s} 0$ in $X_2$. Thus, it is necessary that $\bar{x} = 0$. This means $x_n \xrightarrow{s} 0$ in $X_1$, contradicting (5.28). Hence, our claim holds.

Now, let $\{h_n\} \in L^p(0, T; X_0) \bigcap W^{1,q}([0, T]; X_2)$ be a bounded sequence. Because $1 < p, q < \infty$ and both $X_0$ and $X_2$ are reflexive, we

## §5. Second Order Evolution Systems

have the reflexivity of $L^p(0,T;X_0)$ and $L^q(0,T;X_2)$. Consequently, we may assume that

(5.29)
$$\begin{cases} h_n \xrightarrow{w} h, & \text{in } L^p(0,T;X_0), \\ \dot{h}_n \xrightarrow{w} \dot{h}, & \text{in } L^q(0,T;X_2). \end{cases}$$

By considering $h_n - h$ if necessary, we may assume that $h = 0$. Now, for any $s \in [0,T)$, we have

(5.30) $$h_n(s) = h_n(t) - \int_s^t \dot{h}_n(r)\, dr, \qquad t \in [0,T].$$

Integrate it over $(s, s+\sigma)$ $(\sigma \in (0, T-s])$

(5.31)
$$\begin{aligned} h_n(s) &= \frac{1}{\sigma}\int_s^{s+\sigma} h_n(t)\, dt - \frac{1}{\sigma}\int_s^{s+\sigma}\int_s^t \dot{h}_n(r)\, dr\, dt \\ &= \frac{1}{\sigma}\int_s^{s+\sigma} h_n(t)\, dt - \frac{1}{\sigma}\int_s^{s+\sigma}(s+\sigma-r)\dot{h}_n(r)\, dr \\ &\equiv a_n + b_n. \end{aligned}$$

We observe that

(5.32)
$$\begin{aligned} |b_n|_{X_2} &\leq \frac{1}{\sigma}\int_s^{s+\sigma}(s+\sigma-r)|\dot{h}_n(r)|_{X_2}\, dr \\ &\leq \int_s^{s+\sigma} |\dot{h}_n(r)|_{X_2}\, dr \leq C\sigma^{1-1/q}. \end{aligned}$$

On the other hand, by (5.29) (recall $h=0$), for any fixed $\sigma > 0$, we have

(5.33) $$a_n = \frac{1}{\sigma}\int_s^{s+\sigma} h_n(t)\, dt \xrightarrow{w} 0, \qquad \text{in } X_0.$$

Thus, by the compactness of $X_0 \hookrightarrow\hookrightarrow X_1$, one has the strong convergence $a_n \xrightarrow{s} 0$ in $X_1$. Hence, combining (5.31) and (5.32), we see that $h_n(s) \xrightarrow{s} 0$ in $X_2$. Because $W^{1,q}([0,T];X_2) \hookrightarrow C([0,T];X_2)$ and $h_n$ is bounded in $W^{1,q}([0,T];X_2)$, we have the boundedness of $h_n$ in $C([0,T];X_2)$. Hence, by the Dominated Convergence Theorem, we obtain

(5.34) $$\lim_{n\to\infty} |h_n(\cdot)|_{L^p(0,T;X_2)} = 0.$$

Finally, for any $\delta > 0$, by (5.27), we have

(5.35) $$|h_n|_{L^p(0,T;X_1)} \leq \delta |h_n|_{L^p(0,T;X_0)} + C_\delta |h_n|_{L^p(0,T;X_2)}.$$

The second term on the right-hand side of (5.35) goes to zero as $n \to \infty$, and the first term can be arbitrarily small. Thus, our conclusion follows. □

Now, we are ready to state the following existence theorem.

**Theorem 5.10.** *Let* (W1)–(W6) *hold. Let there exist a minimizing sequence* $\{(y_k(\cdot), u_k(\cdot))\} \subset \mathcal{A}_{ad}$, *such that for some constant* $C > 0$,

$$\int_0^T |f(t, y_k(t), u_k(t))|^2 \, dt \leq C, \qquad \forall k \geq 1. \tag{5.36}$$

*Then Problem* (W) *admits at least one optimal pair.*

*Proof.* For each $k \geq 1$, from (5.2), (5.10), and the boundedness of $S$, we have that

$$\begin{aligned}
|\dot{y}_k(t)|^2 + \alpha |y_k(t)|_V^2 &\leq \Big\{ [|\dot{y}_k(0)|^2 + \langle Ay_k(0), y_k(0) \rangle]^{1/2} \\
&\quad + \int_0^t |f(s, y_k(s), u_k(s))| \, ds \Big\}^2 \leq C, \qquad \forall t \in [0, T], \quad k \geq 1,
\end{aligned} \tag{5.37}$$

for some constant $C$. Thus, we may assume that

$$\begin{cases} \dot{y}_k(\cdot) \overset{*}{\rightharpoonup} \dot{\bar{y}}(\cdot), & \text{in } L^\infty(0, T; H), \\ y_k(\cdot) \overset{*}{\rightharpoonup} \bar{y}(\cdot), & \text{in } L^\infty(0, T; V). \end{cases} \tag{5.38}$$

Because of the compactness of the embedding $V \hookrightarrow H$, by Lemma 5.9, we know that the embedding $L^2(0, T; V) \cap W^{1,2}([0, T]; H) \hookrightarrow L^2(0, T; H)$ is compact. Thus, one may let

$$\begin{cases} y_k(\cdot) \overset{s}{\to} \bar{y}(\cdot), & \text{in } L^2(0, T; H), \\ y_k(t) \overset{s}{\to} \bar{y}(t), & \text{in } H, \quad \text{a.e. } t \in [0, T]. \end{cases} \tag{5.39}$$

On the other hand, by the boundedness of $S$, we may assume

$$\begin{cases} y_k(0) \overset{w}{\rightharpoonup} \bar{y}_0, & \text{in } V, \\ \dot{y}_k(0) \overset{w}{\rightharpoonup} \bar{y}_1, & \text{in } H. \end{cases} \tag{5.40}$$

Also, from (5.36), we may let

$$f(\cdot, y_k(\cdot), u_k(\cdot)) \overset{w}{\rightharpoonup} \bar{f}(\cdot), \qquad \text{in } L^2(0, T; H). \tag{5.41}$$

By Definition 5.2, for any $v(\cdot) \in L^2(0, T; V) \cap W^{1,2}([0, T]; H)$ with $v(T) = 0$, it holds that

$$\begin{aligned}
\int_0^T \langle f(t, y_k(t), u_k(t)), v(t) \rangle \, dt &= -(\dot{y}_k(0), v(0)) \\
&\quad - \int_0^T (\dot{y}_k(t), \dot{v}(t)) \, dt + \int_0^T \langle y_k(t), Av(t) \rangle \, dt.
\end{aligned} \tag{5.42}$$

Let $k \to \infty$, we obtain

$$\begin{aligned}
\int_0^T \langle \bar{f}(t), v(t) \rangle \, dt &= -(\bar{y}_1, v(0)) \\
&\quad - \int_0^T (\dot{\bar{y}}(t), \dot{v}(t)) \, dt + \int_0^T \langle \bar{y}(t), Av(t) \rangle \, dt.
\end{aligned} \tag{5.43}$$

From (5.39), we can find $\sigma \in (0, T)$, such that

(5.44) $$y_k(\sigma) \xrightarrow{s} \bar{y}(\sigma).$$

Then, for any $h \in H$,

(5.45) $$(\bar{y}_0, h) = \lim_{k \to \infty} (y_k(0), h) = \lim_{k \to \infty} \{(y_k(\sigma), h) - \int_0^\sigma (\dot{y}_k(s), h)\, ds\}$$
$$= (\bar{y}(\sigma), h) - \int_0^\sigma (\dot{\bar{y}}(s), h)\, ds = (\bar{y}(0), h).$$

That gives $\bar{y}(0) = \bar{y}_0$, in $H$. By Definition 5.2, we know that

(5.46) $$\begin{cases} \ddot{\bar{y}}(t) + A\bar{y}(t) = \bar{f}(t), & \text{a.e. } t \in [0, T], \text{ in } V', \\ y(0) = y_0, & \text{in } H, \\ \dot{y}(0) = y_1, & \text{in } V'. \end{cases}$$

Then, by a similar method used in the previous section, we can prove that for some $\bar{f}^0(\cdot) \in L^1(0, T)$,

(5.47) $$(\bar{f}^0(t), \bar{f}(t)) \in \overline{\text{co}}\, \mathcal{E}(t, \bar{y}(t)), \quad \text{a.e. } t \in [0, T],$$

and

(5.48) $$\int_0^T \bar{f}^0(t)\, dt \leq \inf_{\mathcal{A}_{ad}} J(y(\cdot), u(\cdot)).$$

Finally, by Corollaries 2.18 and 2.26, we can find some $\bar{u}(\cdot) \in \mathcal{U}[0, T]$, such that

(5.49) $$\begin{cases} \bar{f}^0(t) \geq f^0(t, \bar{y}(t), \bar{u}(t)), \\ \bar{f}(t) = f(t, \bar{y}(t), \bar{u}(t)), & \text{a.e. } t \in [0, T]. \\ \bar{u}(t) \in \Gamma(t, \bar{y}(t)), \end{cases}$$

That is, $(\bar{y}(\cdot), \bar{u}(\cdot)) \in \mathcal{A}_{ad}$. Hence, $(\bar{y}(\cdot), \bar{u}(\cdot))$ is an optimal pair. □

We see that the key point is the strong convergence of the trajectories, which is guaranteed by the suitable compactness conditions. We also note that one can obtain the existence of feasible pairs as we presented in the previous section. Because the idea is the same, we omit the details here.

## §6. Elliptic Partial Differential Equations and Variational Inequalities

In this section, we are going to discuss the existence of optimal controls for elliptic partial differential equations and variational inequalities. Let us first introduce a notion. Let $\beta \subset \mathbb{R} \times \mathbb{R}$, such that

(6.1) $$(y_1 - y_2)(x_1 - x_2) \geq 0, \quad \forall (x_1, y_1), (x_2, y_2) \in \beta.$$

We call such a $\beta$ a *monotone graph* in $\mathbb{R} \times \mathbb{R}$. Furthermore, if for any other monotone graph $\beta' \subset \mathbb{R} \times \mathbb{R}$ containing $\beta$, it holds that $\beta = \beta'$, then we call $\beta$ a *maximal monotone graph* in $\mathbb{R} \times \mathbb{R}$. For any nondecreasing function $h : \mathbb{R} \to \mathbb{R}$, the closure of its graph $\{(x, h(x)) \mid x \in \mathbb{R}\}$ is a maximal monotone graph in $\mathbb{R} \times \mathbb{R}$. In particular, the graph of any continuous nondecreasing function defined on $\mathbb{R}$ is maximal monotone. Let us give two other examples. The first one is the following:

$$(6.2) \qquad \beta = \big((-\infty, 0) \times \{0\}\big) \bigcup \big(\{0\} \times [0, 1]\big) \bigcup \big((0, \infty) \times \{1\}\big).$$

This is the maximal monotone graph containing the graph of the Heaviside function $H(x) = \chi_{[0,\infty)}(x)$. The other one is the following:

$$(6.3) \qquad \beta = \big(\{0\} \times (-\infty, 0]\big) \bigcup \big((0, \infty) \times \{0\}\big).$$

This graph consists of a positive $x$-axis and negative $y$-axis plus the origin.

Next, denote

$$(6.4) \qquad \begin{cases} \mathcal{D}(\beta) = \{x \in \mathbb{R} \mid \exists y \in \mathbb{R}, \text{ with } (x, y) \in \beta\}, \\ \beta(x) = \{y \in \mathbb{R} \mid (x, y) \in \beta\}, \qquad \forall x \in \mathcal{D}(\beta). \end{cases}$$

We call $\mathcal{D}(\beta)$ the *domain* of $\beta$ and $\beta(x)$ defined above the *associated multifunction*. For example, for $\beta$ given by (6.3), $\mathcal{D}(\beta) = [0, \infty)$, and $\beta(x) = 0$, for all $x > 0$ and $\beta(0) = (-\infty, 0]$. It should be pointed out that $\mathbb{R} \times \{0\}$ is a maximal monotone graph for which we have $\beta(x) = 0$, for all $x \in \mathbb{R}$.

Hereafter, we will not distinguish the monotone graph $\beta$ and the associated multifunction $\beta(\cdot)$. We note that if $\beta \subset \mathbb{R} \times \mathbb{R}$ is a maximal monotone graph, then there are at most countably many points $s \in \mathcal{D}(\beta)$, at which $\beta(s)$ consists of more than one point. Also, outside of this countable set, $\beta(s)$ is continuous (as a single valued function). The following result will be useful in sequel.

**Lemma 6.1.** *Let $\beta \subset \mathbb{R} \times \mathbb{R}$ be a maximal monotone graph containing the origin. Then there exists a family of smooth strictly increasing functions $\beta_\varepsilon : \mathbb{R} \to \mathbb{R}$, such that $\beta_\varepsilon(0) = 0$ and*

$$(6.5) \qquad \lim_{\varepsilon \to 0} \beta_\varepsilon(s) \in \beta(s), \qquad \forall s \in \mathcal{D}(\beta).$$

For the cases like (6.2) and (6.3), the readers can readily provide a proof for the above result. The general case is a little bit more technical, and we are not going to get into it. Interested readers can look at many books on maximal monotone operators, see for example Barbu [1].

Now, let us make the following assumptions.

(E1) $\Omega$ is a bounded region in $\mathbb{R}^n$ with a $C^2$ boundary $\partial\Omega$, and $U$ is a Polish space.

## §6. Elliptic PDEs and Variational Inequalities

(E2) $\beta$ is a maximal monotone graph in $\mathbb{R} \times \mathbb{R}$ containing the origin, and $A$ is a second order uniformly elliptic differential operator of divergence form. More precisely,

$$(6.6) \qquad Ay(x) = -\sum_{i,j=1}^{n} \left(a_{ij}(x) y_{x_j}(x)\right)_{x_i}, \qquad \forall y(\cdot) \in H^1(\Omega),$$

with $a_{ij}(\cdot) \in C(\overline{\Omega})$ and for some $\lambda > 0$,

$$(6.7) \qquad \sum_{i,j=1}^{n} a_{ij}(x) \xi_i \xi_j \geq \lambda |\xi|^2, \qquad \forall \xi \in \mathbb{R}^n, \quad x \in \Omega.$$

(E3) $f : \Omega \times \mathbb{R} \times U \to \mathbb{R}$ is Borel measurable in $(x, y, u) \in \Omega \times \mathbb{R} \times U$ and continuous in $(y, u) \in \mathbb{R} \times U$ for almost all $x \in \Omega$,

$$(6.8) \qquad f_y(x, y, u) \leq 0, \qquad \forall (x, y, u) \in \Omega \times \mathbb{R} \times U.$$

Moreover, for any $R > 0$, there exists an $M_R > 0$, such that

$$(6.9) \qquad |f(x, y, u)| + |f_y(x, y, u)| \leq M_R, \qquad \forall (x, u) \in \Omega \times U, \ |y| \leq R.$$

(E4) $f^0 : \Omega \times \mathbb{R} \times U \to \mathbb{R}$ is Borel measurable in $(x, y, u) \in \Omega \times \mathbb{R} \times \mathbb{R}$, and lower semicontinuous in $(y, u) \in \mathbb{R} \times U$ for almost all $x \in \Omega$, and there exists a constant $K \geq 0$, such that

$$(6.10) \qquad f^0(x, y, u) \geq -K, \qquad (x, y, u) \in \Omega \times \mathbb{R} \times U.$$

(E5) $\Gamma : \Omega \times \mathbb{R} \to 2^U$ is pseudo-continuous (see Definition 2.4).

(E6) $Q \subset \mathbb{R}$ is closed.

Our controlled system is the following:

$$(6.11) \qquad \begin{cases} Ay(x) + \beta(y(x)) \ni f(x, y(x), u(x)), & \text{a.e. } x \in \Omega, \\ y|_{\partial \Omega} = 0. \end{cases}$$

We note that in (E1), the assumption that $\beta \ni 0$ is just for convenience. If $\beta$ does not contain the origin, we may make a proper translation and absorb the proper term into the right-hand side term so that the new monotone graph has such a property.

The above is referred to as a variational inequality. Let us look at a special case. Take $\beta$ to be the one given by (6.3). Let $a_{ij}$ be $C^1(\overline{\Omega})$. Then, in order for $y(x)$ to be a solution of (6.10), it is necessary that $\beta(y(x)) \neq \phi$, a.e. $x \in \Omega$, which implies that $y(x) \geq 0$, a.e. $x \in \Omega$. Thus, notifying that $\beta(y) = 0, \forall y > 0$, and $\beta(0) = (-\infty, 0]$, we see that (6.11) is equivalent to the following:

$$\text{(6.12)} \quad \begin{cases} y(x) \geq 0, & \text{a.e. } x \in \Omega, \\ Ay(x) \geq f(x, y(x), u(x)), & \text{a.e. } x \in \Omega, \\ y(x)(Ay(x) - f(x, y(x), u(x))) = 0, & \text{a.e. } x \in \Omega, \\ y|_{\partial\Omega} = 0. \end{cases}$$

This is a typical variational inequality (see Chapter 1, §5).

We note that if $\beta = 0$, then (6.11) becomes a controlled elliptic partial differential equation. Thus, the results presented in this section hold true for optimal control problems of elliptic equations.

Next, we need the following definition.

**Definition 6.2.** A function $y$ is called a solution of (6.11), if $y \in W_0^{1,p}(\Omega)$ for some $p \in [1, \infty)$ and $\beta(y)$ admits a selection $\zeta \in L^p(\Omega)$, i.e.,

$$\text{(6.13)} \quad \zeta(x) \in \beta(y(x)), \quad \text{a.e. } x \in \Omega,$$

and the following holds: For any $\varphi \in W_0^{1,p'}(\Omega)$ $(p' = \frac{p}{p-1})$

$$\text{(6.14)} \quad \int_\Omega \Big(a_{ij}(x)y_{x_j}(x)\varphi_{x_i}(x) + \zeta(x)\varphi(x)\Big)\,dx = \int_\Omega f(x, y(x), u(x))\varphi(x)\,dx.$$

The following result gives the well-posedness of (6.11).

**Proposition 6.3.** *Let (E1)–(E3) hold. Then, for any $u(\cdot) \in \mathcal{U} \equiv \{u(\cdot) : \Omega \to U \mid u(\cdot) \text{ measurable}\}$ and any $2 \leq p < \infty$, there exists a unique solution $y(\cdot) \equiv y(\cdot\,;u(\cdot)) \in W_0^{1,p}(\Omega)$ to (6.11). Moreover,*

$$\text{(6.15)} \quad \begin{cases} \|y(\cdot\,;u(\cdot))\|_{W_0^{1,p}(\Omega)} \leq C_p, \\ \|y(\cdot\,;u(\cdot))\|_{L^\infty(\Omega)} \leq C, \end{cases} \quad \forall u(\cdot) \in \mathcal{U}.$$

*Proof.* Let $\beta_\varepsilon$ be the smooth increasing functions satisfying (6.5). We consider the following approximate equation:

$$\text{(6.16)} \quad \begin{cases} Ay^\varepsilon + \beta_\varepsilon(y^\varepsilon) = f(x, y^\varepsilon, u(x)), & \text{in } \Omega, \\ y^\varepsilon|_{\partial\Omega} = 0. \end{cases}$$

Note that $f_y(x, y, u(x)) - \beta_\varepsilon'(y) \leq 0$. Thus, by Theorem 6.11 of Chapter 2, there exists a unique weak solution $y^\varepsilon \in W_0^{1,p}(\Omega)$. From the proof of Theorem 6.11 of Chapter 2, we further see that

$$\text{(6.17)} \quad \|y^\varepsilon\|_{L^\infty(\Omega)} \leq C,$$

§6. Elliptic PDEs and Variational Inequalities

with the constant being independent of $\varepsilon > 0$ and $u(\cdot) \in \mathcal{U}$. Now, let $F_\varepsilon(s) = |\beta_\varepsilon(s)|^{p-2}\beta_\varepsilon(s)$. Then we have

$$(6.18) \qquad F'_\varepsilon(s) = (p-1)|\beta_\varepsilon(s)|^{p-2}\beta'_\varepsilon(s) \geq 0,$$

and $F_\varepsilon(y^\varepsilon) \in L^\infty(\Omega) \cap W_0^{1,p}(\Omega)$ because (6.17) holds and $\beta_\varepsilon(0) = 0$. Now, by (6.16), (6.18), and (6.7), we have

$$(6.19) \quad \begin{aligned} \int_\Omega |\beta_\varepsilon(y^\varepsilon(x))|^p \, dx & \\ & \leq \int_\Omega \Big( \sum_{i,j=1}^n a_{ij}(x) y^\varepsilon_{x_j}(x) F'_\varepsilon(y^\varepsilon(x)) y^\varepsilon_{x_i}(x) + |\beta_\varepsilon(y^\varepsilon(x))|^p \Big) \, dx \\ &= \langle Ay^\varepsilon + \beta(y^\varepsilon), F_\varepsilon(y^\varepsilon) \rangle \\ &= \int_\Omega f(x, y^\varepsilon(x), u(x)) F_\varepsilon(y^\varepsilon(x)) \, dx \\ &\leq \int_\Omega |f(x, y^\varepsilon(x), u(x))| |\beta_\varepsilon(y^\varepsilon(x))|^{p-1} \, dx \\ &\leq \Big( \int_\Omega |f(x, y^\varepsilon(x), u(x))|^p \, dx \Big)^{1/p} \Big( \int_\Omega |\beta_\varepsilon(y^\varepsilon(x))|^p \, dx \Big)^{1/p'}. \end{aligned}$$

Thus, from (6.17) and (6.9) we obtain

$$(6.20) \qquad \|\beta_\varepsilon(y^\varepsilon)\|_{L^p(\Omega)} \leq C, \qquad \forall \varepsilon > 0, \ u(\cdot) \in \mathcal{U}.$$

Then, from (6.16) and Theorem 6.11 of Chapter 2, we have

$$(6.21) \qquad \|y^\varepsilon\|_{W_0^{1,p}(\Omega)} \leq C, \qquad \forall \varepsilon > 0, \ u(\cdot) \in \mathcal{U}.$$

Now, we may assume that

$$(6.22) \quad \begin{aligned} y^\varepsilon &\rightharpoonup y, \qquad \text{weakly in } W_0^{1,p}(\Omega), \\ y^\varepsilon &\to y, \qquad \text{strongly in } L^p(\Omega), \qquad \text{a.e. } x \in \Omega. \end{aligned}$$

Also, from (6.20), we may let

$$(6.23) \qquad \beta_\varepsilon(y^\varepsilon) \rightharpoonup \zeta, \qquad \text{weakly in } L^p(\Omega).$$

Since, for any $\varphi \in W_0^{1,p'}(\Omega)$,

$$(6.24) \quad \begin{aligned} \int_\Omega \Big( a_{ij}(x) y^\varepsilon_{x_j}(x) \varphi_{x_i}(x) + \beta_\varepsilon(y^\varepsilon(x)) \varphi(x) \Big) \, dx \\ = \int_\Omega f(x, y^\varepsilon(x), u(x)) \varphi(x) \, dx, \end{aligned}$$

by letting $\varepsilon \to 0$, we see that (6.14) holds. Now, for any $z \in \mathcal{D}(\beta)$, and any measurable set $E \subset \Omega$, by the monotonicity of $\beta_\varepsilon$, we have

$$
\begin{aligned}
0 &\leq \int_E (\beta_\varepsilon(y^\varepsilon(x)) - \beta_\varepsilon(z))(y^\varepsilon(x) - z)\, dx \\
&\to \int_E (\zeta(x) - \xi)(y(x) - z)\, dx, \qquad (\xi \in \beta(z)).
\end{aligned}
\tag{6.25}
$$

Thus, we see that

$$
(\zeta(x) - \xi)(y(x) - z) \geq 0, \qquad \forall (z, \xi) \in \beta, \text{ a.e. } x \in \Omega.
\tag{6.26}
$$

Here, we have used the fact that except for at most a countable set, $\beta(s)$ is single valued and continuous. Then, by the maximality of $\beta$, we must have (6.13). This proves the existence of a solution. Now, suppose there is another solution $\bar{y}$ with the associated $\bar{\zeta} \in \beta(\bar{y})$. Then, for any $\varphi \in W_0^{1,p'}(\Omega)$, it holds that

$$
\int_\Omega \left( a_{ij}(x) \bar{y}_{x_j}(x) \varphi_{x_i}(x) + \bar{\zeta}(x) \varphi(x) \right) dx \\
= \int_\Omega f(x, \bar{y}(x), u(x)) \varphi(x)\, dx,
\tag{6.27}
$$

We make the difference of (6.14) and (6.27) and take $\varphi = y - \bar{y}$. This is allowed because $p \geq 2$, $W_0^{1,p}(\Omega) \hookrightarrow W_0^{1,p'}(\Omega)$. By (6.8) and (6.7),

$$
\begin{aligned}
0 &\geq \int_\Omega (f(x, y, u) - f(x, \bar{y}, u))(y - \bar{y})\, dx \\
&= \int_\Omega \left( a_{ij}(y - \bar{y})_{x_j}(y - \bar{y})_{x_i} + (\zeta - \bar{\zeta})(y - \bar{y}) \right) dx \\
&\geq \lambda \int_\Omega |\nabla(y - \bar{y})|^2\, dx.
\end{aligned}
\tag{6.28}
$$

Thus, $\bar{y} = y$. Then, from (6.14) and (6.27), we obtain $\bar{\zeta} = \zeta$. $\square$

In our problem, the control and the state constraints are given by

$$u(x) \in \Gamma(x, y(x)), \qquad \text{a.e. } x \in \Omega, \tag{6.29}$$

$$y(x) \in Q, \qquad \text{a.e. } x \in \Omega. \tag{6.30}$$

We let $p \geq 2$. A pair $(y(\cdot), u(\cdot)) \in W_0^{1,p}(\Omega) \times \mathcal{U}$ is said to be *feasible* if (6.11) and (6.29) are satisfied and is said to be *admissible* if (6.11), (6.29) and (6.30) hold and $f^0(\cdot, y(\cdot), u(\cdot)) \in L^1(\Omega)$. We let $\mathcal{A}_{ad}$ be the set of all admissible pairs $(y(\cdot), u(\cdot))$. Next, for any $(y(\cdot), u(\cdot)) \in \mathcal{A}_{ad}$, we define the cost functional to be the following:

$$J(y(\cdot), u(\cdot)) = \int_\Omega f^0(x, y(x), u(x))\, dx. \tag{6.31}$$

## §6. Elliptic PDEs and Variational Inequalities

Then, our optimal control problem can be stated as follows:

**Problem (E).** Find $(\bar{y}(\cdot), \bar{u}(\cdot)) \in \mathcal{A}_{ad}$, such that

$$(6.32) \qquad J(\bar{y}(\cdot), \bar{u}(\cdot)) = \inf_{\mathcal{A}_{ad}} J(y(\cdot), u(\cdot)).$$

If a pair $(\bar{y}(\cdot), \bar{u}(\cdot)) \in \mathcal{A}_{ad}$ exists satisfying (6.32), we call $(\bar{y}(\cdot), \bar{u}(\cdot))$, $\bar{y}(\cdot)$, and $\bar{u}(\cdot)$ an *optimal pair, state,* and *control,* respectively.

To establish the existence of an optimal pair for Problem (E), let us introduce the following set:

$$(6.33) \qquad \mathcal{E}(x,y) = \{(z^0, z) \in \mathbb{R}^2 \mid z^0 \geq f^0(x,y,u), z = f(x,y,u), \\ u \in \Gamma(x,y)\}, \quad \forall (x,y) \in \Omega \times \mathbb{R}.$$

We further introduce the following assumption (compare (P6) and (W6) in §§4 and 5).

(E6) For a.e. $x \in \Omega$, the map $\mathcal{E}(x, \cdot)$ has the Cesari property on $Q$.

Now, we are ready to state and prove the main result of this section.

**Theorem 6.4.** *Let (E1)–(E6) hold and $\mathcal{A}_{ad} \neq \phi$. Then Problem (E) admits at least one optimal pair.*

*Proof.* Let $(y_k(\cdot), u_k(\cdot)) \in \mathcal{A}_{ad}$ be any minimizing sequence. By Proposition 6.3 (let $p > n$),

$$(6.34) \qquad \begin{cases} |y_k(\cdot)|_{W_0^{1,p}(\Omega)} \leq C_p, \\ |y_k(\cdot)|_{L^\infty(\Omega)} \leq C, \end{cases} \quad \forall k \geq 1.$$

Thus, we may let

$$(6.35) \qquad \begin{aligned} y_k(\cdot) &\to \bar{y}(\cdot), \quad \text{weakly in } W^{1,p}(\Omega), \\ &\quad\quad\quad\quad\,\, \text{strongly in } C^\alpha(\bar{\Omega}), \end{aligned}$$

for some $\bar{y}(\cdot) \in W_0^{1,p}(\Omega)$ and $\alpha > 0$. On the other hand, by (6.9) and (6.34), we know that $f(\cdot, y_k(\cdot), u_k(\cdot))$ are uniformly bounded. Hence, we may let

$$(6.36) \qquad f(\cdot, y_k(\cdot), u_k(\cdot)) \to \bar{f}(\cdot), \quad \text{weakly in } L^p(\Omega),$$

for some $\bar{f}(\cdot) \in L^\infty(\Omega)$. Then, by the Mazur Theorem (Corollary 2.8 of Chapter 2), one can find $\alpha_{ij} \geq 0$, $\sum_{i \geq 1} \alpha_{ij} = 1$, such that

$$(6.37) \qquad \psi_j \equiv \sum_{i \geq 1} \alpha_{ij} f(\cdot, y_{i+j}(\cdot), u_{i+j}(\cdot)) \to \bar{f}(\cdot), \quad \text{strongly in } L^p(\Omega).$$

Set

$$(6.38) \qquad \psi_j^0(\cdot) = \sum_{i \geq 1} \alpha_{ij} f^0(\cdot, y_{i+j}(\cdot), u_{i+j}(\cdot)),$$

(6.39) $$\bar{f}^0(x) = \lim_{j\to\infty} \psi_j^0(x) \geq -K, \quad \text{a.e. } x \in \Omega.$$

By (6.35), we see that for any $\varepsilon > 0$, there exists a $j_0$, such that for $j \geq j_0$,

(6.40) $$(\psi_j^0(x), \psi_j(x)) \in \mathrm{co}\,\mathcal{E}(x, \mathcal{O}_\varepsilon(\bar{y}(x))), \quad \text{a.e. } x \in \Omega.$$

Thus,

(6.41) $$(\bar{f}^0(x), \bar{f}(x)) \in \overline{\mathrm{co}}\,\mathcal{E}(x, \mathcal{O}_\varepsilon(\bar{y}(x))), \quad \text{a.e. } x \in \Omega.$$

Then, by (E6), we see that

(6.42) $$(\bar{f}^0(x), \bar{f}(x)) \in \mathcal{E}(x, \bar{y}(x)), \quad \text{a.e. } x \in \Omega.$$

Hence, similar to the previous sections, we can find $\bar{u}(\cdot) \in \mathcal{U}$, such that

(6.43) $$\begin{cases} \bar{f}^0(x) \geq f^0(x, \bar{y}(x), \bar{u}(x)), \\ \bar{f}(x) = f(x, \bar{y}(x), \bar{u}(x)), \\ \bar{u}(x) \in \Gamma(x, \bar{y}(x)), \end{cases} \quad \text{a.e. } x \in \Omega.$$

On the other hand, from the admissibility of $(y_k(\cdot), u_k(\cdot))$, the convergence (6.35) and (6.36), and (6.43), we have

(6.44) $$\begin{cases} A\bar{y}(x) + \beta(\bar{y}(x)) \ni f(x, \bar{y}(x), \bar{u}(x)), \quad \text{a.e. } x \in \Omega, \\ \bar{y}|_{\partial\Omega} = 0. \end{cases}$$

By (E5), we have

(6.45) $$\bar{y}(x) \in Q, \quad \text{a.e. } x \in \Omega.$$

Hence, $(\bar{y}(\cdot), \bar{u}(\cdot)) \in \mathcal{A}_{ad}$. Finally, by Fatou's Lemma,

(6.46) $$\begin{aligned} J(\bar{y}(\cdot), \bar{u}(\cdot)) &= \int_\Omega f^0(x, y(x), \bar{u}(x))\, dx \\ &\leq \lim_{k\to\infty} J(y_k(\cdot), u_k(\cdot)) = \inf_{\mathcal{A}_{ad}} J(y(\cdot), u(\cdot)). \end{aligned}$$

Thus, $(\bar{y}(\cdot), \bar{u}(\cdot))$ is an optimal pair. □

**Remark 6.5.** In the above, the operator $A$ is not necessarily of second order. Also, we see that if $\beta = 0$, the the (6.11) is reduced to a semilinear elliptic partial differential system with the leading operator of divergence form and coercive. Thus, the above result covers the existence of optimal controls for elliptic equations.

**Remark 6.6.** From the above, we see that the existence of an optimal pair follows essentially from the relative compactness of the minimizing sequence in suitable spaces and the type of convexity conditions. The multivaluedness of term $Ay + \beta(y)$ does not cause any difficulty. However, if this multivalued operator also depends on the control variable, then the situation becomes technically difficult.

*Remark 6.7.* It is also possible to discuss the existence of optimal controls for evolutionary variational inequalities.

## Remarks

For finite dimensional systems, the existence of optimal controls is very closely related to the classical calculus of variation. Along this line, many results are available. We refer the readers to the books by Berkovitz [1], Cesari [5], and Warga [1]. For nonlinear infinite dimensional systems, the study of existence theory for optimal controls probably started from the work of Cesari [1,2], to our best knowledge. Later, many authors contributed in this aspect. Among them, we mention the works by Ahmed [1], Ahmed-Teo [3], Berkovitz [2], Cesari [1-4], Fattorini [9], Hou [1], Nababan-Teo [1], Papageorgiou [1-3], and Suryanarayana [2].

The material in this chapter is based on the work of Yong [7]. §1 introduces the theory of Souslin space. This is selected from Bourbaki [1] with some modifications, so that the presentation is shorter and self-contained. §2 discusses multifunctions and measurable selection theorems. This is based on the paper by Himmelberg-Jacobs-Van Vleck [1]. By using the results of §1, the presentation of this section is self-contained also. For an excellent survey on the measurable selection theorems, see Wagner [1]. §§3–6 provide an existence theory of optimal controls for several interesting cases.

# Chapter 4
# Necessary Conditions for Optimal Controls — Abstract Evolution Equations

## §1. Formulation of the Problem

In this chapter we present some necessary conditions of optimal controls for evolutionary systems. These necessary conditions are usually referred to as *Pontryagin's maximum principle*.

Let us begin with the following hypothesis.

(H1) Let $X$ be a Banach space, $S$ be a closed and convex subset of $X \times X$, $U$ be a separable metric space, and $T > 0$ be a constant. Let $e^{At}$ be a $C_0$ semigroup on $X$ with generator $A : \mathcal{D}(A) \subset X \to X$.

(H2) Let $f : [0,T] \times X \times U \to X$ and $f^0 : [0,T] \times X \times U \to \mathbb{R}$ such that $f(t,y,u)$ and $f^0(t,y,u)$ are strongly measurable in $t \in [0,T]$, and continuously Fréchet differentiable in $y \in X$ with $f(t,\cdot,\cdot)$, $f_y(t,\cdot,\cdot)$, $f^0(t,\cdot,\cdot)$ and $f_y^0(t,\cdot,\cdot)$, continuous. Moreover, for some constant $L > 0$,

(1.1) $$|f_y(t,y,u)|, |f_y^0(t,y,u)|, |f(t,0,u)|, |f^0(t,0,u)| \leq L,$$
$$\forall (t,y,u) \in [0,T] \times X \times U.$$

We consider the following evolution equation:

(1.2) $$y(t) = e^{At}y(0) + \int_0^t e^{A(t-s)} f(s, y(s), u(s))\, ds, \qquad t \in [0,T],$$

where $u(\cdot) \in \mathcal{U}[0,T] \equiv \{u : [0,T] \to U \mid u(\cdot) \text{ measurable}\}$. We know that (1.2) is the mild form of the following equation:

(1.3) $$\dot{y}(t) = Ay(t) + f(t, y(t), u(t)), \qquad t \in [0,T].$$

It is clear that under (H2), for any $(y(0), u(\cdot)) \in X \times \mathcal{U}[0,T]$, there exists a unique solution $y(\cdot) \in C([0,T]; X)$ of (1.2). Also, we point out that condition (1.1) is only used for the simplicity of presentation and it can be slightly relaxed.

Recall that any $(y(\cdot), u(\cdot)) \in C([0,T]; X) \times \mathcal{U}[0,T]$ satisfying (1.2) is called a *feasible pair*. The set of all feasible pairs is denoted by $\mathcal{A}$. Next, we let $\mathcal{A}_{ad}$ be the set of all feasible pairs $(y(\cdot), u(\cdot)) \in \mathcal{A}$, such that

(1.4) $$(y(0), y(T)) \in S,$$

and $f^0(\cdot, y(\cdot), u(\cdot)) \in L^1(0,T)$. Any $(y(\cdot), u(\cdot)) \in \mathcal{A}_{ad}$ is called an *admissible pair*. Now, for any $(y(\cdot), u(\cdot)) \in \mathcal{A}_{ad}$, we define

(1.5) $$J(y(\cdot), u(\cdot)) = \int_0^T f^0(s, y(s), u(s))\, ds.$$

## §1. Formulation of the Problem

Our optimal control problem can be stated as follows:

**Problem (C).** Find $(\bar{y}(\cdot), \bar{u}(\cdot)) \in \mathcal{A}_{ad}$, such that

$$J(\bar{y}(\cdot), \bar{u}(\cdot)) = \inf_{\mathcal{A}_{ad}} J(y(\cdot), u(\cdot)). \tag{1.6}$$

If such a pair exists, we call it an *optimal pair* and refer to $\bar{y}(\cdot)$ and $\bar{u}(\cdot)$ as an *optimal trajectory* and *control*, respectively.

We will assume that there exists an optimal pair $(\bar{y}(\cdot), \bar{u}(\cdot))$ for Problem (C). Our goal is to derive necessary conditions for the pair $(\bar{y}(\cdot), \bar{u}(\cdot))$.

In what follows, we let $C_{w^*}([0,T]; X^*)$ be the set of all weak* continuous functions from $[0,T]$ to $X^*$ and (see Chapter 2, §3.2).

**Definition 1.1.** Let (H1) and (H2) hold and $(\bar{y}(\cdot), \bar{u}(\cdot)) \in \mathcal{A}_{ad}$ be an optimal pair. We say that $(\bar{y}(\cdot), \bar{u}(\cdot))$ satisfies *Pontryagin's maximum principle*, if there exists a nontrivial pair $(\psi^0, \psi(\cdot)) \in \mathbb{R} \times C_{w^*}([0,T]; X^*)$, i.e., $(\psi^0, \psi(\cdot)) \neq 0$, such that

$$\psi^0 \leq 0, \tag{1.7}$$

$$\begin{aligned}\psi(t) =& e^{A^*(T-t)}\psi(T) + \int_t^T e^{A^*(s-t)} f_y(s, \bar{y}(s), \bar{u}(s))^* \psi(s)\, ds \\&+ \int_t^T e^{A^*(s-t)} \psi^0 f_y^0(s, \bar{y}(s), \bar{u}(s))\, ds, \quad \text{a.e. } t \in [0,T],\end{aligned} \tag{1.8}$$

$$\langle \psi(0), x_0 - \bar{y}(0) \rangle - \langle \psi(T), x_1 - \bar{y}(T) \rangle \leq 0, \quad \forall (x_0, x_1) \in S, \tag{1.9}$$

$$\begin{aligned}H(t, \bar{y}(t), \bar{u}(t), \psi^0, \psi(t)) =& \max_{u \in U} H(t, \bar{y}(t), u, \psi^0, \psi(t)), \\& \text{a.e. } t \in [0,T],\end{aligned} \tag{1.10}$$

where

$$\begin{aligned}H(t, y, u, \psi^0, \psi) =& \psi^0 f^0(t, y, u) + \langle \psi, f(t, y, u) \rangle, \\& \forall (t, y, u, \psi^0, \psi) \in [0,T] \times X \times U \times \mathbb{R} \times X^*.\end{aligned} \tag{1.11}$$

In the above, we refer to $\psi(\cdot)$ as the *costate* or the *adjoint variable*. Thus, (1.8) is called the *adjoint system* along the optimal pair $(\bar{y}(\cdot), \bar{u}(\cdot))$. Also, we call (1.9) the *transversality condition* and (1.10) the *maximum condition*. The function $H$ defined by (1.11) is called the *Hamiltonian*.

*Remark 1.2.* It is clear that by taking $(\psi^0, \psi(\cdot)) = 0$, (1.7)–(1.10) hold trivially. Thus, it is crucial to require the pair $(\psi^0, \psi(\cdot))$ to be nontrivial.

We note that in the case where $X = \mathbb{R}^n$, without loss of generality, we may let $A = 0$. Then system (1.2) is the same as

$$\dot{y}(t) = f(t, y(t), u(t)), \quad \text{a.e. } t \in [0,T], \tag{1.12}$$

and the adjoint system (1.8) is equivalent to the following:

(1.13) $\quad \dot{\psi}(t) = -f_y(t, \bar{y}(t), \bar{u}(t))^* \psi(t) + \psi^0 f_y^0(t, \bar{y}(t), \bar{u}(t)),$ a.e. $t \in [0, T].$

For the finite dimensional case, we have the following classical result.

**Theorem 1.3.** (Pontryagin) *Let (H1) and (H2) hold with $X = \mathbb{R}^n$ and $A = 0$. Let $(\bar{y}(\cdot), \bar{u}(\cdot))$ be an optimal pair for Problem (C). Then, $(\bar{y}(\cdot), \bar{u}(\cdot))$ satisfies Pontryagin's maximum principle.*

For the infinite dimensional case, Theorem 1.3 is not necessarily true if we do not make some further assumptions. To see this, let us present the following example.

*Example 1.4.* Let $X$ be a Hilbert space of infinite dimension. Let $A = A^* : \mathcal{D}(A) \subset X \to X$ be a self-adjoint operator with the following properties: There exist sequences $\{\varphi_k\}_{k \geq 1} \subset X$ and $\{\lambda_k\}_{k \geq 1} \subset \mathbb{R}$ with the properties that $\{\varphi_k\}_{k \geq 1}$ forms an orthonormal basis of $X$ and $\{e^{\lambda_k \cdot}\}_{k \geq 1}$ forms a basis of $L^2(0, 1)$, such that

(1.14) $\quad \begin{cases} 0 < \lambda_1 \leq \lambda_2 \leq \cdots, & \lim\limits_{k \to \infty} \lambda_k = +\infty, \\ A\varphi_k = -\lambda_k \varphi_k, & k \geq 1. \end{cases}$

This can be easily achieved. Now, we let

(1.15) $\quad \begin{cases} b \in X, & b_k \equiv \langle b, \varphi_k \rangle \neq 0, \quad k \geq 1; \\ a \in X, & c \in \mathbb{R}, \quad c \neq \langle b, A^{-1}a \rangle; \\ q = (e^A - I)A^{-1}b; & U = [-2, 2]. \end{cases}$

Consider the following system:

(1.16) $\quad \dot{y}(t) = Ay(t) + bu(t), \qquad t \in [0, 1],$

with $u(\cdot) \in \mathcal{U} \equiv \{u(\cdot) : [0, 1] \to U \mid u(\cdot) \text{ measurable }\}$. Our constraint for the endpoints of the state is

(1.17) $\quad y(0) = 0, \qquad y(1) = q,$

i.e., $S = \{0\} \times \{q\}$ and the cost functional is given by

(1.18) $\quad J(u(\cdot)) = \int_0^1 \Big[ \langle a, y(t) \rangle + cu(t) \Big] dt$

Solutions to system (1.16) are understood to be mild solutions. Thus, for given $u(\cdot) \in \mathcal{U}$ and the initial state $y(0) = 0$, we have

(1.19) $\quad y(t) = \int_0^t e^{A(t-s)} bu(s) \, ds, \qquad t \in [0, 1].$

## §1. Formulation of the Problem

If we let $\bar{u}(t) \equiv 1$, then the corresponding trajectory, denoted by $\bar{y}(\cdot) \equiv y(\cdot\,; \bar{u}(\cdot))$, satisfies

$$\bar{y}(1) = \int_0^1 e^{As} b \, ds = (e^A - I) A^{-1} b = q. \tag{1.20}$$

Thus, $(\bar{y}(\cdot), \bar{u}(\cdot))$ is admissible. If $\widehat{u}(\cdot) \in \mathcal{U}$ is such that $(y(\cdot\,; \widehat{u}(\cdot)), \widehat{u}(\cdot))$ is also admissible, then

$$\begin{aligned}
0 = y(1; \widehat{u}(\cdot)) - \bar{y}(1) &= \sum_{k \geq 1} \Big[ \int_0^1 e^{-\lambda_k(1-s)} \big(\widehat{u}(s) - \bar{u}(s)\big) \, ds \Big] b_k \varphi_k \\
&= \sum_{k \geq 1} \Big[ \int_0^1 e^{\lambda_k s} \big(\widehat{u}(s) - \bar{u}(s)\big) \, ds \Big] e^{-\lambda_k} b_k \varphi_k.
\end{aligned} \tag{1.21}$$

Hence,

$$\int_0^1 e^{\lambda_k s} \big(\widehat{u}(s) - \bar{u}(s)\big) \, ds = 0, \qquad k \geq 1. \tag{1.22}$$

Because $\{e^{\lambda_k \cdot}\}_{k \geq 1}$ forms a basis of $L^2(0,1)$, it follows from (1.22) that $\widehat{u}(s) = \bar{u}(s)$, a.e. $s \in [0,1]$. This means that $(\bar{y}(\cdot), \bar{u}(\cdot))$ is the only admissible pair and hence it is optimal. Now, suppose the pair $(\bar{y}(\cdot), \bar{u}(\cdot))$ satisfies Pontryagin's maximum principle; that is, there exists a pair $(\psi^0, \psi(\cdot)) \neq 0$, with $\psi^0 \leq 0$ and

$$\begin{aligned}
\psi(t) &= e^{A(1-t)} \psi(1) + \psi^0 \int_t^1 e^{A(s-t)} a \, ds \\
&= e^{A(1-t)} \psi(1) + \psi^0 (e^{A(1-t)} - I) A^{-1} a,
\end{aligned} \tag{1.23}$$

such that

$$H(\bar{y}(t), \bar{u}(t), \psi^0, \psi(t)) = \max_{|u| \leq 2} H(\bar{y}(t), u, \psi^0, \psi(t)), \quad \text{a.e. } t \in [0,1], \tag{1.24}$$

where

$$\begin{aligned}
H(y, u, \psi^0, \psi) &= \psi^0 \big(\langle a, y \rangle + cu\big) + \langle \psi, b \rangle u, \\
&\forall (y, u, \psi^0, \psi) \in X \times U \times \mathbb{R} \times X^*.
\end{aligned} \tag{1.25}$$

Then (1.24) implies (note that $\bar{u}(t) \equiv 1$)

$$\big[\psi^0 c + \langle \psi(t), b \rangle\big] = \max_{|u| \leq 2} \big[\psi^0 c + \langle \psi(t), b \rangle\big] u, \quad \text{a.e. } t \in [0,1]. \tag{1.26}$$

In order for this to be true, we must have (note the continuity of $\psi(\cdot)$)

$$\psi^0 c + \langle \psi(t), b \rangle \equiv 0, \qquad t \in [0,1]. \tag{1.27}$$

Combining the above with (1.23), we have

$$
\begin{aligned}
0 = \psi^0 c + \langle \psi(t), b \rangle &= \psi^0 c + \langle \psi(1), e^{A(1-t)} b \rangle \\
&\quad + \psi^0 \langle (e^{A(1-t)} - I) A^{-1} a, b \rangle \\
&= \psi^0 \big[ c - \langle A^{-1} a, b \rangle \big] + \sum_{k \geq 1} \langle \psi(1) + \psi^0 A^{-1} a, \varphi_k \rangle e^{-\lambda_k (1-t)} b_k.
\end{aligned}
\tag{1.28}
$$

Thus, it follows that

$$
\begin{cases} \psi^0 \big[ c - \langle A^{-1} a, b \rangle \big] = 0, \\ \langle \psi(1) + \psi^0 A^{-1} a, \varphi_k \rangle = 0, \qquad k \geq 1, \end{cases}
\tag{1.29}
$$

which gives

$$
\begin{cases} \psi^0 \big[ c - \langle A^{-1} a, b \rangle \big] = 0, \\ \psi(1) + \psi^0 A^{-1} a = 0. \end{cases}
\tag{1.30}
$$

By (1.15), we derive from (1.30) that

$$
\psi^0 = 0, \quad \psi(1) = 0.
\tag{1.31}
$$

But, this gives the pair $(\psi^0, \psi(\cdot)) = 0$, which leads to a contradiction.

The above example shows that, in general, for an infinite dimensional control problem, the usual Pontryagin type maximum principle does not necessarily hold. To ensure that Pontryagin's maximum principle does hold, we need some additional assumptions. Now, we again let $(\bar{y}(\cdot), \bar{u}(\cdot)) \in \mathcal{A}_{ad}$ be optimal and define

$$
\begin{aligned}
\mathcal{R} = \Big\{ \xi(T) \in X \mid \xi(t) &= \int_0^t e^{A(t-s)} f_y(s, \bar{y}(s), \bar{u}(s)) \xi(s) \, ds \\
&\quad + \int_0^t e^{A(t-s)} [f(s, \bar{y}(s), u(s)) - f(s, \bar{y}(s), \bar{u}(s))] \, ds, \\
&\qquad t \in [0, T], \ u(\cdot) \in \mathcal{U}[0, T] \Big\},
\end{aligned}
\tag{1.32}
$$

$$
Q = \Big\{ y_1 - \eta(T) \mid \eta(t) = e^{At} y_0 + \int_0^t e^{A(t-s)} f_y(s, \bar{y}(s), \bar{u}(s)) \eta(s) \, ds, \\
t \in [0, T], \ (y_0, y_1) \in S \Big\}.
\tag{1.33}
$$

The equation satisfied by $\xi(\cdot)$ in (1.32) is called the *variational system* along the optimal pair $(\bar{y}(\cdot), \bar{u}(\cdot))$ and the set $\mathcal{R}$ is called the corresponding *reachable set*. For simplicity, we will call $\mathcal{R}$ the reachable set associated with $(\bar{y}(\cdot), \bar{u}(\cdot))$ hereafter. The set $Q$ is referred to as the *modified endpoint constraint set*. The following definition is necessary in sequel.

**Definition 1.5.** Let $Z$ be a Banach space. A subset $S$ of $Z$ is said to be *finite codimensional* in $Z$, denoted by codim $_Z S < \infty$, if there exists a point $z_0 \in \overline{co}\, S$, such that span $\{S - z_0\} \triangleq$ the closed subspace spanned by $\{z - z_0 \mid z \in S\}$ is a finite codimensional subspace of $Z$ and $\overline{co}(S - z_0)$ has a nonempty interior in this subspace.

Our main result in this chapter is the following maximum principle.

**Theorem 1.6.** *Let $X$ be a Banach space. Let (H1) and (H2) hold. Let $(\bar{y}(\cdot), \bar{u}(\cdot))$ be an optimal pair of Problem (C) with the associated reachable set $\mathcal{R}$ and the modified endpoint constraint set $Q$. Suppose further that the following (H3) and (H4) hold:*

(H3) $\mathcal{R} - Q \equiv \{r - q \mid r \in \mathcal{R}, q \in Q\}$ *is finite codimensional in $X$.*

(H4) *The dual $X^*$ of $X$ is strictly convex.*

*Then the pair $(\bar{y}(\cdot), \bar{u}(\cdot))$ satisfies Pontryagin's maximum principle.*

*Remark 1.7.* (H4) is actually very general. From Chapter 2, §2.2, we see that for the case where $X$ is separable or reflexive, one can find an equivalent norm under which (H4) holds. Thus, if our state space is one of $L^p(\Omega)$, $W^{m,p}(\Omega)$ $(1 \leq p < \infty)$, or $C(\overline{\Omega})$, $C([-r, 0]; \mathbb{R}^n)$, then (H4) can be assumed. These cases are fairly common in applications.

*Remark 1.8.* (H3) is crucial for the nontriviality of the pair $(\psi^0, \psi(\cdot))$. In §3, we will see that if one of $\mathcal{R}$ and $Q$ is finite codimensional in $X$, then so is $\mathcal{R} - Q$. Also, we will see some interesting cases covered by this condition.

## §2. Ekeland Variational Principle

In this section, we give the following result, due to Ekeland, which is very important in sequel.

**Lemma 2.1.** *(Ekeland's Variational Principle) Let $(V, d)$ be a complete metric space and let $F : V \to (-\infty, +\infty]$ be a proper (i.e., $F \not\equiv +\infty$), lower semicontinuous function bounded from below. Let $v_0 \in \mathcal{D}(F) \equiv \{v \in V \mid F(v) < \infty\}$ and $\delta > 0$ be fixed. Then there exists a $\bar{v} \in V$, such that*

$$(2.1) \qquad F(\bar{v}) + \delta d(\bar{v}, v_0) \leq F(v_0),$$

$$(2.2) \qquad F(\bar{v}) < F(v) + \delta d(v, \bar{v}), \qquad \forall v \neq \bar{v}.$$

*Proof.* By changing the metric to $\delta d(\cdot, \cdot)$ if necessary, we need only consider $\delta = 1$. Also, by considering $F(\cdot) - \inf_{v \in V} F(v)$ instead of $F(\cdot)$, we may assume that $F(\cdot)$ is nonnegative valued. Next, we define

$$(2.3) \qquad G(v) = \{w \in V \mid F(w) + d(w, v) \leq F(v)\}.$$

Then, for any $v \in V$, $G(v)$ is a closed set in $V$ because $F$ is lower semicontinuous. Also, trivially, we have

$$(2.4) \qquad v \in G(v), \qquad \forall v \in V.$$

Next, we claim that

(2.5) $$w \in G(v) \Rightarrow G(w) \subseteq G(v).$$

In fact, (2.5) is true if $F(v) = +\infty$, since $G(v) = V$ in this case. Now let $F(v) < \infty$. Then $w \in G(v)$ implies

(2.6) $$F(w) + d(w, v) \leq F(v),$$

and for any $u \in G(w)$, we have

(2.7) $$F(u) + d(u, w) \leq F(w).$$

Thus, combining (2.6) and (2.7), using the triangle inequality, we have

(2.8) $$F(u) + d(u, v) \leq F(u) + d(u, w) + d(w, v) \leq F(v).$$

Thus, (2.5) holds. Next, we define

(2.9) $$f(v) = \inf_{w \in G(v)} F(w), \qquad \forall v \in \mathcal{D}(F).$$

Then, for any $w \in G(v)$, $f(v) \leq F(w) \leq F(v) - d(w, v)$. Hence,

(2.10) $$d(w, v) \leq F(v) - f(v).$$

Thus, the diameter $\operatorname{diam} G(v)$ of the set $G(v)$ satisfies

(2.11) $$\operatorname{diam} G(v) \equiv \sup_{w, u \in G(v)} d(w, u) \leq 2(F(v) - f(v)).$$

Now, we define a sequence in the following way: $v_{n+1} \in G(v_n)$, $n \geq 0$, such that

(2.12) $$F(v_{n+1}) \leq f(v_n) + \frac{1}{2^n}.$$

By (2.5), we know that $G(v_{n+1}) \subseteq G(v_n)$. Thus,

(2.13) $$f(v_n) \leq f(v_{n+1}), \qquad n \geq 0.$$

On the other hand, $f(w) \leq F(w)$ because $w \in G(w)$ (see (2.4)). Thus, together with (2.12) and (2.13), we obtain

(2.14) $$0 \leq F(v_{n+1}) - f(v_{n+1}) \leq f(v_n) + \frac{1}{2^n} - f(v_{n+1}) \leq \frac{1}{2^n}.$$

Then, by (2.11), we see that the diameter of $G(v_n)$ goes to 0 as $n \to \infty$. Becuse $G(v_n)$ is a sequence of nested closed sets in $V$ (i.e., $G(v_{n+1}) \subseteq G(v_n)$, $\forall n \geq 0$) and $V$ is complete, we must have some point $\bar{v} \in V$, such that

(2.15) $$\bigcap_{n \geq 0} G(v_n) = \{\bar{v}\}.$$

In particular, $\bar{v} \in G(v_0)$, which gives (2.1) with $\delta = 1$. Also, because $\bar{v} \in G(v_n)$ for any $n \geq 0$, we have from (2.5) that

$$G(\bar{v}) \subseteq \bigcap_{n \geq 0} G(v_n) = \{\bar{v}\}. \tag{2.16}$$

This implies that $G(\bar{v}) = \{\bar{v}\}$. Hence, for any $v \neq \bar{v}$, we have $v \notin G(\bar{v})$, which gives (2.2) (with $\delta = 1$). □

**Corollary 2.2.** *Let the assumption of Lemma 2.1 hold. Let $\varepsilon > 0$ and $v_0 \in V$ be such that*

$$F(v_0) \leq \inf_{v \in V} F(v) + \varepsilon. \tag{2.17}$$

*Then there exists a $v_\varepsilon \in V$, such that*

$$F(v_\varepsilon) \leq F(v_0), \qquad d(v_\varepsilon, v_0) \leq \sqrt{\varepsilon}, \tag{2.18}$$

*and for all $v \in V$,*

$$-\sqrt{\varepsilon} d(v, v_\varepsilon) \leq F(v) - F(v_\varepsilon). \tag{2.19}$$

*Proof.* We take $\delta = \sqrt{\varepsilon}$. Then, by Lemma 2.1, there exists a $v_\varepsilon \in V$, such that (note (2.17))

$$F(v_\varepsilon) + \sqrt{\varepsilon} d(v_\varepsilon, v_0) \leq F(v_0) \leq \inf_{v \in V} F(v) + \varepsilon \leq F(v_\varepsilon) + \varepsilon, \tag{2.20}$$

and

$$F(v_\varepsilon) < F(v) + \sqrt{\varepsilon} d(v, v_\varepsilon), \qquad \forall v \neq v_\varepsilon. \tag{2.21}$$

Then (2.18) follows from (2.20), and (2.19) follows from (2.21). □

## §3. Other Preliminary Results

In this section, we present some other preliminary results. Some of them are necessary for the proof of our main result (Theorem 1.6) and the others are illustrations of the conditions we imposed in our main theorem.

### §3.1. Finite codimensionality

We first collect some results concerning the finite codimensionality of subsets in Banach spaces.

**Proposition 3.1.** *Let $X$ be a Banach space and $Q \subset X$. Then, for any $y \in Q$, $z \in \overline{co}\, Q$,*

$$\operatorname{span}(Q - y) = \operatorname{span} \overline{co}(Q - z). \tag{3.1}$$

*Moreover, $\operatorname{codim}_X Q < \infty$ if and only if there exists a $y_0 \in \overline{co}\, Q$, such that $X_0 \equiv \operatorname{span} \overline{co}(Q - y_0)$ is a finite codimensional subspace of $X$ and*

$$0 \in \operatorname{Int}_{X_0} \overline{co}(Q - y_0), \tag{3.2}$$

where $\text{Int}_{X_0} \overline{\text{co}}(Q - y_0)$ is the interior of $\overline{\text{co}}(Q - y_0)$ in $X_0$.

*Proof.* For any $y \in Q$ and $z \in \overline{\text{co}} \, Q$, we have

(3.3) $\quad Q - z = Q - y - (z - y) \subseteq Q - y - \overline{\text{co}}(Q - y) \subseteq \text{span}\,(Q - y).$

On the other hand,

(3.4) $\quad Q - y = (Q - z) + z - y \subset \overline{\text{co}}(Q - z) - (y - z) \subset \text{span}\,\overline{\text{co}}(Q - z).$

Thus, (3.1) follows from (3.3) and (3.4). For the second assertion, we need only to prove the necessity. Thus, let $Q$ be finite codimensional in $X$. Then, by Definition 1.5, there exists a $z \in \overline{\text{co}}\, Q$, such that $X_0 \equiv \text{span}\, \overline{\text{co}}(Q - z)$ is a finite codimensional subspace of $X$ and there exists a $z_0$, such that

(3.5) $\qquad\qquad z_0 \in \text{Int}_{X_0} \overline{\text{co}}(Q - z).$

Hence, by setting $y_0 = z + z_0$, we obtain (3.2). $\qquad\square$

We note that in the case where $Q$ is convex, one can drop $\overline{\text{co}}$ in the above. Next, let us take a closer look at the finite codimensionality.

**Proposition 3.2.** *Let $X_0$ be a subspace of some Banach space $X$. Then the following are equivalent:*

(i) *$X_0$ is finite codimensional in $X$, i.e., $\text{codim}_X X_0 < \infty$.*

(ii) *There exist finitely many bounded linear functionals $f_1, \cdots, f_m \in X^*$, such that*

(3.6) $$X_0 = \bigcap_{i=1}^m \mathcal{N}(f_i),$$

*where $\mathcal{N}(f_i) = \{x \in X \mid f_i(x) = 0\}$.*

(iii) *There exists a compact set $K \subset X$, such that*

(3.7) $\qquad\qquad \text{Int}\,(X_0 + K) \neq \phi,$

*where $X_0 + K = \{x_0 + k \mid x_0 \in X_0, k \in K\}$.*

*Proof.* (i) $\Rightarrow$ (ii). By definition, there exist $x_1, \cdots, x_m \in X \setminus X_0$, linearly independent, such that

(3.8) $\qquad\qquad X_0 + \text{span}\,\{x_1, \cdots, x_m\} = X.$

Clearly, we have

(3.9) $\quad x_i \notin X_0 + \text{span}\,\{x_1, \cdots, x_{i-1}, x_{i+1}, \cdots, x_m\} \equiv L_i, \quad 1 \leq i \leq m.$

Thus, by the Hahn–Banach Theorem, we can find $f_i \in X^*$, such that

(3.10) $\qquad f_i(x_i) = 1, \quad |f_i|_{X^*} = 1, \quad \mathcal{N}(f_i) \supseteq L_i, \quad 1 \leq i \leq m.$

## §3. Other Preliminary Results

Then we see that

$$(3.11) \qquad X_0 \subseteq \bigcap_{i=1}^m L_i \subseteq \bigcap_{i=1}^m \mathcal{N}(f_i).$$

On the other hand, for any $x \in \bigcap_{i=1}^m \mathcal{N}(f_i)$, by (3.8), we have

$$(3.12) \qquad x = x_0 + \sum_{i=1}^m \alpha_i x_i, \qquad x_0 \in X_0, \alpha_i \in \mathbb{R}.$$

Now, applying $f_j$ to the above, we see that

$$(3.13) \qquad 0 = f_j(x) = f_j(x_0) + \sum_{i=1}^m \alpha_i f_j(x_i) = \alpha_j, \qquad 1 \le j \le m.$$

Hence,

$$(3.14) \qquad x = x_0 \in X_0, \qquad \forall x \in \bigcap_{i=1}^m \mathcal{N}(f_i).$$

Combining (3.11) and (3.14), we obtain (3.6). This proves (ii).

(ii) $\Rightarrow$ (i). Without loss of generality, we may assume that $f_1, \cdots, f_m$ are linearly independent. By Corollary 1.35 of Chapter 2, there exist $x_1, x_2, \cdots, x_m \in X$, such that

$$(3.15) \qquad f_j(x_i) = \delta_{ij}, \qquad 1 \le i, j \le m.$$

Now, for any $x \in X$, we have the following decomposition:

$$(3.16) \qquad x = (x - \sum_{i=1}^m f_i(x)x_i) + \sum_{i=1}^m f_i(x)x_i.$$

Because $x - \sum_{i=1}^m f_i(x)x_i \in \bigcap_{i=1}^m \mathcal{N}(f_i) = X_0$, we obtain (i).

(i) $\Rightarrow$ (iii). Again, let $x_i$'s and $f_i$'s be as in the proof of (i) $\Rightarrow$ (ii). Then define

$$(3.17) \qquad K = \{\sum_{i=1}^m a_i x_i \mid a_i \in \mathbb{R}, |a_i| \le 1, \ 1 \le i \le m\}.$$

Clearly, $K$ is a compact set. Now, for any $x \in B_1(0) \equiv \{x \in X \mid |x| \le 1\}$, we have

$$(3.18) \qquad x = (x - \sum_{i=1}^m f_i(x)x_i) + \sum_{i=1}^m f_i(x)x_i \equiv x_0 + \sum_{i=1}^m a_i x_i,$$

with $x_0 = x - \sum_{i=1}^m f_i(x)x_i \in X_0$ (note the proved (ii)) and $|a_i| = |f_i(x)| \le |x| \le 1$. Thus,

$$(3.19) \qquad B_1(0) \subset X_0 + K.$$

Hence, (iii) follows.

(iii)⇒(i). Without loss of generality, we assume that (3.19) holds with some compact set $K$. Suppose $X_0$ is not finite codimensional in $X$. Then we can find $|x_j| = 1$, $j = 1, 2, \cdots$, linearly independent, such that $x_j \notin X_0$ for all $j \geq 1$. Then, by the Hahn–Banach Theorem, there exist $f_i \in X^*$, $i \geq 1$, such that

$$(3.20) \qquad f_i(x_j) = \delta_{ij}, \quad \mathcal{N}(f_i) \supset X_0, \quad i, j \geq 1.$$

On the other hand, from (3.19), we have

$$(3.21) \qquad x_j = x_j^0 + y_j, \quad x_j^0 \in X_0, \quad y_j \in K, \quad \forall j \geq 1.$$

Thus, it follows that

$$(3.22) \quad \begin{aligned} |y_i - y_j| &\geq f_i(y_i - y_j) = f_i(x_i - x_i^0 - x_j + x_j^0) \\ &= f_i(x_i - x_j) = f_i(x_i) = 1, \quad \forall i \neq j. \end{aligned}$$

This contradicts the compactness of $K$. Hence, (i) holds. □

**Corollary 3.3.** *Let $Q$ be a subset of some Banach space $X$. Then the following are true:*

*(i) If $\operatorname{codim}_X Q < \infty$, then there exists a compact set $K \subset X$, such that*

$$(3.23) \qquad \operatorname{Int}(\overline{\operatorname{co}} Q + K) \neq \phi.$$

*(ii) If (3.23) holds for some compact set $K \subset X$, then*

$$(3.24) \qquad \operatorname{codim}_X \operatorname{span}(Q - q) < \infty, \quad \forall q \in \overline{\operatorname{co}} Q.$$

*Proof.* Without loss of generality, we assume that $Q$ is convex and $0 \in Q$. Let $X_0 = \operatorname{span} Q$. Now, if $\operatorname{codim}_X Q < \infty$, by definition, we may assume, without loss of generality, that

$$(3.25) \qquad B_\delta(0) \bigcap X_0 \subset Q,$$

for some $0 < \delta < 1$. We define $x_i$'s and $f_i$'s as in the proof of (i) ⇒ (ii) of Proposition 3.2. Then set $K$ as in (3.17), which is compact, and moreover, similar to (3.18) and (3.19), we have $B_\delta(0) \subset Q + K$. This proves (i). Now, to prove (ii), we note that

$$(3.26) \qquad \operatorname{Int}(Q + K) \subset \operatorname{Int}(X_0 + K).$$

Then the conclusion follows from Proposition 3.2. □

We conjecture that under some mild conditions, $\operatorname{codim}_X Q < \infty$ if and only if there exists a compact set $K$ such that $\operatorname{Int}(Q + K) \neq \phi$. Proposition 3.2 says that this is true for the case where $Q$ is a subspace. We are not able to prove the general case at the present time.

## §3. Other Preliminary Results

**Proposition 3.4.** *Let $X$ be a Banach space and $S_1, S_2 \subset X$. Let $S_1$ be finite codimensional in $X$. Then, for any $\alpha \in \mathbb{R} \setminus \{0\}$ and $\beta \in \mathbb{R}$, the set*

$$(3.27) \qquad \alpha S_1 - \beta S_2 \equiv \{\alpha x_1 - \beta x_2 \mid x_1 \in S_1, \ x_2 \in S_2\}$$

*is finite codimensional in $X$.*

*Proof.* It is clear that $S_1$ is finite codimensional in $X$, if and only if for any $\alpha \in \mathbb{R}\setminus\{0\}$, the set $\alpha S_1$ is finite codimensional in $X$. Thus, the set $\alpha S_1 - \beta S_2$ is finite codimensional in $X$ if and only if $S_1 - \frac{\beta}{\alpha} S_2$ is so. Hence, we see that to prove our result, it suffices to assume $\alpha = \beta = 1$. Now, by Proposition 3.1, we have $y_1 \in \overline{co}\, S_1$, such that the subspace $X_1 \overset{\Delta}{=} \text{span}\,(S_1 - y_1)$ is finite codimensional and for some $\delta > 0$,

$$(3.28) \qquad B_\delta(0) \cap X_1 \subset \overline{co}(S_1 - y_1).$$

We let $y_2 \in S_2$. Then

$$(3.29) \qquad z_0 \overset{\Delta}{=} y_1 - y_2 \in \overline{co}\, S_1 - S_2 \subseteq \overline{co}(S_1 - S_2),$$

and

$$(3.30) \qquad \begin{aligned} X_0 &\overset{\Delta}{=} \text{span}\,(S_1 - S_2 - z_0) = \text{span}\,((S_1 - y_1) - (S_2 - y_2)) \\ &\supseteq \text{span}\,(S_1 - y_1) = X_1. \end{aligned}$$

Thus, $X_0$ is a finite codimensional subspace of $X$. Now, if $S_2 - y_2 \subset X_1$, by (3.28), we are done. Otherwise, we can find $x_1, x_2, \cdots, x_k \in (S_2 - y_2) \setminus X_1$, which are linearly independent such that

$$(3.31) \qquad X_0 = \text{span}\{X_1, x_1, x_2, \cdots, x_k\}.$$

Because $0 \in S_2 - y_2$, we see that

$$(3.32) \qquad \sum_{i=1}^k \lambda_i x_i \in \overline{co}(S_2 - y_2), \qquad \forall \lambda_i \geq 0, \ \sum_{i=1}^k \lambda_i \leq 1.$$

Thus, $\forall x_0 \in B_\delta(0) \cap X_1$, $\lambda_i \geq 0$, $\sum_{i=1}^k \lambda_i \leq 1$, $\lambda, \mu \geq 0$, $\lambda + \mu = 1$,

$$(3.33) \quad \lambda x_0 - \mu \sum_{i=1}^k \lambda_i x_i \in \lambda \overline{co}(S_1 - y_1) - \mu \overline{co}(S_2 - y_2) \subset \overline{co}(S_1 - S_2 - z_0).$$

Consequently, for some $\varepsilon > 0$, we have (with $\bar{x} = \frac{1}{2}\sum_{i=1}^k x_i$)

$$(3.34) \qquad B_\varepsilon(\bar{x}) \cap X_0 \subset \overline{co}(S_1 - S_2 - z_0).$$

This proves our result. $\square$

The above result implies that if one of $S_1$ and $S_2$ is finite codimensional in $X$, then so is $S_1 - S_2$. The next result will be used to link the finite dimensionality of some sets in the space $X$ and $X \times X$. This result will be used in the proof of our main result.

**Proposition 3.5.** *Let $X$ be a Banach space, $\mathcal{R} \subset X$, $S \subset X \times X$, and $G \in \mathcal{L}(X)$. For any $r > 0$, let*

(3.35)
$$\widetilde{\mathcal{R}}_r = \{ \begin{pmatrix} \eta \\ z \end{pmatrix} \in X \times X \mid z = G\eta + \xi, \ \xi \in \mathcal{R}, \ \eta \in B_r(0) \}$$
$$Q = \{ y_1 - Gy_0 \mid (y_0, y_1) \in S \}.$$

*Then $\mathcal{R} - Q$ is finite codimensional in $X$ if and only if $\widetilde{\mathcal{R}}_r - S$ is so in $X \times X$.*

*Proof.* We note that

(3.36)
$$\widetilde{\mathcal{R}}_r = \begin{pmatrix} I \\ G \end{pmatrix} B_r(0) + \begin{pmatrix} 0 \\ \mathcal{R} \end{pmatrix}, \qquad Q = (-G, I)S,$$

with $I : X \to X$ being the identity. Thus, we have

(3.37)
$$\begin{pmatrix} I & 0 \\ -G & I \end{pmatrix} (\widetilde{\mathcal{R}}_r - S) = \begin{pmatrix} B_r(0) - (I, 0)S \\ \mathcal{R} + (G, -I)S \end{pmatrix} = \begin{pmatrix} B_r(0) - (I, 0)S \\ \mathcal{R} - Q \end{pmatrix}.$$

Clearly, $\mathcal{R} - Q$ is finite codimensional in $X$ if and only if $B_r(0) \times (\mathcal{R} - Q)$ is finite codimensional in $X \times X$. As the operator $\begin{pmatrix} I & 0 \\ -G & I \end{pmatrix} : X \times X \to X \times X$ is invertible, it does not change the finite codimensionality of the sets in $X \times X$. Thus, our conclusion follows. □

As we said in Remark 1.8, the finite codimensionality of the set $\mathcal{R} - Q$ in $X$ guarantees the nontriviality of $(\psi^0, \psi(\cdot))$. This will be achieved essentially through the following result.

**Lemma 3.6.** *Let $Q$ be finite codimensional in $X$. Let $\{f_n\}_{n \geq 1} \subset X^*$ with*

(3.38)
$$|f_n|_{X^*} \geq \delta > 0, \qquad f_n \overset{*}{\rightharpoonup} f \in X^*,$$

*and*

(3.39)
$$\langle f_n, x \rangle \geq -\varepsilon_n, \qquad \forall x \in Q, \ n \geq 1,$$

*where $\varepsilon_n \to 0$. Then $f \neq 0$.*

*Proof.* First of all, it is clear that (3.39) holds for any $x \in \overline{\mathrm{co}}\, Q$. Thus, we may assume that $Q$ is convex. By Corollary 3.3, there exists a compact set $K$, such that (3.23) holds. Then, by translating the set $K$, if necessary, we may assume that there exists an $\alpha > 0$, such that

(3.40)
$$B_\alpha(0) \subset Q + K.$$

§3. Other Preliminary Results

Now, for any $|y| < \alpha$, by (3.40), we have $y = x + z$ for some $x \in Q$ and $z \in K$. Thus, by (3.39),

(3.41)  $\langle f_n, y \rangle = \langle f_n, x \rangle + \langle f_n, z \rangle \geq -\varepsilon_n + \langle f_n, z \rangle.$

Suppose $f = 0$, i.e., $f_n \xrightarrow{w} 0$. Then, note the compactness of $K$, we have

(3.42)  $0 < \delta \leq |f_n| \leq \dfrac{1}{\alpha}\left(\varepsilon_n + \sup_{z \in K}|\langle f_n, z \rangle|\right) \to 0, \quad n \to \infty.$

This is a contradiction. Thus $f \neq 0$. □

## §3.2. Preliminaries for spike perturbation

We note that the control domain $U$ is just a metric space and does not necessarily have any algebraic structure. Thus, we should not talk about the convexity of $U$. Consequently, in the derivation of optimality conditions, the control variation is restricted to a very special type. Let us describe it as follows: Let $u \in \mathcal{U}$ be a feasible control. Pick any other $v \in \mathcal{U}$ and a measurable set $E \subset [0,T]$; we define $u_E = v$ on the set $E$ and $u_E = u$ on the set $[0,T] \setminus E$. That is, we only change the values of $u$ on the set $E$. Clearly, $u_E \in \mathcal{U}$. We call $u_E$ a *spike perturbation* of $u$ (sometimes it is also called a *"needlelike"* perturbation. In this subsection, we present some preliminaries related to the spike perturbation of the controls. Let us recall that for a finite measure space $(\Omega, \mathcal{F}, \mu)$, we say that $\mu$ is *nonatomic* if for any $E \in \mathcal{F}$ and any $\rho \in (0,1)$, there exists a set $F \in \mathcal{F}$ with $F \subset E$ and $\mu(F) = \rho\mu(E)$. The following result is very crucial in sequel. It tells us how we can "approximate" the function 1 by the "oscillatory" function $\frac{1}{\rho}\chi_E$ for any given $\rho \in (0,1)$.

**Lemma 3.7.** *Let $X$ be a Banach space and $(\Omega, \mathcal{F}, \mu)$ be a finite measure space where $\mu$ is nonatomic. For any $\rho \in (0,1)$, let $\mathcal{E}_\rho = \{E \in \mathcal{F} \mid \mu(E) = \rho\mu(\Omega)\}$. Then, for any $h \in L^1(\Omega; X)$,*

(3.43)  $\displaystyle\inf_{E \in \mathcal{E}_\rho} \left| \int_\Omega (\dfrac{1}{\rho}\chi_E - 1)h \, d\mu \right| = 0.$

*Proof.* For any $\varepsilon > 0$, by the Bochner integrability of the function $h$, there exists a simple function $g = \sum_{k=1}^m a_k \chi_{F_k}$, with $a_k \in X$, $F_k \in \mathcal{F}$ ($1 \leq k \leq m$), $F_k$'s being mutually disjoint, and $\Omega = \bigcup_{k=1}^m F_k$, such that

(3.44)  $\displaystyle\int_\Omega |h - g| \, d\mu < \varepsilon.$

Because $\mu$ is nonatomic, for each $F_k$ there exists an $F_k^\rho \subset F_k$, such that

(3.45)  $\mu(F_k^\rho) = \rho\mu(F_k).$

We set $E = \bigcup_{k=1}^m F_k^\rho$. Clearly, $E \in \mathcal{E}_\rho$. On the other hand,

(3.46) $$\int_\Omega (\frac{1}{\rho}\chi_E - 1)g\, d\mu = \sum_{k=1}^m a_k[\frac{1}{\rho}\mu(F_k^\rho) - \mu(F_k)] = 0.$$

Hence,

(3.47) $$\left|\int_\Omega (\frac{1}{\rho}\chi_E - 1)h\, d\mu\right| \leq \left|\int_\Omega (\frac{1}{\rho}\chi_E - 1)g\, d\mu\right| + \int_\Omega |g - h|\, d\mu$$
$$+ \frac{1}{\rho}\int_E |g - h|\, d\mu < (1 + \frac{1}{\rho})\varepsilon.$$

Because $\varepsilon > 0$ is arbitrary, (3.43) follows. □

**Corollary 3.8.** *Let $X$ be a Banach space and for any $\rho \in (0,1)$, let $\mathcal{E}_\rho = \{E \subset [0,T] \mid |E| = \rho T\}$, where $|E|$ is the Lebesgue measure of $E$. Then, for any $h \in C([0,T]; L^1(0,T;X))$,*

(3.48) $$\inf_{E \in \mathcal{E}_\rho} \left|\int_0^T (\frac{1}{\rho}\chi_E(s) - 1)h(\cdot, s)\, ds\right|_{C[0,T]} = 0.$$

*Proof.* For any $\varepsilon > 0$, by the compactness of $[0,T]$ and the continuity of $h$, we can find a $\delta > 0$, such that

(3.49) $$\int_0^T |h(t,s) - h(\bar{t},s)|\, ds < \varepsilon, \qquad \forall |t - \bar{t}| \leq \delta,\ t, \bar{t} \in [0,T].$$

Let $0 = t_0 < t_1 < t_2 < \cdots < t_k = T$, $|t_i - t_{i-1}| < \delta$, and let

(3.50) $$g(s) = (h(t_0, s), h(t_1, s), \cdots, h(t_k, s)), \qquad s \in [0,T].$$

Then, $g(\cdot) \in L^1(0,T; X^{k+1})$. By Lemma 3.7, for any $\varepsilon > 0$, there exists an $E \in \mathcal{E}_\rho$, such that

(3.51) $$\left|\int_0^T (\frac{1}{\rho}\chi_E(s) - 1)g(s)\, ds\right| < \varepsilon.$$

Now, for any $t \in [0,T]$, we can find an $i$, with $|t - t_i| < \delta$. Then

(3.52) $$\left|\int_0^T (\frac{1}{\rho}\chi_E(s) - 1)h(t,s)\, ds\right| \leq \left|\int_0^T (\frac{1}{\rho}\chi_E(s) - 1)h(t_i, s)\, ds\right|$$
$$+ \int_0^T |h(t,s) - h(t_i, s)|\, ds + \frac{1}{\rho}\int_E |h(t,s) - h(t_i, s)|\, ds < (2 + \frac{1}{\rho})\varepsilon.$$

Then, (3.48) follows. □

**Corollary 3.9.** *Let $\overline{\Delta} = \{(t,s) \in [0,T] \times [0,T] \mid 0 \leq s \leq t \leq T\}$ and $X$ be a Banach space. Let $h: \overline{\Delta} \to X$ have the following properties: There exists a $\varphi(\cdot) \in L^1(0,T)$, such that*

(3.53) $$|h(t,s)| \leq \varphi(s), \qquad \forall t \in [0,T],\ \text{a.e. } s \in [0,t],$$

## §3. Other Preliminary Results

and for almost all $s \in [0,T]$, $h(\cdot,s) : [s,T] \to X$ is continuous. Then, for any $\rho \in (0,1)$, there exists an $E_\rho \subset [0,T]$, such that $|E_\rho| = \rho T$ and

$$(3.54) \qquad \sup_{t \in [0,T]} \left| \rho \int_0^t h(t,s)\,ds - \int_{E_\rho \cap [0,t]} h(t,s)\,ds \right| = o(\rho).$$

*Proof.* Let

$$(3.55) \qquad g(t,s) = \begin{cases} h(t,s), & 0 \le s \le t \le T, \\ 0, & 0 \le t < s \le T. \end{cases}$$

Then, for any $t, \bar{t} \in [0,T]$, we have

$$(3.56) \qquad \int_0^T |g(t,s) - g(\bar{t},s)|\,ds \le \int_0^{t \wedge \bar{t}} |h(t,s) - h(\bar{t},s)|\,ds + \int_{t \wedge \bar{t}}^{t \vee \bar{t}} \varphi(s)\,ds.$$

Thus, $g(\cdot,\cdot) \in C([0,T]; L^1(0,T;X))$ and Corollary 3.8 applies. □

Next, let us introduce the following:

$$(3.57) \qquad \bar{d}(u(\cdot), v(\cdot)) = |\{t \in [0,T] \mid u(t) \neq v(t)\}|, \quad \forall u(\cdot), v(\cdot) \in \mathcal{U}[0,T].$$

It is not hard to show that $\bar{d}$ is a metric. Thus, $(\mathcal{U}[0,T], \bar{d})$ is a metric space. We will apply the Ekeland variational principle to some functionals defined on this space. Thus, we need the completeness of this space, which is provided by the following result.

**Proposition 3.10.** *Let $U$ be a measurable space. Then $(\mathcal{U}[0,T], \bar{d})$ is a complete metric space.*

*Proof.* Let $\{u_n(\cdot)\}$ be a Cauchy sequence in $\mathcal{U}[0,T]$, i.e.,

$$(3.58) \qquad \bar{d}(u_n(\cdot), u_m(\cdot)) \equiv |\{t \in [0,T] \mid u_n(t) \neq u_m(t)\}| \to 0, \quad n, m \to \infty.$$

Then there exists a subsequence $\{u_{n_k}(\cdot)\}$, such that

$$(3.59) \qquad \bar{d}(u_{n_k}(\cdot), u_{n_{k+1}}(\cdot)) \le 2^{-k}, \quad k \ge 1.$$

Let

$$(3.60) \qquad \begin{cases} E_{nm} = \{t \in [0,T] \mid u_n(t) \neq u_m(t)\}, & n, m \ge 1, \\ A_k = \bigcup_{p \ge k} E_{n_p, n_{p+1}}, & k \ge 1. \end{cases}$$

We see that $A_k \supseteq A_{k+1}$, $\forall k \ge 1$, and

$$(3.61) \qquad |A_k| \le \sum_{p=k}^{\infty} 2^{-p} = 2^{1-k}, \quad k \ge 1.$$

Consequently, $\left| \bigcup_{k \ge 1} A_k^c \right| = T$. Now, we define

$$(3.62) \qquad \bar{u}(t) = u_{n_k}(t), \quad t \in A_k^c, \quad k \ge 1.$$

From the definition of $A_k$, we see that $\bar{u}(\cdot)$ is well defined and $\bar{u}(\cdot) \in \mathcal{U}[0,T]$. Moreover,

(3.63) $$\bar{d}(u_{n_k}(\cdot), \bar{u}(\cdot)) \leq |A_k| \leq 2^{1-k} \to 0.$$

Therefore, we have $\bar{d}(u_n(\cdot), \bar{u}(\cdot)) \to 0$, which proves the completeness of $\mathcal{U}[0,T]$. □

It should be pointed out that we only require $U$ to be a measurable space. In particular, the above result applies to the case where $U$ is a metric space (not necessarily complete) with the $\sigma$-field generated by all open sets. Also, $[0,T]$ can be replaced by any measure space with a nonatomic measure; in particular, we can replace $[0,T]$ by any domain $\Omega \subset \mathbb{R}^n$ with the Lebesgue measure.

## §3.3. The distance function

Let $Q$ be a convex and closed subset in some Banach space $Z$. Define

(3.64) $$d_Q(z) = \inf_{z' \in Q} |z - z'|_Z.$$

We call $d_Q$ the *distance function* (to the set $Q$). In proving our main result, this function will play an important role. In this subsection, we are going to prove some relevant results about the distance function, which will be useful later.

To begin with, let us first observe that by (3.64), it can be shown that $d_Q(\cdot)$ is a convex function, i.e., $d_Q(\lambda x + (1-\lambda)y) \leq \lambda d_Q(x) + (1-\lambda)d_Q(y)$, for all $\lambda \in [0,1]$ and $x, y \in Z$. Also, it is Lipschitz continuous with the Lipschitz constant being 1:

(3.65) $$|d_Q(z) - d_Q(z')| \leq |z - z'|_Z.$$

We define the *subdifferential* of the function $d_Q(\cdot)$ as follows:

(3.66) $$\partial d_Q(z) = \{\zeta \in Z^* \mid d_Q(z') - d_Q(z) \geq \langle \zeta, z' - z \rangle, \quad \forall z' \in Z\}.$$

The following proposition collects interesting properties of $\partial d_Q(z)$.

**Proposition 3.11.** *Let $Q$ be a convex and closed subset in some Banach space $Z$. Then*

(i) *The set $\partial d_Q(z)$ is nonempty, convex, and weak\*-compact in $Z^*$. For any $z, \xi \in Z$, the limit*

(3.67) $$\lim_{\rho \downarrow 0} \frac{d_Q(z + \rho\xi) - d_Q(z)}{\rho} \triangleq d_Q^0(z; \xi) = \max\{\langle \zeta, \xi \rangle \mid \zeta \in \partial d_Q(z)\},$$

*exists, which is positively homogeneous and subadditive in $\xi$. Moreover, $\partial d_Q(z)$ can be represented by*

(3.68) $$\partial d_Q(z) = \{\zeta \in Z^* \mid \langle \zeta, \xi \rangle \leq d_Q^0(z; \xi), \quad \forall \xi \in Z \}.$$

## §3. Other Preliminary Results

Further, if $z_\alpha \in Z$ and $\zeta_\alpha \in Z^*$ are two nets with

(3.69) $$z_\alpha \xrightarrow{s} z, \quad \zeta_\alpha \xrightarrow{*} \zeta, \quad \zeta_\alpha \in \partial d_Q(z_\alpha),$$

then $\zeta \in \partial d_Q(z)$. Consequently, the map $z \mapsto \partial d_Q(z)$ is pseudo-continuous on $Z$.

(ii) For any $z \notin Q$,

(3.70) $$|\zeta|_{Z^*} = 1, \quad \forall \zeta \in \partial d_Q(z).$$

*Proof.* (i) First of all, for any $z, \xi \in Z$ and $0 < \rho_1 < \rho_2$, by the convexity of the function $d_Q(\cdot)$, we have

(3.71) $$d_Q(z + \rho_1 \xi) \le \frac{\rho_1}{\rho_2} d_Q(z + \rho_2 \xi) + \frac{\rho_2 - \rho_1}{\rho_2} d_Q(z),$$

which implies

(3.72) $$\frac{d_Q(z + \rho_1 \xi) - d_Q(z)}{\rho_1} \le \frac{d_Q(z + \rho_2 \xi) - d_Q(z)}{\rho_2}.$$

That means that the map $\rho \mapsto \frac{d_Q(z+\rho\xi)-d_Q(z)}{\rho}$ is nondecreasing and bounded by $|\xi|$ (see (3.65)). Thus, the limit

(3.73) $$\lim_{\rho \downarrow 0} \frac{d_Q(z + \rho \xi) - d_Q(z)}{\rho} \triangleq d_Q^0(z; \xi)$$

exists. It is clear that

(3.74) $$d_Q^0(z; \alpha \xi) = \alpha d_Q^0(z; \xi), \quad \forall \alpha \ge 0, \ z, \xi \in Z.$$

Also, for any $\xi_1, \xi_2, z \in Z$ and $0 \le \beta \le \bar\beta$, by (3.72), we have

(3.75) $$\frac{d_Q(z + \beta \xi_1 + \beta \xi_2) - d_Q(z + \beta \xi_1)}{\beta}$$
$$\le \frac{d_Q(z + \beta \xi_1 + \bar\beta \xi_2) - d_Q(z + \beta \xi_1)}{\bar\beta}.$$

Thus, for $0 < \rho < 1$, $\beta = \rho^2$, and $\bar\beta = \rho$, we obtain (note (3.65))

(3.76) $$\begin{aligned}&\frac{d_Q(z + \rho^2 \xi_1 + \rho^2 \xi_2) - d_Q(z + \rho^2 \xi_1)}{\rho^2} \\ &\le \frac{d_Q(z + \rho^2 \xi_1 + \rho \xi_2) - d_Q(z + \rho^2 \xi_1)}{\rho} \\ &= \frac{d_Q(z + \rho \xi_2) - d_Q(z)}{\rho} + \frac{d_Q(z + \rho^2 \xi_1 + \rho \xi_2) - d_Q(z + \rho \xi_2)}{\rho} \\ &\quad + \frac{d_Q(z) - d_Q(z + \rho^2 \xi_1)}{\rho} \\ &\le \frac{d_Q(z + \rho \xi_2) - d_Q(z)}{\rho} + 2\rho |\xi_1|.\end{aligned}$$

Hence,

$$
\begin{aligned}
d_Q^0(z;\xi_1+\xi_2) &= \lim_{\rho\downarrow 0}\frac{d_Q(z+\rho\xi_1+\rho\xi_2)-d_Q(z)}{\rho} \\
&= \lim_{\rho\downarrow 0}\Big\{\frac{d_Q(z+\rho\xi_1+\rho\xi_2)-d_Q(z+\rho\xi_1)}{\rho} \\
&\quad + \frac{d_Q(z+\rho\xi_1)-d_Q(z)}{\rho}\Big\} \\
&= \lim_{\rho\downarrow 0}\frac{d_Q(z+\rho^2\xi_1+\rho^2\xi_2)-d_Q(z+\rho^2\xi_1)}{\rho^2} + d_Q^0(z;\xi_1) \\
&\le \lim_{\rho\downarrow 0}\Big\{\frac{d_Q(z+\rho\xi_2)-d_Q(z)}{\rho}+2\rho|\xi_1|\Big\} + d_Q^0(z;\xi_1) \\
&= d_Q^0(z;\xi_1) + d_Q^0(z;\xi_2).
\end{aligned}
\tag{3.77}
$$

From (3.74) and (3.77), we see that for a given $z\in Z$, the map $d_Q^0(z;\cdot)$ is positively homogeneous and subadditive. In particular, we have

$$0 = d_Q^0(z;\xi-\xi) \le d_Q^0(z;\xi) + d_Q^0(z;-\xi), \tag{3.78}$$

which gives

$$-d_Q^0(z;\xi) \le d_Q^0(z;-\xi), \qquad \forall z,\xi\in Z. \tag{3.79}$$

Now, let us prove (3.68). To this end, let $\zeta\in\partial d_Q(z)$. By the definition (see (3.66))

$$\langle\zeta,\xi\rangle \le \frac{d_Q(z+\rho\xi)-d_Q(z)}{\rho}, \qquad \forall \rho\in\mathbb{R},\ \xi\in Z, \tag{3.80}$$

which gives (by letting $\rho\downarrow 0$) $\langle\zeta,\xi\rangle\le d_Q^0(z;\xi)$, $\forall \xi\in Z$. This means that

$$\partial d_Q(z) \subseteq \{\zeta\in Z^* \mid \langle\zeta,\xi\rangle\le d_Q^0(z;\xi),\ \forall\xi\in Z\}. \tag{3.81}$$

Conversely, let $\zeta\in Z^*$ with the property that

$$\langle\zeta,\xi\rangle \le d_Q^0(z;\xi), \qquad \forall\xi\in Z. \tag{3.82}$$

Hence, by (3.73), we have

$$\langle\zeta,\xi\rangle \le d_Q^0(z;\xi) \le \lim_{\rho\downarrow 0}\frac{d_Q(z+\rho\xi)-d_Q(z)}{\rho} \le d_Q(z+\xi)-d_Q(z). \tag{3.83}$$

This implies $\zeta\in\partial d_Q(z)$. Thus, (3.68) holds. Next, we prove the nonemptiness of $\partial d_Q(z)$ and the second equality in (3.67). For given $z,\xi\in Z$, we define

$$f(\alpha\xi) = \alpha d_Q^0(z;\xi), \qquad \forall\alpha\in\mathbb{R}. \tag{3.84}$$

§3. Other Preliminary Results

Then, trivially, $f$ is a linear bounded functional defined on $Z_0 \equiv \text{span}\{\xi\}$. By (3.74) and (3.79), we see that

(3.85) $$f(z_0) \leq d_Q^0(z; z_0), \qquad \forall z_0 \in Z_0.$$

Because $d_Q^0(z; \cdot)$ is positively homogeneous and subadditive, by the Hahn–Banach Theorem (Chapter 2, Theorem 1.14), there exists a $\zeta_0 \in Z^*$ such that

(3.86) $$\begin{cases} \langle \zeta_0, \alpha \xi \rangle = f(\alpha \xi) \equiv \alpha d_Q^0(z; \xi), & \forall \alpha \in \mathbb{R}, \\ \langle \zeta_0, y \rangle \leq d_Q^0(z; y), & \forall y \in Z. \end{cases}$$

By the proved (3.68), we have $\zeta_0 \in \partial d_Q(z)$. This proves the nonemptiness of $\partial d_Q(z)$ and

(3.87) $$d_Q^0(z; \xi) = \langle \zeta_0, \xi \rangle = \max_{\zeta \in \partial d_Q(z)} \langle \zeta, \xi \rangle.$$

This completes the proof of (3.67).

The convexity and the weak*-closeness of the set $\partial d_Q(z)$ are clear from (3.66). On the other hand, from (3.65), we see that for any $\zeta \in \partial d_Q(z)$,

(3.88) $$\langle \zeta, \xi \rangle \leq d_Q^0(z; \xi) \leq |\xi|_Z, \qquad \forall \xi \in Z.$$

Thus, it follows that

(3.89) $$|\zeta|_{Z^*} \leq 1, \qquad \forall \zeta \in \partial d_Q(z).$$

Then, by the Alaoglu Theorem (Chapter 2, §1.3), $\partial d_Q(z)$ is weak*-compact. Next, from (3.66), we easily see that if (3.69) holds, then $\zeta \in \partial d_Q(z)$. This implies that the graph of the multifunction $\partial d_Q(\cdot)$ is closed. Also, the weak*-compactness of the set $\partial d_Q(z)$ implies its closeness for any $z \in Z$. Thus, by Chapter 3, Proposition 2.5, we have the pseudo-continuity of $\partial d_Q(\cdot)$.

(ii) Now, for $z \notin Q$ and any $0 < \delta < 1$, there exists a $q_\delta \in Q$, such that

(3.90) $$d_Q(z) \geq (1 - \delta)|z - q_\delta|_Z > 0.$$

Thus, for any $\zeta \in \partial d_Q(z)$, by (3.66), we have (note $d_Q(q_\delta) = 0$)

(3.91) $$-d_Q(z) \geq \langle \zeta, q_\delta - z \rangle.$$

Combining (3.90) and (3.91), we obtain

(3.92) $$0 < (1 - \delta)|z - q_\delta|_Z \leq d_Q(z) \leq -\langle \zeta, q_\delta - z \rangle \leq |\zeta|_{Z^*}|z - q_\delta|_Z.$$

Hence, $|\zeta|_{Z^*} \geq 1 - \delta$. By sending $\delta \to 0$, (3.70) follows from this and (3.89). □

**Corollary 3.12.** *Let $Z$ be a Banach space with strictly convex dual $Z^*$. Let $Q$ be a convex and closed subset of $Z$. Then, for any $z \notin Q$, the set $\partial d_Q(z)$ consists of one point.*

*Proof.* By Proposition 3.11, we know that $\partial d_Q(z)$ is convex and any element $\zeta \in \partial d_Q(z)$ satisfies (3.70). Thus, by the strict convexity of the space $Z^*$, we must have that $\partial d_Q(z)$ is a singleton. □

## §4. Proof of the Maximum Principle

This section is devoted to the proof of the maximum principle stated in §1 (Theorem 1.6). The main steps are the following: (i) Define a penalty functional, via which for convenience, we will transform the original problem with endpoint constraint to another one called the approximate problem, which has no endpoint constraint. (ii) Apply the Ekeland variational principle to find an optimal pair for the approximate problem; this pair turns out to be close to the original optimal pair. (iii) Use the spike variation technique to derive the (first order) necessary conditions for the optimal pair of the approximate problem. (iv) Pass to the limit to obtain the necessary conditions for the original optimal pair, the maximum principle.

Let us first introduce some notation. For all $(x, u(\cdot)), (\widehat{x}, \widehat{u}(\cdot)) \in X \times \mathcal{U}[0, T]$, we define

$$(4.1) \qquad \widetilde{d}((x, u(\cdot)), (\widehat{x}, \widehat{u}(\cdot))) = |x - \widehat{x}|_X + \bar{d}(u(\cdot), \widehat{u}(\cdot)),$$

where $\bar{d}$ is the distance introduced in §3 (see (3.57)). Then, by Proposition 3.10, we can easily show that $(X \times \mathcal{U}[0,T], \widetilde{d})$ is a complete metric space. Next, we know that under (H1) and (H2), for any $(y_0, u(\cdot)) \in X \times \mathcal{U}[0,T]$, (1.2) admits a unique solution $y(\cdot)$ with $y(0) = y_0$ and the cost functional (1.5) is also uniquely determined by $(y_0, u(\cdot))$. In sequel, we denote the unique solution of (1.2) with $y(0) = y_0$ by $y(\cdot; y_0, u(\cdot))$ and the corresponding cost functional by $J(y_0, u(\cdot))$. Now, let $(\bar{y}(\cdot), \bar{u}(\cdot)) \in \mathcal{A}_{ad}$ be the optimal pair given in Theorem 1.6, for which we want to derive the necessary conditions. We denote $\bar{y}(0) = \bar{y}_0$. Without loss of generality, we may assume that $J(\bar{y}_0, \bar{u}(\cdot)) = 0$; otherwise, we may consider the optimal control problem with a cost functional of $J(y_0, u(\cdot)) - J(\bar{y}_0, \bar{u}(\cdot))$.

*Step 1.* Introduction of the penalty functional.

For any $\varepsilon > 0$, we define

$$(4.2) \quad \begin{aligned} J_\varepsilon(y_0, u(\cdot)) &= \left\{ d_S^2(y_0, y(T; y_0, u(\cdot))) + \left[ (J(y_0, u(\cdot)) + \varepsilon)^+ \right]^2 \right\}^{1/2}, \\ &\forall (y_0, u(\cdot)) \in X \times \mathcal{U}[0, T], \end{aligned}$$

where

$$(4.3) \quad \begin{aligned} d_S(y_0, y_1) = d((y_0, y_1), S) &\equiv \inf_{(x_0, x_1) \in S} \left\{ |y_0 - x_0|^2 + |y_1 - x_1|^2 \right\}^{1/2}, \\ &\forall (y_0, y_1) \in X \times X. \end{aligned}$$

## §4. Proof of the Maximum Principle

**Lemma 4.1.** *Let (H1) and (H2) hold. Then there exists a constant $C > 0$, such that for all $(y_0, u(\cdot)), (\widehat{y}_0, \widehat{u}(\cdot)) \in X \times \mathcal{U}[0,T]$,*

(4.4)
$$\begin{cases} \sup_{t\in[0,T]} |y(t; y_0, u(\cdot)) - y(t; \widehat{y}_0, \widehat{u}(\cdot))| \\ \qquad \leq C(1 + |y_0| \vee |\widehat{y}_0|) \widetilde{d}\big((y_0, u(\cdot)), (\widehat{y}_0, \widehat{u}(\cdot))\big), \\ |J(y_0, u(\cdot)) - J(\widehat{y}_0, \widehat{u}(\cdot))| \\ \qquad \leq C(1 + |y_0| \vee |\widehat{y}_0|) \widetilde{d}\big((y_0, u(\cdot)), (\widehat{y}_0, \widehat{u}(\cdot))\big). \end{cases}$$

*Proof.* Denote $y(\cdot) = y(\cdot; y_0, u(\cdot))$ and $\widehat{y}(\cdot) = y(\cdot; \widehat{y}_0, \widehat{u}(\cdot))$. By (1.1) and (1.2) and Gronwall's inequality, we have

(4.5) $\quad |y(t)| \leq C(1 + |y_0|), \quad |\widehat{y}(t)| \leq C(1 + |\widehat{y}_0|), \quad \forall t \in [0,T],$

where the constant $C$ is independent of controls $u(\cdot)$ and $\widehat{u}(\cdot)$. Then it follows that (the constant $C$ below could be different at different places)

(4.6)
$$\begin{aligned} |y(t) - \widehat{y}(t)| &\leq C|y_0 - \widehat{y}_0| + C\int_0^t |y(s) - \widehat{y}(s)|\, ds \\ &\quad + \int_0^t |f(s, y(s), u(s)) - f(s, y(s), \widehat{u}(s))|\, ds \\ &\leq C|y_0 - \widehat{y}_0| + C(1 + |y_0|)\bar{d}(u(\cdot), \widehat{u}(\cdot)) + C\int_0^t |y(s) - \widehat{y}(s)|\, ds. \end{aligned}$$

Thus, by Gronwall's inequality, we obtain the first inequality in (4.4). The second inequality can be proved similarly. □

**Corollary 4.2.** *The functional $J_\varepsilon(y_0, u(\cdot))$ is continuous on the space $(X \times \mathcal{U}[0,T], \widetilde{d})$.*

The proof is clear from the definition of $J_\varepsilon(y_0, u(\cdot))$ (see (4.2)) and Lemma 4.1.

*Step 2.* Application of Ekeland variational principle.

By the definition of $J_\varepsilon(y_0, u(\cdot))$, we see that

(4.7) $\quad \begin{cases} J_\varepsilon(y_0, u(\cdot)) > 0, \quad \forall (y_0, u(\cdot)) \in X \times \mathcal{U}[0,T], \\ J_\varepsilon(\bar{y}_0, \bar{u}(\cdot)) = \varepsilon \leq \inf_{X \times \mathcal{U}[0,T]} J_\varepsilon(y_0, u(\cdot)) + \varepsilon. \end{cases}$

Thus, by Corollary 2.2, there exists a pair $(y_0^\varepsilon, u^\varepsilon(\cdot)) \in X \times \mathcal{U}[0,T]$, such that

(4.8) $\qquad \widetilde{d}\big((x_0^\varepsilon, u^\varepsilon(\cdot)), (\bar{x}_0, \bar{u}(\cdot))\big) \leq \sqrt{\varepsilon},$

(4.9) $\quad \begin{aligned} J_\varepsilon(y_0, u(\cdot)) + \sqrt{\varepsilon}\widetilde{d}\big((y_0, u(\cdot)), (y_0^\varepsilon, u^\varepsilon(\cdot))\big) &\geq J_\varepsilon(y_0^\varepsilon, u^\varepsilon(\cdot)), \\ &\forall (y_0, u(\cdot)) \in X \times \mathcal{U}[0,T]. \end{aligned}$

152    Chapter 4. Necessary Conditions for Evolution Equations

The above implies that if we let $y^\varepsilon(\cdot) = y(\cdot\,; y_0^\varepsilon, u^\varepsilon(\cdot))$, then $(y^\varepsilon(\cdot), u^\varepsilon(\cdot))$ is an optimal pair for the problem where the state equation is (1.2) and the cost functional is the left-hand side of (4.9).

*Step 3.* Derivation of the necessary conditions for $(y^\varepsilon(\cdot), u^\varepsilon(\cdot))$.

We fix any $(\eta, v(\cdot)) \in B_1(0) \times \mathcal{U}[0,T]$ and $\rho \in (0, 1]$. Let

$$(4.10) \qquad h(t,s) = \begin{pmatrix} f^0(s, y^\varepsilon(s), v(s)) - f^0(s, y^\varepsilon(s), u^\varepsilon(s)) \\ e^{A(t-s)}[f(s, y^\varepsilon(s), v(s)) - f(s, y^\varepsilon(s), u^\varepsilon(s))] \end{pmatrix}.$$

Then, by Corollary 3.9, there exists a set $E_\rho \subset [0, T]$, with $|E_\rho| = \rho T$, such that

$$(4.11) \qquad \sup_{t \in [0,T]} \left| \int_0^t h(t,s)\,ds - \frac{1}{\rho} \int_0^t \chi_{E_\rho}(s) h(t,s)\,ds \right| = o(1).$$

Next, we define

$$(4.12) \qquad u_\rho^\varepsilon(t) = \begin{cases} u^\varepsilon(t), & t \in [0, T] \setminus E_\rho, \\ v(t), & t \in E_\rho. \end{cases}$$

It is clear that $u_\rho^\varepsilon(\cdot) \in \mathcal{U}[0,T]$ and

$$(4.13) \qquad \bar{d}(u_\rho^\varepsilon(\cdot), u^\varepsilon(\cdot)) \le |E_\rho| = \rho T.$$

We let $y_\rho^\varepsilon(\cdot) = y(\cdot\,; y_0^\varepsilon + \rho\eta, u_\rho^\varepsilon(\cdot))$ and recall that $y^\varepsilon(\cdot) = y(\cdot\,; y_0^\varepsilon, u^\varepsilon(\cdot))$. The following lemma gives a sort of Taylor expansion of $y_\rho^\varepsilon(\cdot)$ and $J(y_0^\varepsilon + \rho\eta, u_\rho^\varepsilon(\cdot))$ with respect to $\rho$ at $\rho = 0$. This is very important in deriving the necessary conditions for the pair $(y^\varepsilon(\cdot), u^\varepsilon(\cdot))$.

**Lemma 4.3.** *It holds that*

$$(4.14) \qquad \sup_{t \in [0,T]} |y_\rho^\varepsilon(t) - y^\varepsilon(t) - \rho z_\varepsilon(t)| = o(\rho),$$

$$(4.15) \qquad |J(y_0^\varepsilon + \rho\eta, u_\rho^\varepsilon(\cdot)) - J(y_0^\varepsilon, u^\varepsilon(\cdot)) - \rho z_\varepsilon^0| = o(\rho),$$

*where*

$$(4.16) \qquad \begin{cases} z_\varepsilon(t) = e^{At}\eta + \int_0^t e^{A(t-s)} f_y(s, y^\varepsilon(s), u^\varepsilon(s)) z_\varepsilon(s)\,ds \\ \qquad + \int_0^t e^{A(t-s)}[f(s, y^\varepsilon(s), v(s)) - f(s, y^\varepsilon(s), u^\varepsilon(s))]\,ds, \\ \hfill t \in [0, T], \\ z_\varepsilon^0 = \int_0^T f_y^0(s, y^\varepsilon(s), u^\varepsilon(s)) z_\varepsilon(s)\,ds \\ \qquad + \int_0^T [f^0(s, y^\varepsilon(s), v(s)) - f^0(s, y^\varepsilon(s), u^\varepsilon(s))]\,ds. \end{cases}$$

## §4. Proof of the Maximum Principle

*Proof.* Set $z_\rho^\varepsilon(t) = \frac{1}{\rho}(y_\rho^\varepsilon(t) - y^\varepsilon(t))$. Then, by (4.11), we have
(4.17)
$$z_\rho^\varepsilon(t) = e^{At}\eta + \int_0^t e^{A(t-s)} \frac{f(s, y_\rho^\varepsilon(s), u_\rho^\varepsilon(s)) - f(s, y^\varepsilon(s), u_\rho^\varepsilon(s))}{\rho} ds$$
$$+ \int_{E_\rho \cap [0,t]} e^{A(t-s)} \frac{f(s, y^\varepsilon(s), v(s)) - f(s, y^\varepsilon(s), u^\varepsilon(s))}{\rho} ds$$
$$= e^{At}\eta + \int_0^t e^{A(t-s)} \left[ \int_0^1 f_y(s, (1-\sigma)y^\varepsilon(s) + \sigma y_\rho^\varepsilon(s), u_\rho^\varepsilon(s)) \, d\sigma \right] z_\rho^\varepsilon(s) \, ds$$
$$+ \int_0^t e^{A(t-s)} \Big[ f(s, y^\varepsilon(s), v(s)) - f(s, y^\varepsilon(s), u^\varepsilon(s)) \Big] ds + o(1).$$

Hence, we obtain (recall (4.16))

(4.18)
$$|z_\rho^\varepsilon(t) - z_\varepsilon(t)| \leq C \int_0^t |z_\rho^\varepsilon(s) - z_\varepsilon(s)| \, ds$$
$$+ C \int_0^t \left| \left[ \int_0^1 f_y(s, (1-\sigma)y^\varepsilon(s) + \sigma y_\rho^\varepsilon(s), u_\rho^\varepsilon(s)) \, d\sigma \right. \right.$$
$$\left. \left. - f_y(s, y^\varepsilon(s), u^\varepsilon(s)) \right] z_\varepsilon(s) \right| \, ds + o(1)$$
$$\leq C \int_0^t |z_\rho^\varepsilon(s) - z_\varepsilon(s)| \, ds + o(1).$$

Here, we have used (4.13) and the fact that $|y_\rho^\varepsilon(s) - y^\varepsilon(s)| \leq C\rho$ (see Lemma 4.1). Then, by Gronwall's inequality, we obtain

(4.19)
$$\sup_{t \in [0,T]} |z_\rho^\varepsilon(t) - z_\varepsilon(t)| = o(1).$$

This proves (4.14). Now we prove (4.15). It is simpler.
(4.20)
$$\frac{1}{\rho} |J(y_0^\varepsilon + \rho\eta, u_\rho^\varepsilon(\cdot)) - J(y_0^\varepsilon, u^\varepsilon(\cdot)) - \rho z_\varepsilon^0|$$
$$\leq \int_0^T \left| \int_0^1 f_y^0(s, (1-\sigma)y^\varepsilon(s) + \sigma y_\rho^\varepsilon(s), u_\rho^\varepsilon(s)) \, d\sigma \right.$$
$$\left. - f_y^0(s, y^\varepsilon(s), u^\varepsilon(s)) \right| |z_\varepsilon(s)| \, ds$$
$$+ \int_0^T |f_y^0(s, y^\varepsilon(s), u^\varepsilon(s))| \, |z_\rho^\varepsilon(s) - z_\varepsilon(s)| \, ds$$
$$+ \left| \int_0^T (\frac{1}{\rho}\chi_{E_\rho}(s) - 1)\Big[ f^0(s, y^\varepsilon, v(s)) - f^0(s, y^\varepsilon(s), u^\varepsilon(s)) \Big] ds \right|$$
$$= o(1).$$

This proves (4.15). □

From Corollary 4.2 and Lemma 4.3, we see that $J_\varepsilon(y_0^\varepsilon + \rho\eta, u_\rho^\varepsilon(\cdot)) = J_\varepsilon(y_0^\varepsilon, u^\varepsilon(\cdot)) + o(1)$, as $\rho \to 0$. Furthermore, by (4.9),

$$
\begin{aligned}
-\sqrt{\varepsilon}(|\eta|+T) &\leq \frac{1}{\rho}\Big(J_\varepsilon(y_0^\varepsilon+\rho\eta, u_\rho^\varepsilon(\cdot)) - J_\varepsilon(y_0^\varepsilon, u^\varepsilon(\cdot))\Big) \\
&= \frac{1}{2J_\varepsilon(y_0^\varepsilon, u^\varepsilon(\cdot)) + o(1)}\Big\{\frac{1}{\rho}[d_S^2(y_0^\varepsilon+\rho\eta, y_\rho^\varepsilon(T)) - d_S^2(y_0^\varepsilon, y^\varepsilon(T))] \\
&\quad + \frac{1}{\rho}\big\{[(J(y_0^\varepsilon+\rho\eta, u_\rho^\varepsilon(\cdot))+\varepsilon)^+]^2 - [(J(y_0^\varepsilon, u^\varepsilon(\cdot))+\varepsilon)^+]^2\big\}\Big\}.
\end{aligned}
$$
(4.21)

We note that the map $(y_0, y_1) \mapsto d_S^2(y_0, y_1)$ is continuously Fréchet differentiable on $X \times X$ with the Fréchet derivative

$$
Dd_S^2(y_0, y_1) = \begin{cases} 2d_S(y_0,y_1)(a,b), & \{(a,b)\} = \partial d_S(y_0, y_1), \\ & \text{if } (y_0, y_1) \notin S, \\ 0, & \text{if } (y_0, y_1) \in S. \end{cases}
$$
(4.22)

Here, we should note that because $X^*$ is strictly convex, by Proposition 3.11 and Corollary 3.12, as long as $(y_0, y_1) \notin S$, $\partial d_S(y_0, y_1)$ consists of one element with norm 1. Also, the set $\partial d_S(y_0, y_1)$ is always contained in the unit ball of $X \times X$. Thus, it will be no ambiguity to write

$$
\begin{cases} Dd_S^2(y_0, y_1) = 2d_S(y_0, y_1)(a, b), \\ (a,b) \in \partial d_S(y_0, y_1), \qquad |a|_{X^*}^2 + |b|_{X^*}^2 = 1. \end{cases}
$$
(4.23)

Then, by Lemma 4.3, we have

$$
\lim_{\rho \to 0} \frac{d_S^2(y_0^\varepsilon + \rho\eta, y_\rho^\varepsilon(T)) - d_S^2(y_0^\varepsilon, y^\varepsilon(T))}{\rho} \\
= 2d_S(y_0^\varepsilon, y^\varepsilon(T))(\langle a_\varepsilon, \eta\rangle + \langle b_\varepsilon, z_\varepsilon(T)\rangle),
$$
(4.24)

with $(a_\varepsilon, b_\varepsilon) \in \partial d_S(y_0^\varepsilon, y^\varepsilon(T))$ and

$$
|a_\varepsilon|_{X^*}^2 + |b_\varepsilon|_{X^*}^2 = 1.
$$
(4.25)

Similarly, we have

$$
\lim_{\rho \to 0} \frac{1}{\rho}\Big\{[(J(y_0^\varepsilon+\rho\eta, u_\rho^\varepsilon(\cdot))+\varepsilon)^+]^2 - [(J(y_0^\varepsilon, u^\varepsilon(\cdot))+\varepsilon)^+]^2\Big\} \\
= 2(J(y_0^\varepsilon, u^\varepsilon(\cdot))+\varepsilon)^+ z_\varepsilon^0.
$$
(4.26)

Combining (4.25) and (4.26), by sending $\rho \to 0$ in (4.21), we obtain (note Corollary 4.2)

$$
-\sqrt{\varepsilon}(|\eta|+T) \leq \langle \bar\varphi_\varepsilon, \eta\rangle + \langle \bar\psi_\varepsilon, z_\varepsilon(T)\rangle + \bar\psi_\varepsilon^0 z_\varepsilon^0,
$$
(4.27)

## §4. Proof of the Maximum Principle

where

(4.28)
$$\begin{cases} (\bar{\varphi}_\varepsilon, \bar{\psi}_\varepsilon) = \dfrac{d_S(y_0^\varepsilon, y^\varepsilon(T))}{J_\varepsilon(y_0^\varepsilon, u^\varepsilon(\cdot))}(a_\varepsilon, b_\varepsilon), \\[2mm] \bar{\psi}_\varepsilon^0 = \dfrac{(J(y_0^\varepsilon, u^\varepsilon(\cdot)) + \varepsilon)^+}{J_\varepsilon(y_0^\varepsilon, u^\varepsilon(\cdot))}. \end{cases}$$

By (4.25) and (4.2), we have

(4.29) $$|\bar{\varphi}_\varepsilon|_{X^*}^2 + |\bar{\psi}_\varepsilon|_{X^*}^2 + (\bar{\psi}_\varepsilon^0)^2 = 1.$$

On the other hand, by (3.66), we have

(4.30) $$\langle \bar{\varphi}_\varepsilon, x_0 - y_0^\varepsilon \rangle + \langle \bar{\psi}_\varepsilon, x_1 - y^\varepsilon(T) \rangle \leq 0, \qquad \forall (x_0, x_1) \in S.$$

Conditions (4.27)–(4.30) can be regarded as necessary conditions for the pair $(y^\varepsilon(\cdot), u^\varepsilon(\cdot))$. We may proceed further to derive an approximate maximum principle. However, we prefer to pass to the limit first and derive the maximum principle directly for the original optimal pair $(\bar{y}(\cdot), \bar{u}(\cdot))$.

*Step 4.* Pass to the limit.

From (4.8) and Lemma 4.1, similar to (4.18) and (4.20), we have that

(4.31) $$\lim_{\varepsilon \to 0} \left[ |y_0^\varepsilon - \bar{y}_0| + \sup_{t \in [0,T]} |z_\varepsilon(t) - z(t)| + |z_\varepsilon^0 - z^0| \right] = 0,$$

where $z(\cdot)$ and $z^0$ satisfy the following equations:

(4.32) $$z(t) = e^{At}\eta + \int_0^t e^{A(t-s)} f_y(s, \bar{y}(s), \bar{u}(s)) z(s)\, ds \\ + \int_0^t e^{A(t-s)} \left[ f(s, \bar{y}(s), v(s)) - f(s, \bar{y}(s), \bar{u}(s)) \right] ds, \quad t \in [0, T],$$

(4.33) $$z^0 = \int_0^T f_y^0(s, \bar{y}(s), \bar{u}(s)) z(s)\, ds \\ + \int_0^T \left[ f^0(s, \bar{y}(s), v(s)) - f^0(s, \bar{y}(s), \bar{u}(s)) \right] ds.$$

Then, by (4.30), we have $\forall (x_0, x_1) \in S$,

(4.34) $$\langle \bar{\varphi}_\varepsilon, x_0 - \bar{y}_0 \rangle + \langle \bar{\psi}_\varepsilon, x_1 - \bar{y}(T) \rangle \\ \leq \left( |y_0^\varepsilon - \bar{y}_0|^2 + |y^\varepsilon(T) - \bar{y}(T)|^2 \right)^{1/2} \equiv \delta_\varepsilon \to 0, \quad \text{as } \varepsilon \to 0.$$

Hence, combining (4.27) and (4.34), we obtain

(4.35) $$\langle \bar{\varphi}_\varepsilon, \eta - (x_0 - \bar{y}_0) \rangle + \langle \bar{\psi}_\varepsilon, z(T) - (x_1 - \bar{y}(T)) \rangle + \bar{\psi}_\varepsilon^0 z^0 \\ \geq -\sqrt{\varepsilon}(|\eta| + T) - \delta_\varepsilon - |z_\varepsilon(T) - z(T)| - |z_\varepsilon^0 - z^0| \geq -\theta_\varepsilon, \\ \forall (x_0, x_1) \in S.$$

Here, we can see that $\theta_\varepsilon$ is uniform in $v(\cdot) \in \mathcal{U}_{ad}$ and $\eta \in B_1(0)$. Now, we let $G(\cdot, \cdot)$ be the evolution operator generated by $A + f_y(\cdot, \bar{y}(\cdot), \bar{u}(\cdot))$. Then, by Chapter 2, Proposition 5.7, we know that the solution $z(\cdot)$ of (4.32) can be represented by

$$(4.36) \quad z(t) = G(t,0)\eta + \int_0^t G(t,s)\big[f(s,\bar{x}(s),v(s)) - f(s,\bar{x}(s),\bar{u}(s))\big]ds,$$
$$t \in [0,T].$$

Next, we let (recall the definition of $\mathcal{R}$, see (1.32))

$$(4.37) \quad \widetilde{\mathcal{R}} = \left\{ \begin{pmatrix} \eta \\ z \end{pmatrix} \in X \times X \mid z = G(T,0)\eta + \xi, \ \xi \in \mathcal{R}, \eta \in B_1(0) \right\}.$$

On the other hand, by (1.33), we have

$$(4.38) \quad Q = \{x_1 - G(T,0)x_0 \mid (x_0, x_1) \in S\}.$$

By our assumption and Proposition 3.5, we know that $\widetilde{\mathcal{R}} - S$ is finite codimensional in $X \times X$, and therefore so is $\widetilde{\mathcal{R}} - S + \begin{pmatrix} \bar{y}_0 \\ \bar{y}(T) \end{pmatrix}$. We note that (4.35) can be written as follows:

$$(4.39) \quad \left\langle \begin{pmatrix} \bar{\varphi}_\varepsilon \\ \bar{\psi}_\varepsilon \end{pmatrix}, \begin{pmatrix} \zeta_0 \\ \zeta_1 \end{pmatrix} \right\rangle + \bar{\psi}_\varepsilon^0 z^0 \geq -\theta_\varepsilon, \quad \forall \begin{pmatrix} \zeta_0 \\ \zeta_1 \end{pmatrix} \in \widetilde{\mathcal{R}} - S + \begin{pmatrix} \bar{y}_0 \\ \bar{y}(T) \end{pmatrix}.$$

Thus, by Lemma 3.6 and (4.29), we can find a subsequence (still denoted by itself), such that

$$(4.40) \quad (\bar{\varphi}_\varepsilon, \bar{\psi}_\varepsilon, \bar{\psi}_\varepsilon^0) \xrightharpoonup{*} (\bar{\varphi}, \bar{\psi}, \bar{\psi}^0) \neq 0.$$

In fact, if $\bar{\psi}^0 \neq 0$, we are done. Otherwise, $\bar{\psi}_\varepsilon^0 \to 0$. Hence, $|\bar{\varphi}_\varepsilon|_{X^*}^2 + |\bar{\psi}_\varepsilon|_{X^*}^2 \geq \delta > 0$ for $\varepsilon > 0$ small enough. By Lemma 3.6, $(\bar{\varphi}, \bar{\psi}) \neq 0$.

Now, by sending $\varepsilon \to 0$ in (4.35), we obtain

$$(4.41) \quad \langle \bar{\varphi}, \eta - (x_0 - \bar{y}_0) \rangle + \langle \bar{\psi}, z(T) - (x_1 - \bar{y}(T)) \rangle + \bar{\psi}^0 z^0 \geq 0,$$
$$\forall (x_0, x_1) \in S, \ v(\cdot) \in \mathcal{U}[0,T], \ \eta \in B_1(0).$$

Next, let $\psi(\cdot)$ be the unique solution of (1.8) with

$$(4.42) \quad \psi^0 = -\bar{\psi}^0 \leq 0, \quad \psi(T) = -\bar{\psi}.$$

Then, by Chapter 2, Proposition 5.7, we have

$$(4.43) \quad \psi(t) = G^*(T,t)\psi(T) + \psi^0 \int_t^T G^*(s,t)g(s)\,ds, \quad t \in [0,T],$$

where $g(s) = f_y^0(s, \bar{y}(s), \bar{u}(s))$, $s \in [0,T]$. We rewrite (4.41) as follows:

$$(4.44) \quad \langle \bar{\varphi}, x_0 - \bar{y}_0 - \eta \rangle - \langle \psi(T), x_1 - \bar{y}(T) - z(T) \rangle + \psi^0 z^0 \leq 0,$$
$$\forall (x_0, x_1) \in S, \ v(\cdot) \in \mathcal{U}[0,T], \ \eta \in B_1(0).$$

## §4. Proof of the Maximum Principle

From (4.33) and (4.36), we know that

(4.45)
$$\begin{cases} z(t) = G(t,0)\eta + \int_0^t G(t,s)h(s)\,ds, & t \in [0,T], \\ z^0 = \int_0^T [\langle g(s), z(s) \rangle + h^0(s)]\,ds, \end{cases}$$

where

(4.46)
$$\begin{cases} h(s) = f(s, \bar{x}(s), v(s)) - f(s, \bar{x}(s), \bar{u}(s)), \\ h^0(s) = f^0(s, \bar{x}(s), v(s)) - f^0(s, \bar{x}(s), \bar{u}(s)), \end{cases} \quad t \in [0,T].$$

We have the following duality equality: (see Chapter 2, Proposition 5.7)
(4.47)
$$\langle \psi(T), z(T) \rangle - \langle \psi(0), \eta \rangle + \psi^0 z^0$$
$$= \langle \psi(T), G(T,0)\eta + \int_0^T G(T,s)h(s)\,ds \rangle$$
$$- \langle G^*(T,0)\psi(T) + \psi^0 \int_0^T G^*(s,0)g(s)\,ds, \eta \rangle$$
$$+ \psi^0 \int_0^T [\langle g(s), G(s,0)\eta + \int_0^s G(s,\tau)h(\tau)\,d\tau \rangle + h^0(s)]\,ds$$
$$= \int_0^T \{\langle G^*(T,s)\psi(T), h(s) \rangle$$
$$+ \int_s^T \langle \psi^0 G^*(\tau,s)g(\tau)d\tau, h(s) \rangle + \psi^0 h^0(s)\}\,ds$$
$$= \int_0^T [\langle \psi(s), h(s) \rangle + \psi^0 h^0(s)]\,ds,$$
$$= \int_0^T \left[ H(t, \bar{y}(t), v(t), \psi^0, \psi(t)) - H(t, \bar{y}(t), \bar{u}(t), \psi^0, \psi(t)) \right]\,dt,$$

for all $\eta \in B_1(0)$ and $v(\cdot) \in \mathcal{U}_{ad}$, where $(\psi(\cdot), \psi^0, z(\cdot), z^0)$ satisfy (1.8), (4.43), and (4.45). Hence, by setting $\eta = 0$ and $(x_0, x_1) = (\bar{y}_0, \bar{y}(T))$ in (4.44), we obtain

(4.48)
$$\int_0^T \left[ H(t, \bar{y}(t), v(t), \psi^0, \psi(t)) - H(t, \bar{y}(t), \bar{u}(t), \psi^0, \psi(t)) \right]\,dt \leq 0,$$
$$\forall v(\cdot) \in \mathcal{U}_{ad}.$$

As $U$ is separable, there exists a countable dense set $U_0 = \{u_i, i \geq 1\} \subset U$. For each $u_i \in U_0$, we denote

(4.49) $\quad g_i(s) = H(t, \bar{y}(s), u_i, \psi^0, \psi(s)) - H(t, \bar{y}(s), \bar{u}(s), \psi^0, \psi(s)).$

Then $g_i(\cdot) \in L^1(0,T)$. Thus, there exists a measurable set $F_i \subset [0,T]$ with $|F_i| = T$, such that any point in $F_i$ is a Lebesgue point of $g_i$. Namely,

$$(4.50) \qquad \lim_{\delta \to 0} \frac{1}{\delta} \int_{t-\delta}^{t+\delta} |g_i(s) - g_i(t)| \, ds = 0, \qquad \forall t \in F_i.$$

Now, for any $t \in F_i$, we define

$$(4.51) \qquad v_\delta(s) = \begin{cases} \bar{u}(s), & |s-t| > \delta, \\ u_i, & |s-t| \leq \delta. \end{cases}$$

Then, by (4.48), we obtain

$$(4.52) \qquad \int_{t-\delta}^{t+\delta} g_i(s) \, ds \leq 0, \qquad \forall \delta > 0.$$

Dividing by $\delta > 0$ and sending $\delta \to 0$, we obtain $g_i(t) \leq 0$. That means

$$(4.53) \qquad \begin{aligned} H(t, \bar{y}(t), u_i, \psi^0, \psi(t)) &\leq H(t, \bar{y}(t), \bar{u}(t), \psi^0, \psi(t)), \\ &\forall t \in F \equiv \bigcap_{i \geq 1} F_i, \quad u_i \in U_0. \end{aligned}$$

As $U_0$ is countable, we have $|F| = T$. Then, by the continuity of the Hamiltonian $H$ in $u \in U$, we obtain (1.10) from (4.53).

Next, by taking $v(\cdot) = \bar{u}(\cdot)$ and $(x_0, x_1) = (\bar{y}_0, \bar{y}(T))$ in (4.44), using (4.47), we get

$$(4.54) \qquad \langle \bar{\varphi}, \eta \rangle \geq \langle \psi(T), z(T) \rangle + \psi^0 z^0 = \langle \psi(0), \eta \rangle, \qquad \forall \eta \in B_1(0).$$

Thus, $\bar{\varphi} = \psi(0)$. Then, taking $\eta = 0$ and $v(\cdot) = \bar{u}(\cdot)$ in (4.44), we obtain the transversality condition (1.9).

Finally, we claim that $(\psi^0, \psi(\cdot)) \neq 0$. Otherwise, we have, in particular, that

$$(4.55) \qquad \bar{\psi} = -\psi(T) = 0, \qquad \bar{\varphi} = \psi(0) = 0.$$

This, together with $\psi^0 = 0$, gives a contradiction to (4.40). The proof of the maximum principle is completed.

To conclude this section, let us look at some conditions under which $\psi^0 \neq 0$. To this end, we suppose $\psi^0 = 0$. Then (4.44) reads (recall $\bar{\varphi} = \psi(0)$, (4.36), and (4.43))

$$(4.56) \qquad \begin{aligned} 0 &\geq \langle \psi(0), x_0 - \bar{y}_0 - \eta \rangle - \langle \psi(T), x_1 - \bar{y}(T) - z(T) \rangle \\ &= \langle \psi(0), x_0 - \bar{y}_0 \rangle - \langle G^*(T,0)\psi(T), \eta \rangle \\ &\quad - \langle \psi(T), x_1 - \bar{y}(T) \rangle + \langle \psi(T), G(T,0)\eta + \xi \rangle \\ &= \langle \psi(0), x_0 - \bar{y}_0 \rangle - \langle \psi(T), x_1 - \bar{y}(T) \rangle + \langle \psi(T), \xi \rangle, \\ &\quad \forall (x_0, x_1) \in S, \ \xi \in \mathcal{R}. \end{aligned}$$

The above further implies that (note (4.43) again)

(4.57) $\langle -\psi(T), q - \xi - (\bar{y}(T) - G(T,0)\bar{y}_0) \rangle \leq 0,$
$\forall q \in Q \equiv (-G(T,0), I)S, \ \xi \in \mathcal{R}.$

By Theorem 1.6, we know that $(\psi^0, \psi(T)) \neq 0$. Thus, to obtain the cases in which $\psi^0 \neq 0$, we need only to find conditions under which we can deduce from (4.56) or (4.57) that $\psi(T) = 0$. This leads to the following result.

**Proposition 4.4.** *Let (H1)–(H4) hold and $(\bar{y}(\cdot), \bar{u}(\cdot))$ be an optimal pair with $\bar{y}_0 = \bar{y}(0)$. Then, $\psi^0 \neq 0$ under any of the following conditions:*

(i) *$\bar{y}(T) - G(T,0)\bar{y}_0 \in \text{Int}\,(Q - \mathcal{R})$; in particular, $0 \in \text{Int}\,\mathcal{R}$ or $Q = X$.*

(ii) *Suppose $S = Q_0 \times Q_1$ with $Q_0, Q_1 \subset X$ closed and convex. Let $\bar{y}(T) \in \text{Int}\,Q_1$; in particular, $Q_1 = X$.*

(iii) *Suppose $S = Q_0 \times Q_1$ with $Q_0, Q_1 \subset X$ closed and convex. Let $\bar{y}_0 \in \text{Int}\,Q_0$ and $G(T,0)$ be injective; in particular, $Q_0 = X$ and $G(T,0)^{-1} \in \mathcal{L}(X)$.*

(iv) *Let*

(4.58) $\left\{ \bigcup_{\lambda \geq 0} \lambda \partial d_{Q-\mathcal{R}}(\bar{y}(T) - G(T,0)\bar{y}_0) \right\}$
$\cap \left\{ \bar{\psi} \in X^* \ \Big| \ \begin{pmatrix} -G^*(T,0) \\ I \end{pmatrix} \bar{\psi} \in \bigcup_{\lambda \geq 0} \lambda \partial d_S(\bar{y}_0, \bar{y}(T)) \right\} = \{0\}.$

We have seen that in many cases, $\psi^0 \neq 0$. The case where $y_0$ is fixed and $y(T)$ is free is the most classical one. This case amounts to saying that $Q_1 = X$, which is contained in (ii). It is not hard for the readers to identify the cases listed above.

## §5. Applications

First of all, let us point out some important special cases covered by Theorem 1.6.

In what follows, let us assume that (H1) and (H2), and (H4) hold, and let $G(t,s)$ be the evolution operator generated by $A + f_y(t, \bar{y}(t), \bar{u}(t))$ with $(\bar{y}(\cdot), \bar{u}(\cdot))$ being some optimal pair. We consider the different types of endpoint constraints.

*1. The control problem with fixed endpoints.*

In this case, the constraint set of the problem is $S = \{(y_0, y_1)\}$ and the endpoint constraint takes the following form:

(5.1) $\qquad y(0) = y_0, \qquad y(T) = y_1.$

Thus, the modified state constraint set is $Q = \{x_1 - G(T,0)x_0\}$, which is a singleton. Hence, in order for the maximum principle to hold, we need the reachable set $\mathcal{R}$ associated with $(\bar{y}(\cdot), \bar{u}(\cdot))$ to be finite codimensional in $X$. This is the case if $e^{At}$ is a $C_0$ group on $X$, $f(t, y, u) = f_1(t, y) + u$, and $U$ is a ball in $X$. This result is essentially due to Fattorini [6,7].

2. *The control problem with a terminal state constraint.*

Let $x_0 \in X$ and $Q_1 \subset X$ be convex and closed. We consider the problem with the endpoint constraints:

(5.2) $$y(0) = x_0, \qquad y(T) \in Q_1.$$

Thus, the constraint set is $S = \{x_0\} \times Q_1$ and the modified constraint set is given by $Q = Q_1 - G(T,0)x_0$. Clearly, this set is finite codimensional in $X$ if and only if $Q_1$ is so. Hence, if $Q_1$ is finite codimensional in $X$, then the maximum principle holds. This result is due to Li–Yao [3].

3. *The control problem with separated endpoint constraints.*

Let $Q_0, Q_1 \subset X$ be convex and closed sets. We consider the optimal control problem with the endpoint constraints:

(5.3) $$y(0) \in Q_0, \qquad y(T) \in Q_1,$$

Then the constraint set can be written as $S = Q_0 \times Q_1$ and the modified constraint set is given by $Q = Q_1 - G(T,0)Q_0$. Now, by Proposition 3.4, we know that if $Q_1$ or $G(T,0)Q_0$ is finite codimensional in $X$, then so is $Q$. In this case, we have the maximum principle. In the case where $e^{At}$ is a $C_0$ group, which is the case for the wave equation without damping, $G(T, 0)$ is invertible; and we need only to have the finite codimensionality of either $Q_0$ or $Q_1$.

Next, let us discuss an interesting case — The optimal periodic control problem. To study such a problem, let us first assume (H1), (H2), and (H4). In addition, we make the following periodic conditions:

(H5) For any $(t, x, u) \in [0, \infty) \times X \times U$,

(5.4) $$f(t + T, x, u) = f(t, x, u), \quad f^0(t + T, x, u) = f^0(t, x, u).$$

It is required that the state satisfy the following periodic condition:

(5.5) $$y(0) = y(T).$$

Thus, in the present case, the state constraint set is $S = \{(x, x) \mid x \in X\}$. This is the "diagonal" set in $X \times X$. Clearly, this set is *not* finite codimensional in $X \times X$ (if $\dim X = \infty$). The modified constraint set is given by $Q = (I - G(T, 0))X$. Now the question is: When is this set finite codimensional in $X$? From Chapter 2, Corollary 1.33, we know that if $G(T, 0)$ is a compact operator, then the range $\mathcal{R}(I - G(T, 0))$ of $I - G(T, 0)$ is a closed finite codimensional subspace of $X$. Consequently, by Proposition 3.4, we see that (H3) is satisfied and then, the maximum principle holds.

## §5. Applications

The next question is: When is $G(T,0)$ compact? For this, we have the following result.

**Proposition 5.1.** *Let $e^{At}$ be a compact semigroup. Then, for any $F(\cdot) \in L^1(0,T;\mathcal{L}(X))$, the evolution operator $G(\cdot,\cdot)$ generated by $A+F(\cdot)$ has the property that $G(T,0)$ is compact.*

*Proof.* First of all, by definition, we have

$$(5.6) \quad G(t,0)x = e^{At}x + \int_0^t e^{A(t-r)} F(r) G(r,0) x \, dr, \quad \forall x \in X, \ t \in [0,T].$$

By Gronwall's inequality, we see that $\|G(r,0)\|_{\mathcal{L}(X)} \leq C$, for all $r \in [0,T]$. Thus, the function $\widetilde{F}(r) \equiv F(r)G(r,0)$ is in $L^1(0,T;\mathcal{L}(X))$. Because $e^{AT}$ is compact, from (5.6) we see that to prove the compactness of $G(T,0)$, it suffices to show that the following operator is compact:

$$(5.7) \quad \widehat{G}x \equiv \int_0^T e^{A(T-r)} \widetilde{F}(r) x \, dr, \quad \forall x \in X.$$

To this end, we need to prove that $\widehat{G}B_1(0)$ is relatively compact. ($B_1(0)$ is the closed unit ball in $X$.) For any $\varepsilon > 0$, there exists a $\delta > 0$, such that

$$(5.8) \quad \left| \int_{T-\delta}^T e^{A(T-r)} \widetilde{F}(r) x \, dr \right| \leq C \int_{T-\delta}^T \|\widetilde{F}(r)\| \, dr < \varepsilon/2, \quad \forall |x| \leq 1.$$

On the other hand, we have

$$(5.9) \quad \left| \int_0^{T-\delta} e^{A(T-\delta-r)} \widetilde{F}(r) x \, dr \right| \leq C \int_0^T \|\widetilde{F}(r)\| \, dr, \quad \forall x \in B_1(0),$$

Thus, by the compactness of $e^{A\delta}$, we can find finitely many points $y_1, \cdots, y_k$, such that

$$(5.10) \quad Y_0 \equiv \{ e^{A\delta} \int_0^{T-\delta} e^{A(T-\delta-r)} \widetilde{F}(r) x \, dr \mid |x| \leq 1 \} \subset \bigcup_{i=1}^k B_{\varepsilon/2}(y_i).$$

Then it follows that

$$\widehat{G}x = \int_{T-\delta}^T e^{A(T-r)} \widetilde{F}(r) x \, dr + e^{A\delta} \int_0^{T-\delta} e^{A(T-\delta-r)} \widetilde{F}(r) x \, dr$$

(5.11)
$$\in B_{\varepsilon/2}(Y_0) \subset \bigcup_{i=1}^k B_\varepsilon(y_i), \quad \forall x \in B_1(0).$$

This shows that $\widehat{G}B_1(0)$ is relatively compact. □

Now, we can prove the following result.

**Proposition 5.2.** *Let (H1), (H2), (H4), and (H5) hold. Let $e^{At}$ be a compact semigroup. Let $(\bar{y}, \bar{u})$ be an optimal pair for the optimal periodic* control problem. Then there exists a pair $(\psi^0, \psi(\cdot)) \neq 0$, such that (1.7)–(1.10) hold and

(5.12) $$\psi(0) = \psi(T).$$

Moreover, if $\mathcal{N}(I - G^*(T, 0)) = \{0\}$, $\psi^0 \neq 0$.

*Proof.* From Proposition 5.1 and the analysis we made before the statement of Proposition 5.1, we see that the modified constraint set $Q$ is finite codimensional in $X$, and so is $\mathcal{R} - Q$ by Proposition 3.4. Thus, all the assumptions of Theorem 1.6 hold for the present case. Consequently, there exists a nontrivial pair $(\psi^0, \psi(\cdot))$, such that (1.7)–(1.10) hold. Now, we prove (5.12). Note that in the present case, $S = \{(x_0, x_0) \mid x_0 \in X\}$. Thus, by (1.9) (recall that $\bar{y}(T) = \bar{y}_0$, because $\bar{y}(\cdot)$ is periodic with the period $T$),

(5.13) $$\langle \psi(0) - \psi(T), x_0 - \bar{y}_0 \rangle \leq 0, \qquad \forall x_0 \in X.$$

This implies (5.12). Finally, if $\mathcal{N}(I - G^*(T, 0)) = \{0\}$, we must have $\psi^0 \neq 0$. In fact, if $\psi^0 = 0$, then, by (4.43), we have

(5.14) $$\psi(t) = G^*(T, t)\psi(T), \qquad \forall t \in [0, T].$$

Now, by (5.12), we obtain that $(I - G^*(T, 0))\psi(T) = 0$. Thus, by our condition, $\psi(T) = 0$, which implies $\psi(\cdot) = 0$, a contradiction because $(\psi^0, \psi(\cdot)) \neq 0$. Hence, we have $\psi^0 \neq 0$. □

Next, we give a concrete example.

*Example 5.3.* Consider the following controlled diffusion reaction equation:

(5.15) $$\begin{cases} y_t - \Delta y = f(t, x, y)u(t, x) + g(t, x, y), & \text{in } \Omega_T, \\ y|_{\partial \Omega} = 0, \\ y(t + T, x) = y(t, x), \end{cases}$$

where $\Omega \subset \mathbb{R}^3$ with a smooth boundary $\partial \Omega$ and $\Omega_T = \Omega \times [0, T]$. We can regard the above as a controlled chemical reaction process with diffusion or a controlled single species population dynamics with diffusion. The process is desired to be periodic (with period $T$). We seek optimal periodic control, which minimizes the following cost functional:

(5.16) $$J(u) = \int_0^T \int_\Omega \left\{ f^0(t, x, y(t, x))u(x, t) + g^0(t, x, y(t, x)) \right\} dx\, dt.$$

Let us assume that our control $u(x, t)$ takes values in $U = [-1, 1]$ and is periodic in $t$ (with period $T$). We let $X = L^2(\Omega)$, which is a Hilbert space. Thus, (H4) holds. Also, we assume that functions $f, g, f^0, g^0$ are $C^1$ in $y$ with bounded partial derivatives in $y$, and that they are periodic in $t$. Then (H2) and (H5) hold. On the other hand, form Chapter 2, Example 4.23, we know that if we set $A = \Delta$, with $\mathcal{D}(A) = W^{2,2}(\Omega) \cap W_0^{1,2}(\Omega)$, then $e^{At}$

## §5. Applications

is a compact semigroup on $X$. Thus, by Proposition 5.2, the maximum principle holds. We would like to further discuss the optimal control. To this end, let us first give the following lemma.

**Lemma 5.4.** *Let*

(5.17) $$\begin{cases} |f_y(t,x,y)| + g_y(t,x,y) \leq c(t), & \forall (t,x,y) \in [0,T] \times \Omega \times \mathbb{R}, \\ \int_0^T c(t)\, dt < \lambda T, \end{cases}$$

*where $\lambda > 0$, is the first eigenvalue of $-\Delta$ with the Dirichlet boundary condition. Then, $\mathcal{N}(I - G^*(T,0)) = \{0\}$.*

*Proof.* We denote $F(t,x) = f_y(t,x,\bar{y}(t,x))\bar{u}(t,x) + g_y(t,x,\bar{y}(t,x))$. Consider the following problem:

(5.18) $$\begin{cases} z_t - \Delta z = F(t,x)z, & \text{in } \Omega_T, \\ z\big|_{\partial\Omega} = 0, \\ z\big|_{t=0} = z\big|_{t=T}. \end{cases}$$

Clearly, 0 is a solution of (5.18). Suppose $z$ is any other solution. Then, by (5.17), we have

(5.19) $$\begin{aligned} \frac{1}{2}\frac{d}{dt}\|z(t)\|_{L^2(\Omega)}^2 &+ \|\nabla z(t)\|_{L^2(\Omega)}^2 \\ &= \int_\Omega F(t,x)z(t,x)^2\, dx \leq c(t)\|z(t)\|_{L^2(\Omega)}^2. \end{aligned}$$

By the definition of $\lambda$, we have $\|\nabla z(t)\|_{L^2(\Omega)} \geq \lambda\|z(t)\|_{L^2(\Omega)}^2$. Thus, (5.19) implies that

(5.20) $$\frac{d}{dt}\|z(t)\|_{L^2(\Omega)}^2 \leq 2(c(t) - \lambda)\|z(t)\|_{L^2(\Omega)}^2, \qquad \forall t \in [0,\infty).$$

Hence, it follows that

(5.21) $$\|z(0)\|_{L^2(\Omega)}^2 = \|z(T)\|_{L^2(\Omega)}^2 \leq e^{2\int_0^T c(t)dt - 2\lambda T}\|z(0)\|_{L^2(\Omega)}^2.$$

By condition (5.17), we see that the above is possible only if $z(0) = 0$. Thus, by the definition of $G(T,0)$, we see that $\mathcal{N}(I - G(T,0)) = \{0\}$. It is clear that in the present case, $G(T,0)$ is self-adjoint. Thus, our conclusion follows. □

Now, let us assume (5.17) for our control problem. Then, by Proposition 5.2, we can let $\psi^0 = -1$. Also, there exists a periodic solution $\psi(\cdot)$ of the following:

(5.22) $$\begin{cases} \psi_t + \Delta\psi = -\big(f_y(t,x,\bar{y})\bar{u} + g_y(t,x,\bar{y})\big)\psi \\ \qquad\qquad + f_y^0(t,x,\bar{y})\bar{u} + g_y^0(t,x,\bar{y}), & \text{in } \Omega_T, \\ \psi\big|_{\partial\Omega} = 0, \quad \psi\big|_{t=0} = \psi\big|_{t=T}. \end{cases}$$

From Lemma 5.4, it is not hard to see that such a solution is unique. The following maximum condition holds:

$$\int_\Omega \{\psi(t,x)f(t,x,\bar{y}(t,x)) - f^0(t,x,\bar{y}(t,x))\}\bar{u}(t,x)\,dx$$
(5.23)
$$\geq \int_\Omega \{\psi(t,x)f(t,x,\bar{y}(t,x)) - f^0(t,x,\bar{y}(t,x))\}v(t,x)\,dx,$$
$$\forall v \in L^\infty(\Omega_T; U), \text{ a.e. } t \in [0,T].$$

From this, we can deduce that

(5.24) $\bar{u}(t,x) = \begin{cases} 1, & \text{if } \psi(t,x)f(t,x,\bar{y}(t,x)) - f^0(t,x,\bar{y}(t,x)) > 0, \\ -1, & \text{if } \psi(t,x)f(t,x,\bar{y}(t,x)) - f^0(t,x,\bar{y}(t,x)) < 0. \end{cases}$

Thus, on the set $\{(t,x) \mid \psi(t,x)f(t,x,\bar{y}(t,x)) - f^0(t,x,\bar{y}(t,x)) \neq 0\}$, the optimal control $\bar{u}$ only takes value $\pm 1$. This is usually called the *bang-bang principle*.

Next, we give an example involving ordinary functional differential equations.

*Example 5.5.* Let $0 < r < T$, $X = C([-r,0]; \mathbb{R}^n)$. Then $X$ is a separable Banach space. Thus, we may endow a new norm to $X$ so that $X^*$ is strictly convex (see Chapter 2, Theorem 2.18). Consider the following functional differential system:

(5.25) $$\frac{dy(t)}{dt} = f(t, y_t, u(t)),$$

where $f : \mathbb{R} \times X \times \mathbb{R}^m \to \mathbb{R}^n$ is a given map and $y_t \in X$ is defined by $y_t(\theta) = y(t+\theta), \forall \theta \in [-r,0]$. Furthermore, let $f^0 : \mathbb{R} \times X \times \mathbb{R}^m \to \mathbb{R}$ be given. Let us assume that (H2) and (H5) hold for the maps $f$ and $f^0$. Then a periodic optimal control problem can be posed. We may regard (5.25) as a semilinear evolution equation in $X$ with the generator $A$ of the $C_0$ semigroup given by (similar to Chapter 2, Example 4.22)

(5.26) $$\begin{cases} \mathcal{D}(A) = \{\varphi \in C^1([-r,0]; \mathbb{R}^n) \mid \dot\varphi(0) = 0\}, \\ A\varphi(\theta) = \begin{cases} \dot\varphi(\theta), & -r \leq \theta < 0, \\ 0, & \theta = 0. \end{cases} \end{cases}$$

It is clear that $e^{At}$ is a $C_0$ semigroup that is compact for $t > r$ with $r > 0$. We would like to point out that Proposition 5.1 does not apply to this case! From Phillips [1], we know that, in general, if $e^{At}$ is a $C_0$ semigroup that is compact for $t > t_0$ with $t_0 > 0$, then there could be that some bounded operator $B$, such that $e^{(A+B)t}$ is not compact for $t > t_0$ (note that $e^{(A+B)t}$ is always a $C_0$ semigroup). However, the ordinary functional differential equations have their own feature and we can prove the compactness of $G(T, 0)$ directly. As a matter of fact, suppose that $(\bar{y}, \bar{u})$ is an optimal pair; then the evolution operator $G(t, s)$ is generated by the following homogeneous equation:

$$\text{(5.27)} \qquad \frac{dz(t)}{dt} = f_y(t, \bar{y}_t, \bar{u}(t))z_t,$$

where $f_y(t, y_t, u)$ is the Fréchet derivative of $f(t, y_t, u)$ in $y_t$. Then, for $T > r$, we see that $G(T, 0)$ maps from $X = C([-r, 0]; \mathbb{R}^n)$ to $C^1([-r, 0]; \mathbb{R}^n)$, which is compactly embedded into $X$. Thus, $G(T, 0)$ is compact. Then a result similar to Proposition 5.2 holds for the ordinary functional differential equations. In this example, we may also take $X = W^{1,p}([-r, 0]; \mathbb{R}^n)$, or $X = \mathbb{R}^n \times L^p(-r, 0; \mathbb{R}^n)$ with $1 \leq p < \infty$.

## Remarks

The first result on the optimal control problem for infinite dimensional systems was due to Butkovsky–Lerner [1] published in 1960. They studied an optimal control problem where the state equation was the one-dimensional heat equation. Using integral representation of the solutions (in terms of Green's function), they reduced the state equation to a situation in which the Pontryagin maximum principle applied. The optimal control was then determined.

The infinite dimensional version of the Pontryagin maximum principle was first proved by Butkovsky [1] for systems governed by integral equations and by Kharatishvile [1] for systems governed by ordinary delay equations in 1961. Later, A.I. Egorov [1] adopted the method of Pontryagin to derive the maximum principle for nonlinear evolution equations formally. Soon after, Yu.V. Egorov [1,2] constructed an example showing that the maximum principle does not necessarily hold for infinite dimensional systems. His example is as follows. Let $X = \ell^2$. The system and the constraints are as follows

$$\text{(R.1)} \qquad \begin{cases} \dot{y}(t) = u(t), \quad y(0) = 0, \; y(t_1) = (1, \frac{1}{2}, \cdots, \frac{1}{n}, \cdots) \\ u(t) = (u_1(t), u_2(t), \cdots), \quad |u_n(t)| \leq \frac{1}{n} + \frac{1}{n^2}. \end{cases}$$

Consider the associated time optimal control problem. Then $T = 1$ is the minimum time and the optimal control is given by

$$\text{(R.2)} \qquad \bar{u}(t) = (1, \frac{1}{2}, \cdots, \frac{1}{n}, \cdots).$$

It was shown that the maximum principle does not hold. In the same paper, Yu. V. Egorov [1] imposed some conditions and by applying the method of Pontryagin, he proved the maximum principle. In the mid-1960s, Friedman [2] pointed out that there was a gap in the proof of Yu.V. Egorov [1,2]. Later, Yu.V. Egorov made some further comments on the work of Friedman.

In his book, Lions [2, p.270] said: "*We are unable to follow all points of the proof given by the author* (Yu.V. Egorov). *The results are probably all correct.*" In 1967, under the so-called completeness of admissible controls, A.I. Egorov [2] proved the maximum principle for parabolic and hyperbolic systems with the terminal state constrained by finitely many equalities. Such a terminal constraint set is finite codimensional in the state space. But for a long time, people did not realize it.

Because of the counterexample by Yu.V. Egorov, in the 1960s and 1970s, many people working in this area concentrated mainly on the discussion of the time optimal control problem for linear infinite dimensional systems. The major problem occurs when the "bang-bang" principle and maximum principle hold. See the remarks for Chapter 7 for details.

On the other hand, since the later 1970s, Ahmed–Teo [1,2], Wolfersdorf [1], and some others studied the Pontryagin maximum principle for evolutionary partial differential equations with no terminal state constraints and with convex control domain. These results were summarized in the book by Ahmed–Teo [3]. Starting from the beginning of 1980s, Barbu and some other authors studied the maximum principle for evolutionary variational inequalities without endpoint constraint and with convex control domain. Readers are referred to the books by Barbu [1,5] and Tiba [2] for details. About the same time, Yao [1,3] proved the maximum principle for evolution control systems with free terminal state and with the control domain allowed to be nonconvex. His method is essentially adopted from Li–Yao [1,2] (see the remarks for Chapter 7).

In 1985, Li–Yao [3] discussed the optimal control problem for the system

$$(R.3) \quad \begin{cases} y(t) = G(t,t_0)x_0 + \int_{t_0}^t G(t,s)b(x,y(s),y(s-h(s)),u(s))\,ds, \\ y(t_1) \in Q \subset X(\text{a Banach space}), \end{cases}$$

with the Lagrange type cost functional and general control domain. Here, $G(t,s)$ is the strongly continuous evolution operator on $X$. By assuming the finite codimensionality condition on the set $Q$, together with some minor conditions, they proved the maximum principle. The vector measure theory was successfully adopted and the corresponding spike variation technique was developed.

At the same time, Fattorini [6,7] discussed the case where $X$ is a Hilbert space. He used the Ekeland variational principle and proved that the maximum principle holds if the reachable set of the variational system is finite codimensional in $X$.

In 1989, Xu [1] proved that if $X$ is a uniformly convex Banach space and if the set $\mathcal{R} - Q$ is finite codimensional in $X$ (where $\mathcal{R}$ is the reachable set of the variational system and $Q$ is the terminal state constraint set), then the maximum principle holds.

The material in this chapter is based on the work by Li–Yong [1]. In this work, the methods of Li–Yao [3] and Fattorini [6,7] were combined in some sense. The result of Li–Yao [3] was not completely covered.

Corollary 2.2 is due to Ekeland [1,2], the general form, Lemma 2.1 was essentially taken from Aubin–Frankowska [1]. Lemma 3.6 was first proved by Fattorini [6,7] for the case in which underlying space is a Hilbert space and by Li–Yong [1] for the general case. Proposition 3.10 was due to Ekeland [2]. The material of §3.3 has been selected from Clarke [1] with certain modifications; the conclusion (ii) of Proposition 3.11 and Corollary 3.12 are due to Li–Yong [1].

For the control problem discussed by Yu.V. Egorov [1,2], by using the method similar to that of Li–Yong [1], Yong [10] recently proved the maximum principle. With such an approach, the question raised by Friedman [2] can be avoided.

In 1993, Yong [9] derived necessary conditions for infinite dimensional Volterra–Stieltjes evolution equations with state constraints. The motivation comes from infinite dimensional impulse control problems (Yong–Zhang [1]).

Recently, some other relevant works have appeared. Among them, we mention the following: Basile–Mininni [1], Fattorini [8,9], Fattorini–Frankowska [1], Fattorini–Murphy [1], Fattorini–Sritharan [1–3], Frankowska [1,2], Hu–Yong [1], Jin–Li [1], and Yong [3,4,10].

# Chapter 5

# Necessary Conditions for Optimal Controls — Elliptic Partial Differential Equations

In this chapter we present the Pontryagin type maximum principle for optimal control problems where the state equations are second order elliptic partial differential equations and variational inequalities. The existence theory of optimal controls for similar control problems can be found in Chapter 3. Unlike the evolution equations studied in the previous chapter, in this case, we do not have the time variable $t$. Thus, some techniques used in Chapter 4 have to be substantially modified.

## §1. Semilinear Elliptic Equations

### §1.1. Optimal control problem and the maximum principle

We begin with the optimal control problem for which the state is governed by a semilinear second order elliptic partial differential equation with a distributed control. Thus, our system reads

$$(1.1) \quad \begin{cases} Ay(x) = f(x, y(x), u(x)), & \text{in } \Omega, \\ y|_{\partial\Omega} = 0. \end{cases}$$

The cost functional is given by

$$(1.2) \quad J(y, u) = \int_\Omega f^0(x, y(x), u(x))\, dx,$$

where $(y, u)$ is a pair of state and control satisfying (1.1). We make the following assumptions.

(S1) $\Omega \subset \mathbb{R}^n$ is a bounded region with $C^{1,\gamma}$ boundary $\partial\Omega$, for some $\gamma > 0$ and $U$ is a separable metric space.

(S2) Operator $A$ is defined by

$$(1.3) \quad Ay(x) = -\sum_{i,j=1}^n \left(a_{ij}(x)y_{x_j}(x)\right)_{x_i},$$

with $a_{ij} \in C(\overline{\Omega})$, $a_{ij} = a_{ji}$, $1 \leq i, j \leq n$, and for some $\lambda > 0$,

$$(1.4) \quad \sum_{i,j=1}^n a_{ij}(x)\xi_i\xi_j \geq \lambda \sum_{i=1}^n |\xi_i|^2, \qquad \forall x \in \Omega,\ (\xi_1, \xi_2, \cdots, \xi_n) \in \mathbb{R}^n.$$

## §1. Semilinear Elliptic Equations

(S3) The function $f : \Omega \times \mathbb{R} \times U \to \mathbb{R}$ has the following properties: $f(\,\cdot\,, y, u)$ is measurable on $\Omega$, and $f(x, \,\cdot\,, u)$ is in $C^1(\mathbb{R})$ with $f(x, \,\cdot\,, \,\cdot\,)$ and $f_y(x, \,\cdot\,, \,\cdot\,)$ continuous on $\mathbb{R} \times U$. Moreover,

(1.5) $$f_y(x, y, u) \leq 0, \qquad \forall (x, y, u) \in \Omega \times \mathbb{R} \times U,$$

and for any $R > 0$, there exists a constant $M_R > 0$, such that

(1.6) $$|f(x, y, u)| + |f_y(x, y, u)| \leq M_R, \qquad \forall (x, u) \in \Omega \times U, \ |y| \leq R.$$

(S4) The function $f^0 : \Omega \times \mathbb{R} \times U \to \mathbb{R}$ satisfies (S3) except for (1.5).

Next, we set $\mathcal{U} = \{u : \Omega \to U \mid u \text{ is measurable }\}$. Any element $u \in \mathcal{U}$ is referred to as a control. The following result is basic.

**Proposition 1.1.** *Let (S1)–(S3) hold. Then, for any $p \in [1, \infty)$ and any $u \in \mathcal{U}$, there exists a unique $y \equiv y(\,\cdot\,; u) \in W_0^{1,p}(\Omega) \cap L^\infty(\Omega)$ solving (1.1). Furthermore, there exists a constant $C_p > 0$, independent of $u \in \mathcal{U}$, such that*

(1.7) $$\|y(\,\cdot\,; u)\|_{W_0^{1,p}(\Omega) \cap L^\infty(\Omega)} \leq C_p, \qquad \forall u \in \mathcal{U}.$$

The proof is similar to that given in Chapter 2, §6.3. We should note that if we let $p > n$, by the Sobolev Embedding Theorem (see Chapter 2, §6.1), the solution $y$ of (1.1) is actually in $C^\beta(\bar{\Omega})$ for some $\beta \in (0, 1)$. In what follows, we fix a $p \in [1, \infty)$. Any pair $(y, u) \in W_0^{1,p}(\Omega) \times \mathcal{U}$ satisfying (1.1) is called a *feasible pair* and we refer to the corresponding $y$ and $u$ as a *feasible state* and a *feasible control*, respectively. We let $\mathcal{A}$ be the set of all feasible pairs. Clearly, under (S1)–(S3), $\mathcal{U}$ coincides with the set of all feasible controls and for each feasible control $u \in \mathcal{U}$, there corresponds a unique feasible state and the cost functional (1.2) is well defined. Therefore, under (S1)–(S4), we can write $J(y, u)$ as $J(u)$ without any ambiguity. In what follows, we will use $J(u)$ or $J(y, u)$ according to our convenience. Next, we introduce the following:

(S5) $\mathcal{Y}$ is a Banach space with strict convex dual $\mathcal{Y}^*$, $F : W_0^{1,p}(\Omega) \to \mathcal{Y}$ is continuously Fréchet differentiable, and $Q \subset \mathcal{Y}$ is closed and convex.

We see that under (S1)–(S3), for any $u \in \mathcal{U}$, the solution $y(\,\cdot\,; u)$ of (1.1) is in $W_0^{1,p}(\Omega)$. Thus, a state constraint of the following type makes sense:

(1.8) $$F(y) \in Q.$$

We will present many examples of the state constraints covered by (1.8). Now, we let $\mathcal{A}_{ad}$ be the set of all pairs $(y, u) \in \mathcal{A}$, such that the state constraint (1.8) is satisfied. Any $(y, u) \in \mathcal{A}_{ad}$ is called an *admissible pair*. Our optimal control problem can be stated as follows:

**Problem (SD).** Find an admissible pair $(\bar{y}, \bar{u}) \in \mathcal{A}_{ad}$, such that

(1.9) $$J(\bar{y}, \bar{u}) = \inf_{\mathcal{A}_{ad}} J(y, u).$$

We refer to such a pair $(\bar{y}, \bar{u})$, if it exists, as an *optimal pair* and refer to $\bar{y}$ and $\bar{u}$ as an *optimal state* and *control*, respectively.

Next, we let $(\bar{y}, \bar{u}) \in \mathcal{A}_{ad}$ be an optimal pair of Problem (SD). For any $u \in \mathcal{U}$, we let $z = z(\,\cdot\,; u) \in W_0^{1,p}(\Omega)$ be the unique solution of the following problem:

(1.10)
$$\begin{cases} Az(x) = f_y(x, \bar{y}(x), \bar{u}(x))z(x) + f(x, \bar{y}(x), u(x)) \\ \qquad\qquad - f(x, \bar{y}(x), \bar{u}(x)), \qquad \text{in } \Omega, \\ z\big|_{\partial\Omega} = 0. \end{cases}$$

This system is referred to as the *variational system* along the pair $(\bar{y}, \bar{u})$. We define

(1.11)
$$\mathcal{R} = \{z(\,\cdot\,; u) \mid u \in \mathcal{U}\}.$$

This set is called the *reachable set* of the variational system (1.10). It is clear that $\mathcal{R} \subset W_0^{1,p}(\Omega)$. We define the Hamiltonian as follows:

(1.12)
$$H(x, y, u, \psi^0, \psi) = \psi^0 f^0(x, y, u) + \psi f(x, y, u),$$
$$\forall (x, y, u, \psi^0, \psi) \in \Omega \times \mathbb{R} \times U \times \mathbb{R} \times \mathbb{R}.$$

The following are the first order necessary conditions for optimal pairs of our Problem (SD).

**Theorem 1.2.** (Maximum Principle) *Let (S1)–(S5) hold. Let $(\bar{y}, \bar{u}) \in \mathcal{A}_{ad}$ be an optimal pair of Problem (SD). Let $F'(\bar{y})\mathcal{R} - Q$ be finite codimensional in $\mathcal{Y}$. Then there exists a triplet $(\psi^0, \psi, \varphi) \in [-1, 0] \times W_0^{1,p'}(\Omega) \times \mathcal{Y}^*$, such that $(\psi^0, \varphi) \neq 0$,*

(1.13)
$$\langle \varphi, \eta - F(\bar{y}) \rangle_{\mathcal{Y}^*, \mathcal{Y}} \leq 0, \qquad \forall \eta \in Q.$$

(1.14)
$$\begin{cases} A\psi(x) = f_y(x, \bar{y}(x), \bar{u}(x))\psi(x) \\ \qquad\qquad + \psi^0 f_y^0(x, \bar{y}(x), \bar{u}(x)) - F'(\bar{y}(\cdot))^*\varphi, \quad \text{in } \Omega, \\ \psi\big|_{\partial\Omega} = 0. \end{cases}$$

(1.15)
$$H(x, \bar{y}(x), \bar{u}(x), \psi^0, \psi(x)) = \max_{u \in U} H(x, \bar{y}(x), u, \psi^0, \psi(x)),$$
$$\text{a.e. } x \in \Omega.$$

*In addition, if $F'(\bar{y})^*$ is injective (i.e., $\mathcal{N}(F'(\bar{y})^*) = \{0\}$), then*

(1.16)
$$(\psi^0, \psi) \neq 0.$$

We refer to (1.13), (1.14), and (1.15) as the *transversality condition*, the *adjoint system* (along the given optimal pair), and the *maximum condition*, respectively.

§1. Semilinear Elliptic Equations

Let us make some remarks on the above result. First of all, by Proposition 1.1, we know that $\bar{y} \in L^\infty(\Omega)$. Thus, by (S3) and (S4), $f_y(\cdot, \bar{y}(\cdot), \bar{u}(\cdot)), f_y^0(\cdot, \bar{y}(\cdot), \bar{u}(\cdot)) \in L^\infty(\Omega)$. On the other hand, by (S5), $F'(\bar{y}) : W_0^{1,p}(\Omega) \to \mathcal{Y}$. Thus $F'(\bar{y})^* : \mathcal{Y}^* \to W^{-1,p'}(\Omega)$ ($p' = \frac{p}{p-1}$). Consequently, $F'(\bar{y})^*\varphi \in W^{-1,p'}(\Omega)$. Then, similar to Chapter 2, Theorem 6.8, (1.14) admits a unique solution $\psi(\cdot) \in W_0^{1,p'}(\Omega)$ (if $\psi^0$ is specified).

## §1.2. The state constraints

Let us now look at various kinds of state constraints covered by (1.8). Also, we will study the properties of the $\varphi$ appearing in Theorem 1.2. Remember that in applications, the state $y(x)$ of our control system can be regarded as the steady state of temperature distribution, concentration or potential, etc. Thus, the state constraints are really the constraints for these physical quantities (see Chapter 1). We note that by Chapter 4, Proposition 3.4, if $Q$ is finite codimensional in $\mathcal{Y}$, then so is $F'(\bar{y})\mathcal{R} - Q$. In what follows, we will give many kinds of $Q$'s that are finite codimensional in the corresponding state space $\mathcal{Y}$.

(1) Let $Q = \mathcal{Y}$. This corresponds to the case of no state constraint. In such a case, by the transversality condition (1.13), we have $\varphi = 0$. This is also the case if the state constraint is *inactive*, i.e., the set $Q$ has a nonempty interior in which the optimal state $\bar{y}$ lies. In such a case, the proof of Theorem 1.2 can be substantially simplified (we do not need the Ekeland principle here!).

(2) Let $\mathcal{Y} = C(\overline{\Omega})$. We note that under usual norms of continuous functions, the dual $\mathcal{Y}^*$ is not strictly convex. But, because $C(\overline{\Omega})$ is separable, from Chapter 2, Theorem 2.18, we can find an equivalent norm $|\cdot|_0$ on $C(\overline{\Omega})$, under which the dual $(C(\overline{\Omega}), |\cdot|_0)^*$ is strictly convex (see Chapter 4 for similar arguments). It is also clear that under this equivalent norm, we still have the Riesz Representation Theorem: $C(\overline{\Omega})^* = \mathcal{M}(\overline{\Omega}) \equiv \{\text{all the regular Borel measures on}\ \overline{\Omega}\}$. We consider the following type of state constraint:

(1.17) $$h(x, y(x)) \leq 0, \qquad \forall x \in \Omega,$$

where $h \in C(\overline{\Omega} \times \mathbb{R})$ satisfies the following: For each $x \in \overline{\Omega}$, $h(x, \cdot)$ is $C^1$ with $h_y \in C(\overline{\Omega} \times \mathbb{R})$. Moreover,

(1.18) $$h(x, 0) < 0, \qquad \forall x \in \partial\Omega.$$

We define

(1.19) $$\begin{cases} F(y)(x) = h(x, y(x)), & x \in \overline{\Omega},\ y \in C(\overline{\Omega}); \\ Q = \{\eta \in C(\overline{\Omega}) \mid \eta(x) \leq 0,\ \forall x \in \overline{\Omega}\}. \end{cases}$$

Then, (1.17) is of form (1.8). We claim that $Q$ has a nonempty interior in $\mathcal{Y}$. In fact, by taking $y_0(x) \equiv 2$, we see that the ball in $\mathcal{Y}$ centered at $y_0$

with radius 1 is contained in $Q$. This implies that $Q$ is codimension 0 in $\mathcal{Y}$. Let us look at the property that the corresponding $\varphi$ possesses. By the transversality condition (1.13), we have

$$\text{(1.20)} \qquad \int_{\bar{\Omega}} \big(\eta(x) - h(x,\bar{y}(x))\big)\, d\varphi(x) \leq 0, \qquad \forall \eta \in Q.$$

Denote $\Omega_0 = \{x \in \Omega \mid h(x,\bar{y}(x)) = 0\}$. This set is called the *active* set, meaning that the state constraint is active. The complement $\Omega \setminus \Omega_0$ is called the *inactive* set. Because $\bar{y}$ and $h$ are continuous, $\Omega \setminus \Omega_0$ is open. Also, by (1.18), we see that $\Omega_0 \cap \partial\Omega = \phi$. Now, for any open set $E \subset \Omega \setminus \Omega_0$, we let $\zeta \in C_0^\infty(E)$. Then, by the definition of $E$, for all $\varepsilon > 0$ small enough, $\eta_\varepsilon \equiv h(\cdot,\bar{y}(\cdot)) + \varepsilon\zeta(\cdot) \in Q$. Thus, taking $\eta = \eta_\varepsilon$ in (1.20), we obtain

$$\text{(1.21)} \qquad \int_E \zeta(x)\, d\varphi(x) \leq 0, \qquad \forall \zeta \in C_0^\infty(E).$$

This shows that the support $\operatorname{supp}\varphi$ of $\varphi$ is disjoint with $E$. Because $E$ is an arbitrary open set in $\Omega \setminus \Omega_0$, we arrive at

$$\text{(1.22)} \qquad \operatorname{supp}\varphi \subset \Omega_0.$$

Finally, from (1.20), when $x \in \Omega_0$, the integrand is nonpositive, thus $\varphi$ is a nonnegative valued measure supported on $\Omega_0$.

(3) Let $\mathcal{Y} = \mathbb{R}^m$ and $p > n$. Let $x_1, \cdots, x_m \in \Omega$. Define $F : W_0^{1,p}(\Omega) \to \mathbb{R}^m$ by the following:

$$\text{(1.23)} \qquad F(y(\cdot)) = (y(x_1), \cdots, y(x_m)), \qquad \forall y(\cdot) \in W_0^{1,p}(\Omega).$$

Clearly, $F$ is a linear bounded operator. Thus, it is Fréchet differentiable. We let $Q \subset \mathbb{R}^m$ be a (nonempty) convex and closed set. Because $\mathbb{R}^m$ is finite dimensional, $Q$ is of course finite codimensional in $\mathcal{Y}$. In the present case, our state constraint has the following form:

$$\text{(1.24)} \qquad (y(x_1), \cdots, y(x_m)) \in Q.$$

Let us look at the form of $\varphi \in \mathcal{Y}^* \equiv \mathbb{R}^m$. By the transversality condition (1.13), we see that

$$\text{(1.25)} \qquad \begin{cases} \varphi = 0, & \text{if } (\bar{y}(x_1), \cdots, \bar{y}(x_m)) \in \operatorname{Int} Q, \\ \varphi \in N_Q(\bar{y}(x_1), \cdots, \bar{y}(x_m)), & \text{if } (\bar{y}(x_1), \cdots, \bar{y}(x_m)) \in \partial Q, \end{cases}$$

where $N_Q(a_1, \cdots, a_m)$ is the set of all outward normals of $Q$ at the point $(a_1, \cdots, a_m) \in \partial Q$. We would like to see the form $F'(\bar{y}(\cdot))^*\varphi$ as such a term appears in the adjoint equation (1.14). Because $F$ is linear, we must have

$$\text{(1.26)} \quad F'(\bar{y}(\cdot))z(\cdot) = F(z(\cdot)) = (z(x_1), \cdots, z(x_m)), \qquad \forall z(\cdot) \in W_0^{1,p}(\Omega).$$

## §1. Semilinear Elliptic Equations

Then $F'(\bar{y}(\cdot))^* : \mathbb{R}^m \to W^{-1,p'}(\Omega)$ and it has the following form:

$$(1.27) \qquad \langle F'(\bar{y}(\cdot))^* \varphi, z(\cdot) \rangle = \sum_{i=1}^m \varphi_i z(x_i), \qquad \forall z(\cdot) \in W_0^{1,p}(\Omega),$$

where $\varphi = (\varphi_1, \cdots, \varphi_m)$. In other words,

$$(1.28) \qquad F'(\bar{y})^* \varphi = \sum_{i=1}^m \varphi_i \delta(x - x_i),$$

where $\delta(\cdot)$ is the $\delta$-function, by definition, $\langle \delta(\cdot - x_i), z(\cdot) \rangle = z(x_i)$, for all $z \in C(\bar{\Omega})$. Now, let us look at some interesting special cases of the above:

(a) Let $b_1, \cdots, b_m \in \mathbb{R}$ and define

$$(1.29) \qquad Q = \{(s_1, \cdots, s_m) \in \mathbb{R}^m \mid s_i \leq b_i, \ 1 \leq i \leq m\}.$$

Then, our state constraint becomes

$$(1.30) \qquad y(x_i) \leq b_i, \qquad 1 \leq i \leq m.$$

In this case, if we let $\varphi = (\varphi_1, \cdots, \varphi_m)$ and let $I_0 = \{i \mid 1 \leq i \leq m, \ \bar{y}(x_i) = b_i\}$, then (see (1.13))

$$(1.31) \qquad \begin{cases} \varphi_i = 0, & i \notin I_0, \\ \varphi_i \geq 0, & i \in I_0. \end{cases}$$

(b) Let $b_1, \cdots, b_m \in \mathbb{R}$, $\varepsilon_i > 0$ and define

$$(1.32) \qquad Q = \{(s_1, \cdots, s_m) \in \mathbb{R}^m \mid |s_i - b_i| \leq \varepsilon_i, \ 1 \leq i \leq m\}.$$

Then, the state constraint is of the form:

$$(1.33) \qquad |y(x_i) - b_i| \leq \varepsilon_i, \qquad 1 \leq i \leq m.$$

It is not hard for us to see that by setting $I_0 = \{i \mid 1 \leq i \leq m, \ |\bar{y}(x_i) - b_i| = \varepsilon_i\}$, we have that $\varphi_i = 0$ if $t \notin I_0$.

We can carry out a similar analysis for more general cases: Let $x_1, \cdots, x_k \in \Omega$, $h \in C^1(\mathbb{R}^k, \mathbb{R}^m)$ and define

$$(1.34) \qquad F(y) = h(y(x_1), \cdots, y(x_k)), \qquad \forall y(\cdot) \in W_0^{1,p}(\Omega).$$

We list some of the state constraints that can be covered by the above with appropriate choices of $F$ and $Q$.

$$(1.35) \qquad y(x_1) = y(x_2) = \cdots = y(x_m);$$

$$(1.36) \qquad y(x_1) \leq y(x_2) \leq \cdots \leq y(x_m);$$

$$(1.37) \qquad h_i(y(x_i)) = b_i, \qquad 1 \leq i \leq m.$$

It is not hard to come up with many other similar constraints; we leave the details to the interested readers.

(4) Let $\mathcal{Y} = \mathbb{R}^m$, $1 \leq p < \infty$, $h_i \in L^{p'}(\Omega)$, $1 \leq i \leq m$, linearly independent. We define $F : W_0^{1,p}(\Omega) \to \mathbb{R}^m$ as follows:

$$(1.38) \qquad F(y) = \Big( \int_\Omega y(x) h_1(x)\, dx, \cdots, \int_\Omega y(x) h_m(x)\, dx \Big),$$
$$\forall y(\cdot) \in W_0^{1,p}(\Omega).$$

Clearly, $F$ is linear and thus it coincides with its Fréchet derivative. Hence, $F'(\bar{y})^* : \mathbb{R}^m \to W^{-1,p'}(\Omega)$ can be identified by

$$(1.39) \qquad F'(\bar{y})^* \varphi = \sum_{i=1}^m \varphi_i h_i \in L^p(\Omega), \qquad \forall \varphi \in \mathbb{R}^m.$$

On the other hand, as $\mathbb{R}^m$ is finite dimensional, we may take any nonempty convex and closed set $Q$ in $\mathbb{R}^m$ as a constraint set.

We may consider a more general type of $F$, for example,

$$(1.40) \qquad F(y) = \Big( \int_\Omega f_1(x, y(x))\, dx, \cdots, \int_\Omega f_m(x, y(x))\, dx \Big),$$
$$\forall y \in W_0^{1,p}(\Omega),$$

for some differentiable functions $f_i$.

(5) Let $\mathcal{Y} = W^{1,p}(\Omega)$, $1 < p < \infty$. We take $F$ to be the identity operator on $\mathcal{Y}$ and set

$$(1.41) \qquad Q = \{ y(\cdot) \in W^{1,p}(\Omega) \mid \|y(\cdot)\|_{W^{1,p}(\Omega)} \leq 1 \}.$$

Clearly, this $Q$ has a nonempty interior. Thus $\operatorname{codim}_{\mathcal{Y}} Q = 0$. In this case, from (1.13), we have

$$(1.42) \qquad \|\varphi\|_{W^{-1,p'}(\Omega)} = \sup_{\|\eta\|_{W_0^{1,p}(\Omega)} \leq 1} \langle \varphi, \eta \rangle \leq \langle \varphi, \bar{y} \rangle \leq \|\varphi\|_{W^{-1,p'}(\Omega)}.$$

Thus, the equalities hold. Hence, we see that if $\|\bar{y}\|_{W_0^{1,p}(\Omega)} < 1$, then $\varphi = 0$. Now, we let $\|\bar{y}\|_{W_0^{1,p}(\Omega)} = 1$. Because $\varphi \in W^{-1,p'}(\Omega)$, we have the following representation (see Chapter 2, §6.1)

$$(1.43) \qquad \varphi = \varphi_0 - \sum_{i=1}^m \frac{\partial}{\partial x_i} \varphi_i, \qquad \text{in } \mathcal{D}'(\Omega),$$

for some $\varphi_0, \cdots, \varphi_n \in L^{p'}(\Omega)$. Then the equality in (1.42) implies that
(1.44)
$$\begin{cases} \int_\Omega \varphi_0 \bar{y}\, dx = \Big( \int_\Omega |\varphi_0|^{p'}\, dx \Big)^{1/p'} \Big( \int_\Omega |\bar{y}|^p\, dx \Big)^{1/p}; \\ \int_\Omega \varphi_i \bar{y}_{x_i}\, dx = \Big( \int_\Omega |f_i|^{p'}\, dx \Big)^{1/p'} \Big( \int_\Omega |\bar{y}_{x_i}|^p\, dx \Big)^{1/p}, \quad 1 \leq i \leq n. \end{cases}$$

By the condition that the Hölder inequality becomes equality, we know that

(1.45) $$\begin{cases} \varphi_0 = \mu|\bar{y}|^{p-1}\mathrm{sgn}\,(\bar{y}), \\ \varphi_i = \mu|\bar{y}_{x_i}|^{p-1}\mathrm{sgn}\,(\bar{y}_{x_i}), & 1 \leq i \leq n, \end{cases}$$

for some constant $\mu \geq 0$. This gives a representation of $\varphi$ in terms of $\bar{y}$.

It is possible to consider more complicated state constraints. For example,

(1.46) $$\begin{cases} \int_\Omega y(x)h_i(x)\,dx = a_i, & 1 \leq i \leq m, \\ |y(x)| \leq 1, & x \in \Omega. \end{cases}$$

This is a mixture of equality and inequality constraints. Of course, some conditions should be imposed to make the constraints consistent.

## §2. Variation Along Feasible Pairs

As we have seen in the previous chapter, in deriving necessary conditions for optimal pairs, one needs to make certain perturbations for the control, and the corresponding variations of the state and the cost functional need to be determined. This amounts to finding a "Taylor expansion" (of first order) for the state and the cost with respect to the perturbation of the control. We note that because the control domain $U$ is just a metric space, the perturbation of the control has to be of the "spike" type. Thus, to look for the corresponding "Taylor expansion" is a little technical. In the previous chapter, we have treated such a situation for evolution equations. In this section, we present a similar result for elliptic partial differential equations.

Let us first prove a lemma that is of independent interest. Recall that $|S|$ stands for the Lebesgue measure of the set $S$ in some Euclidean space.

**Lemma 2.1.** *Let $h^0 \in L^1(\Omega)$ and $h \in L^p(\Omega)$, $1 < p < \infty$. For any $\rho \in (0,1)$, we define*

(2.1) $$\mathcal{E}_\rho = \{E \subset \Omega \mid E \text{ measurable and } |E| = \rho|\Omega|\,\}.$$

*Let $\mathcal{Y}$ be a Banach space such that the embedding $\mathcal{Y} \hookrightarrow L^{p'}(\Omega)$ is compact $(p' = \frac{p}{p-1})$. Then*

(2.2) $$\inf_{E \in \mathcal{E}_\rho} \left\{ \left| \int_\Omega \left(1 - \frac{1}{\rho}\chi_E(x)\right) h^0(x)\,dx \right| + \left\| \left(1 - \frac{1}{\rho}\chi_E\right) h \right\|_{\mathcal{Y}^*} \right\} = 0.$$

*Proof.* Let $\rho \in (0,1)$ be given and let $\delta > 0$ be arbitrary. Let $B$ be the closed unit ball in $\mathcal{Y}$. By our assumption, $B$ is compact in $L^{p'}(\Omega)$. Thus, there exists a set of finitely many step functions $\Theta \equiv \{\theta_i, 1 \leq i \leq r\}$, such that for any $y \in B$, there exists a $\theta_i \in \Theta$ satisfying

(2.3) $$\|y - \theta_i\|_{L^{p'}(\Omega)} < \delta.$$

Now, let $\eta = (h^0, h\theta_1, \cdots, h\theta_r) \in L^1(\Omega; \mathbb{R}^{r+1})$. By Chapter 4, Lemma 3.7, we have

(2.4) $$\inf_{E \in \mathcal{E}_\rho} \left| \int_\Omega \left(1 - \frac{1}{\rho}\chi_E(x)\right)\eta(x)\,dx \right| = 0.$$

Consequently, for any $\varepsilon > 0$, there exists an $E \in \mathcal{E}_\rho$, such that

(2.5) $$\begin{cases} \left| \int_\Omega \left(1 - \frac{1}{\rho}\chi_E(x)\right)h^0(x)\,dx \right| < \varepsilon, \\ \left| \int_\Omega \left(1 - \frac{1}{\rho}\chi_E(x)\right)h(x)\theta_i(x)\,dx \right| < \varepsilon, \quad 1 \leq i \leq r. \end{cases}$$

Then, for any $y \in B$, by letting $\theta_i \in \Theta$ satisfying (2.3), we have

(2.6) $$\begin{aligned}
& \left| \int_\Omega \left(1 - \frac{1}{\rho}\chi_E(x)\right)h(x)y(x)\,dx \right| \\
& \leq \left| \int_\Omega \left(1 - \frac{1}{\rho}\chi_E(x)\right)h(x)\theta_i(x)\,dx \right| \\
& \quad + \left| \int_\Omega \left(1 - \frac{1}{\rho}\chi_E(x)\right)h(x)\{y(x) - \theta_i(x)\}\,dx \right| \\
& \leq \varepsilon + \left(1 + \frac{1}{\rho}\right)\|h\|_{L^p(\Omega)}\|y - \theta_i\|_{L^{p'}(\Omega)} \leq \varepsilon + \left(1 + \frac{1}{\rho}\right)\delta\|h\|_{L^p(\Omega)}.
\end{aligned}$$

This implies

(2.7) $$\left\|\left(1 - \frac{1}{\rho}\chi_E\right)h\right\|_{Y^*} \leq \varepsilon + \left(1 + \frac{1}{\rho}\right)\delta\|h\|_{L^p(\Omega)}.$$

Hence, together with (2.5), we obtain (2.2). □

The next result gives a sort of "Taylor expansion" formula.

**Theorem 2.2.** *Let $(y, u) \in \mathcal{A}$ be a fixed feasible pair and $v \in \mathcal{U}$ be fixed. Then, for any $\rho \in (0, 1)$, there exists a measurable set $E_\rho \subset \Omega$, with property*

(2.8) $$|E_\rho| = \rho|\Omega|,$$

*such that if we define $u_\rho$ by*

(2.9) $$u_\rho(x) = \begin{cases} u(x), & \text{if } x \in \Omega \setminus E_\rho, \\ v(x), & \text{if } x \in E_\rho, \end{cases}$$

*and let $y_\rho$ be the state corresponding to $u_\rho$, then it holds that*

(2.10) $$\begin{cases} y_\rho = y + \rho z + r_\rho, \\ \lim_{\rho \to 0} \frac{1}{\rho}\|r_\rho\|_{W^{1,p}(\Omega)} = 0; \end{cases}$$

## §2. Variation Along Feasible Pairs

and

(2.11) $$\begin{cases} J(u_\rho) = J(u) + \rho z^0 + r_\rho^0, \\ \lim_{\rho \to 0} \frac{1}{\rho} |r_\rho^0| = 0; \end{cases}$$

where $z$ and $z^0$ satisfy the following:

(2.12) $$\begin{cases} Az(x) = f_y(x, y(x), u(x))z(x) + f(x, y(x), v(x)) \\ \qquad\qquad - f(x, y(x), u(x)), \qquad \text{in } \Omega, \\ z|_{\partial\Omega} = 0. \end{cases}$$

(2.13) $$z^0 = \int_\Omega [f_y^0(x, y(x), u(x))z(x) + f^0(x, y(x), v(x)) \\ - f^0(x, y(x), u(x))]\, dx.$$

Before proving the above result, let us introduce the following: For any $u, v \in \mathcal{U}$, define

(2.14) $$\bar{d}(u, v) = |\{x \in \Omega \mid u(x) \neq v(x)\}|.$$

Using an argument similar to one in Chapter 4, one can show that $(\mathcal{U}, \bar{d})$ is a complete metric space. The following result is concerned with the continuity of the state $y(\,\cdot\,; u)$ with respect to the control $u$ under the above metric.

**Lemma 2.3.** Let $(y, u), (\widehat{y}, \widehat{u}) \in \mathcal{A}$. Then

(2.15) $$\|y - \widehat{y}\|_{W^{1,p}(\Omega)} \leq \begin{cases} C_p \bar{d}(u, \widehat{u})^{\frac{n+p}{np}}, & \text{if } p > \frac{n}{n-1}, \\ C_{p,q} \bar{d}(u, \widehat{u})^{1/q}, & \forall q > 1, \quad \text{if } p = \frac{n}{n-1}, \\ C_p \bar{d}(u, \widehat{u}), & \text{if } 1 \leq p < \frac{n}{n-1}, \end{cases}$$

where the constants $C_p$ and $C_{p,q}$ are independent of $u$ and $\widehat{u}$.

*Proof.* Denote

(2.16) $$\widetilde{a}(x) = -\int_0^1 f_y(x, y(x) + \tau(\widehat{y}(x) - y(x)), \widehat{u}(x))\, d\tau.$$

From (1.1), we see that $\widehat{y} - y$ satisfies the following:

(2.17) $$\begin{cases} A(\widehat{y}(x) - y(x)) + \widetilde{a}(x)(\widehat{y}(x) - y(x)) = f(x, y(x), \widehat{u}(x)) \\ \qquad\qquad - f(x, y(x), u(x)), \qquad \text{in } \Omega, \\ (\widehat{y} - y)|_{\partial\Omega} = 0. \end{cases}$$

By the $L^p$-estimate for the divergence form elliptic equations (see Chapter 2, §6), we obtain

$$\tag{2.18} \|\widehat{y} - y\|_{W^{1,p}(\Omega)} \leq C \|f(\cdot, y, \widehat{u}) - f(\cdot, y, u)\|_{W^{-1,p}(\Omega)}.$$

By Sobolev embedding (Chapter 2, §6.1) and the duality, we have

$$\tag{2.19} \begin{cases} L^{\frac{np}{n+p}}(\Omega) \hookrightarrow W^{-1,p}(\Omega), & \text{for } p > \dfrac{n}{n-1}, \\ L^q(\Omega) \hookrightarrow W^{-1,p}(\Omega), & \text{for } p = \dfrac{n}{n-1}, \quad \forall q > 1, \\ L^1(\Omega) \hookrightarrow W^{-1,p}(\Omega), & \text{for } 1 \leq p < \dfrac{n}{n-1}. \end{cases}$$

This, together with (2.18) and (1.6), gives (2.15). In the above, we should note that $\widehat{y}$ and $y$ are bounded and the constant in the $L^p$-estimate only depends on the modulus of continuity of the leading coefficients, the ellipticity constant, the bounds of the coefficients, and the domain. Thus, the constants appearing in (2.15) are independent of controls $u$ and $\widehat{u}$. □

Now, we are ready to prove Theorem 2.2.

*Proof of Theorem 2.2.* For any $\rho \in (0,1)$, by Lemma 2.1 with $\mathcal{Y} = W^{1,p'}(\Omega)$, we can find an $E_\rho \in \mathcal{E}_\rho$, such that

$$\tag{2.20} \left| \int_\Omega (1 - \frac{1}{\rho} \chi_{E_\rho}(x)) h^0(x) \, dx \right| + \|(1 - \frac{1}{\rho}\chi_{E_\rho})h\|_{W^{-1,p}(\Omega)} \leq \rho,$$

where $h^0$ and $h$ are defined by the following:

$$\tag{2.21} \begin{cases} h(x) = f(x, y(x), v(x)) - f(x, y(x), u(x)), \\ h^0(x) = f^0(x, y(x), v(x)) - f^0(x, y(x), u(x)). \end{cases}$$

Let $u_\rho$ be defined by (2.9) and let $y_\rho$ be the corresponding state. Set

$$\tag{2.22} z_\rho(x) = \frac{y_\rho(x) - y(x)}{\rho}, \qquad x \in \Omega.$$

Then $z_\rho$ satisfies the following:

$$\tag{2.23} \begin{cases} A z_\rho(x) + a^\rho(x) z_\rho(x) = \dfrac{1}{\rho} \chi_{E_\rho}(x) h(x), & \text{in } \Omega, \\ z_\rho\big|_{\partial\Omega} = 0, \end{cases}$$

where

$$\tag{2.24} a^\rho(x) = -\int_0^1 f_y(x, y(x) + \sigma(y_\rho(x) - y(x)), u_\rho(x)) \, d\sigma.$$

Clearly, by Lemma 2.3 and (S1)–(S3), we see that

$$\tag{2.25} a^\rho(x) \to a(x) \equiv -f_y(x, y(x), u(x)), \qquad \text{in } L^p(\Omega).$$

By recalling $z$, the solution of (2.12), we have the following:

(2.26)
$$\begin{cases} A(z_\rho(x) - z(x)) + a^\rho(x)(z_\rho(x) - z(x)) \\ \qquad = (a(x) - a^\rho(x))z(x) - (1 - \dfrac{1}{\rho})\chi_{E_\rho}(x)h(x), \\ (z_\rho - z)\big|_{\partial\Omega} = 0. \end{cases}$$

By the $L^p$-estimate (see Chapter 2, §6.2), we have

(2.27)
$$\begin{aligned} \dfrac{\|r_\rho\|_{W^{1,p}(\Omega)}}{\rho} &= \|z_\rho - z\|_{W^{1,p}(\Omega)} \\ &\le C\{\|(a - a^\rho)z\|_{W^{-1,p}(\Omega)} + \|(1 - \dfrac{1}{\rho}\chi_{E_\rho})h\|_{W^{-1,p}(\Omega)}\} \\ &\le C\{\|(a - a^\rho)z\|_{L^p(\Omega)} + \rho\} = o(1). \end{aligned}$$

This proves (2.10). Finally, we define $z^0$ as in (2.13) and let

(2.28)
$$r_\rho^0 = \dfrac{1}{\rho}(J(u_\rho) - J(u) - z^0).$$

Then, using (2.20) and (2.15), we have

(2.29)
$$\begin{aligned} \dfrac{1}{\rho}|r_\rho^0| &= \left|\dfrac{J(u_\rho) - J(u)}{\rho} - z^0\right| \\ &\le \left|\int_\Omega \Big(\int_0^1 f_y^0(x, y(x) + \sigma(y_\rho(x) - y(x)), u_\rho(x))\, d\sigma z_\rho(x)\right. \\ &\qquad \left.- f_y^0(x, y(x), u(x))z(x)\Big)\, dx\right| \\ &\quad + \left|\int_\Omega (1 - \dfrac{1}{\rho}\chi_{E_\rho}(x))h^0(x)\, dx\right| = o(1). \end{aligned}$$

This proves (2.11). □

## §3. Proof of the Maximum Principle

This section is devoted to a proof of the maximum principle stated in §1. The main idea is the same as that for evolution systems.

*Proof of Theorem 1.2.* Let $(\bar{y}, \bar{u})$ be an optimal pair. For any $u \in \mathcal{U}$, let $y(\,\cdot\,; u)$ be the corresponding state, emphasizing the dependence on the control. Similar to the previous chapter, we may assume, without loss of generality, that $J(\bar{u}) = 0$. For any $\varepsilon > 0$, define

(3.1)
$$J_\varepsilon(u) = \{[(J(u) + \varepsilon)^+]^2 + d_Q^2(F(y(\,\cdot\,; u)))\}^{1/2}.$$

Clearly, this functional is continuous on the (complete) metric space $(\mathcal{U}, \bar{d})$ (recall that $\bar{d}$ is the distance defined in (2.14)). Also, we have

(3.2)
$$\begin{cases} J_\varepsilon(u) > 0, & \forall u \in \mathcal{U}, \\ J_\varepsilon(\bar{u}) = \varepsilon \le \inf_{\mathcal{U}} J_\varepsilon(u) + \varepsilon. \end{cases}$$

Hence, by Ekeland's variational principle (see Chapter 4), we can find a $u^\varepsilon \in \mathcal{U}$, such that

$$(3.3) \quad \begin{cases} \bar{d}(\bar{u}, u^\varepsilon) \leq \sqrt{\varepsilon}, \\ J_\varepsilon(\hat{u}) - J_\varepsilon(u^\varepsilon) \geq -\sqrt{\varepsilon}\,\bar{d}(\hat{u}, u^\varepsilon), \quad \forall \hat{u} \in \mathcal{U}. \end{cases}$$

We let $v \in \mathcal{U}$ and $\varepsilon > 0$ be fixed and let $y^\varepsilon = y(\,\cdot\,;u^\varepsilon)$. By Theorem 2.2, we know that for any $\rho \in (0,1)$, there exists a measurable set $E_\rho^\varepsilon \subset \Omega$ with the property $|E_\rho^\varepsilon| = \rho|\Omega|$, such that if we define

$$(3.4) \quad u_\rho^\varepsilon(x) = \begin{cases} u^\varepsilon(x), & \text{if } x \in \Omega \setminus E_\rho^\varepsilon, \\ v(x), & \text{if } x \in E_\rho^\varepsilon, \end{cases}$$

and let $y_\rho^\varepsilon = y(\,\cdot\,;u_\rho^\varepsilon)$ be the corresponding state, then

$$(3.5) \quad \begin{cases} y_\rho^\varepsilon = y^\varepsilon + \rho z^\varepsilon + r_\rho^\varepsilon, \\ J(u_\rho^\varepsilon) = J(u^\varepsilon) + \rho z^{0,\varepsilon} + r_\rho^{0,\varepsilon}, \end{cases}$$

where $z^\varepsilon$ and $z^{0,\varepsilon}$ satisfy the following:

$$(3.6) \quad \begin{cases} Az^\varepsilon - f_y(x, y^\varepsilon(x), u^\varepsilon(x))z^\varepsilon = h^\varepsilon(x), & \text{in } \Omega, \\ z^\varepsilon\big|_{\partial\Omega} = 0. \end{cases}$$

$$(3.7) \quad z^{0,\varepsilon} = \int_\Omega [f_y^0(x, y^\varepsilon(x), u^\varepsilon(x))z^\varepsilon(x) + h^{0,\varepsilon}(x)]dx,$$

with

$$(3.8) \quad \begin{cases} h^\varepsilon(x) = f(x, y^\varepsilon(x), v(x)) - f(x, y^\varepsilon(x), u^\varepsilon(x)), \\ h^{0,\varepsilon}(x) = f^0(x, y^\varepsilon(x), v(x)) - f^0(x, y^\varepsilon(x), u^\varepsilon(x)). \end{cases}$$

and

$$(3.9) \quad \lim_{\rho \to 0}\frac{1}{\rho}\|r_\rho^\varepsilon\|_{W_0^{1,p}(\Omega)} = \lim_{\rho \to 0}\frac{1}{\rho}|r_\rho^{0,\varepsilon}| = 0.$$

Now, in the second relation of (3.3), we take $\hat{u} = u_\rho^\varepsilon$. It follows that

$$(3.10)\quad\begin{aligned} -\sqrt{\varepsilon}|\Omega| &\leq \frac{J_\varepsilon(u_\rho^\varepsilon) - J_\varepsilon(u^\varepsilon)}{\rho} \\ &= \frac{1}{J_\varepsilon(u_\rho^\varepsilon) + J_\varepsilon(u^\varepsilon)}\Big\{\frac{[(J(u_\rho^\varepsilon)+\varepsilon)^+]^2 - [(J(u^\varepsilon)+\varepsilon)^+]^2}{\rho} \\ &\qquad + \frac{d_Q^2(F(y_\rho^\varepsilon)) - d_Q^2(F(y^\varepsilon))}{\rho}\Big\} \\ &\to \frac{(J(u^\varepsilon)+\varepsilon)^+}{J_\varepsilon(u^\varepsilon)}z^{0,\varepsilon} + \langle\frac{d_Q(F(y^\varepsilon))\xi_\varepsilon}{J_\varepsilon(u^\varepsilon)},\ F'(y^\varepsilon)z^\varepsilon\rangle, \quad (\rho \to 0), \end{aligned}$$

§3. Proof of the Maximum Principle

where (note that $\mathcal{Y}^*$ is strictly convex)

(3.11) $$\xi_\varepsilon = \begin{cases} \nabla d_Q(F(y^\varepsilon)), & \text{if } F(y^\varepsilon) \notin Q, \\ 0, & \text{if } F(y^\varepsilon) \in Q. \end{cases}$$

We note that because $F : W_0^{1,p}(\Omega) \to \mathcal{Y}$, to obtain the convergence in (3.10), the expansion (3.5) *in the space* $W_0^{1,p}(\Omega)$ is necessary.

Next, we define $(\varphi^{0,\varepsilon}, \varphi^\varepsilon) \in [0,1] \times \mathcal{Y}^*$ as follows:

(3.12) $$\varphi^{0,\varepsilon} = \frac{(J(u^\varepsilon) + \varepsilon)^+}{J_\varepsilon(u^\varepsilon)}, \qquad \varphi^\varepsilon = \frac{d_Q(F(y^\varepsilon))\xi_\varepsilon}{J_\varepsilon(u^\varepsilon)}.$$

Then, (3.10) becomes

(3.13) $$-\sqrt{\varepsilon}|\Omega| \leq \varphi^{0,\varepsilon} z^{0,\varepsilon} + \langle \varphi^\varepsilon, F'(y^\varepsilon) z^\varepsilon \rangle.$$

By (3.1) and Chapter 4, Proposition 3.11, we have

(3.14) $$|\varphi^{0,\varepsilon}|^2 + \|\varphi^\varepsilon\|_{\mathcal{Y}^*}^2 = 1.$$

On the other hand, by the definition of the subdifferential (for the distance function), we have

(3.15) $$\langle \varphi^\varepsilon, \eta - F(y^\varepsilon) \rangle \leq 0, \qquad \forall \eta \in Q.$$

Next, from the first relation in (3.3) and Lemma 2.3,

(3.16) $$\|y^\varepsilon - \bar{y}\|_{W_0^{1,p}(\Omega)} \to 0, \qquad (\varepsilon \to 0).$$

Consequently,

(3.17) $$\lim_{\varepsilon \to 0} \|F'(y^\varepsilon) - F'(\bar{y})\|_{\mathcal{L}(W_0^{1,p}(\Omega), \mathcal{Y})} = 0.$$

From equations (3.6) and (3.7), we have

(3.18) $$\begin{cases} z^\varepsilon \to z, & \text{in } W_0^{1,p}(\Omega), \\ z^{0,\varepsilon} \to z^0, \end{cases} \qquad (\varepsilon \to 0),$$

where $z$ is the solution of the following variational system:

(3.19) $$\begin{cases} Az = f_y(x, \bar{y}(x), \bar{u}(x))z \\ \qquad + f(x, \bar{y}(x), v(x)) - f(x, \bar{y}(x), \bar{u}(x)), & \text{in } \Omega, \\ z|_{\partial\Omega} = 0, \end{cases}$$

and

(3.20) $$z^0 = \int_\Omega f_y^0(x, \bar{y}(x), \bar{u}(x)) z(x) \, dx \\ + \int_\Omega [f^0(x, \bar{y}(x), v(x)) - f^0(x, \bar{y}(x), \bar{u}(x))] \, dx.$$

We note that the solution $z$ of (3.19) and the quantity $z^0$ defined by (3.20) depend on the choice of $v \in \mathcal{U}$. Thus, we denote them by $z(\cdot\,;v)$ and $z^0(v)$, respectively. Then, (3.13), (3.15), and (3.18) give

$$(3.21) \quad \varphi^{0,\varepsilon} z^0(v) + \langle\, \varphi^\varepsilon, F'(\bar y) z(\cdot\,;v) - \eta + F(\bar y)\,\rangle \geq -\delta_\varepsilon, \quad \forall v \in \mathcal{U},\ \eta \in Q,$$

with $\delta_\varepsilon \to 0$ as $\varepsilon \to 0$. Because $F'(\bar y)\mathcal{R} - Q$ is finite codimensional in $\mathcal{Y}$, by Chapter 4, Lemma 3.6, one can extract some subsequence, still denoted by itself, such that

$$(3.22) \qquad\qquad (\varphi^{0,\varepsilon}, \varphi^\varepsilon) \xrightarrow{*} (\varphi^0, \varphi) \neq 0.$$

Then, taking limits in (3.21), we obtain

$$(3.23) \quad \varphi^0 z^0(v) + \langle\, \varphi, F'(\bar y) z(\cdot\,;v) - \eta + F(\bar y)\,\rangle \geq 0, \quad \forall v \in \mathcal{U},\ \eta \in Q.$$

Now, let $\psi^0 = -\varphi^0 \in [-1, 0]$. Then $(\psi^0, \varphi) \neq 0$. We rewrite (3.23) as follows:

$$(3.24)\ \psi^0 z^0(v) + \langle\, \varphi, \eta - F(\bar y)\,\rangle - \langle\, F'(\bar y)^*\varphi, z(\cdot\,;v)\,\rangle \leq 0, \quad \forall u \in \mathcal{U},\ \eta \in Q.$$

Take $v = \bar u$, we obtain the transversality condition (1.13). Next, we let $\eta = F(\bar y)$ to get

$$(3.25) \qquad \psi^0 z^0(v) - \langle\, F'(\bar y)^*\varphi, z(\cdot\,;v)\,\rangle \leq 0, \qquad \forall v \in \mathcal{U}.$$

Because $F'(\bar y)^*\varphi \in W^{-1,p'}(\Omega)$, for the given $\psi^0$, there exists a unique solution $\psi(\cdot) \in W_0^{1,p'}(\Omega)$ of the adjoint equation (1.14). Then, from (3.19), (3.20), and (3.25), we have
(3.26)
$$\begin{aligned}
0 \geq\ & \psi^0 z^0 - \langle\, F'(\bar y)^*\varphi, z\,\rangle \\
=\ & \psi^0 \int_\Omega f_y^0(x, \bar y(x), \bar u(x)) z(x)\, dx \\
& + \int_\Omega \psi^0 \{f^0(x, \bar y(x), v(x)) - f^0(x, \bar y(x), \bar u(x))\}\, dx \\
& + \langle\, A\psi - f_y(\cdot, \bar y, \bar u)\psi - \psi^0 f_y^0(\cdot, \bar y, \bar u), z\,\rangle \\
=\ & \int_\Omega \Big\{ \psi^0 [f^0(x, \bar y(x), v(x)) - f^0(x, \bar y(x), \bar u(x))] \\
& \qquad + \langle\, \psi(x), f(x, \bar y(x), v(x)) - f(x, \bar y(x), \bar u(x))\,\rangle \Big\}\, dx \\
=\ & \int_\Omega \Big\{ H(x, \bar y(x), v(x), \psi^0, \psi(x)) - H(x, \bar y(x), \bar u(x), \psi^0, \psi(x)) \Big\}\, dx.
\end{aligned}$$

Therefore, (1.15) follows. Finally, by (1.14), if $(\psi^0, \psi) = 0$, then $F'(\bar y)^*\varphi = 0$. Thus, in the case where $\mathcal{N}(F'(\bar y)^*) = \{0\}$, we must have $(\psi^0, \psi) \neq 0$, because $(\psi^0, \varphi) \neq 0$. $\square$

## §4. Variational Inequalities

This section is devoted to the study of a problem governed by semilinear variational inequalities with distributed controls. Thus, the state equation is the following:

(4.1) $$\begin{cases} Ay(x) + \beta(y(x)) \ni f(x, y(x), u(x)), & \text{in } \Omega, \\ y|_{\partial\Omega} = 0. \end{cases}$$

Let us make some assumptions. First, we let (V1)–(V5) be the same as (S1)–(S5) with $p = 2$. Further, we assume

(V6) $\beta \subset \mathbb{R} \times \mathbb{R}$ is a maximal monotone graph with $0 \in \mathcal{D}(\beta)$.

As in Chapter 3, §6, we have the following result.

**Proposition 4.1.** *Let (V1)–(V3) and (V6) hold. Then, for any $p \in [2, \infty)$ and any $u \in \mathcal{U}$, (4.1) admits a unique solution $y \equiv y(\cdot\,; u) \in W_0^{1,p}(\Omega) \cap L^\infty(\Omega)$ and*

(4.2) $$\|y(\cdot\,; u)\|_{W_0^{1,p}(\Omega) \cap L^\infty(\Omega)} \leq C_p, \qquad \forall u \in \mathcal{U},$$

*where $C_p$ is a constant independent of $u \in \mathcal{U}$.*

As before, we let $\mathcal{A}$ be the set of all feasible pairs $(y, u) \in W_0^{1,p}(\Omega) \times \mathcal{U}$ and let $\mathcal{A}_{ad}$ be the set of all pairs $(y, u) \in \mathcal{A}$ with the state constraint (1.8) satisfied. The cost functional $J(y, u)$ is taken to be (1.2). Similar to §1, we will use $J(u)$ or $J(y, u)$ as we wish when (V1)–(V4) and (V6) hold. Our optimal control problem can be stated as follows.

**Problem (V).** Find $(\bar{y}, \bar{u}) \in \mathcal{A}_{ad}$, such that

(4.3) $$J(\bar{y}, \bar{u}) = \inf_{\mathcal{A}_{ad}} J(y, u).$$

We should note that although the statement of Problem (V) is the same as that of Problem (SD), they are different because the state equations are different. The main feature of Problem (V) is that the state equation is *not* smooth. Thus, in deriving necessary conditions for the optimal pair, we need to approximate the state equation by some smooth one. In order for such an approximation to work properly, we need to impose some further conditions on the original problem. This leads to the following subsection.

### §4.1. Stability of the optimal cost

Our next assumption is concerned with the stability of the optimal cost with respect to the variation of the state constraint.

(V7) For any $(y_k, u_k) \in \mathcal{A}$, with

(4.4) $$d_Q(F(y_k)) \equiv \inf_{\eta \in Q} \|F(y_k) - \eta\|_Y \to 0, \quad k \to \infty,$$

it holds that

$$\lim_{k \to \infty} J(y_k, u_k) \geq \inf_{\mathcal{A}_{ad}} J(y, u). \tag{4.5}$$

The above (V7) is technically necessary in sequel because we have to regularize the state equation in the presence of the state constraint. This condition is referred to as the *stability of the optimal cost*. Also, if an optimal control problem has a state constraint with property (V7), we say that the problem is *stable*. Such a condition restricts the generality of our result. However, for many interesting problems, (V7) actually holds. Let us point out some of them. The easiest one is the case where $Q = \mathcal{Y}$, i.e., where there are no state constraints. In this case, $\mathcal{A} = \mathcal{A}_{ad}$, and thus, (4.5) trivially holds. Second, let $U$ be a bounded convex set in $\mathbb{R}^m$, $f^0(x, y, u)$ be convex in $u \in U$, and

$$f(x, y, u) = f_1(x, y) + f_2(x, y)u, \qquad \forall (x, y, u) \in \Omega \times \mathbb{R} \times U. \tag{4.6}$$

Then, for any $(y_k, u_k) \in \mathcal{A}$ where (4.4) holds, we may assume

$$\begin{cases} u_k \xrightarrow{*} \widehat{u}, & \text{in } L^\infty(\Omega, \mathbb{R}^m), \\ y_k \xrightarrow{w} \widehat{y}, & \text{in } W^{1,p}(\Omega), \quad p \geq 1, \\ y_k \xrightarrow{s} \widehat{y}, & \text{in } C(\bar{\Omega}), \end{cases} \tag{4.7}$$

and $(\widehat{y}, \widehat{u}) \in \mathcal{A}_{ad}$. Thus, by the convexity of $f^0(x, y, \cdot)$ and Mazur's Theorem (Chapter 2, Corollary 2.8), together with Fatou's Lemma, we see that

$$\lim_{k \to \infty} J(y_k, u_k) \geq J(\widehat{y}, \widehat{u}) \geq \inf_{\mathcal{A}_{ad}} J(y, u). \tag{4.8}$$

Thus, (V7) holds for this case. Finally, let us give the following more general result, whose proof can be carried out similar to that of Chapter 3, Theorem 6.4.

**Proposition 4.2.** *Let (V1)–(V6) hold. Let*

$$\Lambda(x, y) = \{(\lambda^0, \lambda) \in \mathbb{R}^2 \,|\, \lambda^0 \geq f^0(x, y, u), \; \lambda = f(x, y, u), \\ \text{for some } u \in U\}. \tag{4.9}$$

*Assume that for any $y \in W_0^{1,p}(\Omega)$, $x \mapsto \Lambda(x, y(x))$ is a measurable multifunction (this is the case if $f(x, y, u)$ and $f^0(x, y, u)$ are continuous, see Chapter 3 for more general cases) taking convex and closed set values. Then (V7) holds.*

The condition assumed in the above proposition is essentially the same as that assumed in the result for the existence of optimal pairs (see Chapter 3). This shows that (V7) is very closely related to the existence theory and thus, in some sense, it is a reasonable hypothesis. We will see the role played by (V7) in the proof of our main result of this section.

## §4.2. Approximate control problems

In this subsection, we introduce a family of approximate control problems. These problems have regular state equations and the state constraint sets are some small perturbation of the original set $Q$.

Let $\varepsilon > 0$ and consider

(4.10) $$\begin{cases} Ay_\varepsilon(x) + \beta_\varepsilon(y_\varepsilon(x)) = f(x, y_\varepsilon(x), u(x)), & \text{in } \Omega, \\ y_\varepsilon|_{\partial\Omega} = 0. \end{cases}$$

Here, $\beta_\varepsilon : \mathcal{D}(\beta_\varepsilon) \subset \mathbb{R} \to \mathbb{R}$ is a smooth nondecreasing function satisfying the following: (see Chapter 3, §6)

(4.11) $$\begin{cases} \mathcal{D}(\beta_\varepsilon) \supseteq \mathcal{D}(\beta), & \beta_\varepsilon(0) = 0, \\ \lim_{\varepsilon \to 0} \beta_\varepsilon(s) \in \beta(s), & \forall s \in \mathcal{D}(\beta). \end{cases}$$

By Proposition 6.3 of Chapter 3, we have that under conditions (V1)–(V4) and (V6), for any $u \in \mathcal{U}$ and $p \in [2, \infty)$, there exists a unique $y_\varepsilon \equiv y_\varepsilon(\cdot\,; u) \in W_0^{1,p}(\Omega) \cap L^\infty(\Omega)$ solving (4.10) and satisfying the following estimates:

(4.12) $$\begin{cases} \|y_\varepsilon\|_{W_0^{1,p}(\Omega) \cap L^\infty(\Omega)} \leq C_p \\ \|\beta_\varepsilon(y_\varepsilon)\|_{L^p(\Omega)} \leq C_p, \end{cases} \quad \forall u \in \mathcal{U}.$$

We let $\mathcal{A}^\varepsilon$ be the set of all pairs $(y_\varepsilon, u) \in W_0^{1,p}(\Omega) \times \mathcal{U}$ that satisfy (4.10). The following result is an improvement of the proof for Proposition 6.3 in Chapter 3.

**Lemma 4.3.** *Let (V1)–(V3) and (V6) hold. Let $u_\varepsilon \in \mathcal{U}$ be any sequence, $(y_\varepsilon, u_\varepsilon) \in \mathcal{A}^\varepsilon$, and $(y^\varepsilon, u_\varepsilon) \in \mathcal{A}$. Then, for any $p \in [1, \infty)$ and some subsequence,*

(4.13) $$\lim_{\varepsilon \to 0} \|y_\varepsilon - y^\varepsilon\|_{W_0^{1,p}(\Omega)} = 0.$$

*Proof.* It suffices to prove (4.13) for $p \geq 2$. First of all, by the definitions of $(y_\varepsilon, u_\varepsilon) \in \mathcal{A}^\varepsilon$ and $(y^\varepsilon, u_\varepsilon) \in \mathcal{A}$, we have that

(4.14) $$Ay_\varepsilon + \beta_\varepsilon(y_\varepsilon) = f(x, y_\varepsilon, u_\varepsilon), \quad y_\varepsilon|_{\partial\Omega} = 0,$$

(4.15) $$Ay^\varepsilon + \zeta^\varepsilon = f(x, y^\varepsilon, u_\varepsilon), \quad y^\varepsilon|_{\partial\Omega} = 0, \quad \zeta^\varepsilon \in \beta(y^\varepsilon).$$

Clearly, (4.12) holds for $y_\varepsilon$ and the first inequality in (4.12) holds for $y^\varepsilon$. Thus, we may assume that (let $p > \max\{n, 2\}$)

(4.16) $$\begin{cases} y_\varepsilon \xrightarrow{w} \widehat{y}, \quad y^\varepsilon \xrightarrow{w} y & \text{in } W_0^{1,p}(\Omega), \\ y_\varepsilon \xrightarrow{s} \widehat{y}, \quad y^\varepsilon \xrightarrow{s} y & \text{in } C^\alpha(\bar{\Omega}), \quad (0 < \alpha < 1 - \tfrac{n}{p}), \\ \beta_\varepsilon(y_\varepsilon) \xrightarrow{w} \widehat{\zeta}, & \text{in } L^p(\Omega), \\ \beta_\varepsilon(y_\varepsilon) \xrightarrow{s} \widehat{\zeta}, & \text{in } W^{-1,p}(\Omega). \end{cases}$$

By the convergence of $y_\varepsilon$ to $\widehat{y}$ in $C^\alpha(\bar{\Omega})$ and the maximal monotonicity of $\beta$, we see that $\widehat{\zeta} \in \beta(\widehat{y})$. Next, we take the difference between (4.14) and (4.15) to get

(4.17)
$$\begin{cases} A(y_\varepsilon - y^\varepsilon) + (\beta_\varepsilon(y_\varepsilon) - \zeta^\varepsilon) = f(x, y_\varepsilon, u_\varepsilon) - f(x, y^\varepsilon, u_\varepsilon), \\ (y_\varepsilon - y^\varepsilon)|_{\partial\Omega} = 0. \end{cases}$$

Multiplying the above by $y_\varepsilon - y^\varepsilon$, using (V1)–(V3) and (V6), we have

$$\lambda \|\nabla(y_\varepsilon - y^\varepsilon)\|_{L^2(\Omega)} \leq \langle A(y_\varepsilon - y^\varepsilon), y_\varepsilon - y^\varepsilon \rangle$$
$$= -\int_\Omega (\beta_\varepsilon(y_\varepsilon) - \zeta^\varepsilon)(y_\varepsilon - y^\varepsilon)\, dx$$
$$+ \int_\Omega \big(f(x, y_\varepsilon, u_\varepsilon) - f(x, y^\varepsilon, u_\varepsilon)\big)(y_\varepsilon - y^\varepsilon)\, dx$$

(4.18)
$$\leq -\int_\Omega \Big\{(\beta_\varepsilon(y_\varepsilon) - \widehat{\zeta})(y_\varepsilon - y^\varepsilon) + (\widehat{\zeta} - \zeta^\varepsilon)(\widehat{y} - y^\varepsilon)$$
$$+ (\widehat{\zeta} - \zeta^\varepsilon)(y_\varepsilon - \widehat{y})\Big\}\, dx$$
$$\leq \|\beta_\varepsilon(y_\varepsilon) - \widehat{\zeta}\|_{W^{-1,2}(\Omega)} \|y_\varepsilon - y^\varepsilon\|_{W^{1,2}(\Omega)}$$
$$+ \big(\|\widehat{\zeta}\|_{L^2(\Omega)} + \|\zeta^\varepsilon\|_{L^2(\Omega)}\big)\|y_\varepsilon - \widehat{y}\|_{L^2(\Omega)}$$
$$\leq C\big\{\|\beta_\varepsilon(y_\varepsilon) - \widehat{\zeta}\|_{W^{-1,2}(\Omega)} + \|y_\varepsilon - \widehat{y}\|_{L^2(\Omega)}\big\} \to 0.$$

In the above, we note that the monotonicity of $\beta$ implies

(4.19)
$$\int_\Omega (\widehat{\zeta} - \zeta^\varepsilon)(\widehat{y} - y^\varepsilon)\, dx \geq 0.$$

Then, combining with (4.17), we see that

(4.20)
$$\lim_{\varepsilon \to 0} \|\beta_\varepsilon(y_\varepsilon) - \zeta^\varepsilon\|_{W^{-1,2}(\Omega)} = 0.$$

For any $p \geq 2$, $L^p(\Omega) \hookrightarrow\hookrightarrow W^{-1,p}(\Omega) \hookrightarrow W^{-1,2}(\Omega)$; thus, by the proof of Lemma 5.9 (see (5.27), in particular) in Chapter 3, we have that for any $\delta > 0$, there exists a $C_\delta > 0$, such that

(4.21)
$$\|\beta_\varepsilon(y_\varepsilon) - \zeta^\varepsilon\|_{W^{-1,p}(\Omega)} \leq \delta \|\beta_\varepsilon(y_\varepsilon) - \zeta^\varepsilon\|_{L^p(\Omega)}$$
$$+ C_\delta \|\beta_\varepsilon(y_\varepsilon) - \zeta^\varepsilon\|_{W^{-1,2}(\Omega)}.$$

Next, by the boundedness of $\beta_\varepsilon(y_\varepsilon)$ and $\zeta^\varepsilon$ in $L^p(\Omega)$ and (4.20), we see that

(4.22)
$$\lim_{\varepsilon \to 0} \|\beta_\varepsilon(y_\varepsilon) - \zeta^\varepsilon\|_{W^{-1,p}(\Omega)} = 0, \quad \forall p \geq 2.$$

Hence, returning to (4.17), using the $L^p$-estimate and (4.18), we have

(4.23)
$$\|y_\varepsilon - y^\varepsilon\|_{W^{1,p}(\Omega)} \leq C\Big\{\|\beta_\varepsilon(y_\varepsilon) - \zeta^\varepsilon\|_{W^{-1,p}(\Omega)}$$
$$+ \|f(\cdot, y_\varepsilon, u_\varepsilon) - f(\cdot, y^\varepsilon, u_\varepsilon)\|_{L^p(\Omega)}\Big\} \to 0.$$

## §4. Variational Inequalities

This proves (4.13). □

Now, we let $(\bar{y}, \bar{u}) \in \mathcal{A}_{ad}$ be an optimal pair of Problem (V) and let $(\bar{y}_\varepsilon, \bar{u}) \in \mathcal{A}^\varepsilon$. By Lemma 4.3, we have that for any $p \in [2, \infty)$,

$$\lim_{\varepsilon \to 0} \|\bar{y}_\varepsilon - \bar{y}\|_{W^{1,p}(\Omega)} = 0. \tag{4.24}$$

Consequently, $\|F(\bar{y}_\varepsilon) - F(\bar{y})\|_{\mathcal{Y}} \to 0$. Thus,

$$\lim_{\varepsilon \to 0} d_Q(F(\bar{y}_\varepsilon)) = 0. \tag{4.25}$$

Hence, it is possible to construct a family $\{Q_\varepsilon\}$ of convex and closed subsets of $\mathcal{Y}$, such that

$$\begin{cases} Q \subseteq Q_\varepsilon, \quad F(\bar{y}_\varepsilon) \in Q_\varepsilon, \quad \forall \varepsilon > 0, \\ \rho_H(Q, Q_\varepsilon) \equiv \max\{\sup_{\eta \in Q} d_{Q_\varepsilon}(\eta), \sup_{\eta \in Q_\varepsilon} d_Q(\eta)\} \to 0, \quad (\varepsilon \to 0). \end{cases} \tag{4.26}$$

Set $\mathcal{A}^\varepsilon_{ad} = \{(y_\varepsilon, u) \in \mathcal{A}^\varepsilon \mid F(y_\varepsilon) \in Q_\varepsilon\}$ and introduce the following approximating problem:

**Problem (V)$_\varepsilon$.** Find a pair $(\tilde{y}_\varepsilon, \tilde{u}) \in \mathcal{A}^\varepsilon_{ad}$, such that

$$J(\tilde{y}_\varepsilon, \tilde{u}) = \inf_{\mathcal{A}^\varepsilon_{ad}} J(y_\varepsilon, u). \tag{4.27}$$

The following result is crucial in sequel.

**Proposition 4.4.** *Let (V1)–(V7) hold. Then the following is true.*

$$\lim_{\varepsilon \to 0} \inf_{\mathcal{A}^\varepsilon_{ad}} J(y_\varepsilon, u) = \inf_{\mathcal{A}_{ad}} J(y, u). \tag{4.28}$$

*Proof.* By the construction of $Q_\varepsilon$, we see that $(\bar{y}_\varepsilon, \bar{u}) \in \mathcal{A}^\varepsilon_{ad}$. Thus,

$$\bar{J}^\varepsilon \triangleq \inf_{\mathcal{A}^\varepsilon_{ad}} J(y_\varepsilon, u) \le J(\bar{y}_\varepsilon, \bar{u}). \tag{4.29}$$

It follows that (by Lemma 4.3)

$$\overline{\lim_{\varepsilon \to 0}} \bar{J}^\varepsilon \le J(\bar{y}, \bar{u}) = \inf_{\mathcal{A}_{ad}} J(y, u) \triangleq \bar{J}. \tag{4.30}$$

On the other hand, for any $\varepsilon > 0$, one can find $(y_\varepsilon, u_\varepsilon) \in \mathcal{A}^\varepsilon_{ad}$, such that

$$J(y_\varepsilon, u_\varepsilon) \le \bar{J}^\varepsilon + \varepsilon. \tag{4.31}$$

Let $(y^\varepsilon, u_\varepsilon) \in \mathcal{A}$ (i.e., $y^\varepsilon$ is the solution of (4.1) corresponding to $u_\varepsilon$). Then, by Lemma 4.3, we have

$$\lim_{\varepsilon \to 0} \|y^\varepsilon - y_\varepsilon\|_{W^{1,p}(\Omega)} = 0. \tag{4.32}$$

On the other hand, $F(y_\varepsilon) \in Q_\varepsilon$. Thus, by (4.26),

(4.33) $$\lim_{\varepsilon \to 0} d_Q(F(y^\varepsilon)) = 0.$$

Then, by (V7), we obtain

(4.34) $$\lim_{\varepsilon \to 0} J(y^\varepsilon, u_\varepsilon) \geq \bar{J}.$$

Using the continuity of $f^0$, (4.31), (4.32), and (4.34), we obtain

(4.35) $$\lim_{\varepsilon \to 0} \bar{J}_\varepsilon \geq \lim_{\varepsilon \to 0} J(y_\varepsilon, u_\varepsilon) \geq \bar{J}.$$

Then, (4.28) follows from (4.30) and (4.35). $\square$

We have seen that to obtain inequality (4.30), we do not need the stability condition (V7). However, in proving the other inequality (4.35), condition (V7) is essential.

### §4.3. Maximum principle and its proof

With the preparations made in the previous subsections, we are now ready to state and prove the maximum principle for Problem (V).

**Theorem 4.5.** (Maximum Principle) *Let (V1)–(V7) hold and $Q$ be finite codimensional in $\mathcal{Y}$. Let $(\bar{y}, \bar{u}) \in \mathcal{A}_{ad}$ be an optimal pair for Problem (V). Then there exists a pair $(\psi^0, \varphi) \in ([-1,0] \times \mathcal{Y}^*) \setminus \{0\}$, a $\psi \in W_0^{1,2}(\Omega)$, and a $\mu \in W^{-1,2}(\Omega)$, such that*

(4.36) $$\begin{cases} A\psi(x) + \mu = f_y(x, \bar{y}(x), \bar{u}(x))\psi(x) \\ \qquad\qquad + \psi^0 f^0(x, \bar{y}(x), \bar{u}(x)) - F'(\bar{y})^*\varphi, & \text{in } \Omega, \\ \psi|_{\partial\Omega} = 0. \end{cases}$$

(4.37) $$\langle \varphi, \eta - F(\bar{y}) \rangle_{\mathcal{Y}^*, \mathcal{Y}} \leq 0, \qquad \forall \eta \in Q.$$

(4.38) $$H(x, \bar{y}(x), \bar{u}(x), \psi^0, \psi(x)) = \max_{u \in U} H(x, \bar{y}(x), u, \psi^0, \psi(x)),$$
$$\text{a.e. } x \in \Omega,$$

*with the Hamiltonian $H$ given by (1.12). Moreover, if the following holds:*

(4.39) $$\operatorname{supp} \mu \bigcap \operatorname{supp}(F'(\bar{y})^*\varphi) = \phi, \qquad \mathcal{N}(F'(\bar{y})^*) = \{0\},$$

*then*

(4.40) $$(\psi^0, \psi(\cdot)) \neq 0.$$

## §4. Variational Inequalities

*Proof.* Let $\varepsilon > 0$ be fixed. For any $u \in \mathcal{U}$, let $y_\varepsilon \equiv y_\varepsilon(\cdot\,; u(\cdot))$ be the solution of (4.10) corresponding to $u$, which is unique. Then let

$$(4.41) \quad J_\varepsilon(u) = \left\{ [(J(y_\varepsilon, u) - \bar{J}^\varepsilon + \varepsilon)^+]^2 + d_{Q_\varepsilon}(F(y_\varepsilon))^2 \right\}^{1/2}, \quad \forall u \in \mathcal{U},$$

where $\bar{J}^\varepsilon$ is defined in (4.29). Again, as in §2, $J_\varepsilon$ is continuous on $(\mathcal{U}, \bar{d})$ and

$$(4.42) \quad J_\varepsilon(u) > 0, \quad \forall u \in \mathcal{U},$$

From Proposition 4.4, we have

$$(4.43) \quad \begin{aligned} 0 &< J_\varepsilon(\bar{u}) = [J(\bar{y}_\varepsilon, \bar{u}) - \bar{J}^\varepsilon + \varepsilon]^+ \stackrel{\Delta}{=} \sigma(\varepsilon) \\ &\leq |J(\bar{y}_\varepsilon, \bar{u}) - \bar{J}| + |\bar{J} - \bar{J}^\varepsilon| + \varepsilon \to 0, \quad (\varepsilon \to 0). \end{aligned}$$

Hence, by Ekeland's variational principle, there exists a $u^\varepsilon \in \mathcal{U}$, such that

$$(4.44) \quad \begin{cases} \bar{d}(u^\varepsilon, \bar{u}) \leq \sqrt{\sigma(\varepsilon)}, \\ J_\varepsilon(\hat{u}) - J_\varepsilon(u^\varepsilon) \geq -\sqrt{\sigma(\varepsilon)}\,\bar{d}(u^\varepsilon, \bar{u}), \quad \forall \hat{u} \in \mathcal{U}. \end{cases}$$

Then, almost the same as in the previous section, we can obtain the following (note that $\varepsilon > 0$ is fixed and let $(y^\varepsilon, u^\varepsilon) \in \mathcal{A}^\varepsilon$).

$$(4.45) \quad \begin{cases} \varphi^{0,\varepsilon} z^{0,\varepsilon}(v) + \langle \varphi^\varepsilon, F'(y^\varepsilon) z^\varepsilon(\cdot\,; v) - \eta + F(y^\varepsilon) \rangle \\ \quad \geq -\sqrt{\sigma(\varepsilon)}\,|\Omega|, \quad \forall v \in \mathcal{U},\ \eta \in Q_\varepsilon, \\ |\varphi^{0,\varepsilon}|^2 + \|\varphi^\varepsilon\|^2_{\mathcal{Y}^*} = 1, \end{cases}$$

where $z^\varepsilon(\cdot\,; v)$ is the unique solution of the following problem:

$$(4.46) \quad \begin{cases} A z^\varepsilon(x) + \beta'_\varepsilon(y^\varepsilon(x)) z^\varepsilon(x) = f_y(x, y^\varepsilon(x), u^\varepsilon(x)) z^\varepsilon(x) \\ \qquad + f(x, y^\varepsilon(x), v(x)) - f(x, y^\varepsilon(x), u^\varepsilon(x)), \quad \text{in } \Omega, \\ z^\varepsilon|_{\partial\Omega} = 0, \end{cases}$$

and $z^{0,\varepsilon}(v)$ is defined by

$$(4.47) \quad \begin{aligned} z^{0,\varepsilon}(v) = &\int_\Omega f^0_y(x, y^\varepsilon(x), u^\varepsilon(x)) z^\varepsilon(x)\, dx \\ &+ \int_\Omega [f^0(x, y^\varepsilon(x), v(x)) - f^0(x, y^\varepsilon(x), u^\varepsilon(x))]\, dx. \end{aligned}$$

Now, we let $\psi^0_\varepsilon = -\varphi^{0,\varepsilon} \in [-1, 0]$ and $\psi_\varepsilon \in W^{1,p}_0(\Omega)$ be the unique solution of the following problem:

$$(4.48) \quad \begin{cases} A\psi_\varepsilon(x) + \beta'_\varepsilon(y^\varepsilon(x))\psi_\varepsilon(x) = f_y(x, y^\varepsilon(x), u^\varepsilon(x))\psi_\varepsilon(x) \\ \qquad + \psi^0_\varepsilon f^0_y(x, y^\varepsilon(x), u^\varepsilon(x)) - F'(y^\varepsilon)^* \varphi^\varepsilon, \quad \text{in } \Omega, \\ \psi_\varepsilon|_{\partial\Omega} = 0. \end{cases}$$

Then we can obtain that

$$
(4.49) \quad \sqrt{\sigma(\varepsilon)}\,|\Omega| \geq \int_{\Omega} \bigl[ H(x, y^\varepsilon(x), v(x), \psi_\varepsilon^0, \psi_\varepsilon(x)) \\
- H(x, y^\varepsilon(x), u^\varepsilon(x), \psi_\varepsilon^0, \psi_\varepsilon(x)) \bigr]\, dx, \quad \forall v \in \mathcal{U},
$$

with the Hamiltonian $H$ defined by (1.12). This can be regarded as an approximate maximum condition. Our next goal is to take the limits to get the final result.

To this end, we first note that by (4.44) and the proof of Lemma 4.3, we have

$$
(4.50) \quad \lim_{\varepsilon \to 0} \|y^\varepsilon - \bar{y}\|_{W_0^{1,p}(\Omega)} = 0.
$$

Thus,

$$
(4.51) \quad \lim_{\varepsilon \to 0} \|F'(y^\varepsilon)^* - F'(\bar{y})^*\|_{\mathcal{L}(\mathcal{Y}^*, W^{-1,2}(\Omega))} = 0.
$$

Consequently, $F'(y^\varepsilon)^* \varphi^\varepsilon$ is bounded in $W^{-1,2}(\Omega)$. Then, multiplying (4.48) by $\psi_\varepsilon$ and integrating it over $\Omega$, we have (note that $\beta'_\varepsilon(y^\varepsilon) \geq 0$ and $f_y(x, y^\varepsilon, u^\varepsilon) \leq 0$)

$$
(4.52) \quad \lambda \|\nabla \psi_\varepsilon\|_{L^2(\Omega)}^2 + \int_\Omega \beta'_\varepsilon(y^\varepsilon) \psi_\varepsilon^2 \, dx \leq C \|\psi_\varepsilon\|_{W^{1,2}(\Omega)}, \quad \forall \varepsilon > 0.
$$

This implies that

$$
(4.53) \quad \|\psi_\varepsilon\|_{W^{1,2}(\Omega)} \leq C, \quad \forall \varepsilon > 0.
$$

Then, from (4.48), we get

$$
(4.54) \quad \|\beta'_\varepsilon(y^\varepsilon) \psi_\varepsilon\|_{W^{-1,2}(\Omega)} \leq C.
$$

Hence, we may let

$$
(4.55) \quad \begin{cases} \psi_\varepsilon \xrightarrow{w} \psi, & \text{in } W^{1,2}(\Omega), \\ \beta'_\varepsilon(y^\varepsilon) \psi_\varepsilon \xrightarrow{w} \mu, & \text{in } W^{-1,2}(\Omega), \\ \psi_\varepsilon^0 \to \psi^0, & \\ \varphi^\varepsilon \xrightarrow{*} \varphi, & \text{in } \mathcal{Y}^*. \end{cases}
$$

Clearly, $\psi$ satisfies (4.36). In (4.45), taking $v = u^\varepsilon$, we have

$$
(4.56) \quad \langle \varphi^\varepsilon, F(\bar{y}) - \eta \rangle \geq -\delta_\varepsilon, \quad \forall \eta \in Q,
$$

with $\delta_\varepsilon \to 0$. Hence, by the finite codimensionality of $Q$ and $|\varphi^{0,\varepsilon}|^2 + \|\varphi^\varepsilon\|_{\mathcal{Y}^*}^2 = 1$, we know that the pair $(\psi^0, \varphi) \in \mathbb{R} \times \mathcal{Y}^*$ is not zero. Now, letting $\varepsilon \to 0$ in (4.49) and (4.56), we obtain (4.38) and (4.37). Finally, if (4.39) holds and $(\psi^0, \psi(\cdot)) = 0$, then, by (4.36), we have $\mu = -F'(\bar{y})^* \varphi$.

Hence, by (4.39), we must have $\mu = 0$ and $F'(\bar{y})^*\varphi = 0$. Consequently, $\varphi = 0$. This implies $(\psi^0, \varphi) = 0$, a contradiction. □

It is noted that in the last step of the proof, we need some estimates on $\psi_\varepsilon$ and $\beta'_\varepsilon(y^\varepsilon)\psi_\varepsilon$ uniform in $\varepsilon$. We are only able to obtain such estimates, like (4.52)–(4.54) using $L^2$ theory. This is the reason why we have to restrict ourselves in the case $F: W_0^{1,2}(\Omega) \to \mathcal{Y}$. Also, we see that (V7) ensures the convergence $\bar{J}_\varepsilon \to \bar{J}$, which leads to $\sigma(\varepsilon) \to 0$ in (4.43). This is crucial in applying the Ekeland variational principle.

In the case where $\beta \equiv 0$, the state equation (4.1) is reduced to (1.1), a semilinear elliptic equation. In this case, we can see that $\mu = 0$. Thus, Theorem 4.5 will be a special case of Theorem 1.2 as the former only allows $p = 2$. In this case, condition (4.39) coincides with $\mathcal{N}(F'(\bar{y})^*) = \{0\}$, which appeared in Theorem 1.2.

Now, let us give an example to illustrate condition (4.39). We let $n = 1$ and $\mathcal{Y} = W_0^{1,2}(\Omega) \supset C(\bar{\Omega})$, $F = I$, the identity operator. Let

(4.57) $$Q = \{y \in \mathcal{Y} \mid y(x_i) = a_i, \quad 1 \le i \le m\},$$

with $x_i \in \Omega \subset \mathbb{R}$ and $a_i \in \mathbb{R}$ $(1 \le i \le m)$. We know that $\varphi \in \mathcal{Y}^* \subset C(\bar{\Omega})^*$ is a signed measure with

(4.58) $$\operatorname{supp} \varphi \subset \{x_i \mid 1 \le i \le m\}.$$

The proof of this fact can be carried out similarly as in §1.2. Thus, (4.39) means that

(4.59) $$\operatorname{supp} \varphi \bigcap \operatorname{supp} \mu = \phi.$$

We also see that, by (4.55) and the convergence $y^\varepsilon \to \bar{y}$ in $C(\bar{\Omega})$,

(4.60) $$\operatorname{supp} \mu \subset \{x \in \Omega \mid \bar{y}(x) \in \operatorname{Jump}(\beta)\},$$

where $\operatorname{Jump}(\beta)$ is the set of all points in $\mathcal{D}(\beta)$ at which $\beta$ is not single valued. Thus, (4.59) is implied by

(4.61) $$\{x_i \mid 1 \le i \le m\} \bigcap \{x \in \Omega \mid \bar{y}(x) \in \operatorname{Jump}(\beta)\} = \phi,$$

or

(4.62) $$\bar{y}(x_i) \notin \operatorname{Jump}(\beta), \quad 1 \le i \le m.$$

## §5. Quasilinear Equations

### §5.1. The state equation and the optimal control problem

In this section we study the control problem whose state equation is a quasilinear second order elliptic partial differential equation of the following form:

(5.1) $$\begin{cases} -\nabla \cdot a(x, \nabla y(x)) = f(x, y(x), u(x)), & \text{in } \Omega, \\ y|_{\partial \Omega} = 0. \end{cases}$$

We make the following assumptions. Let (Q1) and (Q3)–(Q4) be the same as (S1) and (S3)–(S4).

(Q2) The function $a : \overline{\Omega} \times \mathbb{R}^n \to \mathbb{R}^n$ is continuous with

(5.2) $$a(x, 0) = 0, \quad \forall x \in \overline{\Omega}.$$

For each $x \in \overline{\Omega}$, $a(x, \cdot)$ is differentiable and $a_\zeta(\cdot, \cdot)$ is continuous (we use $\zeta$ as the dummy argument for $\nabla y$). Moreover there exist constants $\alpha > 1$, $0 < \sigma \leq 1$, $\Lambda \geq \lambda > 0$, such that for all $x, \widehat{x} \in \Omega$, $\zeta, \xi \in \mathbb{R}^n$,

(5.3) $$\lambda(1 + |\zeta|)^{\alpha-2}|\xi|^2 \leq \langle a_\zeta(x, \zeta)\xi, \xi \rangle \leq \Lambda(1 + |\zeta|)^{\alpha-2}|\xi|^2,$$

(5.4) $$|a(x, \zeta) - a(\widehat{x}, \zeta)| \leq \Lambda(1 + |\zeta|)^{\alpha-1}|x - \widehat{x}|^\sigma.$$

A typical example of $a(x, \zeta)$ is the following:

(5.5) $$a(x, \zeta) = |\zeta|^{\alpha-2}\zeta, \quad (x, \zeta) \in \overline{\Omega} \times \mathbb{R}^n.$$

The corresponding state equation appears in the study of fluids. The number $\alpha$ is the characteristic of the medium. A medium with $\alpha > 2$ is called a *dilatant fluid*, one with $1 < \alpha < 2$ is called a *pseudo-plastic fluid*, and one with $\alpha = 2$ is called a *Newtonian fluid* (see Diaz [1]).

Let us recall that (see Chapter 2, §6.1) $C_0(\Omega)$ is the set of all continuous functions on $\overline{\Omega}$ that vanish on $\partial \Omega$ and $C^{1,\beta}(\overline{\Omega})$ is the set of all continuously differentiable functions on $\overline{\Omega}$ for which the first order partial derivatives are Hölder continuous with the exponent $\beta \in (0, 1)$. Now, let us give the following simple result.

**Proposition 5.1.** *Let (Q2) hold. Then, for any $y, z \in W_0^{1,\alpha}(\Omega)$, we have*

(5.6) $$\int_\Omega a(x, \nabla y(x)) \cdot \nabla z(x)\, dx \leq \frac{\Lambda}{\alpha - 1}\left(|\Omega|^{\frac{\alpha-1}{\alpha}} + \|\nabla y\|_{L^\alpha(\Omega)}\right)\|\nabla z\|_{L^\alpha(\Omega)},$$

(5.7) $$\int_\Omega \bigl(a(x, \nabla y) - a(x, \nabla z)\bigr) \cdot \nabla(y - z)\, dx \\ \geq \begin{cases} \dfrac{\lambda}{4^{\alpha-1}}\bigl\{\|\nabla(y - z)\|_{L^\alpha(\Omega)}^\alpha + \|\nabla(y - z)\|_{L^2(\Omega)}^2\bigr\}, & \alpha \geq 2, \\[2mm] \dfrac{\lambda\|\nabla(y - z)\|_{L^\alpha(\Omega)}^2}{\|(1 + |\nabla y| + |\nabla z|)\|_{L^\alpha(\Omega)}^{2-\alpha}}, & 1 < \alpha < 2. \end{cases}$$

## §5. Quasilinear Equations

*In particular,*

$$
(5.8) \quad \int_\Omega a(x,\nabla y)\cdot \nabla y\, dx \geq \begin{cases} \dfrac{\lambda}{4^{\alpha-1}}\|\nabla y\|^\alpha_{L^\alpha(\Omega)}, & \alpha \geq 2, \\[2mm] \dfrac{\lambda \|\nabla y\|^2_{L^\alpha(\Omega)}}{\|(1+|\nabla y|)\|^{2-\alpha}_{L^\alpha(\Omega)}}, & 1 < \alpha < 2. \end{cases}
$$

*Proof.* Denote $\xi(x) = \nabla y(x)$ and $\eta(x) = \nabla z(x)$. By (5.2) and (5.3),

$$
\left|\int_\Omega a(x,\xi(x))\eta(x)\, dx\right| = \left|\int_\Omega (a(x,\xi(x)) - a(x,0))\eta(x)\, dx\right|
$$

$$
= \left|\int_\Omega \int_0^1 \langle a_\zeta(x,\sigma\xi(x))\xi(x), \eta(x)\rangle\, d\sigma\, dx\right|
$$

(5.9)
$$
\leq \Lambda \int_\Omega \int_0^1 (1+\sigma|\xi(x)|)^{\alpha-2}|\xi(x)|\,|\eta(x)|\, d\sigma\, dx
$$

$$
= \frac{\Lambda}{\alpha-1}\int_\Omega \left\{(1+|\xi(x)|)^{\alpha-1} - 1\right\}|\eta(x)|\, dx
$$

$$
\leq \frac{\Lambda}{\alpha-1}\left\{\int_\Omega (1+|\xi(x)|)^\alpha dx\right\}^{\frac{\alpha-1}{\alpha}} \|\eta\|_{L^\alpha(\Omega)} < \infty.
$$

This implies (5.6). Next, let $\alpha \geq 2$. Then, for any $\bar\xi, \bar\eta \in \mathbb{R}^n$, we may let, say, $|\bar\xi| \geq |\bar\eta|$ (the other case is similar). Then $|\bar\xi - \bar\eta| \leq 2|\bar\xi|$. Hence,

$$
\int_0^1 (1+|\sigma\bar\xi + (1-\sigma)\bar\eta|)^{\alpha-2}\, d\sigma
$$

(5.10)
$$
\geq \int_{3/4}^1 (1+\sigma|\bar\xi| - (1-\sigma)|\bar\eta|)^{\alpha-2}\, d\sigma
$$

$$
\geq \int_{3/4}^1 \left(1+\frac{1}{2}|\bar\xi|\right)^{\alpha-2} d\sigma \geq \frac{1}{4}\left(1+\frac{1}{4}|\bar\xi-\bar\eta|\right)^{\alpha-2}.
$$

Consequently (for $\alpha \geq 2$),

$$
\int_\Omega (a(x,\xi) - a(x,\eta))\cdot(\xi-\eta)\, dx
$$

(5.11)
$$
= \int_\Omega \int_0^1 \langle a_\zeta(x,\sigma\xi + (1-\sigma)\eta)(\xi-\eta), \xi-\eta\rangle\, d\sigma\, dx
$$

$$
\geq \lambda \int_\Omega \int_0^1 (1+|\sigma\xi + (1-\sigma)\eta|)^{\alpha-2}|\xi-\eta|^2\, d\sigma\, dx
$$

$$
\geq \frac{\lambda}{4^{\alpha-1}}\int_\Omega (1+|\xi-\eta|)^{\alpha-2}|\xi-\eta|^2\, dx.
$$

Thus, the relation in (5.7) for $\alpha \geq 2$ follows.

To get the estimate for the case where $1 < \alpha < 2$, we first observe the following: For any $g, h \in L^\alpha(\Omega)$ with $g > 0$ and $h \geq 0$,

(5.12)
$$\left\{\int_\Omega h^\alpha \, dx\right\}^{\frac{2}{\alpha}} = \left\{\int_\Omega \frac{h^\alpha}{g^{\frac{(2-\alpha)\alpha}{2}}} g^{\frac{(2-\alpha)\alpha}{2}} \, dx\right\}^{2/\alpha}$$
$$\leq \left\{\int_\Omega \frac{h^2}{g^{2-\alpha}} \, dx\right\} \left\{\int_\Omega g^\alpha \, dx\right\}^{\frac{2-\alpha}{\alpha}}.$$

Consequently,

(5.13)
$$\int_\Omega \frac{h^2}{g^{2-\alpha}} \, dx \geq \|g\|_{L^\alpha(\Omega)}^{\alpha-2} \|h\|_{L^\alpha(\Omega)}^2.$$

Thus (taking $g = 1 + |\xi| + |\eta|$ and $h = |\xi - \eta|$ below),

(5.14)
$$\int_\Omega (a(x,\xi) - a(x,\eta)) \cdot (\xi - \eta) \, dx \geq \lambda \int_\Omega \frac{|\xi - \eta|^2}{(1 + |\xi| + |\eta|)^{2-\alpha}} \, dx$$
$$\geq \frac{\lambda \|\xi - \eta\|_{L^\alpha(\Omega)}^2}{\|(1 + |\xi| + |\eta|)\|_{L^\alpha(\Omega)}^{2-\alpha}}.$$

Thus, the relation in (5.7) for $1 < \alpha < 2$ also holds. Finally, (5.8) follows easily by taking $z = 0$ in (5.7). □

The above suggests that we introduce the following definition for the weak solution of (5.1).

**Definition 5.2.** A function $y \in W_0^{1,\alpha}(\Omega)$ is called a *weak solution* of (5.1) if for any $z \in W_0^{1,\alpha}(\Omega)$,

(5.15)
$$\int_\Omega a(x, \nabla y(x)) \cdot \nabla z(x) \, dx = \int_\Omega f(x, y(x), u(x)) z(x) \, dx.$$

Based on Proposition 5.1, similar to the case of semilinear equations, we have the following basic result concerning the state equation.

**Proposition 5.3.** *Let (Q1)–(Q3) hold. Then, for any $u \in \mathcal{U}$, (5.1) admits a unique weak solution $y \equiv y(\cdot\,;u) \in C^{1,\beta}(\overline{\Omega}) \cap C_0(\Omega)$ for some $\beta \in (0, \min\{\sigma, \gamma\})$. Furthermore, there exists a constant $C > 0$, independent of $u \in \mathcal{U}$, such that*

(5.16)
$$\|y(\cdot\,;u)\|_{C^{1,\beta}(\overline{\Omega})} \leq C, \qquad \forall u \in \mathcal{U}.$$

*Sketch of the Proof.* First of all, we truncate $f$: For any $m > 0$, let

(5.17)
$$f_m(x,y,u) = \begin{cases} f(x,y,u), & \text{if } |y| \leq m, \\ f(x,-m,u), & \text{if } y < -m, \\ f(x,m,u), & \text{if } y > m. \end{cases}$$

## §5. Quasilinear Equations

Then we consider the following truncated problem:

(5.18) $$\begin{cases} -\nabla \cdot a(x, \nabla y(x)) = f_m(x, y(x), u(x)), & \text{in } \Omega, \\ y|_{\partial\Omega} = 0. \end{cases}$$

From Proposition 5.1, we know that the operator defined by

(5.19) $$A(y)(x) = -\nabla \cdot a(x, \nabla y(x)) - f_m(x, y(x), u(x)), \qquad x \in \Omega,$$

is from $W_0^{1,\alpha}(\Omega)$ to $W^{-1,\alpha'}(\Omega)$ ($\alpha' = \frac{\alpha}{\alpha-1}$) satisfies the following condition:

(5.20) $$\langle A(y) - A(z), y - z \rangle \geq \begin{cases} C_\alpha \|y - z\|_{W_0^{1,\alpha}(\Omega)}^\alpha, & \alpha \geq 2, \\[2mm] \dfrac{C\|y - z\|_{W_0^{1,\alpha}(\Omega)}^2}{\left(1 + \|y\|_{W_0^{1,\alpha}(\Omega)} + \|z\|_{W_0^{1,\alpha}(\Omega)}\right)^{\alpha-2}}, & 1 < \alpha < 2. \end{cases}$$

This implies that $A$ is *strictly monotone*. Some more careful arguments will enable us to obtain the existence and the uniqueness of a weak solution $y_m \in W_0^{1,\alpha}(\Omega)$ to (5.1) and the fact that there exists an absolute constant $C > 0$ independent of $m$ and $u \in \mathcal{U}$, such that

(5.21) $$\|y_m(\cdot\,;u)\|_{C^{1,\beta}(\overline{\Omega})} \leq C, \qquad \forall m > 0, \ u \in \mathcal{U}.$$

Consequently, for $m > C$, we obtain that $y_m = y$ is a weak solution of (5.1). This proves our proposition. $\square$

A complete proof of the above result is very technical and we prefer not to get into it. Interested readers are referred to Rakotoson [1] and Lieberman [1].

In what follows, any pair $(y, u) \in (C^{1,\beta}(\overline{\Omega}) \cap C_0(\Omega)) \times \mathcal{U}$ satisfying (5.1) in the weak sense is called a *feasible pair*. We let $\mathcal{A}$ be the set of all feasible pairs. Next, let us introduce another map $h : \overline{\Omega} \times \mathbb{R} \to \mathbb{R}$. We assume the following:

(Q5) The map $h$ is continuous, and $h_y(\cdot\,,\cdot)$ exists and is continuous on $\overline{\Omega} \times \mathbb{R}$. Moreover we assume that

(5.22) $$h(x, 0) < 0, \qquad \forall x \in \partial\Omega.$$

From above, we know that under (Q1)–(Q3), for any $u \in \mathcal{U}$, the corresponding feasible state $y$ is in $C^{1,\beta}(\overline{\Omega})$. Thus, we may talk about the state constraint of the form

(5.23) $$h(x, y(x)) \leq 0, \qquad \forall x \in \overline{\Omega}.$$

As before, we let $\mathcal{A}_{ad}$ be the set of all pairs, called *admissible pairs*, that satisfy the constraint (5.23). Note that in the present case, we are working in the space $C^{1,\beta}(\overline{\Omega})$, instead of $W_0^{1,p}(\Omega)$. This is due to the fact that we need the state to have a continuous gradient in order to have the well-posedness of the linearized state equation (or the variational equation). It is possible to consider more general state constraints as in §1. However, we prefer to treat the above case for a change and simplicity. Finally, let us again take the cost functional to be (1.2). Then our optimal control problem can be stated as follows:

**Problem (Q).** Find $(\bar{y}, \bar{u}) \in \mathcal{A}_{ad}$, such that

(5.24) $$J(\bar{y}, \bar{u}) = \inf_{\mathcal{A}_{ad}} J(y, u).$$

## §5.2. The maximum principle

We want to present a corresponding result of Theorem 1.2 for Problem (Q). To this end, let us make some preliminaries. We let $\mathcal{M}(\overline{\Omega})$ be the space of all real Borel measures on $\overline{\Omega}$ and $\mathcal{M}(\Omega) = \{\mu|_\Omega \mid \mu \in \mathcal{M}(\overline{\Omega})\}$. Because $C_0(\Omega)$ is a separable Banach space, by Chapter 2, Theorem 2.18, we know that there exists a norm, denoted by $|\cdot|_0$, which is equivalent to the norm $\|\cdot\|_{C_0(\Omega)}$, such that the dual of $(C_0(\Omega), |\cdot|_0)$ is strictly convex. It is clear that any element $\mu \in (C_0(\Omega), |\cdot|_0)^*$ can still be identified with an element of $\mathcal{M}(\Omega)$, such that

(5.25) $$\langle \mu, \eta \rangle = \int_\Omega \eta(x) \, d\mu(x), \qquad \forall \eta \in C_0(\Omega).$$

In the rest of this section, whenever we write $C_0(\Omega)$, the corresponding norm is taken to be the above $|\cdot|_0$, and the dual space is identified to be $\mathcal{M}(\Omega)$ with the corresponding dual norm $|\cdot|_*$, which is equivalent to the usual norm $\|\cdot\|_{\mathcal{M}(\overline{\Omega})}$. Now, we define

(5.26) $$Q = \{\eta \in C_0(\Omega) \mid \eta(x) \leq 0, \quad \forall x \in \overline{\Omega}\}.$$

Clearly, $Q$ is convex, closed, and has a nonempty interior in $C_0(\Omega)$. Let

(5.27) $$d_Q(\eta) = \inf_{\tilde{\eta} \in Q} |\eta - \tilde{\eta}|_0, \qquad \forall \eta \in C_0(\Omega).$$

Then $d_Q : C_0(\Omega) \to \mathbb{R}$ convex and Lipschitz continuous (the Lipschitz constant is 1). Some analysis about this function can be found in Chapter 4, §3.3.

The following result is comparable with Theorem 1.2. Also the proof is very similar to that given in §3; thus, we will omit it.

**Theorem 5.4.** (Maximum Principle) *Let (Q1)–(Q5) hold. Let $(\bar{y}, \bar{u})$ be an optimal pair for Problem (Q). Then there exist a $\psi^0 \leq 0$, a $\psi \in W_0^{1,p'}(\Omega)$ with $p' < n/(n-1)$, and a $\varphi \in \mathcal{M}(\Omega)$, such that*

(5.28) $$(\psi^0, \varphi) \neq 0.$$

$$
(5.29) \quad \begin{cases} -\sum_{i,j=1}^{n} \left(a_{i,\zeta_j}(x, \nabla \bar{y}(x))\psi_{x_j}(x)\right)_{x_i} = f_y(x, \bar{y}(x), \bar{u}(x))\psi(x) \\ \qquad\qquad + \psi^0 f_y^0(x, \bar{y}(x), \bar{u}(x)) - h_y(x, \bar{y}(x))\varphi, \quad \text{in } \Omega, \\ \psi\big|_{\partial\Omega} = 0. \end{cases}
$$

$$
(5.30) \quad \int_\Omega (\eta(x) - h(x, \bar{y}(x)))\, d\varphi(x) \leq 0,
$$
$$
\forall \eta \in C_0(\Omega), \text{ with } \eta(x) \leq 0, \quad \forall x \in \overline{\Omega}.
$$

$$
(5.31) \quad H(x, \bar{y}(x), \bar{u}(x), \psi^0, \psi(x)) = \max_{u \in U} H(x, \bar{y}(x), u, \psi^0, \psi(x)),
$$
$$
\text{a.e. } x \in \Omega,
$$

where the Hamiltonian $H$ is given by (1.12).

## §6. Minimax Control Problem

### §6.1. Statement of the problem

In this section, we consider control system (1.1) with the following cost functional:

$$
(6.1) \quad J(u) = \operatorname{ess\,sup}_{x \in \Omega} f^0(x, y(x; u), u(x)).
$$

We make the following assumptions. Recall (S1)–(S5) stated in §1. Let (M2), (M3), and (M5) be the same as (S2), (S3), and (S5), and

(M1) The conditions in (S1) hold. Moreover, $U$ is a Polish space (see Chapter 3, §1.1).

(M4) $f^0(x, y, u)$ is continuous and satisfies (S4).

We note that in (S1), $U$ is any metric space and in (S4), $f^0$ is not assumed to be continuous. These are the only differences. It is clear that under (M1)–(M4), by Proposition 1.1, the above cost functional is well defined. A motivation of the above cost functional is the following: Suppose we would like to control the state $y$, which is subject to (1.1) together with state constraint (1.8), so that the largest deviation of the state from the desired one, say $y_0$, is minimized. In this case, we could take $f^0(x, y, u) = |y - y_0(x)|^2$. The above problem is referred to as a *minimax* control problem as our goal is to minimize a "maximum" (see Chapter 1).

Now, let us state our minimax control problem more precisely. As before, we let $\mathcal{A}$ and $\mathcal{U}$ be the sets of all feasible pairs and controls; and $\mathcal{A}_{ad}$ and $\mathcal{U}_{ad}$ be the sets of all admissible pairs and controls. Then, our optimal control problem can be stated as follows.

**Problem (M).** Find $\bar{u} \in \mathcal{U}_{ad}$, such that

(6.2) $$J(\bar{u}) = \inf_{\mathcal{U}_{ad}} J(u).$$

Following the basic idea of Chapter 3, the readers can prove the existence of optimal controls under suitable conditions (like the Cesari property for some multifunctions). Our purpose here is to derive some necessary conditions on the optimal pairs. It is seen that in addition to the difficulties we have had before, i.e., the state constraint exists and the control domain has no convexity, we also have a cost functional that is not smooth (in some sense). A similar difficulty is encountered in discussing the problem of elliptic variational inequalities (see §4), in which the state equation is nonsmooth. There, we regularized (or approximated) the state equation. It is thus very natural that we should regularize the cost functional for the present case.

Before going further, let us make some reductions. First of all, by scaling, we may assume that

(6.3) $$|\Omega| \equiv \operatorname{meas} \Omega = 1.$$

Next, from (M1)–(M3) and Proposition 1.1, it follows that $y$ is uniformly bounded, independent of $u \in \mathcal{U}$. Thus, by (M4), we see that in our problem, the values of $f^0(x, y, u)$ for large $y$ are irrelevant. Hence, without changing the original Problem (M), we may redefine $f^0(x, y, u)$ for large $y$ suitably so that (M4) remains true with the $M_R$ replaced by some fixed constant $M$. This means that we may assume without loss of generality, that

(6.4) $$|f^0(x, y, u)| \leq M, \quad \forall (x, y, u) \in \Omega \times \mathbb{R} \times U.$$

Once (6.4) holds, we may further set

(6.5) $$\widehat{f}^0(x, y, u) = \frac{f^0(x, y, u) + M + 1}{2(M+1)}, \quad \forall (x, y, u) \in \Omega \times \mathbb{R} \times U.$$

This yields

(6.6) $$0 < \frac{1}{2(M+1)} \leq \widehat{f}^0(x, y, u) \leq \frac{2M+1}{2(M+1)} < 1,$$
$$\forall (x, y, u) \in \Omega \times \mathbb{R} \times U.$$

On the other hand, it is clear that minimizing $J(u)$ is equivalent to minimizing

$$\widehat{J}(u) \equiv \operatorname{ess\,sup}_{x \in \Omega} \widehat{f}^0(x, y(x; u), u(x)).$$

Hence, again without loss of generality, we may assume at the beginning that

(6.7) $$0 < a \leq f^0(x, y, u) \leq b < 1, \quad \forall (x, y, u) \in \Omega \times \mathbb{R} \times U,$$

## §6. Minimax Control Problem

for some constants $a, b$. We will use assumptions (6.3) and (6.7) in the rest of the section. Clearly, under (6.7),

(6.8) $\quad J(u) \equiv \text{ess sup}_{x \in \Omega} f^0(x, y(x), u(x)) = \|f^0(\cdot, y(\cdot), u(\cdot))\|_{L^\infty(\Omega)}.$

### §6.2. Regularization of the cost functional

In this subsection, we are going to introduce a regularization of the cost functional. Then we will prove the stability of the optimal cost, which will be crucial in proving the maximum principle later. The basic idea is very simple; it comes from the following result of real analysis.

**Lemma 6.1.** *Let $h \in L^\infty(\Omega)$. Then*

(6.9) $\quad \lim_{r \to \infty} \|h\|_{L^r(\Omega)} = \|h\|_{L^\infty(\Omega)}.$

The proof of the above result is very simple and we leave it to the readers. From the above lemma, we immediately come up with the idea that the original cost functional (6.8) should be approximated by the following:

(6.10) $\quad J_r(u) = \left\{ \int_\Omega (f^0(x, y(x; u), u(x)))^r \, dx \right\}^{1/r}, \quad \forall u \in \mathcal{U}.$

In what follows, we denote

(6.11) $\quad m_r = \inf_{u \in \mathcal{U}_{ad}} J_r(u).$

Thus, $m_r$ is the optimal cost for Problem (M) where cost functional $J(u)$ is replaced by (6.10). As in §4 (see (V7)), we impose the following condition:

(M6) For any $(y_k, u_k) \in \mathcal{A}$, with

(6.12) $\quad \lim_{k \to \infty} d_Q(F(y_k)) = 0,$

it holds that

(6.13) $\quad \lim_{k \to \infty} J(u_k) \geq \inf_{\mathcal{U}_{ad}} J(u) \equiv \bar{m}.$

The following result is the stability of the optimal cost value, and is comparable with Proposition 4.4.

**Theorem 6.2.** *Let (M1)–(M6) hold. Then it holds that*

(6.14) $\quad \lim_{r \to \infty} m_r = \bar{m}.$

To prove this result, we need the following lemmas.

**Lemma 6.3.** *There exists a nondecreasing continuous function $\widehat{\omega} : [0, \infty) \to [0, \infty)$, with $\widehat{\omega}(0) = 0$, such that for any $(y, u) \in \mathcal{A}$ and $\alpha \in \mathbb{R}$, there exists a $(\tilde{y}, \tilde{u}) \in \mathcal{A}$, with the property that*

(6.15) $\quad f^0(x, \tilde{y}(x), \tilde{u}(x)) \leq \alpha + \widehat{\omega}(|M_\alpha|), \quad \text{a.e. } x \in \Omega,$

*where*

(6.16) $$M_\alpha = \{x \in \Omega \mid f^0(x, y(x), u(x)) > \alpha\},$$

*and*

(6.17) $$E \equiv \{x \in \Omega \mid u(x) \neq \widetilde{u}(x)\} \subseteq M_\alpha.$$

*Proof.* If $|M_\alpha| = 0$ or $|M_\alpha| = 1$ (recall $|\Omega| = 1$), (6.15) is trivially true by choosing suitable $\widehat{\omega}(\cdot)$ with $\widehat{\omega}(1) \geq 1 - \alpha$ (recall (6.7)). Now, we let $0 < |M_\alpha| < 1$. Let $\delta > 0$ be such that

(6.18) $$|M_\alpha| < |\mathcal{O}_\delta(0)| < 2|M_\alpha|,$$

where $\mathcal{O}_\delta(x)$ is an open ball centered at $x$ with radius $\delta$. Then we can choose $x_i \in \Omega$, such that

(6.19) $$\bigcup_{i \geq 1} \mathcal{O}_\delta(x_i) \supset \overline{\Omega}.$$

By (6.18), we know that for each $i \geq 1$, there exists an $\widetilde{x}_i \in \mathcal{O}_\delta(x_i) \setminus M_\alpha$. For such $\widetilde{x}_i$, it holds that

(6.20) $$f^0(\widetilde{x}_i, y(\widetilde{x}_i), u(\widetilde{x}_i)) \leq \alpha.$$

Then, we define

(6.21) $$\widetilde{u}(x) = \begin{cases} u(x), & x \in \Omega \setminus M_\alpha, \\ u(\widetilde{x}_i), & x \in \left[\mathcal{O}_\delta(x_i) \setminus \bigcup_{j=1}^{i-1} \mathcal{O}_\delta(x_j)\right] \bigcap M_\alpha. \end{cases}$$

Clearly, (6.17) holds. On the other hand, by $L^p$-theory, we know that there exists a constant $C$, independent of $u$ and $\widetilde{u}$, such that ($\widetilde{y} = y(\cdot\,;\widetilde{u})$) and

(6.22) $$\|y - \widetilde{y}\|_{W^{1,p}(\Omega)} \leq C|E|^{1/p} \leq C|M_\alpha|^{1/p}.$$

(See Lemma 2.3 for a stronger result.) Now, for any $x \in \Omega \setminus M_\alpha$, we have (let $\omega_0(\cdot)$ be the modulus of continuity for $f^0$ and let $p > n$)

(6.23) $$\begin{aligned} f^0(x, \widetilde{y}(x), \widetilde{u}(x)) &= f^0(x, \widetilde{y}(x), u(x)) \\ &\leq \alpha + \omega_0(|y(x) - \widetilde{y}(x)|) \leq \alpha + \omega_0(C|M_\alpha|^{1/p}). \end{aligned}$$

On the other hand, for $x \in M_\alpha \bigcap [\mathcal{O}_\delta(x_i) \setminus \bigcup_{j=1}^{i-1} \mathcal{O}_\delta(x_j)]$, we have (note (6.18))

(6.24) $$\begin{aligned} f^0(x, \widetilde{y}(x), \widetilde{u}(x)) &= f^0(x, \widetilde{y}(x), u(\widetilde{x}_i)) \\ &\leq \alpha + \omega_0(|x - \widetilde{x}_i| + |y(x) - \widetilde{y}(x)| + |y(\widetilde{x}_i) - y(x)|) \\ &\leq \alpha + \omega_0(\delta + C|M_\alpha|^{1/p} + C\delta^\beta) \leq \alpha + \widehat{\omega}(|M_\alpha|). \end{aligned}$$

## §6. Minimax Control Problem

Hence, (6.15) follows with a proper choice of $\widehat{\omega}$ (which is independent of $u \in \mathcal{U}$). Here, we have used the fact that $\|y(\cdot\,;u)\|_{C^\beta(\overline{\Omega})}$ is bounded uniformly in $u \in \mathcal{U}$ ($0 < \beta < 1 - \frac{n}{p}$ for large $p$). □

**Lemma 6.4.** *Let (M1)–(M6) hold. Then, for any sequence $(y_r, u_r) \in \mathcal{A}$ with*

(6.25) $$\lim_{r \to \infty} d_Q(F(y_r)) = 0,$$

*it holds*

(6.26) $$\lim_{r \to \infty} J_r(u_r) \geq \bar{m} \equiv \inf_{\mathcal{U}_{ad}} J(u).$$

*Proof.* Suppose that (6.26) does not hold. Then, for some $\varepsilon > 0$ and some subsequence (still denoted by itself) $(y_r, u_r) \in \mathcal{A}$ satisfying (6.25), we have

(6.27) $$\left( \int_\Omega (f^0(x, y_r(x), u_r(x)))^r \, dx \right)^{1/r} \leq \bar{m} - 2\varepsilon, \qquad \forall r \geq r_0.$$

Let

(6.28) $$\widehat{M}_r = \{ x \in \Omega \mid f^0(x, y_r(x), u_r(x)) > \bar{m} - \varepsilon \}.$$

Then (6.27) implies

(6.29) $$\bar{m} - 2\varepsilon \geq \left( \int_\Omega (f^0(x, y_r(x), u_r(x)))^r \, dx \right)^{1/r} \geq (\bar{m} - \varepsilon)|\widehat{M}_r|^{1/r}.$$

This yields

(6.30) $$|\widehat{M}_r| \leq \left( \frac{\bar{m} - 2\varepsilon}{\bar{m} - \varepsilon} \right)^r \to 0, \qquad (r \to \infty).$$

On the other hand, by Lemma 6.3, there exists a $(\widetilde{y}_r, \widetilde{u}_r) \in \mathcal{A}$, such that

(6.31) $$f^0(x, \widetilde{y}_r(x), \widetilde{u}_r(x)) \leq \bar{m} - \varepsilon + \widehat{\omega}(|\widehat{M}_r|), \qquad \text{a.e. } x \in \Omega,$$

and

(6.32) $$|\{ x \in \Omega \mid u_r(x) \neq \widetilde{u}_r(x) \}| \leq |\widehat{M}_r| \to 0.$$

Thus,

(6.33) $$\varlimsup_{r \to \infty} \operatorname{ess\,sup}_{x \in \Omega} f^0(x, \widetilde{y}_r(x), \widetilde{u}_r(x)) \leq \bar{m} - \varepsilon.$$

Also, (6.32) implies (note that $F$ is Fréchet differentiable)

(6.34) $$\begin{aligned} d_Q(F(\widetilde{y}_r)) &\leq C \|y_r - \widetilde{y}_r\|_{W^{1,p}(\Omega)} + d_Q(F(y_r)) \\ &\leq C |\widehat{M}_r|^{1/p} + d_Q(F(y_r)) \to 0. \end{aligned}$$

Hence, by (M6), we obtain

(6.35) $$\varlimsup_{r\to\infty} \text{ess sup}_{x\in\Omega} f^0(x,\widetilde{y}_r(x),\widetilde{u}_r(x)) \geq \bar{m}.$$

This contradicts (6.33). Thus, (6.26) holds. □

*Proof of Theorem 6.2.* First of all, by the Hölder inequality, we have (recall $|\Omega| = 1$ and (6.11), the definition of $m_r$)

(6.36) $$m_r \leq J_r(u) \leq J(u), \qquad \forall u \in \mathcal{U}_{ad}.$$

Thus, it follows that

(6.37) $$\varlimsup_{r\to\infty} m_r \leq \bar{m}.$$

On the other hand, there exists a pair $(y_r, u_r) \in \mathcal{A}_{ad}$, such that

(6.38) $$J_r(u_r) \leq m_r + \frac{1}{r}, \qquad r \geq 1.$$

Because $F(y_r) \in Q$, by Lemma 6.4, we have

(6.39) $$\lim_{r\to\infty} J_r(u_r) \geq \bar{m}.$$

Hence, our conclusion follows from (6.37)–(6.39). □

To conclude this subsection, let us make some comments on (M6) (compare those for (V7) made in §4.1). First of all, if $Q = \mathcal{Y}$, i.e., if there is no state constraint, then, (M6) holds. Second, if for each $(x,y) \in \Omega \times \mathbb{R}$, the set $f(x,y,U)$ is convex and closed, and $f^0(x,y,u) \equiv f^0(x,y)$, then (M6) holds. In fact, in this case, if $(y_k, u_k) \in \mathcal{A}$ satisfying (6.25), then, we can show that there exists a pair $(\widetilde{y}, \widetilde{u}) \in \mathcal{A}_{ad}$, such that for some subsequence,

(6.40) $$\|y_k - \widetilde{y}\|_{W^{1,p}(\Omega)} \to 0.$$

Thus,

(6.41) $$\bar{m} \leq J(\widetilde{u}) \leq \lim_{k\to\infty} J(u_k).$$

It is not hard to see that actually, under certain convexity conditions, that ensuring the existence of optimal pairs, (M6) holds. We leave the exact statement of such a result to the readers.

## §6.3. Necessary conditions for optimal controls

In this subsection, we state and prove the following necessary conditions for the optimal controls.

## §6. Minimax Control Problem

**Theorem 6.5.** (Maximum Principle) *Let (M1)–(M6) hold. Let $(\bar{y}, \bar{u}) \in \mathcal{A}_{ad}$ be an optimal pair for Problem (M). Moreover, let Q be finite codimensional in $\mathcal{Y}$. Then there exist $(\psi^0, \varphi) \in [-1, 0] \times \mathcal{Y}^* \setminus \{0\}$, $\mu, \lambda \in (L^\infty(\Omega))^*$, $\psi \in W_0^{1,p'}(\Omega)$, $p' = \frac{p}{p-1} \in (1, \frac{n}{n-1})$, and $N \subset \Omega$ with $|N| = 0$, such that*

(6.42) $\begin{cases} A\psi(x) = f_y(x, \bar{y}(x), \bar{u}(x))\psi(x) \\ \qquad\qquad + \psi^0 f_y^0(x, \bar{y}(x), \bar{u}(x))\mu - F'(\bar{y})^*\varphi - \lambda, & \text{in } \Omega, \\ \psi|_{\partial\Omega} = 0, \end{cases}$

(6.43) $\qquad\qquad \langle \varphi, \eta - F(\bar{y}) \rangle \leq 0, \qquad \forall \eta \in Q,$

(6.44) $\psi(x) f(x, \bar{y}(x), \bar{u}(x)) = \max_{u \in U(x)} \psi(x) f(x, \bar{y}(x), u), \quad \text{a.e. } x \in \overline{\Omega}_0,$

*where*

(6.45) $\begin{cases} \Omega_0 = \{x \in \Omega \mid f^0(x, \bar{y}(x), \bar{u}(x)) < \|f^0(\cdot, \bar{y}(\cdot), \bar{u}(\cdot))\|_{L^\infty(\Omega)}\}, \\ U(x) = \{u \in U \mid f^0(x, \bar{y}(x), u) \leq \|f^0(\cdot, \bar{y}(\cdot), \bar{u}(\cdot))\|_{L^\infty(\Omega)}\}, \\ \qquad\qquad\qquad\qquad\qquad\qquad\qquad\qquad x \in \Omega, \end{cases}$

*for any Lebesgue measurable set $G \subset N^c$,*

(6.46) $\qquad\qquad \lambda(G) \equiv \langle \lambda, \chi_G \rangle = 0,$

(6.47) $\qquad\qquad \mu(\Omega) \equiv \langle \mu, \chi_\Omega \rangle \geq a,$

*and in the case $|\Omega_0| > 0$, for any $0 < \varepsilon < |\Omega_0|$, there exists a measurable set $S_\varepsilon \subset \Omega_0$ with $|S_\varepsilon| \geq \varepsilon$, such that*

(6.48) $\qquad\qquad \mu(S_\varepsilon) = 0.$

*If $\mathcal{N}(F'(\bar{y})^*) = \{0\}$, it holds that $(\psi^0, \psi) \neq 0$.*

*If $f^0(x, y, u) \equiv f^0(x, y)$, $\Omega_0$ can be replaced by $\Omega$, $U(x) = U$, and $U$ can be any separable metric space (does not have to be a Polish space).*

We see that if $\psi \neq 0$, then (6.44) gives a necessary condition for the optimal control $\bar{u}$. Whereas, if $\psi = 0$, then (6.44) is trivial. In this case, (6.42) tells us that

(6.49) $\qquad\qquad \psi^0 f^0(x, \bar{y}(x), \bar{u}(x))\mu - F'(\bar{y})^*\varphi - \lambda = 0.$

This gives (implicitly, if $f^0$ is independent of $u$) a necessary condition for $\bar{u}$. By $(\psi^0, \varphi) \neq 0$, we know that (6.49) is a nontrivial condition. Also, if $|\overline{\Omega}_0| = 0$, (6.44) tells us nothing. But, in this case, we must have

(6.50) $\qquad f^0(x, \bar{y}(x), \bar{u}(x)) = \|f^0(\cdot, \bar{y}(\cdot), \bar{u}(\cdot))\|_{L^\infty(\Omega)}, \qquad \text{a.e. } x \in \Omega,$

This has already given us some information about the optimal pair $(\bar{y}, \bar{u})$.

We note that in general, the above $\mu$ is only a finitely additive measure and is not necessarily in $\mathcal{M}(\bar{\Omega})$. However, if $\mu$ happens to be in $\mathcal{M}(\bar{\Omega})$, then there exists a measurable set $S \subset \Omega_0$ with $|\Omega_0 \setminus S| = 0$, such that

$$\mu(S) = 0. \tag{6.51}$$

In another word, the support of $\mu$ is disjoint with $\Omega_0$.

*Proof of Theorem 6.5.* We first assume that $f^0$ does depend on $u$. Let $\Omega_0 \neq \phi$ (otherwise, there is nothing to prove). For $r > 1$, define

$$\widehat{J}_r(u) = \left\{ \left[ (J_r(u) - m_r + \frac{1}{r})^+ \right]^2 + d_Q(F(y(\cdot\,;u)))^2 \right\}^{1/2}, \tag{6.52}$$

$$\forall u \in \mathcal{U}.$$

Then, it is seen that (note Theorem 6.2)

$$\begin{cases} \widehat{J}_r(u) > 0, & \forall u \in \mathcal{U}, \\ \widehat{J}_r(\bar{u}) \leq |J_r(\bar{u}) - \bar{m}| + |\bar{m} - m_r| + \dfrac{1}{r} \equiv \sigma_r \to 0. \end{cases} \tag{6.53}$$

Thus, by the Ekeland variational principle, there exists a $u_r \in \mathcal{U}$, such that (recall the distance $\bar{d}$, see (2.14))

$$\begin{cases} \bar{d}(u_r, \bar{u}) \leq \sqrt{\sigma_r}, \\ -\sqrt{\sigma_r}\, \bar{d}(u_r, \widehat{u}) \leq \widehat{J}_r(\widehat{u}) - \widehat{J}_r(u_r), & \forall \widehat{u} \in \mathcal{U}. \end{cases} \tag{6.54}$$

Next, we let $s > 0$ such that (noting $\Omega_0 \neq \phi$)

$$\Omega_s \triangleq \{ x \in \Omega \mid f^0(x, \bar{y}(x), \bar{u}(x)) \leq \bar{m} - s \} \neq \phi. \tag{6.55}$$

Then denote

$$U_s(x) = \{ u \in U \mid f^0(x, \bar{y}(x), u) \leq \bar{m} - s \}. \tag{6.56}$$

From Chapter 3, Theorem 2.20, $U_s : \Omega_s \to 2^U$ is a measurable multifunction. Consequently, we see that the multifunction $\Gamma_s : \Omega \to 2^U$ defined by

$$\Gamma_s(x) = \begin{cases} U_s(x), & x \in \Omega_s, \\ \{\bar{u}(x)\}, & x \in \Omega \setminus \Omega_s \end{cases} \tag{6.57}$$

is measurable and takes closed set values almost everywhere. Thus, by Chapter 3, Theorem 2.23, there exists a measurable selection $v \in \mathcal{U}$ with $v(x) \in \Gamma_s(x)$, a.e. $x \in \Omega$. Here, we have used the assumption that $U$ is a Polish space (see Chapter 3). We let $\mathcal{V}_s$ be the set of all such selections and let $\mathcal{U}_s^r$ be the set of all $u \in \mathcal{U}$ given by

$$u(x) = \begin{cases} v(x), & x \in \Omega_s, \\ u_r(x), & x \in \Omega \setminus \Omega_s, \end{cases} \tag{6.58}$$

## §6. Minimax Control Problem

with $v \in \mathcal{V}_s$. Thus, any element $u \in \mathcal{U}_s^r$ can only be possibly different from $u_r$ on $\Omega_s$. Next, we fix $v \in \mathcal{V}_s$ and let $u \in \mathcal{U}_s^r$ be defined by (6.58). For any $\rho \in (0,1)$, by Theorem 2.2, there exists a measurable set $E_\rho \subset \Omega$ with $|E_\rho| = \rho|\Omega| = \rho$, such that if we define

$$(6.59) \qquad u_r^\rho(x) = \begin{cases} u_r(x), & x \in \Omega \setminus E_\rho, \\ u(x), & x \in E_\rho, \end{cases}$$

and let $y_r$ and $y_r^\rho$ be the states corresponding to the controls $u_r$ and $u_r^\rho$, respectively, then,

$$(6.60) \qquad \begin{cases} y_r^\rho = y_r + \rho z_r + \theta_r^\rho, \\ \|\theta_r^\rho\|_{W^{1,p}(\Omega)} = o(\rho), & (\rho \to 0), \end{cases}$$

and

$$(6.61) \qquad \int_\Omega (1 - \frac{1}{\rho}\chi_{E_\rho}(x))\{(f^0(x, y_r(x), u(x)))^r - (f^0(x, y_r(x), u_r(x)))^r\} \, dx = o(1), \qquad (\rho \to 0),$$

where $z_r$ is the solution of the following problem:

$$(6.62) \qquad \begin{cases} Az_r(x) = f_y(x, y_r(x), u_r(x))z_r(x) \\ \qquad\quad + f(x, y_r(x), u(x)) - f(x, y_r(x), u_r(x)), & \text{in } \Omega, \\ z_r|_{\partial\Omega} = 0. \end{cases}$$

The first inequality in (6.54) implies that as $r \to \infty$,

$$(6.63) \qquad \begin{cases} z_r \overset{w}{\to} z, & \text{in } W^{1,p}(\Omega), \\ z_r \overset{s}{\to} z, & \text{in } C(\bar{\Omega}), \end{cases}$$

where $z$ is the solution of

$$(6.64) \qquad \begin{cases} Az(x) = f_y(x, \bar{y}(x), \bar{u}(x))z(x) \\ \qquad\quad + \{f(x, \bar{y}(x), v(x)) - f(x, \bar{y}(x), \bar{u}(x))\}\chi_{\Omega_s}(x), & \text{in } \Omega, \\ z|_{\partial\Omega} = 0. \end{cases}$$

Here, we should note that by (6.58),

$$(6.65) \qquad \begin{aligned} & f(x, y_r(x), u(x)) - f(x, y_r(x), u_r(x)) \\ &= \{f(x, y_r(x), v(x)) - f(x, y_r(x), u_r(x))\}\chi_{\Omega_s}(x) \\ &\to \{f(x, \bar{y}(x), v(x)) - f(x, \bar{y}(x), \bar{u}(x))\}\chi_{\Omega_s}(x). \end{aligned}$$

Now, taking $\widehat{u} = u_r^\rho$ in (6.54), we obtain

(6.66)
$$\begin{aligned}
-\sqrt{\sigma_r} &\le \frac{\widehat{J}_r(u_r^\rho) - \widehat{J}_r(u_r)}{\rho} \\
&\to \frac{1}{\widehat{J}_r(u_r)}\Big\{(J_r(u_r) - m_r + \tfrac{1}{r})^+ \\
&\quad \int_\Omega \Big(\frac{(f^0(x, y_r(x), u_r(x)))^{r-1} f_y^0(x, y_r(x), u_r(x))}{(J_r(u_r))^{r-1}} z_r(x) \\
&\quad + \frac{(f^0(x, y_r(x), u(x)))^r - (f^0(x, y_r(x), u_r(x)))^r}{r(J_r(u_r))^{r-1}}\Big) dx \\
&\quad + \langle d_Q(F(y_r)) \nabla d_Q(F(y_r)), F'(y_r) z_r \rangle \Big\} \\
&\equiv \psi_r^0 \int_\Omega \Big(\mu_r(x) f_y^0(x, y_r(x), u_r(x)) z_r(x) + \frac{\delta h_r(x)}{r(J_r(u_r))^{r-1}}\Big) dx \\
&\quad + \langle F'(y_r)^* \varphi_r, z_r \rangle,
\end{aligned}$$

where

(6.67)
$$\begin{cases}
\psi_r^0 = \dfrac{(J_r(u_r) - m_r + \tfrac{1}{r})^+}{\widehat{J}_r(u_r(\cdot))}, \\
\mu_r(x) = \Big(\dfrac{f^0(x, y_r(x), u_r(x))}{J_r(u_r)}\Big)^{r-1}, \\
\delta h_r(x) = (f^0(x, y_r(x), u(x)))^r - (f^0(x, y_r(x), u_r(x)))^r, \\
\varphi_r = \dfrac{d_Q(F(y_r)) \nabla d_Q(F(y_r))}{\widehat{J}_r(u_r)}.
\end{cases}$$

We should note that there is no ambiguity in the definition of $\varphi_r$ because of the strict convexity of $\mathcal{Y}^*$ (similar to §3). It is also clear that (see Chapter 4, Proposition 3.11)

(6.68)
$$(\psi_r^0)^2 + \|\varphi_r\|_{\mathcal{Y}^*}^2 = 1.$$

Next, by (6.7), it is seen that for any $r > 1$, function $\mu_r(x) \ge 0$ satisfies

(6.69)
$$\begin{aligned}
\int_\Omega \mu_r(x) \, dx &= \int_\Omega \Big(\frac{f^0(x, y_r(x), u_r(x))}{J_r(u_r)}\Big)^{r-1} dx \\
&\ge \frac{1}{(J_r(u_r))^{r-1}} \int_\Omega (f^0(x, y_r(x), u_r(x)))^r \, dx = J_r(u_r) \ge a,
\end{aligned}$$

and by the Hölder inequality,

(6.70)
$$\int_\Omega \mu_r(x) \, dx \le \|\mu_r\|_{L^{r/(r-1)}(\Omega)} = 1.$$

## §6. Minimax Control Problem

Thus, we may assume

(6.71) $\quad \mu_r \stackrel{*}{\rightharpoonup} \mu, \quad \mu_r f_y^0(\cdot, y_r(\cdot), u_r(\cdot)) \stackrel{*}{\rightharpoonup} \widetilde{\mu}, \quad \text{in } (L^\infty(\Omega))^*,$

for some $\mu, \widetilde{\mu} \in (L^\infty(\Omega))^*$. Clearly, by (6.69), $\mu$ satisfies (6.47). By (6.53) and (6.54), we have

(6.72) $\quad 0 < \widehat{J}_r(u_r) \leq \widehat{J}_r(\bar{u}) + \sqrt{\sigma_r}\, \bar{d}(u_r, \bar{u}) \leq 2\sigma_r \to 0, \qquad (r \to \infty).$

Hence, it follows from the definition of $\widehat{J}_r(u)$ and Theorem 6.2 that

(6.73) $\quad \varlimsup_{r \to \infty} J_r(u_r) \leq \bar{m}, \qquad \lim_{r \to \infty} d_Q(F(y_r)) = 0.$

Thus, Lemma 6.4 applies; namely, (6.26) holds. Then, combining the first inequality in (6.73) with (6.26), one obtains

(6.74) $\quad \lim_{r \to \infty} J_r(u_r) = \bar{m}.$

Now, by the definition of $u$ (see (6.58) and (6.7)), we obtain

(6.75) $\quad \begin{aligned}\int_\Omega \frac{\delta h_r(x)}{r(J_r(u_r))^{r-1}}\, dx &\leq \frac{J_r(u_r)}{r} \int_{\Omega_s} \left(\frac{f^0(x, y_r(x), v(x))}{J_r(u_r)}\right)^r dx \\ &\leq \frac{b}{r} \int_{\Omega_s} \left(\frac{f^0(x, \bar{y}(x), v(x)) + \varepsilon_r}{\bar{m} - \varepsilon_r}\right)^r dx \\ &\leq \frac{b}{r} \int_\Omega \left(\frac{\bar{m} - s + \varepsilon_r}{\bar{m} - \varepsilon_r}\right)^r dx \to 0, \qquad (r \to \infty)\end{aligned}$

because $\varepsilon_r \to 0$ and $s > 0$ is fixed. Next, by the convexity of $Q$, we have

(6.76) $\quad \langle \varphi_r, \eta - F(y_r) \rangle \leq 0, \qquad \forall \eta \in Q.$

From (6.68) and the finite dimensionality of $Q$, one may assume that

(6.77) $\quad (\psi_r^0, \varphi_r) \stackrel{*}{\rightharpoonup} (\bar{\psi}^0, \varphi) \neq 0, \qquad \text{in } \mathbb{R} \times \mathcal{Y}^*.$

Now, we take limits in (6.76) to obtain the transversality condition (6.43). Next, we take limits in (6.66) to obtain

(6.78) $\quad 0 \leq \bar{\psi}^0 \langle \widetilde{\mu}, z(\cdot) \rangle + \langle F'(\bar{y})^* \varphi, z \rangle.$

Here, $z$ is the solution of (6.64) with $v \in \mathcal{V}_s$. On the other hand, by the Egorov Theorem, for any $\varepsilon > 0$, there exists an $N_\varepsilon \subset \Omega$ with $|N_\varepsilon| < \varepsilon$, such that $f_y^0(x, y_r(x), u_r(x)) \to f_y^0(x, \bar{y}(x), \bar{u}(x))$ uniformly in $x \in N_\varepsilon^c$. Thus, for any $\eta \in C(\bar{\Omega})$, and any $G \subset N_\varepsilon^c$, by (6.71), we have

(6.79) $\quad \langle \widetilde{\mu}, \eta \chi_G \rangle = \langle f_y^0(\cdot, \bar{y}(\cdot), \bar{u}(\cdot))\mu, \eta \chi_G \rangle.$

Consequently, there exists an $N \subset \Omega$ with $|N| = 0$, such that

(6.80) $\quad \lambda \stackrel{\Delta}{=} \psi^0(f_y^0(\cdot, \bar{y}(\cdot), \bar{u}(\cdot))\mu - \widetilde{\mu}),$

with $\psi^0 = -\bar{\psi}^0$, satisfying (6.46). Also, we can rewrite (6.78) as follows

(6.81) $\qquad 0 \geq \langle \psi^0 f_y^0(\cdot, \bar{y}(\cdot), \bar{u}(\cdot))\mu - \lambda, z(\cdot) \rangle - \langle F'(\bar{y})^* \varphi, z \rangle.$

Let $\psi \in W_0^{1,p'}(\Omega)$ be the solution of (6.42). Then, combining (6.81) with some straightforward computations, we can obtain

(6.82) $$0 \geq \int_{\Omega_s} \psi(x)[f(x, \bar{y}(x), v(x)) - f(x, \bar{y}(x), \bar{u}(x))]\, dx,$$

$$\forall v \in \mathcal{V}_s,\ s > 0.$$

Thus, the maximum condition (6.44) follows. Now, we show (6.48). If it is not true, then there exists an $\varepsilon_0 \in (0, |\Omega_0|)$, such that for any measurable set $S \subset \Omega_0$ with $|S| \geq \varepsilon_0$,

(6.83) $\qquad \mu(S) \equiv \langle \mu, \chi_S \rangle > 0.$

Because $\bar{d}(u_r, \bar{u}) \to 0$ (see (6.54)), we see that for some $r_0 \geq 1$, $|\Omega_0 \setminus \{u_r \neq \bar{u}\}| > \varepsilon_0$, for $r \geq r_0$. Thus, by the definition of $\Omega_0$, we can find an $S \subset \Omega_0 \setminus \{u_r \neq \bar{u}\}$ with $|S| \geq \varepsilon_0$ and some $\varepsilon > 0$, such that

(6.84) $\qquad f^0(x, \bar{y}(x), \bar{u}(x)) \leq \bar{m} - 3\varepsilon, \qquad x \in S.$

Since $y_r \to \bar{y}$ in $C(\overline{\Omega})$, we obtain (if necessary, we enlarge $r_0$)

(6.85) $\qquad f^0(x, y_r(x), u_r(x)) \leq \bar{m} - 2\varepsilon, \qquad \forall x \in S,\ r \geq r_0.$

On the other hand, by (6.74), for $r$ large enough, one has $J_r(u_r) \geq \bar{m} - \varepsilon$. Hence, by the definition of $\mu$ and $\mu_r$,

(6.86) $$\mu(S) = \lim_{r \to \infty} \int_S \mu_r(x)\, dx = \lim_{r \to \infty} \int_S \left( \frac{f^0(x, y_r(x), u_r(x))}{J_r(u_r)} \right)^{r-1} dx$$

$$\leq \left( \frac{\bar{m} - 2\varepsilon}{\bar{m} - \varepsilon} \right)^{r-1} |S| \to 0.$$

This contradicts (6.83). Thus, (6.48) holds. Similar as before, we can easily prove that $\mathcal{N}(F'(\bar{y})^*) = \{0\}$ implies $(\psi^0, \psi) \neq 0$.

Finally, in the case where $f^0$ is independent of $u$, $\delta h_r = 0$ and we can carry out the proof without considering $\Omega_s$ and $\mathcal{V}_s$, etc. Thus, the final conclusion of Theorem 6.5 follows. $\qquad \square$

## §7. Boundary Control Problems

In this section, we will derive necessary conditions for optimal control problems of semilinear elliptic equations involving boundary controls.

### §7.1. Formulation of the problem

Consider the following state equation:

(7.1) $\qquad \begin{cases} Ay(x) = f(x, y(x), u(x)), & \text{in } \Omega, \\ \dfrac{\partial y}{\partial \nu_A} = g(x, y(x), v(x)), & \text{on } \partial\Omega. \end{cases}$

## §7. Boundary Control Problems

Here, $A$ is a second order elliptic differential operator of form (1.3) and $\frac{\partial y}{\partial \nu_A}$ is the *conormal derivative* associated with the operator $A$ on the boundary $\partial\Omega$ (see Chapter 2, §6.2); $y$ is the state, and $u$ and $v$ are control actions applied in $\Omega$ and on the boundary $\partial\Omega$, respectively. The cost functional is given by

$$(7.2) \quad J(y,u,v) = \int_\Omega f^0(x,y(x),u(x))\,dx + \int_{\partial\Omega} g^0(x,y(x),v(x))\,dx.$$

We make the following assumptions.

(DB1) $\Omega \subset \mathbb{R}^n$ is a bounded region with $C^{1,\gamma}$ boundary $\partial\Omega$, for some $\gamma > 0$, and $U$ and $V$ are separable metric spaces.

(DB2) Operator $A$ is defined by (1.3) with $a_{ij} \in L^\infty(\Omega)$, $a_{ij} = a_{ji}$, $1 \leq i,j \leq n$, and for some $\lambda > 0$, (1.4) holds.

(DB3) The function $f : \Omega \times \mathbb{R} \times U \to \mathbb{R}$ satisfies (S3) stated in §1 with (1.5) replaced by

$$(7.3) \quad f_y(x,y,u) \leq -a, \quad \forall (x,y,u) \in \Omega \times \mathbb{R} \times U,$$

for some constant $a \geq 0$. The function $g : \partial\Omega \times \mathbb{R} \times V \to \mathbb{R}$ has the same properties as $f$ with $u, U, \Omega$ replaced by $v, V, \partial\Omega$ and (7.3) replaced by

$$(7.4) \quad g_y(x,y,v) \leq -b, \quad \forall (x,y,v) \in \partial\Omega \times \mathbb{R} \times V,$$

for some constant $b \geq 0$. We further assume that

$$(7.5) \quad a+b > 0.$$

(DB4) The functions $f^0 : \Omega \times \mathbb{R} \times U \to \mathbb{R}$ and $g^0 : \partial\Omega \times \mathbb{R} \times V \to \mathbb{R}$ satisfy the same conditions as those for $f$ and $g$ stated in (DB3), respectively, except (7.3) and (7.4).

We define $\mathcal{U}$ as in §1 and set $\mathcal{V} = \{v : \partial\Omega \to V \mid v \text{ is measurable }\}$. Any element $u \in \mathcal{U}$ is referred to as a *distributed control* and any $v \in \mathcal{V}$ a *boundary control*. The following result is basic.

**Proposition 7.1.** *Let (DB1)–(DB3) hold. Then there exists an $\alpha \in (0,1)$, such that for any $(u,v) \in \mathcal{U} \times \mathcal{V}$, there exists a unique $y(\cdot) \equiv y(\cdot\,;u,v) \in W^{1,2}(\Omega) \cap C^\alpha(\overline{\Omega})$ solving (7.1). Furthermore, there exists a constant $C_\alpha > 0$, independent of $(u,v) \in \mathcal{U} \times \mathcal{V}$, such that*

$$(7.6) \quad \|y(\cdot\,;u,v)\|_{W^{1,2}(\Omega) \cap C^\alpha(\overline{\Omega})} \leq C_\alpha, \quad \forall (u,v) \in \mathcal{U} \times \mathcal{V}.$$

The proof is similar to that given in Chapter 2, §6.3. Hereafter, (DB1)–(DB4) will always be assumed. Consequently, from the above, $(u,v) \in \mathcal{U} \times \mathcal{V}$ uniquely determine the state $y(\cdot) \equiv y(\cdot\,;u,v)$ and the cost functional (7.2) is well defined. Thus, $y$ in $J(y,u,v)$ will be suppressed below, for notational simplicity. Also, in what follows, we will fix the $\alpha \in (0,1)$ obtained in the above proposition. Next, let us introduce the following pointwise state constraint:

(7.7) $$h(x, y(x)) \leq 0, \quad x \in \Omega,$$

with the function $h : \Omega \times \mathbb{R} \to \mathbb{R}$ satisfying the following

(DB5) For each $x \in \Omega$, $h(x, \cdot)$ is $C^1$ with $h, h_y \in C(\overline{\Omega} \times \mathbb{R})$.

We let $\mathcal{A}$ be the set of all triplets $(y, u, v) \in (W^{1,2}(\Omega) \cap C^\alpha(\overline{\Omega})) \times \mathcal{U} \times \mathcal{V}$ that satisfy the state equation (7.1) and let $\mathcal{A}_{ad}$ be the set of all triplets $(y, u, v) \in \mathcal{A}$ satisfying the state constraint (7.7). Also, the natural projection of the set $\mathcal{A}_{ad}$ onto $\mathcal{U} \times \mathcal{V}$ is denoted by $(\mathcal{U} \times \mathcal{V})_{ad}$. In other words, $(u, v) \in (\mathcal{U} \times \mathcal{V})_{ad}$ if and only if the corresponding state $y$ satisfies (7.7). Any triplet in $\mathcal{A}$ (resp. $\mathcal{A}_{ad}$) is called a *feasible triplet* (resp. an *admissible triplet*) and any pair in $(\mathcal{U} \times \mathcal{V})_{ad}$ is called an *admissible control pair*. Now, our optimal control problem can be stated as follows.

**Problem (DB).** Find $(\bar{u}, \bar{v}) \in (\mathcal{U} \times \mathcal{V})_{ad}$, such that

(7.8) $$J(\bar{u}, \bar{v}) = \inf_{(\mathcal{U} \times \mathcal{V})_{ad}} J(u, v).$$

Any $(\bar{u}, \bar{v}) \in (\mathcal{U} \times \mathcal{V})_{ad}$ satisfying (7.9) is called an *optimal control pair*, and the corresponding state $\bar{y}$ and the triplet $(\bar{y}, \bar{u}, \bar{v})$ are referred to as an *optimal state* and an *optimal triplet*, respectively. As a very good exercise, we suggest that the interested readers impose suitable conditions on the functions involved in the problem to ensure the existence of an optimal control pair for Problem (DB) (using the technique of Chapter 3). Our goal here is to establish a set of necessary conditions for any given optimal triplet. To this end, we need some more conditions, which we shall explain below.

## §7.2. Strong stability and the qualified maximum principle

We have seen the notion of the stability of the optimal control problem in §4.2. In this subsection, we will introduce a slightly stronger notion. To this end, we consider the state constraint:

(7.9) $$h(x, y(x)) \leq \delta, \quad x \in \overline{\Omega},$$

where $\delta \in [0, \infty)$. Thus, when $\delta = 0$, (7.9) is reduced to the original state constraint (7.7). Let $\mathcal{A}_{ad}^\delta$ be the set of all triplets in $\mathcal{A}$ that satisfy the constraint (7.9) and let $(\mathcal{U} \times \mathcal{V})_{ad}^\delta$ be the corresponding admissible control pair set. Then, we can formulate the optimal control problem as Problem (DB) with $(\mathcal{U} \times \mathcal{V})_{ad}$ replaced by $(\mathcal{U} \times \mathcal{V})_{ad}^\delta$. Such a problem is referred to as Problem (BD)$^\delta$. Next, we define

(7.10) $$V(\delta) = \inf_{(\mathcal{U} \times \mathcal{V})_{ad}^\delta} J(u(\cdot), v(\cdot)).$$

## §7. Boundary Control Problems

Thus, $V(0)$ is the optimal cost for the original Problem (DB). Now, we introduce the following notion.

**Definition 7.2.** Problem $(BD)^\delta$ is said to be *strongly stable* at $\delta = \delta_0$ if there exist constants $C, \bar{\delta} > 0$ (depending on $\delta_0$), such that

$$(7.11) \qquad V(\delta_0) - V(\delta) \leq C(\delta - \delta_0), \qquad \forall \delta \in [\delta_0, \delta_0 + \bar{\delta}].$$

It is not hard to see that (7.11) holds if $V(\delta)$ is differentiable at $\delta_0$. Since $(\mathcal{U} \times \mathcal{V})_{ad}^\delta$ is increasing in $\delta$, $V(\delta)$ is decreasing in $\delta$. Thus, $V(\delta)$ is differentiable for almost all $\delta \in [0, \infty)$. Our further assumption is the following:

(DB6) Problem $(DB)^\delta$ is *strongly stable* at $\delta = 0$, i.e., there exist constants $C, \bar{\delta} > 0$, such that

$$(7.12) \qquad V(0) - V(\delta) \leq C\delta, \qquad \forall \delta \in [0, \bar{\delta}].$$

The following gives a sufficient condition for Problem (DB) to be strongly stable.

**Proposition 7.3.** *Let (DB1)–(BD5) hold and let there exist a constant $C > 0$, such that*

$$(7.13) \qquad J(u,v) - V(0) \geq -C\|(h(\cdot, y(\cdot)))^+\|_{C(\overline{\Omega})}, \qquad \forall (y, u, v) \in \mathcal{A}.$$

*Then (DB6) holds.*

*Proof.* For any $(u, v) \in (\mathcal{U} \times \mathcal{V})_{ad}^\delta$, (7.9) holds, which implies $\|(h(\cdot, y(\cdot)))^+\|_{C(\overline{\Omega})} \leq \delta$. Thus, (7.13) implies that

$$(7.14) \qquad J(u, v) - V(0) \geq -C\delta, \qquad \forall (u, v) \in (\mathcal{U} \times \mathcal{V})_{ad}^\delta.$$

Taking the infimum in the above with respect to $(u, v) \in (\mathcal{U} \times \mathcal{V})_{ad}^\delta$, we obtain (7.12) with any $\bar{\delta} > 0$. □

Now, let us make a comparison between the stability (introduced in §4.1) and the strong stability introduced above. In terms of §4.1, our Problem (BD) is stable if and only if for any $(y_k, u_k, v_k) \in \mathcal{A}$ with the property that

$$(7.15) \qquad \lim_{k \to \infty} \|(h(\cdot, y_k(\cdot)))^+\|_{C(\overline{\Omega})} = 0,$$

it holds that

$$(7.16) \qquad \varliminf_{k \to \infty} J(u_k, v_k) \geq V(0).$$

We have the following simple result.

**Proposition 7.4.** *Let (DB1)–(DB5) hold. Suppose that Problem (DB) is strongly stable. Then it is stable.*

*Proof.* Let $(y_k, u_k, v_k) \in \mathcal{A}$ satisfy (7.15). Set

(7.17) $$\delta_k = \|(h(\cdot, y_k(\cdot)))^+\|_{C(\bar{\Omega})}, \qquad k \geq 1.$$

Then $\delta_k \to 0$ and $(u_k, v_k) \in (\mathcal{U} \times \mathcal{V})_{ad}^{\delta_k}$. Thus, by (7.12), there exist $C > 0$ and $k_0 > 0$, such that

(7.18) $$J(u_k, v_k) \geq V(\delta_k) \geq V(0) - C\delta_k, \qquad k \geq k_0.$$

Hence, (7.16) follows. $\qquad\square$

The above tells us that strong stability is stronger than stability. Roughly speaking, the (strong) stability of the optimal control problem means that under small perturbation (in some sense) of the state constraint, the change of the optimal cost is small. This will be made more precise in §7.4.

Now, let $(\bar{y}, \bar{u}, \bar{v}) \in \mathcal{A}_{ad}$ be an optimal triplet of Problem (DB). Our goal is to derive a set of necessary conditions for it. To state the main result, we need to introduce the Hamiltonians:

(7.19) $$\begin{cases} H(x, y, u, \psi) = -f^0(x, y, u) + \psi f(x, y, u), \\ G(x, y, v, \varphi) = -g^0(x, y, v) + \varphi g(x, y, v). \end{cases}$$

The main result of this section is the following maximum principle.

**Theorem 7.5.** *(Maximum Principle) Let (DB1)–(DB6) hold. Let $(\bar{y}, \bar{u}, \bar{v}) \in \mathcal{A}_{ad}$ be an optimal triplet of Problem (DB). Then, for all $s \in (1, \frac{n}{n-1})$, there exist $\psi(\cdot) \in W^{1,s}(\Omega)$ and $\mu \in \mathcal{M}(\bar{\Omega})$, such that*

(7.20) $$\begin{cases} A\psi = f_y(x, \bar{y}(x), \bar{u}(x))\psi - f_y^0(x, \bar{y}(x), \bar{u}(x)) \\ \qquad\qquad - h_y(x, \bar{y}(x))\mu\big|_\Omega, \qquad \text{in } \Omega, \\ \dfrac{\partial \psi}{\partial \nu_A} = g_y(x, \bar{y}(x), \bar{v}(x))\psi - g_y^0(x, \bar{y}(x), \bar{v}(x)) \\ \qquad\qquad - h_y(x, \bar{y}(x))\mu\big|_{\partial\Omega}, \qquad \text{on } \partial\Omega, \end{cases}$$

(7.21) $$\mu \geq 0, \qquad \int_{\bar{\Omega}} h(x, \bar{y}(x)) \, d\mu(x) = 0,$$

(7.22) $\quad H(x, \bar{y}(x), \bar{u}(x), \psi(x)) = \max_{u \in U} H(x, \bar{y}(x), u, \psi(x)), \qquad$ a.e. $x \in \Omega$,

(7.23) $\quad G(x, \bar{y}(x), \bar{v}(x), \psi(x)) = \max_{v \in V} G(x, \bar{y}(x), v, \psi(x)), \qquad$ a.e. $x \in \partial\Omega$,

*where $\mu\big|_\Omega$ and $\mu\big|_{\partial\Omega}$ in (7.20) are the restriction of $\mu$ on $\Omega$ and $\partial\Omega$, respectively.*

## §7. Boundary Control Problems

It is seen that for problems involving boundary controls, we have a maximum condition on the boundary (see (7.23)). This is pretty natural and it makes the above result significantly different from those for problems without the boundary controls (see previous sections). We also note that comparing with the maximum principle proved before, in the above, the corresponding $\psi^0 = -1$. We will see that this is due to the strong stability assumption (DB6). In the case where $\psi^0 = -1$, we say that the corresponding maximum principle is *qualified*. Condition (7.21) is actually equivalent to the usual transversality condition:

$$（7.24） \quad \int_\Omega \{\eta(x) - h(x, \bar{y}(x))\}\, d\mu(x) \leq 0,$$
$$\forall \eta \in C(\overline{\Omega}),\ \eta(x) \leq 0,\ x \in \Omega.$$

We leave the proof to the readers. Finally, in the above, the adjoint equation (7.20) is unusual as measures appear both in the interior of the domain $\Omega$ and on the boundary $\partial\Omega$. Thus, the meaning as well as the well-posedness of (7.20) should be specified carefully. We shall study such a problem in the next subsection.

### §7.3. Neumann problem with measure data

In this subsection, we study the following problem

$$（7.25） \quad \begin{cases} A\psi + a_0(x)\psi = \mu_\Omega, & \text{in } \Omega, \\ \dfrac{\partial \psi}{\partial \nu_A} + b_0(x)\psi = \mu_{\partial\Omega}, & \text{on } \partial\Omega. \end{cases}$$

Here

$$（7.26） \quad \begin{cases} \mu \in \mathcal{M}(\overline{\Omega}),\quad \mu_\Omega = \mu\big|_\Omega,\quad \mu_{\partial\Omega} = \mu\big|_{\partial\Omega}, \\ a_0(\cdot) \in L^\infty(\Omega),\quad a_0(x) \geq a,\ \text{a.e. } x \in \Omega, \\ b_0(\cdot) \in L^\infty(\partial\Omega),\quad b_0(x) \geq b,\ \text{a.e. } x \in \partial\Omega, \end{cases}$$

where $a, b \geq 0$ satisfy (7.5). Clearly, adjoint equation (7.20) is of the above form. In what follows, we let $s \in (1, \frac{n}{n-1})$ and $s' = \frac{s}{s-1}$. Clearly, $s' > n$. A function $\psi \in W^{1,s}(\Omega)$ is called a *weak solution* of (7.25) if the following holds:

$$（7.27） \quad \int_\Omega \Big\{\sum_{i,j=1}^n a_{ij}(x)\psi_{x_i}(x) z_{x_j}(x) + a_0(x)\psi(x) z(x)\Big\} dx$$
$$+ \int_{\partial\Omega} b_0(x)\psi(x) z(x)\, dx = \langle \mu, z \rangle_{\overline{\Omega}},\quad \forall z \in W^{1,s'}(\Omega),$$

where

$$（7.28） \quad \langle \mu, z(\cdot)\rangle_{\overline{\Omega}} = \int_{\overline{\Omega}} z(x)\, d\mu(x),\quad \forall z \in C(\overline{\Omega}).$$

Hereafter, we will not distinguish (7.25) and (7.27). Our goal is to establish the well-posedness of (7.27). The main result is the following.

**Theorem 7.6.** *For any $\mu \in \mathcal{M}(\bar{\Omega})$, (7.25) admits a unique weak solution $\psi(\cdot) \in W^{1,s}(\Omega)$. Moreover, there exists a constant $C > 0$, independent of $\mu \in \mathcal{M}(\bar{\Omega})$, such that*

$$\|\psi\|_{W^{1,s}(\Omega)} \leq C\|\mu\|_{\mathcal{M}(\bar{\Omega})}, \qquad \forall \mu \in \mathcal{M}(\bar{\Omega}). \tag{7.29}$$

*Proof.* Let us first derive the *a priori* estimate (7.29). To this end, let $\psi \in W^{1,s}(\Omega)$ be a weak solution of (7.25). Take $\varphi_i \in C_0^\infty(\Omega)$, $i = 0, 1, \cdots, n$, and set $\varphi = (\varphi_1, \cdots, \varphi_n)$. Consider the following boundary value problem:

$$\begin{cases} Az + a_0(x)z = -\nabla \cdot \varphi + \varphi_0, & \text{in } \Omega, \\ \dfrac{\partial z}{\partial \nu_A} + b_0(x)z = 0, & \text{on } \partial\Omega. \end{cases} \tag{7.30}$$

The weak formulation of (7.30) is the following:

$$\begin{aligned} \int_\Omega \Big\{ \sum_{i,j=1}^n a_{ij}(x) z_{x_j}(x) \psi_{x_i}(x) + a_0(x) z(x) \psi(x) \Big\} \, dx \\ + \int_{\partial\Omega} b_0(x) z(x) \psi(x) dx = \int_\Omega \big\{ -\nabla \cdot \varphi(x) + \varphi_0(x) \big\} \psi(x) \, dx, \\ \forall \psi \in W^{1,s}(\Omega). \end{aligned} \tag{7.31}$$

From Chapter 2, Theorem 6.10, we know that (7.30) admits a unique weak solution $z \in W^{1,s'}(\Omega)$ and for some constant $C > 0$, independent of $\varphi_i$'s, it holds that

$$\|z\|_{C(\bar{\Omega})} \leq C \sum_{i=0}^n \|\varphi_i\|_{L^{s'}(\Omega)}. \tag{7.32}$$

Combining (7.27) and (7.31), we have

$$\begin{aligned} \int_\Omega (\nabla\psi \cdot \varphi + \psi\varphi_0) \, dx &= \int_\Omega \psi(-\nabla \cdot \varphi + \varphi_0) \, dx \\ &= \int_\Omega \Big\{ \sum_{i,j=1}^n a_{ij}(x) z_{x_j}(x) \psi_{x_i}(x) + a_0(x) z(x) \psi(x) \Big\} \, dx \\ &\quad + \int_{\partial\Omega} b_0(x) z(x) \psi(x) \, dx = \langle \mu, z \rangle_{\bar{\Omega}}. \end{aligned} \tag{7.33}$$

Hence, by (7.32),

$$\begin{aligned} \Big| \int_\Omega (\nabla\psi \cdot \varphi + \psi\varphi_0) \, dx \Big| &= |\langle \mu, z \rangle_{\bar{\Omega}}| \\ &\leq \|\mu\|_{\mathcal{M}(\bar{\Omega})} \|z\|_{C(\bar{\Omega})} \leq C\|\mu\|_{\mathcal{M}(\bar{\Omega})} \sum_{i=0}^n \|\varphi_i\|_{L^{s'}(\Omega)}. \end{aligned} \tag{7.34}$$

## §7. Boundary Control Problems

This yields (7.29) because $C_0^\infty(\Omega)$ is dense in $L^{s'}(\Omega)$. Consequently, the uniqueness of solutions follows. Now, we prove the existence of the weak solution to (7.25). As $C^\infty(\overline{\Omega})$ is dense in $\mathcal{M}(\overline{\Omega})$, we can take $\eta_k \in C^\infty(\overline{\Omega})$, such that $\eta_k \xrightarrow{s} \mu$ in $\mathcal{M}(\overline{\Omega})$. Now, consider (7.25) with $\mu$ replaced by $\eta_k$. By Theorem 6.10 of Chapter 2, we see that there exists a unique solution $\psi^k \in W^{1,2}(\Omega) \cap C^\alpha(\overline{\Omega})$. It follows from (7.29) that this $\psi^k$ satisfies the following:

$$(7.35) \qquad \|\psi^k - \psi^\ell\|_{W^{1,s}(\Omega)} \leq C\|\eta_k - \eta_\ell\|_{\mathcal{M}(\overline{\Omega})}, \qquad \forall k, \ell \geq 1.$$

Therefore, by the convergence of $\eta_k$ to $\mu$ in $\mathcal{M}(\overline{\Omega})$, we see that the sequence $\{\psi^k\}$ is a Cauchy sequence in $W^{1,s}(\Omega)$. Consequently, there exists a $\psi \in W^{1,s}(\Omega)$, such that $\psi^k \xrightarrow{s} \psi$ in $W^{1,s}(\Omega)$. Clearly, this $\psi$ is a weak solution of (7.25). □

It follows from Theorem 7.6 that the adjoint equation (7.20) is well-posed.

### §7.4. Exact penalization and a proof of the maximum principle

In this subsection, we are going to give a proof the maximum principle stated in §7.2. The following method is referred to as the *exact penalization*. We will see that the strong stability assumption (BD6) is crucial in order for the method of exact penalization to be applicable. For notational simplicity, we suppress the argument $x$ in the functions $f, f^0, g, g^0$, and $h$ (we will see that argument $x$ in these functions does not play any role in the proof). In what follows, we let $(\bar{y}, \bar{u}, \bar{v}) \in \mathcal{A}_{ad}$ be the given optimal triplet. Let us prove the following lemma first.

**Lemma 7.7.** *Let (DB1)–(DB6) hold and $(\bar{u}, \bar{v}) \in (\mathcal{U} \times \mathcal{V})_{ad}$ be an optimal control pair for Problem (DB). Then there exists a constant $q > 0$, such that*

$$(7.36) \qquad J(\bar{u}, \bar{v}) = \inf_{\mathcal{A}} \left\{ J(u, v) + q\|(h(y))^+\|_{C(\overline{\Omega})} \right\}.$$

It is important to note that the infimum on the right-hand side of (7.36) is taken over the whole set $\mathcal{A}$ of feasible triplets. The above means that the given optimal control pair $(\bar{u}, \bar{v})$ is a solution of the optimal control problem governed by the state equation (7.1) with the cost functional

$$(7.37) \qquad J^q(u, v) \equiv J(u, v) + q\|(h(y))^+\|_{C(\overline{\Omega})},$$

*without* the state constraint! This will give us a big advantage in obtaining the necessary conditions for the given optimal triplet.

*Proof of Lemma 7.7.* Suppose the contrary. Then, for each $k \geq 1$, there exists a triplet $(y_k, u_k, v_k) \in \mathcal{A}$, such that

$$(7.38) \qquad J(u_k, v_k) + k\|(h(y_k))^+\|_{C(\overline{\Omega})} < J(\bar{u}, \bar{v}), \qquad k \geq 1.$$

By the optimality of $(\bar{y}, \bar{u}, \bar{v})$, we see that $h(y_k) \notin Q$. Hence,

(7.39)
$$0 < \delta_k \stackrel{\Delta}{=} \max_{x \in \bar{\Omega}} h(y_k(x)) = \|(h(y_k))^+\|_{C(\bar{\Omega})}$$
$$< \frac{1}{k}\{J(\bar{u}, \bar{v}) - J(u_k, v_k)\} \to 0, \qquad k \to \infty.$$

The last convergence in (7.39) holds because (DB1)–(DB4) implies that $J(u_k, v_k)$ is uniformly bounded. Next, by (DB6), we can find some constant $C > 0$, such that

(7.40) $$V(0) - V(\delta_k) \leq C\delta_k, \qquad \forall k \geq 1.$$

Then, together with (7.38), we obtain (note that $(u_k, v_k) \in (\mathcal{U} \times \mathcal{V})_{ad}^{\delta_k}$)

(7.41)
$$C\delta_k \geq V(0) - V(\delta_k) \geq V(0) - J(u_k, v_k)$$
$$> k\|(h(y_k))^+\|_{C(\bar{\Omega})} = k\delta_k, \qquad \forall k \geq 1.$$

This is impossible. □

Lemma 7.7 suggests that in deriving necessary conditions for $(\bar{y}, \bar{u}, \bar{v})$, we may regard it as a solution of the control problem with the cost $J^q(u, v)$ and with no state constraints. However, we note that the second term in (7.37) is not differentiable. Thus, we need to approximate it by some better functional. To this end, we let $r \in [1, \infty)$ be fixed and introduce the following regularization of the penalty functional $J^q(u, v)$:

(7.42) $$J_r^q(u, v) = J(u, v) + q\{r^{-r} + \|(h(y))^+\|_{L^r(\Omega)}^r\}^{1/r}.$$

Denote

(7.43) $$m_r^q = \inf_{\mathcal{U} \times \mathcal{V}} J_r^q(u, v).$$

The following result will be useful later.

**Proposition 7.8.** *Let (DB1)–(DB6) hold. Then,*

(7.44) $$\lim_{r \to \infty} m_r^q = J(\bar{u}, \bar{v}) = \inf_{\mathcal{U} \times \mathcal{V}} J^q(u, v).$$

*Proof.* First of all, we have

(7.45) $$m_r^q \leq J_r^q(\bar{u}, \bar{v}) = J(\bar{u}, \bar{v}) + \frac{q}{r}.$$

This yields

(7.46) $$\varlimsup_{r \to \infty} m_r^q \leq J(\bar{u}, \bar{v}).$$

Next, for any $r > 0$, we can find $(y_r, u_r, v_r) \in \mathcal{A}$, such that

(7.47) $$J_r^q(u_r, v_r) \leq m_r^q + \frac{1}{r}.$$

§7. Boundary Control Problems

By Proposition 7.1, we know that the set $\{y_r, \ r > 1\}$ is relatively compact in $C(\overline{\Omega})$, and thus so is the set $\{(h(y_r))^+, \ r > 1\}$. On the other hand, for any $\eta \in C(\overline{\Omega})$, we have

(7.48) $$\lim_{r \to \infty} \|\eta\|_{L^r(\Omega)} = \|\eta\|_{C(\overline{\Omega})}.$$

Thus, it is easy to show that

(7.49) $$\lim_{r \to \infty} \left\{ \left[ (r^{-r} + \|(h(y_r))^+\|^r_{L^r(\Omega)} \right]^{1/r} - \|(h(y_r))^+\|_{C(\overline{\Omega})} \right\} = 0.$$

This implies that

(7.50) $$\begin{aligned} & \overline{\lim_{r \to \infty}} J^q_r(u_r, v_r) - \underline{\lim_{r \to \infty}} J^q(u_r, v_r) \\ & \geq \overline{\lim_{r \to \infty}} \left\{ J^q_r(u_r, v_r) - J^q(u_r, v_r) \right\} \\ & = \overline{\lim_{r \to \infty}} \left\{ q[r^{-r} + \|(h(y_r))^+\|^r_{L^r(\Omega)}]^{1/r} - q\|(h(y_r))^+\|_{C(\overline{\Omega})} \right\} = 0. \end{aligned}$$

Now, by (7.46), (7.47), (7.50), and (7.37), we obtain

(7.51) $$\begin{aligned} \overline{\lim_{r \to \infty}} J^q_r(u_r, v_r) & \leq \overline{\lim_{r \to \infty}} m^q_r \leq J(\bar{u}, \bar{v}) \\ & = J^q(\bar{u}, \bar{v}) \leq \underline{\lim_{r \to \infty}} J^q(u_r, v_r) \leq \overline{\lim_{r \to \infty}} J^q_r(u_r, v_r). \end{aligned}$$

Thus, all the equalities in (7.51) hold. The above is true for the arbitrary sequence $r \to \infty$; thus, our conclusion follows. $\square$

Now, we are ready to prove the maximum principle.

*Proof of Theorem 7.5.* First of all, we see that $J^q_r(u, v)$ is continuous on $(\mathcal{U} \times \mathcal{V}, \bar{d})$. Here, $\bar{d}$ is the distance on the space $\mathcal{U} \times \mathcal{V}$:

(7.52) $$\begin{aligned} & \bar{d}((u, v), (\tilde{u}, \tilde{v})) \\ & = |\{x \in \Omega \mid u(x) \neq \tilde{u}(x)\}| + |\{x \in \partial\Omega \mid v(x) \neq \tilde{v}(x)\}|, \\ & \forall (u, v), (\tilde{u}, \tilde{v}) \in \mathcal{U} \times \mathcal{V}. \end{aligned}$$

We note that the second term on the right-hand side of (7.52) is understood as the $(n-1)$-dimensional *Hausdorff measure*, a natural generalization of the Lebesgue measure. Those who are not quite familiar with the Hausdorff measure can just think of it as the Lebesgue measure on an $(n-1)$-dimensional hypersurface. Similar to Chapter 4, we can show that $(\mathcal{U} \times \mathcal{V}, \bar{d})$ is a complete metric space. Next, we consider the following:

(7.53) $$\begin{aligned} J^q_r(\bar{u}, \bar{v}) & = J(\bar{u}, \bar{v}) + \frac{q}{r} \\ & \leq m^q_r + \{|J(\bar{u}, \bar{v}) - m^q_r| + \frac{q}{r}\} \equiv \inf_{\mathcal{U} \times \mathcal{V}} J^q_r(u, v) + \delta_r, \end{aligned}$$

with $\delta_r \equiv |J(\bar{u}, \bar{v}) - m_r^q| + \frac{q}{r} \to 0$ as $r \to \infty$ because of Proposition 7.8. By Ekeland's variational principle (see Chapter 4), we have $(u^r, v^r) \in \mathcal{U} \times \mathcal{V}$, such that

(7.54)
$$\begin{cases} \bar{d}((u^r, v^r), (\bar{u}, \bar{v})) \leq \sqrt{\delta_r}, \\ -\sqrt{\delta_r}\bar{d}((\hat{u}, \hat{v}), (u^r, v^r)) \leq J_r^q(\hat{u}, \hat{v}) - J_r^q(u^r, v^r), \\ \qquad\qquad\qquad\qquad\qquad\qquad\qquad \forall (\hat{u}, \hat{v}) \in \mathcal{U} \times \mathcal{V}. \end{cases}$$

Next, we fix $(u, v) \in \mathcal{U} \times \mathcal{V}$, $\rho \in (0, 1)$ and define

(7.55)
$$u_\rho^r(x) = \begin{cases} u^r(x), & x \in \Omega \setminus E_\rho, \\ u(x), & x \in E_\rho; \end{cases}$$

(7.56)
$$v_\rho^r(x) = \begin{cases} v^r(x), & x \in \partial\Omega \setminus F_\rho, \\ v(x), & x \in F_\rho. \end{cases}$$

Here, $E_\rho \subset \Omega$, and $F_\rho \subset \partial\Omega$ with $|E_\rho| = \rho|\Omega|$ and $|F_\rho| = \rho|\partial\Omega|$ (again, $|F_\rho|$ is understood as the $(n-1)$-dimensional Hausdorff measure). Let $y_\rho^r$ be the state corresponding to $(u_\rho^r, v_\rho^r)$. Then, similar to Lemma 2.3, it is not hard to show that

(7.57)
$$\|y_\rho^r - y^r\|_{C^\alpha(\bar{\Omega})} \leq \omega(\rho),$$

for some continuous function $\omega : [0, \infty) \to [0, \infty)$ with $\omega(0) = 0$. Denote $z_\rho^r = \frac{1}{\rho}(y_\rho^r - y^r)$. We have

(7.58)
$$\begin{cases} Az_\rho^r + a_\rho^r z_\rho^r = \dfrac{1}{\rho}\bigl(f(y^r, u) - f(y^r, u^r)\bigr)\chi_{E_\rho}, & \text{in } \Omega, \\ \dfrac{\partial z_\rho^r}{\partial \nu_A} + b_\rho^r z_\rho^r = \dfrac{1}{\rho}\bigl(g(y^r, v) - g(y^r, v^r)\bigr)\chi_{F_\rho}, & \text{on } \partial\Omega, \end{cases}$$

where

(7.59)
$$\begin{cases} a_\rho^r(x) = -\displaystyle\int_0^1 f_y(y^r(x) + \sigma(y_\rho^r(x) - y^r(x)), u_\rho^r(x))\, d\sigma, & x \in \Omega, \\ b_\rho^r(x) = -\displaystyle\int_0^1 g_y(y^r(x) + \sigma(y_\rho^r(x) - y^r(x)), v_\rho^r(x))\, d\sigma, & x \in \partial\Omega. \end{cases}$$

Now, we take $(\hat{u}, \hat{v}) = (u_\rho^r, v_\rho^r)$ in (7.54).

(7.60)
$$\begin{aligned} -\sqrt{\delta_r} &\leq \frac{1}{\rho}\{J_r^q(u_\rho^r, v_\rho^r) - J_r^q(u^r, v^r)\} \\ &= \frac{1}{\rho}\{J(u_\rho^r, v_\rho^r) - J(u^r, v^r)\} + \frac{q}{\rho}\Bigl\{\bigl[r^{-r} + \|(h(y_\rho^r))^+\|_{L^r(\Omega)}^r\bigr]^{1/r} \\ &\quad - \bigl[r^{-r} + \|(h(y^r))^+\|_{L^r(\Omega)}^r\bigr]^{1/r}\Bigr\} \equiv I_1 + I_2. \end{aligned}$$

## §7. Boundary Control Problems

Clearly, we have

(7.61)
$$I_1 = \int_\Omega \{f_y^0(y^r, u_\rho^r) + \omega_1(\rho)\} z_\rho^r + \frac{1}{\rho} \int_{E_\rho} \{f^0(y^r, u) - f^0(y^r, u^r)\}$$
$$+ \int_{\partial\Omega} \{g_y^0(y^r, u_\rho^r) + \omega_2(\rho)\} z_\rho^r + \frac{1}{\rho} \int_{F_\rho} \{g^0(y^r, v) - g^0(y^r, v^r)\}.$$

Here and below, we denote $\omega_k(\rho)$ ($k \geq 1$) to be the remainder terms that converge to zero as $\rho \to 0$ uniformly in $x \in \overline{\Omega}$. To compute $I_2$, we denote

(7.62)
$$\theta_r(\eta) = q\left\{r^{-r} + \|h(\eta)^+\|_{L^r(\Omega)}^r\right\}^{1/r}, \qquad \forall \eta \in L^r(\Omega).$$

Then $\theta_r(\cdot)$ is Fréchet differentiable and for any $\zeta \in L^{r'}(\Omega)$ ($r' = r/(r-1)$)

(7.63)
$$\langle \nabla \theta_r(\eta), \zeta \rangle = q\left(r^{-r} + \|(h(\eta))^+\|_{L^r(\Omega)}^r\right)^{\frac{1}{r}-1}$$
$$\cdot \int_\Omega \left((h(\eta))^+\right)^{r-1} h'(\eta) \zeta \, dx.$$

By the mean value theorem, there exists a measurable function $\eta_\rho^r : \Omega \to L^\infty(\Omega)$, with the property that for almost all $x \in \Omega$, the value $\eta_\rho^r(x)$ lies between $y_\rho^r(x)$ and $y^r(x)$, such that

(7.64)
$$I_2 \equiv \frac{1}{\rho}\{\theta_r(y_\rho^r) - \theta_r(y^r)\} = \langle \nabla \theta_r(\eta_\rho^r), z_\rho^r \rangle$$
$$= \int_\Omega q\left(r^{-r} + \|(h(\eta_\rho^r))^+\|_{L^r(\Omega)}^r\right)^{\frac{1}{r}-1} \{(h(\eta_\rho^r))^+\}^{r-1} h'(\eta_\rho^r) z_\rho^r \, dx$$
$$\equiv \int_\Omega \mu_\rho^r h'(\eta_\rho^r) z_\rho^r \, dx,$$

where

(7.65)
$$\mu_\rho^r = q\left(r^{-r} + \|(h(\eta_\rho^r))^+\|_{L^r(\Omega)}^r\right)^{\frac{1}{r}-1} \{(h(\eta_\rho^r))^+\}^{r-1}.$$

Substituting (7.61) and (7.64) into (7.60), we obtain

(7.66)
$$-\sqrt{\delta_r} \leq \int_\Omega \{f_y^0(y^r, u_\rho^r) + \omega_1(\rho) + \mu_\rho^r h'(\eta_\rho^r)\} z_\rho^r \, dx$$
$$+ \int_{\partial\Omega} \{g_y^0(y^r, u_\rho^r) + \omega_2(\rho)\} z_\rho^r \, dx$$
$$+ \frac{1}{\rho} \int_{E_\rho} (f^0(y^r, u) - f^0(y^r, u^r)) \, dx$$
$$+ \frac{1}{\rho} \int_{F_\rho} (g^0(y^r, v) - g^0(y^r, v^r)) \, dx \equiv \widehat{I}_1 + \widehat{I}_2 + \widehat{I}_3 + \widehat{I}_4.$$

We introduce the following:

(7.67)
$$\begin{cases} A\psi_\rho^r + a_\rho^r \psi_\rho^r = -f_y^0(y^r, u^r) - \omega_1(\rho) - \mu_\rho^r h'(\eta_\rho^r), & \text{in } \Omega, \\ \dfrac{\partial \psi_\rho^r}{\partial \nu_A} + b_\rho^r \psi_\rho^r = -g_y^0(y^r, v^r) - \omega_2(\rho), & \text{on } \partial\Omega. \end{cases}$$

By Chapter 2, §6.2, we know that the above admits a unique solution $\psi_\rho^r \in W^{1,2}(\Omega) \cap C^\alpha(\overline{\Omega})$. Using the function $\psi_\rho^r$, we can transform the first integral $\widehat{I}_1$ on the right-hand side of (7.66) as follows:

(7.68)
$$\begin{aligned} \widehat{I}_1 &= -\langle A\psi_\rho^r + a_\rho^r \psi_\rho^r, z_\rho^r \rangle \\ &= -\int_\Omega \Big\{ \sum_{i,j=1}^n a_{ij}(\psi_\rho^r)_{x_i}(z_\rho^r)_{x_j} + a_\rho^r \psi_\rho^r z_\rho^r \Big\} dx \\ &\quad + \int_{\partial\Omega} \big\{ -g_y^0(y^r, v^r) - b_\rho^r \psi_\rho^r - \omega_2(\rho) \big\} z_\rho^r\, dx \\ &= -\langle A z_\rho^r + a_\rho^r z_\rho^r, \psi_\rho^r \rangle \\ &\quad + \dfrac{1}{\rho}\int_{\partial\Omega} (g(y^r, v^r) - g(y^r, v)) \chi_{F_\rho} \psi_\rho^r\, dx - \widehat{I}_2 \\ &= \dfrac{1}{\rho}\int_{E_\rho} \big\{ \psi_\rho^r (f(y^r, u^r) - f(y^r, u)) \big\} dx \\ &\quad + \dfrac{1}{\rho}\int_{F_\rho} \big\{ \psi_\rho^r (g(y^r, v^r) - g(y^r, v)) \big\} dx - \widehat{I}_2. \end{aligned}$$

Hence, we obtain from (7.66) that

(7.69)
$$\begin{aligned} -\sqrt{\delta_r} &\leq \dfrac{1}{\rho}\int_{E_\rho} \big\{ H(y^r, u^r, \psi_\rho^r) - H(y^r, u, \psi_\rho^r) \big\} dx \\ &\quad + \dfrac{1}{\rho}\int_{F_\rho} \big\{ G(y^r, v^r, \psi_\rho^r) - G(y^r, v, \psi_\rho^r) \big\} dx + \widetilde{\omega}(\rho), \end{aligned}$$

where $\widetilde{\omega}(\rho) \to 0$ as $\rho \to 0$, uniformly in the choice of $E_\rho$ and $F_\rho$. On the other hand, by (7.57), we must have

(7.70)
$$\lim_{\rho \to 0} \|\eta_\rho^r - y^r\|_{L^\infty(\Omega)} = 0.$$

Thus, by Chapter 2, Theorem 6.10, we can show that

(7.71)
$$\|\psi_\rho^r - \psi^r\|_{W^{1,2}(\Omega) \cap C(\overline{\Omega})} \to 0, \qquad \rho \to 0,$$

where $\psi^r$ is the weak solution of the following problem:

(7.72)
$$\begin{cases} A\psi^r = f_y(y^r, u^r)\psi^r - f_y^0(y^r, u^r) - \mu^r h'(y^r), & \text{in } \Omega, \\ \dfrac{\partial \psi^r}{\partial \nu_A} = g_y(y^r, v^r)\psi^r - g_y^0(y^r, v^r), & \text{on } \partial\Omega, \end{cases}$$

## Remarks

with (see (7.65) for the definition of $\mu_\rho^r$)

$$(7.73) \qquad \mu^r = q\bigl(r^{-r} + \|h(y^r)^+\|_{L^r(\Omega)}^r\bigr)^{\frac{1}{r}-1}\bigl\{(h(y^r))^+\bigr\}^{r-1}.$$

We further note that almost all $x \in \Omega$ (resp. $x \in \partial\Omega$) are Lebesgue points of $H(y^r, u^r, \psi^r) - H(y^r, u, \psi^r)$ (resp. $G(y^r, v^r, \psi^r) - G(y^r, v, \psi^r)$). Thus, suitably choosing $E_\rho$ and $F_\rho$, and letting $\rho \to 0$ in (7.69), we obtain the following:

$$(7.74) \qquad \begin{cases} H(y^r, u^r, \psi^r) - H(y^r, u, \psi^r) \geq -\sqrt{\delta_r}, & \text{a.e. } x \in \Omega; \\ G(y^r, v^r, \psi^r) - G(y^r, v, \psi^r) \geq -\sqrt{\delta_r}, & \text{a.e. } x \in \partial\Omega. \end{cases}$$

Our final goal is sending $r \to \infty$ to obtain the maximum principle. To this end, let us note that

$$(7.75) \qquad \|\mu^r\|_{L^1(\Omega)} \leq q, \qquad \forall r > 1.$$

Consequently, it follows from Theorem 7.6 that

$$(7.76) \qquad \|\psi^r\|_{W^{1,s}(\Omega)} \leq C, \qquad \forall r > 1.$$

Hence, there exist $\mu \in \mathcal{M}(\overline{\Omega})$ and $\psi \in W^{1,s}(\Omega)$, such that for some sequence $r \to \infty$,

$$(7.77) \qquad \begin{aligned} \mu^r &\overset{*}{\rightharpoonup} \mu, \\ \psi^r &\to \psi, \qquad \text{weakly in } W^{1,s}(\Omega), \text{ strongly in } L^s(\Omega). \end{aligned}$$

Clearly, $\psi$ is the weak solution of (7.20). Now, we let $r \to \infty$ in (7.74). Our maximum conditions (7.22) and (7.23) follow. Because $\mu^r \geq 0$ for all $r > 1$, we must have $\mu \geq 0$. Finally, by the first relation in (7.54),

$$(7.78) \qquad (h(y^r))^+ \overset{s}{\to} (h(\bar{y}))^+ = 0, \qquad \text{in } C(\overline{\Omega}).$$

On the other hand, by (7.73), we have $h(y^r)\mu^r = h(y^r)^+\mu^r$. Thus, together with (7.75), one has

$$(7.79) \qquad \begin{aligned} \left|\int_\Omega h(\bar{y}) \, d\mu\right| &= \lim_{r \to \infty} \left|\int_\Omega h(y^r)\mu^r \, dx\right| \\ &\leq \lim_{r \to \infty} \|(h(y^r))^+\|_{C(\overline{\Omega})} \|\mu^r\|_{L^1(\Omega)} = 0. \end{aligned}$$

This gives (7.21). $\qquad \square$

### Remarks

In the mid-1960s, Lions [1] studied the necessary conditions for optimal control of elliptic partial differential equations. In his case, the equation is linear with the quadratic cost functional, there is no state constraint, and the control $u$ takes values in some convex and closed subset of $L^2(\Omega)$. The first order necessary conditions of such problems are of variational inequality form. In 1976, Mignot [1] started the discussion of optimal control for elliptic variational inequalities. Later, many authors made contributions to this topic; for example, Saguez [1], Barbu [1,5], Mignot–Puel [1], Tiba [1,2], Friedman [5,6], Friedman-Huang-Yong [1,2], Friedman–Hoffman [1], Barbu–Friedman [1], and Shi [1]. The problems studied in these works have no state constraint. The study of optimal control problems for partial differential equations with pointwise state constraints probably started from the work by A.I. Egorov [2], in which the system was evolutionary and finitely many equality constraints were imposed on the state at final time. Later, Mossino [1], Mackenroth [1] and Casas [1] continued to study the problem for linear equations with slightly more general form of pointwise state constraints. Afterwards, many works appeared concerning the necessary conditions for optimal control of (semilinear, quasilinear) elliptic equations and variational inequalities. We mention a few here: Bonnans–Casas [1–3], Bonnans–Tiba [1], Cases [2,3], Casas–Fernández [1,2], Casas–Yong [1], and Yong [8,11],. For parabolic equations, similar results were also obtained in Hu–Yong [1].

In this chapter, Sections 1–4 are based on the work of Yong [8] with some significant simplification. The main idea is the same as that of Chapter 4. The key point is that we have proved Lemma 2.1. This result was not that explicit in Yong [8], and had a complicated proof in Casas–Yong [1]. A similar result was proved in Hu–Yong [1] with a simpler and elementary proof using a suggestion of Casas. Here, the proof is further simplified by the idea of vector measure theory used in Chapter 4, (see Lemma 3.7). Section 5 is taken from Casas–Yong [1], which is a natural extension of the result for semilinear equations to quasilinear equations. Such a result was also established for parabolic semilinear and quasilinear equations (Hu–Yong [1]). Section 6 is based on the work of Yong [11]. The key result here is Lemma 6.3. This is an extension of the result in Barron [1]. See Neustadt [1] for the original formulation and related results for the finite dimensional case. Section 7 is an extension of Bonnans–Casas [3].

# Chapter 6

# Dynamic Programming Method for Evolution Systems

This chapter is devoted to the study of another important approach to optimal control problems. This method was originated by Bellman for finite dimensional optimal control problems and is called the *dynamic programming method*. Recent works by Crandall–Lions on the *viscosity solutions* for Hamilton-Jacobi-Bellman equations are a breakthrough in this direction. These works are of key importance in this chapter.

## §1. Optimality Principle and Hamilton-Jacobi-Bellman Equations

We consider the following state equation:

(1.1) $$\begin{cases} \dot{y}(t) = Ay(t) + f(t, y(t), u(t)), & t \in [0, T], \\ y(0) = x, \end{cases}$$

where $A : \mathcal{D}(A) \subset X \to X$ is the generator of some $C_0$ semigroup $e^{At}$ on a separable Hilbert space $X$, with $X^* = X$, where $f : [0,T] \times X \times U \to X$ is a given map with $U$ a metric space in which the control $u(\cdot)$ takes values. As before, for any initial state $x \in X$ and control $u(\cdot) \in \mathcal{U}[0,T] \equiv \{u : [0,T] \to U \mid u(\cdot) \text{ measurable}\}$, the corresponding trajectory $y(\cdot)$ is the mild solution of (1.1). We will assume that $f$ is Lipschitz continuous in $y$, uniformly in $(t,u) \in [0,T] \times U$ (see §2 for details). Thus, the (mild) solution to (1.1) is uniquely determined by the initial state and the control. Our cost functional is given by the following:

(1.2) $$J(u(\cdot)) = \int_0^T f^0(t, y(t), u(t)) \, dt + h(y(T)),$$

where $f^0 : [0,T] \times X \times U \to \mathbb{R}$ and $h : X \to \mathbb{R}$ are given functions. The optimal control problem is stated as follows:

**Problem (C).** Find $\bar{u}(\cdot) \in \mathcal{U}[0,T]$, such that

(1.3) $$J(\bar{u}(\cdot)) = \inf_{u(\cdot) \in \mathcal{U}[0,T]} J(u(\cdot)).$$

Now, let us describe the *Bellman's dynamic programming method*.

Instead of considering Problem (C) with the state equation (1.1) and the cost functional (1.2), we consider the following family of optimal control problems: For any given $(t, x) \in [0, T) \times X$, let us consider the following state equation:

$$(1.4) \quad y_{t,x}(s) = e^{A(s-t)}x + \int_t^s e^{A(s-r)} f(r, y_{t,x}(r), u(r))\, dr, \quad s \in [t, T],$$

with $u(\cdot) \in \mathcal{U}[t, T]$ and the cost functional

$$(1.5) \quad J_{t,x}(u(\cdot)) = \int_t^T f^0(r, y_{t,x}(r), u(r))\, dr + h(y_{t,x}(T)).$$

Here, the subscripts $t$ and $x$ are used to emphasize the dependence of the trajectory and the cost functional on the initial condition $(t, x)$. Of course, they also depend on the control, which should be clear from the context. Next, we define the function $V : [0, T] \times X \to \mathbb{R}$ by the following:

$$(1.6) \quad V(t, x) = \inf_{u(\cdot) \in \mathcal{U}[t,T]} J_{t,x}(u(\cdot)), \quad V(T, x) = h(x).$$

The function $V$ is called the *value function* of Problem (C). The goal of this chapter is to characterize this value function. Furthermore, we will use it to "find" optimal feedback controls (in some generalized sense).

Our first result is the following theorem, which is called the Bellman *optimality principle*, or the *dynamic programming principle*:

**Theorem 1.1.** (Bellman Optimality Principle) *Let $(t, x) \in [0, T) \times X$. Then, for any $s \in [t, T]$,*

$$(1.7) \quad V(t, x) = \inf_{u(\cdot) \in \mathcal{U}[t,s]} \left\{ \int_t^s f^0(r, y_{t,x}(r), u(r))\, dr + V(s, y_{t,x}(s)) \right\}.$$

Before proving the above theorem, let us first make some observations on (1.7). Suppose (1.7) holds and for a given $(t, x) \in [0, T) \times X$, there exists an optimal control $\bar{u}(\cdot)$, and $\bar{y}_{t,x}(\cdot)$ is the corresponding optimal trajectory. Then

$$(1.8) \quad \begin{aligned} V(t, x) &= \int_t^T f^0(r, \bar{y}_{t,x}(r), \bar{u}(r))\, dr + h(\bar{y}_{t,x}(T)) \\ &= \int_t^s f^0(r, \bar{y}_{t,x}(r), \bar{u}(r))\, dr + J_{s, \bar{y}_{t,x}(s)}(\bar{u}|_{[s,T]}(\cdot)) \\ &\geq \int_t^s f^0(r, \bar{y}_{t,x}(r), \bar{u}(r))\, dr + V(s, \bar{y}_{t,x}(s)) \\ &\geq \inf_{u(\cdot) \in \mathcal{U}[t,s]} \left\{ \int_t^s f^0(r, y_{t,x}(r), u(r))\, dr + V(s, y_{t,x}(s)) \right\} \\ &= V(t, x). \end{aligned}$$

Hence, the equalities in the middle of (1.8) hold. This implies that

$$(1.9) \quad J_{s, \bar{y}_{t,x}(s)}(\bar{u}|_{[s,T]}(\cdot)) = V(s, \bar{y}_{t,x}(s)).$$

## §1. Optimality Principle and HJB Equations

In other words, $\bar{u}|_{[s,T]}(\cdot)$ is an optimal control of the problem starting from $(s, \bar{y}_{t,x}(s))$ with the optimal trajectory $\bar{y}_{t,x}|_{[s,T]}(\cdot)$. This says that:

$$\text{globally optimal} \Rightarrow \text{locally optimal},$$

which is the essence of Bellman's dynamic programming method.

Now, let us give a proof of Theorem 1.1.

*Proof of Theorem 1.1.* First of all, for any $u(\cdot) \in \mathcal{U}[s, T]$ and any $u(\cdot) \in \mathcal{U}[t, s]$, by putting them concatenatively, we obtain $u(\cdot) \in \mathcal{U}[t, T]$. Thus, by the definition of the value function,

(1.10) $$V(t, x) \leq \int_t^s f^0(r, y_{t,x}(r), u(r))\, dr + J_{s, y_{t,x}(s)}(u(\cdot)).$$

By taking the infimum over $u(\cdot) \in \mathcal{U}[s, T]$, we obtain

(1.11) $$V(t, x) \leq \int_t^s f^0(r, y_{t,x}(r), u(r))\, dr + V(s, y_{t,x}(s)).$$

Consequently,

(1.12) $$V(t, x) \leq W(t, s, x) \equiv \text{the right-hand side of (1.7)}.$$

Next, for any $\varepsilon > 0$, there exists a $u^\varepsilon(\cdot) \in \mathcal{U}[t, T]$, such that

(1.13) $$\begin{aligned} V(t, x) + \varepsilon &\geq J_{t,x}(u^\varepsilon(\cdot)) \\ &= \int_t^s f^0(r, y_{t,x}(r), u^\varepsilon(r))\, dr + J_{s, y_{t,x}(s)}(u^\varepsilon(\cdot)) \\ &\geq \int_t^s f^0(r, y_{t,x}(r), u^\varepsilon(r))\, dr + V(s, y_{t,x}(s)) \\ &\geq W(t, s, x). \end{aligned}$$

Hence, (1.7) follows. □

Our next goal is to derive the so-called *Hamilton-Jacobi-Bellman equation* for the value function $V$.

**Proposition 1.2.** *Let the value function $V$ be $C^1([0, T] \times X)$. Let the functions $f$, $f^0$ and $h$ be continuous. Then $V$ satisfies the following Hamilton-Jacobi-Bellman equation:*

(1.14) $$\begin{cases} V_t + \langle V_x, Ax \rangle + H(t, x, V_x) = 0, & (t, x) \in [0, T] \times \mathcal{D}(A), \\ V|_{t=T} = h(x), & x \in X, \end{cases}$$

*where*

(1.15) $$H(t, x, p) = \inf_{u \in U}\{\langle p, f(t, x, u) \rangle + f^0(t, x, u)\},$$

$$(t, x, p) \in [0, T] \times X \times X.$$

*Proof.* First of all, by definition, $V(T,x) = h(x)$ is satisfied. Next, let us fix a $u \in U$ and $x \in \mathcal{D}(A)$. By (1.7), we have

$$\begin{aligned}
0 &\leq V(s, y_{t,x}(s)) - V(t,x) + \int_t^s f^0(r, y_{t,x}(r), u)\, dr \\
&= V_t(t,x)(s-t) + \langle V_x(t,x), y_{t,x}(s) - x \rangle \\
&\quad + \int_t^s f^0(r, y_{t,x}(r), u)\, dr + o(|s-t| + |y_{t,x}(s) - x|).
\end{aligned} \tag{1.16}$$

We note that because $x \in \mathcal{D}(A)$, we have

$$\begin{aligned}
\frac{1}{s-t}(y_{t,x}(s) - x) &= \frac{1}{s-t}(e^{A(s-t)} - I)x \\
&\quad + \frac{1}{s-t}\int_t^s e^{A(s-r)} f(r, y_{t,x}(r), u)\, dr \\
&\to Ax + f(t,x,u), \qquad \text{as } s \downarrow t.
\end{aligned} \tag{1.17}$$

Hence, dividing by $(s-t)$ in (1.16) and sending $s \downarrow t$, we obtain

$$0 \leq V_t(t,x) + \langle V_x(t,x), Ax + f(t,x,u) \rangle + f^0(t,x,u), \quad \forall u \in U. \tag{1.18}$$

Thus, it follows that

$$0 \leq V_t(t,x) + \langle V_x(t,x), Ax \rangle + H(t,x, V_x(t,x)). \tag{1.19}$$

On the other hand, let $x \in \mathcal{D}(A)$ be fixed. For any $\varepsilon > 0$ and $s > t$, by (1.7), there exists a $\tilde{u}(\cdot) \equiv u^{\varepsilon,s}(\cdot) \in \mathcal{U}[t,s]$, such that

$$\begin{aligned}
\varepsilon(s-t) &\geq V(s, y_{t,x}(s)) - V(t,x) + \int_t^s f^0(r, y_{t,x}(r), \tilde{u}(r))\, dr \\
&= V_t(t,x)(s-t) + \langle V_x(t,x), (e^{A(s-t)} - I)x \rangle \\
&\quad + \langle V_x(t,x), \int_t^s e^{A(s-r)} f(r, y_{t,x}(r), \tilde{u}(r))\, dr \rangle \\
&\quad + \int_t^s f^0(r, y_{t,x}(r), \tilde{u}(r))\, dr + o(|s-t|) \\
&= V_t(t,x)(s-t) + \langle V_x(t,x), (e^{A(s-t)} - I)x \rangle \\
&\quad + \int_t^s \{ \langle V_x(t,x), f(t,x,\tilde{u}(r)) \rangle + f^0(t,x,\tilde{u}(r)) \}\, dr + o(|s-t|) \\
&\geq V_t(t,x)(s-t) + \langle V_x(t,x), (e^{A(s-t)} - I)x \rangle \\
&\quad + H(t,x, V_x(t,x))(s-t) + o(|s-t|).
\end{aligned} \tag{1.20}$$

Then, dividing through by $(s-t)$ and letting $s - t \to 0$, we obtain

$$\varepsilon \geq V_t(t,x) + \langle V_x(t,x), Ax \rangle + H(t,x, V_x(t,x)). \tag{1.21}$$

Combining with (1.19), we obtain the desired result. □

*Remark 1.3.* We derive the Hamilton-Jacobi-Bellman equation (1.14) by assuming the value function $V$ to be $C^1([0,T] \times X)$. This assumption, however, is not necessarily true in most cases. We will provide an example below to illustrate this point. Hence, the conclusion of Proposition 1.2 has lack of applicability. The purpose of the following sections is to introduce proper notions of solutions to the equation (1.14) so that the value function $V$ is the unique "solution" of (1.14).

*Remark 1.4.* In the above we have assumed that $X$ is a Hilbert space, for the sake of uniformity with later discussions. It is clear that the results of this section remain true if we replace the Hilbert space by any Banach space.

To conclude this section, let us present an example where the value function is not in $C^1([0,T] \times X)$.

*Example 1.5.* Consider in $\mathbb{R}$ the following system:

$$(1.22) \quad \begin{cases} \dot{y}_{t,x}(s) = u(s)y_{t,x}(s), & s \in (t,T], \\ y_{t,x}(t) = x, \end{cases}$$

with the control domain $U = [0,1]$ and the cost functional

$$(1.23) \qquad J_{t,x}(u(\cdot)) = y_{t,x}(T).$$

Then it is not hard to see that the value function is given by

$$(1.24) \qquad V(t,x) = \begin{cases} x, & x \geq 0, \\ xe^{T-t}, & x < 0. \end{cases}$$

Clearly, $V(t,x)$ is just Lipschitz continuous and is not $C^1$.

It is possible to construct many other examples in which the value function is not $C^1$. For some cases, the value function can even be discontinuous. We will not deal with such cases.

## §2. Properties of the Value Functions

In this section, we present some basic properties of the value function associated with our optimal control problem. As in §1 and hereafter, we let $X$ be a separable Hilbert space with the inner product $\langle \cdot, \cdot \rangle$ and the induced norm $|\cdot|$. We also let $U$ be a metric space in which the control takes values.

### §2.1. Continuity

We first study the continuity of value functions. In what follows, by a *modulus of continuity*, we mean a continuous function $\omega : \mathbb{R}^+ \to \mathbb{R}^+$, with $\omega(0) = 0$ and subadditive: $\omega(\rho_1 + \rho_2) \leq \omega(\rho_1) + \omega(\rho_2)$, for all $\rho_1, \rho_2 \geq 0$; by a *local modulus of continuity*, we mean a continuous function $\omega : \mathbb{R}^+ \times \mathbb{R}^+ \to$

$\mathbb{R}^+$, with the property that for each $r \geq 0$, $\rho \mapsto \omega(\rho, r)$ is a modulus of continuity. In what follows, in different places, $\omega$ will represent a different (local) modulus of continuity. Next, let us make the following assumptions:

(A1) The linear, densely defined operator $A : \mathcal{D}(A) \subset X \to X$ generates a $C_0$ contraction semigroup $e^{At}$ on the space $X$. Thus,

$$\|e^{At}\| \leq 1, \qquad \forall t \geq 0. \tag{2.1}$$

(A2) $f : [0,T] \times X \times U \to X$ is continuous, such that for some constant $L > 0$ and local modulus of continuity $\omega$,

$$|f(t,x,u) - f(\bar{t},\bar{x},u)| \leq L|x - \bar{x}| + \omega(|t - \bar{t}|, |x| \vee |\bar{x}|), \tag{2.2}$$
$$\forall t, \bar{t} \in [0,T], \ x, \bar{x} \in X, \ u \in U,$$

$$|f(t,0,u)| \leq L, \qquad \forall (t,u) \in [0,T] \times U. \tag{2.3}$$

(A3) $f^0 : [0,T] \times X \times U \to \mathbb{R}$ and $h : X \to \mathbb{R}$ are continuous, and there exists a local modulus of continuity $\omega$, such that

$$\begin{cases} |f^0(t,x,u) - f^0(\bar{t},\bar{x},u)| \leq \omega(|x-\bar{x}| + |t-\bar{t}|, |x| \vee |\bar{x}|), \\ |h(x) - h(\bar{x})| \leq \omega(|x-\bar{x}|, |x| \vee |\bar{x}|), \\ \qquad \forall t \in [0,T], \ x, \bar{x} \in X, \ u \in U. \end{cases} \tag{2.4}$$

$$|f^0(t,0,u)|, |h(0)| \leq L, \qquad \forall (t,u) \in [0,T] \times U, \tag{2.5}$$

for some constant $L > 0$ (here, we take it to be the same as that in (A2) just for simplicity).

(A2)' In (A2), replace (2.2) by the following:

$$|f(t,x,u) - f(\bar{t},\bar{x},u)| \leq L(|x - \bar{x}| + |t - \bar{t}|), \tag{2.6}$$
$$\forall t, \bar{t} \in [0,T], \ x, \bar{x} \in X, \ u \in U,$$

(A3)' In (A3), replace (2.4) by the following:

$$\begin{cases} |f^0(t,x,u) - f^0(\bar{t},\bar{x},u)| \leq L(|x-\bar{x}| + |t-\bar{t}|), \\ |h(x) - h(\bar{x})| \leq L|x - \bar{x}|, \\ \qquad \forall t \in [0,T], \ x, \bar{x} \in X, \ u \in U. \end{cases} \tag{2.7}$$

We know that for a general $C_0$ semigroup $e^{At}$, one always has $\|e^{At}\| \leq M e^{\omega_0 t}$ for some $M \geq 1$ and $\omega_0 \in \mathbb{R}$ (see Chapter 2, §4). As we are considering semilinear evolution equations, $\omega_0$ can be taken to be 0, without loss of generality. Thus, (2.1) is restrictive only in that $M = 1$. However, it is not hard to see that all the results in this subsection remain true for general cases.

It is clear that under (A1) and (A2), for any $(t,x) \in [0,T] \times X$ and $u(\cdot) \in \mathcal{U}[t,T]$, the state equation (1.4) admits a unique trajectory $y_{t,x}(\cdot)$.

§2. Properties of the Value Functions

To study the boundedness and the continuity of the value function $V$, we first need to look at some properties of the trajectory $y_{t,x}(\cdot)$. We collect these properties in the following lemma. In what follows, $C$ is an absolute constant, that can be different in different places.

**Lemma 2.1.** *Let (A1) and (A2) hold. Then, for any $0 \leq t \leq \bar{t} \leq T$, $x, \bar{x} \in X$, and $u(\cdot) \in \mathcal{U}[t,T]$,*

(2.8) $$|y_{t,x}(s)| \leq C(1+|x|), \qquad s \in [t,T].$$

(2.9) $$|y_{t,x}(s) - y_{t,\bar{x}}(s)| \leq C|x - \bar{x}|, \qquad s \in [t,T].$$

(2.10) $\quad |y_{t,x}(s) - y_{\bar{t},x}(s)| \leq C|(e^{A(\bar{t}-t)} - I)x| + C(1+|x|)(\bar{t}-t), \quad s \in [\bar{t},T].$

(2.11) $$|y_{t,x}(s) - e^{A(s-t)}x| \leq C(1+|x|)(s-t), \qquad s \in [t,T].$$

*Proof.* By Chapter 2, Proposition 5.3, we see that (2.8) and (2.9) hold. Let us show (2.10) and (2.11). To this end, we take $0 \leq t \leq \bar{t} \leq T$ and $x \in X$. From (1.4) and (2.8), we have

(2.12)
$$\begin{aligned}
|y_{t,x}(s) - y_{\bar{t},x}(s)| &\leq |e^{A(s-t)}x - e^{A(s-\bar{t})}x| + \int_t^{\bar{t}} L(1 + |y_{t,x}(r)|)\, dr \\
&\quad + L\int_{\bar{t}}^s |y_{t,x}(r) - y_{\bar{t},x}(r)|\, dr \\
&\leq |(e^{A(\bar{t}-t)} - I)x| + L[1 + C(1+|x|)](\bar{t}-t) \\
&\quad + L\int_{\bar{t}}^s |y_{t,x}(r) - y_{\bar{t},x}(r)|\, dr.
\end{aligned}$$

Thus, by Gronwall's inequality, we obtain (2.10). Finally, from (1.4) and (2.8), we have

(2.13) $\quad |y_{t,x}(s) - e^{A(s-t)}x| \leq \int_t^s L(1 + |y_{t,x}(r)|)\, dr \leq C(1+|x|)(s-t),$

proving (2.11). $\square$

We have seen that the estimates in (2.8)–(2.11) are uniform in the control $u(\cdot)$. This is crucial in obtaining the properties of the value function $V(t,x)$. The next result contains the local boundedness and various kinds of continuities of the value function.

**Theorem 2.2.** *Let (A1)–(A3) hold. Then, for some increasing function $C_0$ and some local modulus of continuity $\widehat{\omega}$,*

(2.14) $$|V(t,x)| \leq C_0(|x|), \qquad \forall (t,x) \in [0,T] \times X,$$

(2.15) $\quad |V(t,x) - V(t,\bar{x})| \leq \widehat{\omega}(|x - \bar{x}|, |x| \vee |\bar{x}|), \qquad \forall t \in [0,T],\ x, \bar{x} \in X.$

(2.16) $$|V(t,x) - V(\bar{t},x)| \leq \widehat{\omega}(|\bar{t} - t| + |(e^{A|t-\bar{t}|} - I)x|, |x|),$$
$$\forall t, \bar{t} \in [0,T], \ x \in X.$$

(2.17) $|V(t,x) - V(\bar{t}, e^{A(\bar{t}-t)}x)| \leq \widehat{\omega}(\bar{t} - t, |x|), \quad \forall 0 \leq t \leq \bar{t} \leq T, \ x \in X.$

*Consequently,*

(2.18) $|V(t,x) - h(e^{A(T-t)}x)| \leq \widehat{\omega}(T - t, |x|), \quad \forall (t,x) \in [0,T] \times X.$

*In the case where (A1), (A2)′, and (A3)′ hold, we have some constant $C > 0$, such that*

(2.19) $|V(t,x)| \leq C(1 + |x|), \quad \forall (t,x) \in [0,T] \times X,$

(2.20) $|V(t,x) - V(t,\bar{x})| \leq C|x - \bar{x}|, \quad \forall t \in [0,T], \ x, \bar{x} \in X.$

(2.21) $$|V(t,x) - V(\bar{t},x)| \leq C\Big((1+|x|)|\bar{t} - t| + |(e^{A|t-\bar{t}|} - I)x|\Big),$$
$$\forall t, \bar{t} \in [0,T], \ x \in X.$$

(2.22) $$|V(t,x) - V(\bar{t}, e^{A(\bar{t}-t)}x)| \leq C(1 + |x|)|t - \bar{t}|,$$
$$\forall 0 \leq t \leq \bar{t} \leq T, \ x \in X.$$

*Proof.* For any $t \in [0,T]$, $x, \bar{x} \in X$, and any control $u(\cdot) \in \mathcal{U}[t,T]$, by (2.4), (2.5), and (2.8), we have

(2.23) $$|V(t,x)| \leq |J_{t,x}(u(\cdot))|$$
$$= \Big|\int_t^T f^0(r, y_{t,x}(r), u(r)) \, dr + h(y_{t,x}(T))\Big|$$
$$\leq \int_t^T \big\{L + \omega\big(|y_{t,x}(r)|, |y_{t,x}(r)|\big)\big\} \, dr$$
$$+ L + \omega\big(|y_{t,x}(T)|, |y_{t,x}(T)|\big)$$
$$\leq (T + 1)\big\{L + \omega\big(C(1 + |x|), C(1 + |x|)\big)\big\}.$$

This gives (2.14). Now, let $t \in [0,T]$, $x, \bar{x} \in X$, and $u(\cdot) \in \mathcal{U}[t,T]$. By (2.4), (2.8), and (2.9), we have

(2.24) $$|J_{t,x}(u(\cdot)) - J_{t,\bar{x}}(u(\cdot))|$$
$$\leq \int_t^T \omega\big(|y_{t,x}(r) - y_{t,\bar{x}}(r)|, |y_{t,x}(r)| \vee |y_{t,\bar{x}}(r)|\big) \, dr$$
$$+ \omega\big(|y_{t,x}(T) - y_{t,\bar{x}}(T)|, |y_{t,x}(T)| \vee |y_{t,\bar{x}}(T)|\big)$$
$$\leq (T + 1)\omega\big(C|x - \bar{x}|, C(1 + |x| \vee |\bar{x}|)\big).$$

§2. Properties of the Value Functions

Thus, taking the infimum in $u(\cdot) \in \mathcal{U}[t,T]$, we obtain (2.15). Next, we let $0 \leq t \leq \bar{t} \leq T$, $x \in X$. By (2.4), (2.5), (2.8), and (2.10), for any $u(\cdot) \in \mathcal{U}[t,T]$, we have

(2.25)
$$\begin{aligned}
&|J_{t,x}(u(\cdot)) - J_{t,x}(\bar{t}, u(\cdot))| \\
&\leq \int_t^{\bar{t}} \Big[L + \omega\big(|y_{t,x}(r)|, |y_{t,x}(r)|\big)\Big]\, dr \\
&\quad + \int_t^T \omega\big(|y_{t,x}(r) - y_{\bar{t},x}(r)|, |y_{t,x}(r)| \vee |y_{\bar{t},x}(r)|\big)\, dr \\
&\quad + \omega\big(|y_{t,x}(T) - y_{\bar{t},x}(T)|, |y_{t,x}(T)| \vee |y_{\bar{t},x}(T)|\big) \\
&\leq \{L + \omega(C(1+|x|), C(1+|x|))\}(\bar{t}-t) \\
&\quad + (T+1)\omega\big(C|(e^{A(\bar{t}-t)} - I)x| \\
&\qquad + C(1+|x|)(\bar{t}-t), C(1+|x|)\big).
\end{aligned}$$

Then we can define $\widehat{\omega}$ such that (2.16) holds. Finally, let $0 \leq t \leq \bar{t} \leq T$ and $x \in X$. For any $u(\cdot) \in \mathcal{U}[t,T]$, by (2.11),

(2.26)
$$\begin{aligned}
&|J_{t,x}(u(\cdot)) - J_{\bar{t},e^{A(\bar{t}-t)}x}(u(\cdot))| \\
&\leq \int_t^{\bar{t}} \{L + \omega(|y_{t,x}(r)|, |y_{t,x}(r)|)\}\, dr \\
&\quad + \int_{\bar{t}}^T \omega\big(|y_{t,x}(r) - y_{\bar{t},e^{A(\bar{t}-t)}x}(r)|, |y_{t,x}(r)| \vee |y_{\bar{t},e^{A(\bar{t}-t)}x}(r)|\big)\, dr \\
&\quad + \omega\big(|y_{t,x}(T) - y_{\bar{t},e^{A(\bar{t}-t)}x}(T)|, |y_{t,x}(T)| \vee |y_{\bar{t},e^{A(\bar{t}-t)}x}(T)|\big) \\
&\leq \{L + \omega(C(1+|x|), C(1+|x|))\}(\bar{t}-t) \\
&\quad + (T+1)\omega\big(C|y_{t,x}(\bar{t}) - e^{A(\bar{t}-t)}x|, C(1+|x|)\big) \\
&\leq \{L + \omega(C(1+|x|), C(1+|x|))\}(\bar{t}-t) \\
&\quad + (T+1)\omega\big(C(1+|x|)(\bar{t}-t), C(1+|x|)\big).
\end{aligned}$$

Hence, (2.17) holds. If we take $\bar{t} = T$, we obtain (2.18). The conclusion under (A1), (A2)', and (A3)' can be proved similarly. □

We note that $V(t,x)$ is not necessarily Lipschitz continuous in $t$; also, if in (A2) and (A3), $\omega(\sigma, r) \equiv \omega(\sigma)$, then $\widehat{\omega}(\sigma, r) \equiv \widehat{\omega}(\sigma)$ in (2.15)–(2.18).

## §2.2. B-continuity

In this subsection, we discuss another kind of continuity, which will play an important role later. Let us make a further assumption:

(A4) $B \in \mathcal{L}(X)$ is a positive self-adjoint operator, such that $\mathcal{R}(B) \subset \mathcal{D}(A^*)$ (thus $A^*B \in \mathcal{L}(X)$), and for some constant $c_0 \geq 0$,

(2.27) $\quad \langle A^*Bx, x \rangle \leq c_0 \langle Bx, x \rangle - |x|^2, \qquad \forall x \in X.$

We define the seminorm induced by $B$ as follows:

(2.28) $$|x|_B = \langle Bx, x \rangle^{1/2}, \quad \forall x \in X.$$

**Definition 2.3.** Let $B \in \mathcal{L}(X)$ be self-adjoint and positive. A function $v : X \to \mathbb{R}$ is said to be $B$-*continuous* at $x_0 \in X$, if for any $x_n \in X$ with $x_n \overset{w}{\rightharpoonup} x_0$ and $|Bx_n - Bx_0| \to 0$, it holds that $v(x_n) \to v(x_0)$.

Note that if $B$ is compact, then $v$ is $B$-continuous if and only if $v$ is sequentially weakly continuous. This observation will be useful later.

Now, let us give an example to illustrate (A4). Consider the following semilinear heat equation:

(2.29) $$\begin{cases} y_t - \Delta y = f(y, u), & \text{in } \Omega, \\ y|_{\partial \Omega} = 0, \\ y|_{t=0} = y_0, \end{cases}$$

where $\Omega \subset \mathbb{R}^n$ with a smooth boundary $\partial \Omega$. Let $X = L^2(\Omega)$, which is a Hilbert space; and let $A = \Delta$, which is a self-adjoint operator with the domain $\mathcal{D}(A) = H^2(\Omega) \cap H_0^1(\Omega)$ compactly embedded into $L^2(\Omega)$. By Chapter 2, Theorem 6.8, we know that $B \equiv (I - A)^{-1} : L^2(\Omega) \to \mathcal{D}(A)$ exists, which is clearly compact (as an operator on $L^2(\Omega)$). On the other hand, we have

(2.30) $$\langle A^* By, y \rangle = \langle A(I - A)^{-1} y, y \rangle = \langle By, y \rangle - |y|^2, \quad \forall y \in X.$$

Thus, (A4) holds for this case with $c_0 = 1$ and with $B$ compact. We may come up with some more general parabolic equations satisfying (A4).

**Lemma 2.4.** *Let (A1) and (A4) hold. Then*

(2.31) $$|e^{At}x|_B^2 + 2t|e^{At}x|^2 \leq e^{2c_0 t}|x|_B^2, \quad \forall (t, x) \in [0, \infty) \times X.$$

*Proof.* For any $x \in \mathcal{D}(A)$, we denote $y(t) = e^{At}x$. Then

(2.32) $$\begin{aligned} \frac{d}{dt}|y(t)|_B^2 &= \langle BAy(t), y(t) \rangle + \langle By(t), Ay(t) \rangle \\ &= 2\langle A^* By(t), y(t) \rangle \leq 2c_0|y(t)|_B^2 - 2|y(t)|^2. \end{aligned}$$

By Gronwall's inequality, we obtain

(2.33) $$|y(t)|_B^2 \leq e^{2c_0 t}|x|_B^2 - 2\int_0^t e^{2c_0(t-r)}|y(r)|^2 \, dr.$$

On the other hand, because $e^{At}$ is a contraction semigroup, we see that $|y(r)|^2$ is nonincreasing in $r$. Thus, (2.33) implies

(2.34) $$|y(t)|_B^2 \leq e^{2c_0 t}|x|_B^2 - 2|y(t)|^2 \int_0^t e^{2c_0(t-r)} \, dr \leq e^{2c_0 t}|x|_B^2 - 2t|y(t)|^2.$$

§2. Properties of the Value Functions

This proves (2.31) for the case $x \in \mathcal{D}(A)$. Because $\mathcal{D}(A)$ is dense in $X$, we can obtain the general case. □

**Lemma 2.5.** *Let (A1), (A2), and (A4) hold. Then there exists a constant $C > 0$, such that for all $s \in [t, T]$,*

$$(2.35) \quad |y_{t,x}(s) - y_{t,\bar{x}}(s)|_B^2 + \int_t^s |y_{t,x}(r) - y_{t,\bar{x}}(r)|^2 \, dr \leq C|x - \bar{x}|_B^2.$$

*Proof.* First of all, by Chapter 2, Proposition 5.5, for any $\varphi \in C^1(X)$ with $A^*\nabla\varphi(\cdot) \in C(X)$, it holds that

$$(2.36) \quad \varphi(y_{t,x}(s)) = \varphi(x) + \int_t^s \big\{ \langle A^*\nabla\varphi(y_{t,x}(r)), y_{t,x}(r) \rangle$$
$$+ \langle \nabla\varphi(y_{t,x}(r)), f(r, y_{t,x}(r), u(r)) \rangle \big\} \, dr, \quad 0 \leq t \leq s \leq T.$$

Take $\varphi(x) = |x|_B^2 \equiv \langle Bx, x \rangle$. Then $\nabla\varphi(x) = 2Bx$. Thus, by (2.9), (A4), and similar to (2.36), we have

$$|y_{t,x}(s) - y_{t,\bar{x}}(s)|_B^2 = |x - \bar{x}|_B^2$$
$$+ 2 \int_t^s \langle A^*B(y_{t,x}(r) - y_{t,\bar{x}}(r)), y_{t,x}(r) - y_{t,\bar{x}}(r) \rangle \, dr$$
$$+ 2 \int_t^s \langle f(r, y_{t,x}(r), u(r)) - f(r, y_{t,\bar{x}}(r), u(r)),$$
$$B[y_{t,x}(r) - y_{t,\bar{x}}(r)] \rangle \, dr$$

$$(2.37) \quad \leq |x - \bar{x}|_B^2$$
$$+ 2 \int_t^s [c_0|y_{t,x}(r) - y_{t,\bar{x}}(r)|_B^2 - |y_{t,x}(r) - y_{t,\bar{x}}(r)|^2] \, dr$$
$$+ 2 \int_t^s L\|B\|^{1/2}|y_{t,x}(r) - y_{t,\bar{x}}(r)|_B|y_{t,x}(r) - y_{t,\bar{x}}(r)| \, dr$$
$$\leq |x - \bar{x}|_B^2 + \int_t^s \big\{ (2c_0 + L^2\|B\|)|y_{t,x}(r) - y_{t,\bar{x}}(r)|_B^2$$
$$- |y_{t,x}(r) - y_{t,\bar{x}}(r)|^2 \big\} \, dr.$$

Hence, by Gronwall's inequality, we obtain (2.35). □

The next result gives some more continuity of the value function.

**Theorem 2.6.** *Let (A1)–(A4) hold. Then, for any $\varepsilon > 0$, the value function $V$ is uniformly $B$-continuous in $x \in B_{1/\varepsilon}(0)$, uniform in $t \in [0, T - \varepsilon]$. Namely, there exists a modulus of continuity $\omega_\varepsilon(\cdot)$, such that*

$$(2.38) \quad |V(t, x) - V(t, \bar{x})| \leq \omega_\varepsilon(|x - \bar{x}|_B), \quad \forall t \in [0, T - \varepsilon], \ |x|, |\bar{x}| \leq \frac{1}{\varepsilon}.$$

*Proof.* Let $R > 0$. Note that $\omega(\cdot, R)$ is a modulus of continuity for the function $f^0(t, \cdot, u)$ and $h(\cdot)$ with the argument $|x| \leq R$ (see (A3)), for any

$\delta > 0$, there exists a constant $C(\delta, R)$, such that

(2.39) $$\omega(r, R) \leq \delta + C(\delta, R)r, \qquad \forall 0 \leq r \leq R.$$

We leave the proof of this fact to the readers. Now, let $\varepsilon > 0$, $0 \leq t \leq T - \varepsilon$ and $|x| \leq R$. For any $u(\cdot) \in \mathcal{U}[t, T]$, we have

(2.40)
$$\begin{aligned}
|J_{t,x}(u(\cdot)) &- J_{t,\bar{x}}(u(\cdot))| \\
&\leq \int_t^T (\delta + C(\delta, R)|y_{t,x}(r) - y_{t,\bar{x}}(r)|) \, dr \\
&\quad + \delta + C(\delta, R)|y_{t,x}(T) - y_{t,\bar{x}}(T)| \\
&\leq C\delta + C(\delta, R) \int_t^T |y_{t,x}(r) - y_{t,\bar{x}}(r)| \, dr \\
&\quad + C(\delta, R)|y_{t,x}(T) - y_{t,\bar{x}}(T)|.
\end{aligned}$$

On the other hand, by (2.35), we have

(2.41)
$$\begin{aligned}
\int_t^T |y_{t,x}(r) - y_{t,\bar{x}}(r)| \, dr &\leq \sqrt{T-t} \Big\{ \int_t^T |y_{t,x}(r) - y_{t,\bar{x}}(r)|^2 \, dr \Big\}^{1/2} \\
&\leq C|x - \bar{x}|_B.
\end{aligned}$$

Also, for $t \in [0, T - \varepsilon]$, by Lemma 2.4 and (2.41), we have

(2.42)
$$\begin{aligned}
|y_{t,x}(T) - y_{t,\bar{x}}(T)| &\leq |e^{A(T-t)}(x - \bar{x})| \\
&\quad + L \int_t^T |y_{t,x}(r) - y_{t,\bar{x}}(r)| \, dr \leq \frac{C}{\sqrt{\varepsilon}} |x - \bar{x}|_B + C|x - \bar{x}|_B.
\end{aligned}$$

Then, combining (2.40)–(2.42), we see that (taking $R = 1/\varepsilon$)

(2.43)
$$|V(t, x) - V(t, \bar{x})| \leq C\delta + \widetilde{C}(\delta, \frac{1}{\varepsilon})|x - \bar{x}|_B,$$
$$\forall \delta > 0, \ |x|, |\bar{x}| \leq \frac{1}{\varepsilon}, \ 0 \leq t \leq T - \varepsilon.$$

This implies that for any $\varepsilon > 0$, in the region $0 \leq t \leq T - \varepsilon$, $|x| \leq 1/\varepsilon$, the value function $V(t, x)$ is uniformly continuous in $x$ with respect to the seminorm $|\cdot|_B$. Hence, the modulus of continuity $\omega_\varepsilon(\cdot)$ can be constructed so that (2.38) holds. □

We see that if the modulus of continuity $\omega(\sigma, r)$ appearing in (A3) satisfies $\omega(\sigma, r) \equiv \omega(\sigma)$, then (2.38) holds for all $x, \bar{x} \in X$. But, we still have to keep $t$ away from $T$.

### §2.3. Semiconcavity

Next, we will give another interesting property of the value function $V$. To this end, let us introduce the following notion.

## §2. Properties of the Value Functions

**Definition 2.7.** Let $\Omega \subset X$ be convex and $\varphi : \Omega \to \mathbb{R}$. We say that $\varphi$ is *weakly semiconcave* if there exists a local modulus of continuity $\omega$, such that for any $\lambda \in [0,1]$ and $x, y \in \Omega$,

$$
\begin{aligned}
&\lambda\varphi(x) + (1-\lambda)\varphi(y) - \varphi(\lambda x + (1-\lambda)y) \\
&\qquad \leq \lambda(1-\lambda)|x-y|\omega(|x-y|, |x| \vee |y|).
\end{aligned} \tag{2.44}
$$

If $\omega(s,r) \leq Cs$, $\forall s, r \geq 0$, then we say that $\varphi$ is *(strongly) semiconcave*.

It is clear that if $\varphi$ is semiconcave, then for some constant $C > 0$, the function $\psi(x) \equiv \varphi(x) - C|x|^2$ is *concave* in the usual sense:

$$
\psi(\lambda x + (1-\lambda)y) \geq \lambda\psi(x) + (1-\lambda)\psi(y), \quad \forall \lambda \in [0,1], \; x, y \in \Omega. \tag{2.45}
$$

From calculus, we know that any concave (or convex) functions are locally Lipschitz continuous. This implies that any (strongly) semiconcave functions are locally Lipschitz continuous. The following result tells us that the same is true for much more general functions.

**Lemma 2.8.** *Let $\varphi : X \to \mathbb{R}$ be a function that is bounded in any bounded set and satisfies the following property:*

$$
\begin{aligned}
&\lambda\varphi(x) + (1-\lambda)\varphi(y) - \varphi(\lambda x + (1-\lambda)y) \\
&\qquad \leq \lambda(1-\lambda)|x-y|\overline{C}(|x| \vee |y|), \quad \forall x, y \in X, \; \lambda \in [0,1],
\end{aligned} \tag{2.46}
$$

*where $\overline{C}(\cdot)$ is some nondecreasing function. Then it holds that*

$$
|\varphi(x) - \varphi(y)| \leq \{2 \sup_{|z| \leq R+1} |\varphi(z)| + \overline{C}(R+1)\}|x-y|, \tag{2.47}
$$

$$\forall |x|, |y| \leq R, \quad R > 0.$$

*That is, $\varphi(\cdot)$ is locally Lipschitz continuous. In particular, it is the case if $\varphi$ is weakly semiconcave.*

*Proof.* Fixing any $x, y \in X$ with $|x|, |y| \leq R$, we want to show (2.47). The case $x = y$ is trivial. Thus, we let $x \neq y$ and define $\xi = \frac{x-y}{|x-y|}$. Set

$$
\theta(t) = \varphi(y + (t-1)\xi), \quad t \in [0, |x-y|+2]. \tag{2.48}
$$

Clearly, it holds that

$$
\theta(1) = \varphi(y), \quad \theta(|x-y|+1) = \varphi(x). \tag{2.49}
$$

By (2.46), for any $s, t \in [0, |x-y|+2]$ and $\lambda \in [0,1]$,

$$
\begin{aligned}
&\lambda\theta(t) + (1-\lambda)\theta(s) - \theta(\lambda t + (1-\lambda)s) \\
&\quad = \lambda\varphi(y + (t-1)\xi) + (1-\lambda)\varphi(y + (s-1)\xi) \\
&\qquad - \varphi(y + [\lambda t + (1-\lambda)s - 1]\xi) \\
&\quad \leq \lambda(1-\lambda)|t-s|\overline{C}(R+1).
\end{aligned} \tag{2.50}
$$

Here, we note that

$$(2.51)\quad \begin{cases} |y+(t-1)\xi| \le |y|+|t-1| \le R+1, \quad t \in [0,2], \\ |y+(t-1)\xi| = \left|y+\dfrac{t-2}{|x-y|}(x-y)+\xi\right| \le R+1, \\ \qquad t \in [2, |x-y|+2]. \end{cases}$$

The same thing holds for $|y+(s-1)\xi|$. Now, for any $0 \le t_1 < t_2 < t_3 \le |x-y|+2$, we have

$$(2.52)\quad t_2 = \frac{t_3-t_2}{t_3-t_1}t_1 + \frac{t_2-t_1}{t_3-t_1}t_3.$$

Thus, by (2.50), (take $\lambda = \frac{t_3-t_2}{t_3-t_1}$, $t=t_1$, $s=t_3$),

$$(2.53)\quad \theta(t_2) \ge \frac{t_3-t_2}{t_3-t_1}\theta(t_1) + \frac{t_2-t_1}{t_3-t_1}\theta(t_3) - \frac{(t_3-t_2)(t_2-t_1)}{t_3-t_1}\overline{C}(R+1).$$

This implies that

$$(2.54)\quad \begin{aligned}\frac{\theta(t_2)-\theta(t_1)}{t_2-t_1} &\ge \frac{\theta(t_3)-\theta(t_1)}{t_3-t_1} - \frac{t_3-t_2}{t_3-t_1}\overline{C}(R+1) \\ &\ge \frac{\theta(t_3)-\theta(t_1)}{t_3-t_1} - \overline{C}(R+1),\end{aligned}$$

and

$$(2.55)\quad \begin{aligned}\frac{\theta(t_3)-\theta(t_2)}{t_3-t_2} &\le \frac{\theta(t_3)-\theta(t_1)}{t_3-t_1} + \frac{t_2-t_1}{t_3-t_1}\overline{C}(R+1) \\ &\le \frac{\theta(t_3)-\theta(t_1)}{t_3-t_1} + \overline{C}(R+1).\end{aligned}$$

Hence, from (2.54), we have (taking $t_1=1$, $t_2=|x-y|+1$, $t_3=|x-y|+2$)

$$(2.56)\quad \begin{aligned}\frac{\varphi(x)-\varphi(y)}{|x-y|} &= \frac{\theta(|x-y|+1)-\theta(1)}{|x-y|} \\ &\ge \frac{\theta(|x-y|+2)-\theta(1)}{|x-y|+1} - \overline{C}(R+1) \\ &\ge -|\varphi(x+\xi)-\varphi(y)| - \overline{C}(R+1) \\ &\ge -2\sup_{|z|\le R+1}|\varphi(z)| - \overline{C}(R+1).\end{aligned}$$

§2. Properties of the Value Functions

From (2.55), we obtain (taking $t_1 = 0$, $t_2 = 1$, $t_3 = |x - y| + 1$)

$$
\begin{aligned}
\frac{\varphi(x) - \varphi(y)}{|x - y|} &= \frac{\theta(|x - y| + 1) - \theta(1)}{|x - y|} \\
&\leq \frac{\theta(|x - y| + 1) - \theta(0)}{|x - y| + 1} + \overline{C}(R + 1) \\
&\leq |\varphi(x) - \varphi(y - \xi)| + \overline{C}(R + 1) \\
&\leq 2 \sup_{|z| \leq R+1} |\varphi(z)| + \overline{C}(R + 1).
\end{aligned}
\tag{2.57}
$$

Hence, (2.47) follows. $\square$

Our next result gives the weak semiconcavity of the value function.

**Theorem 2.9.** *Let (A1), (A2), and (A3)' hold and let $f(t,x,u)$ be $C^1$ in $x$ such that for some local modulus of continuity $\omega(\cdot,\cdot)$,*

$$|f_x(t, x, u) - f_x(t, \bar{x}, u)| \leq \omega(|x - \bar{x}|, |x| \vee |\bar{x}|). \tag{2.58}$$

*Further, let $f^0(t, x, u)$ and $h(x)$ be weakly semiconcave in $x$, uniformly in $(t, u)$. Then the value function $V(t, x)$ is weakly semiconcave in $x$ uniformly in $t \in [0, T]$. Moreover, if instead of (2.58), we have*

$$|f_x(t, x, u) - f_x(t, \bar{x}, u)| \leq C|x - \bar{x}|, \qquad \forall x, \bar{x} \in X. \tag{2.59}$$

*and $f^0(t, x, u)$ and $h(x)$ are (strongly) semiconcave in $x$ uniformly in $(t, u)$, then $V(t, x)$ is semiconcave in $x$, uniformly in $t \in [0, T]$.*

*Proof.* Let $x_0, x_1 \in X$ with $|x_0|, |x_1| \leq K$. By (2.8) and (2.9), there exists a constant $R = R_K$, such that for all $|x|, |\bar{x}| \leq K$, $s \in [t, T]$ and $u(\cdot) \in \mathcal{U}[t, T]$,

$$|y_{t,x}(s)| \leq R, \qquad |y_{t,x}(s) - y_{t,\bar{x}}(s)| \leq R|x - \bar{x}|. \tag{2.60}$$

Now, let $\lambda \in [0, 1]$ and denote $x_\lambda = \lambda x_1 + (1 - \lambda)x_0$. For any $\varepsilon > 0$, there exists a $u_\varepsilon(\cdot) \in \mathcal{U}[t, T]$, such that

$$J_{t, x_\lambda}(u_\varepsilon) < V(t, x_\lambda) + \varepsilon. \tag{2.61}$$

In what follows, for convenience, we use $\omega(\cdot, \cdot)$ to represent different local moduli of continuity at different places.

$$
\begin{aligned}
\lambda V(t, x_1) &+ (1-\lambda)V(t, x_0) - V(t, x_\lambda) - \varepsilon \\
&\leq \lambda J_{t,x_1}(u_\varepsilon) + (1-\lambda) J_{t,x_0}(u_\varepsilon) - J_{t,x_\lambda}(u_\varepsilon) \\
&= \int_t^T [\lambda f^0(r, y_{t,x_1}(r), u_\varepsilon(r)) + (1-\lambda) f^0(r, y_{t,x_0}(r), u_\varepsilon(r)) \\
&\qquad\qquad - f^0(r, y_{t,x_\lambda}(r), u_\varepsilon(r))] \, dr \\
&\quad + \lambda h(y_{t,x_1}(T)) + (1-\lambda) h(y_{t,x_0}(T)) - h(y_{t,x_\lambda}(T)) \\
&\leq \lambda(1-\lambda) \int_t^T \omega(|y_{t,x_1}(r) - y_{t,x_0}(r)|, R) \, dr \\
&\quad + \lambda(1-\lambda) \omega(|y_{t,x_1}(T) - y_{t,x_0}(T)|, R) \\
&\quad + \int_t^T |f^0(r, \lambda y_{t,x_1}(r) + (1-\lambda) y_{t,x_0}(r), u_\varepsilon(r)) \\
&\qquad\qquad - f^0(r, y_{t,x_\lambda}(r), u_\varepsilon(r))| \, dr \\
&\quad + |h(\lambda y_{t,x_1}(T) + (1-\lambda) y_{t,x_0}(T)) - h(y_{t,x_\lambda}(T))| \\
&\leq \lambda(1-\lambda) \omega(R|x_1 - x_0|, R) \\
&\quad + L \int_t^T |\lambda y_{t,x_1}(r) + (1-\lambda) y_{t,x_0}(r) - y_{t,x_\lambda}(r)| \, dr \\
&\quad + L|\lambda y_{t,x_1}(T) + (1-\lambda) y_{t,x_0}(T) - y_{t,x_\lambda}(T)|.
\end{aligned}
$$
(2.62)

Note that under assumption (2.58), we have (denoting $y_\lambda = \lambda y_1 + (1-\lambda) y_0$)

$$
\begin{aligned}
|\lambda f(t, y_1, u) &+ (1-\lambda) f(t, y_0, u) - f(t, y_\lambda, u)| \\
&= \Big|\lambda \int_0^1 f_x(t, y_\lambda + \sigma(1-\lambda)(y_1 - y_0), u) \, d\sigma (1-\lambda)(y_1 - y_0) \\
&\quad + (1-\lambda) \int_0^1 f_x(t, y_\lambda + \sigma\lambda(y_0 - y_1), u) \, d\sigma \lambda(y_0 - y_1) \Big| \\
&= \lambda(1-\lambda)|y_1 - y_0| \Big| \int_0^1 f_x(t, y_\lambda + \sigma(1-\lambda)(y_1 - y_0), u) \\
&\qquad\qquad - f_x(t, y_\lambda + \sigma\lambda(y_0 - y_1), u) \, d\sigma \Big| \\
&\leq \lambda(1-\lambda)|y_1 - y_0|\omega(|y_1 - y_0|, R).
\end{aligned}
$$
(2.63)

## §3. Viscosity Solutions

Thus,

(2.64)
$$\begin{aligned}
&|\lambda y_{t,x_1}(s) + (1-\lambda)y_{t,x_0}(s) - y_{t,x_\lambda}(s)| \\
&\leq \int_t^s |\lambda f(r, y_{t,x_1}(r), u_\varepsilon(r)) + (1-\lambda)f(r, y_{t,x_0}(r), u_\varepsilon(r)) \\
&\qquad - f(r, \lambda y_{t,x_1}(r) + (1-\lambda)y_{t,x_0}(r), u_\varepsilon(r))|\, dr \\
&\quad + L\int_t^s |\lambda y_{t,x_1}(r) + (1-\lambda)y_{t,x_0}(r) - y_{t,x_\lambda}(r)|\, dr \\
&\leq \lambda(1-\lambda)\int_t^s |y_{t,x_1}(r) - y_{t,x_0}(r)|\omega(|y_{t,x_1}(r) - y_{t,x_0}(r)|, R)\, dr \\
&\quad + L\int_t^s |\lambda y_{t,x_1}(r) + (1-\lambda)y_{t,x_0}(r) - y_{t,x_\lambda}(r)|\, dr \\
&\leq \lambda(1-\lambda)|x_1 - x_0|\omega(|x_1 - x_0|, R) \\
&\quad + L\int_t^s |\lambda y_{t,x_1}(r) + (1-\lambda)y_{t,x_0}(r) - y_{t,x_\lambda}(r)|\, dr.
\end{aligned}$$

Then, by Gronwall's inequality,

(2.65)
$$\begin{aligned}
&|\lambda y_{t,x_1}(s) + (1-\lambda)y_{t,x_0}(s) - y_{t,x_\lambda}(s)| \\
&\leq \lambda(1-\lambda)|x_1 - x_0|\omega(|x_1 - x_0|, R), \qquad s \in [t, T].
\end{aligned}$$

Hence, by (2.62) and (2.65), we obtain our assertion. $\qquad\square$

## §3. Viscosity Solutions

In this section, we are going to introduce the notion of viscosity solutions. To begin with, let us introduce the classes of *test functions*. To this end, we recall that a function $\varphi : [0,T] \times X \to \mathbb{R}$ is *weakly sequentially lower semicontinuous* if for any weakly convergent sequence $(t_n, x_n) \xrightarrow{w} (t, x)$, one has

(3.1)
$$\varphi(t, x) \leq \varliminf_{n \to \infty} \varphi(t_n, x_n).$$

Now, we define

(3.2)
$$\begin{aligned}
\Phi &= \{\varphi \in C^1([0,T] \times X) \mid \varphi \text{ is weakly sequentially lower} \\
&\qquad \text{semicontinuous, } A^*\nabla\varphi \in C([0,T] \times X) \,\}, \\
\mathcal{G} &= \{g \in C^1([0,T] \times X) \mid \exists g_0,\, \rho \in C^1(\mathbb{R}), a \in [0, \infty), \text{ with} \\
&\qquad g_0'(r) \geq 0,\, \rho'(r) \geq 0,\, \rho'(0) = 0, \text{ such that} \\
&\qquad g(x, t) = g_0(\rho(|x|) - at), \quad \forall (t, x) \in [0, T] \times X \,\}.
\end{aligned}$$

In the above and hereafter, $\nabla$ stands for the Fréchet derivative in $x$. Note that $\varphi(x) = |x|_B^2 \equiv \langle Bx, x \rangle$ is a typical test function in $\Phi$. For such a $\varphi$, we have $\nabla\varphi(x) = 2Bx$. By our assumption (A4), $A^*\nabla\varphi(x) = 2A^*Bx$ is a very nice continuous function.

Next, let $g \in \mathcal{G}$, $g(x,t) = g_0(\rho(|x|) - at)$. Then we see that

(3.3) $$\nabla g(t,x) = \begin{cases} g_0'(\rho(|x|) - at)\rho'(|x|)\dfrac{x}{|x|}, & x \neq 0, \\ 0, & x = 0. \end{cases}$$

This observation is very crucial in sequel. Now, we consider the following Cauchy problem of the Hamilton-Jacobi-Bellman equation:

(3.4) $$\begin{cases} v_t + \langle \nabla v, Ax \rangle + H(t,x,\nabla v) = 0, & (t,x) \in [0,T] \times \mathcal{D}(A), \\ v(T,x) = h(x), & x \in X, \end{cases}$$

where

(3.5) $$H(t,x,p) = \inf_{u \in U}\{\langle p, f(t,x,u)\rangle + f^0(t,x,u)\},$$
$$(t,x,p) \in [0,T] \times X \times X^*.$$

The above (3.4) and (3.5) coincide with (1.14) and (1.15). We rewrite them here just for convenience. Let us now introduce the following definition.

**Definition 3.1.** Let $v \in C([0,T] \times X)$. We call $v$ a *viscosity subsolution* (resp. *supersolution*) of (3.4) if the terminal condition $v(T,x) \leq h(x)$ (resp. $v(T,x) \geq h(x)$) is satisfied and for any $\varphi \in \Phi$ and $g \in \mathcal{G}$, whenever the function $v - \varphi - g$ attains a local maximum (resp. the function $v + \varphi + g$ attains a local minimum) at $(t,x) \in [0,T) \times X$, we have

(3.6) $$\varphi_t(t,x) + g_t(t,x) + \langle A^*\nabla\varphi(t,x), x \rangle$$
$$+ H(t,x,\nabla\varphi(t,x) + \nabla g(t,x)) \geq 0,$$

(respectively,

(3.7) $$-\varphi_t(t,x) - g_t(t,x) - \langle A^*\nabla\varphi(t,x), x \rangle$$
$$+ H(t,x,-\nabla\varphi(t,x) - \nabla g(t,x)) \leq 0.)$$

In the case where the function $v$ is both a viscosity subsolution and supersolution of (3.4), we call it a *viscosity solution* of (3.4).

Our main result of this section is the following theorem.

**Theorem 3.2.** *Let (A1)–(A3) hold. Then the value function $V(t,x)$ is a viscosity solution of (3.4).*

To prove this theorem, we need two lemmas.

**Lemma 3.3.** *Let (A1)–(A3) hold. Let $\varphi \in \Phi$ and $(t,x) \in [0,T) \times X$. Then the following convergence holds uniformly in $u(\cdot) \in \mathcal{U}[t,T]$.*

(3.8) $$\lim_{s \downarrow t}\left\{\frac{1}{s-t}[\varphi(s,y_{t,x}(s)) - \varphi(t,x)] - \varphi_t(t,x) - \langle A^*\nabla\varphi(t,x), x\rangle \right.$$
$$\left. - \frac{1}{s-t}\int_t^s \langle \nabla\varphi(t,x), f(t,x,u(r))\rangle\, dr\right\} = 0.$$

## §3. Viscosity Solutions

*Proof.* By Chapter 2, Proposition 5.5, we have

$$
\begin{aligned}
\varphi(s, y_{t,x}(s)) = {}& \varphi(t,x) \\
& + \int_t^s \{\varphi_t(r, y_{t,x}(r)) + \langle A^*\nabla\varphi(r, y_{t,x}(r)), y_{t,x}(r) \rangle \\
& + \langle \nabla\varphi(r, y_{t,x}(r)), f(r, y_{t,x}(r), u(r)) \rangle\} \, dr.
\end{aligned}
\tag{3.9}
$$

Then, because $y_{t,x}(s)$ converges to $x$ as $s \to t$, uniformly in $u(\cdot) \in \mathcal{U}[t,T]$ (see (2.11)) and $\nabla\varphi$ and $A^*\nabla\varphi$ are continuous, we obtain our conclusion. □

**Lemma 3.4.** *Let (A1)–(A3) hold. Let $(t,x) \in [0,T) \times X$. Then, for any $g \in \mathcal{G}$ with $g(t,x) = 0$, the following holds:*

$$
\begin{aligned}
\frac{1}{s-t} g(s, y_{t,x}(s)) \leq {}& g_t(t,x) \\
& + \frac{1}{s-t} \int_t^s \langle \nabla g(t,x), f(t,x,u(r)) \rangle \, dr + o(1),
\end{aligned}
\tag{3.10}
$$

*where $o(1)$ is uniformly in $u(\cdot) \in \mathcal{U}[t,T]$.*

*Proof.* First, let $x \neq 0$. Then, provided $s - t > 0$ is small enough, we have $y_{t,x}(s) \neq 0$ for all $u(\cdot) \in \mathcal{U}[t,T]$. Now, by (1.4),

$$
\begin{aligned}
|y_{t,x}(s)| = {}& \langle \frac{y_{t,x}(s)}{|y_{t,x}(s)|}, e^{A(s-t)} x \rangle \\
& + \int_t^s \langle e^{A^*(s-r)} \frac{y_{t,x}(s)}{|y_{t,x}(s)|}, f(r, y_{t,x}(r), u(r)) \rangle \, dr \\
\leq {}& |x| + \int_t^s \langle \frac{x}{|x|}, f(t,x,u(r)) \rangle \, dr + o(s-t).
\end{aligned}
\tag{3.11}
$$

In the above, we have used the assumptions (2.1)–(2.2) and the fact that

$$
\lim_{s \downarrow t} \left| e^{A^*(s-t)} \frac{y_{t,x}(s)}{|y_{t,x}(s)|} - \frac{x}{|x|} \right| = 0, \quad \text{uniformly in } u(\cdot) \in \mathcal{U}[t,T].
\tag{3.12}
$$

The above fact can be easily proved by using (2.1) and (2.11). We leave the details to the readers. Now, for $g \in \mathcal{G}$, we have (noting (3.3) and our assumption $g(t,x) = g_0(\rho(|x|) - at) = 0$)

$$g(s, y_{t,x}(s)) = g_0(\rho(|y_{t,x}(s)|) - as)$$
$$\leq g_0\left(\rho(|x| + \int_t^s \langle \frac{x}{|x|}, f(t,x,u(r)) \rangle \, dr + o(s-t)) - as\right)$$
(3.13)
$$= g_0(\rho(|x|) + \rho'(|x|) \int_t^s \langle \frac{x}{|x|}, f(t,x,u(r)) \rangle \, dr + o(s-t) - as)$$
$$= g_0(\rho(|x|) - at) + g_0'(\rho(|x|) - at)\left\{\rho'(|x|) \int_t^s \langle \frac{x}{|x|}, f(t,x,u(r)) \rangle \, dr\right.$$
$$\left. + o(s-t) - a(s-t)\right\} + o(s-t)$$
$$= \int_t^s \langle \nabla g(t,x), f(t,x,u(r)) \rangle \, dr + g_t(t,x)(s-t) + o(s-t).$$

This gives (3.10) for $x \neq 0$. Now, if $x = 0$, then, by (3.3), $\nabla g(t,x) = 0$. Thus, we need only to show that

(3.14) $\qquad |\frac{1}{s-t} g(s, y_{t,x}(s)) - g_t(t,x)| = o(1), \qquad$ as $s \downarrow t$.

In fact, we have (note that $\rho'(0) = 0$)

$$g(s, y_{t,0}(s)) = g_0(\rho(|y_{t,0}(s)|) - as)$$
(3.15) $\quad = g_0(\rho(|\int_t^s f(t,0,u(r)) \, dr| + o(s-t)) - as)$
$$= g_0(\rho(0) + \rho'(0)(|\int_t^s f(t,0,u(r)) \, dr| + o(s-t)) + o(s-t) - as)$$
$$= g_0(\rho(0) - as + o(s-t))$$
$$= g(t,0) + g_t(t,0)(s-t) + o(s-t).$$

Hence, our conclusion follows. □

Now, let us prove Theorem 3.2.

*Proof of Theorem 3.2.* First, we let $(t,x) \in [0,T) \times X$ be a local maximum of the function $V - \varphi - g$ with $\varphi \in \Phi$ and $g \in \mathcal{G}$. Without loss of generality, by adding some constants to $\varphi$ and $g$ if necessary, we may assume that

(3.16) $\qquad\qquad g(t,x) = 0, \quad V(t,x) = \varphi(t,x).$

Then, for $s - t > 0$ small, we have that

(3.17) $\quad V(s, y_{t,x}(s)) - \varphi(s, y_{t,x}(s)) - g(s, y_{t,x}(s))$
$\qquad\qquad \leq V(t,x) - \varphi(t,x) - g(t,x) = 0.$

## §3. Viscosity Solutions

Thus, for fixed $u \in U$, by (1.7),

$$\varphi(t,x) = V(t,x) \le \int_t^s f^0(r, y_{t,x}(r), u)\, dr + V(s, y_{t,x}(s)) \tag{3.18}$$

$$\le \int_t^s f^0(r, y_{t,x}(r), u)\, dr + \varphi(s, y_{t,x}(s)) + g(s, y_{t,x}(s)).$$

Then

$$0 \le \frac{1}{s-t} \int_t^s f^0(r, y_{t,x}(r), u)\, dr + \frac{1}{s-t}[\varphi(s, y_{t,x}(s)) - \varphi(t,x)] \tag{3.19}$$

$$+ \frac{1}{s-t} g(s, y_{t,x}(s)).$$

Now, applying Lemmas 3.3 and 3.4, we obtain that

$$0 \le \varphi_t(t,x) + g_t(t,x) + \langle A^*\nabla\varphi(t,x), x \rangle \tag{3.20}$$
$$+ \langle \nabla\varphi(t,x) + \nabla g(t,x), f(t,x,u) \rangle + f^0(t,x,u), \qquad \forall u \in U.$$

Taking the infimum in $u \in U$, we see that $V$ is a viscosity subsolution of (3.4).

Next, we let $(t,x) \in [0,T) \times X$ be a local minimum of the function $V + \varphi + g$ with $\varphi \in \Phi$ and $g \in \mathcal{G}$. Similar to (3.16), we may assume that

$$g(t,x) = 0, \quad V(t,x) + \varphi(t,x) = 0. \tag{3.21}$$

Then, for $s - t > 0$ small,

$$V(s, y_{t,x}(s)) + \varphi(s, y_{t,x}(s)) + g(s, y_{t,x}(s)) \tag{3.22}$$
$$\ge V(t,x) + \varphi(t,x) + g(t,x) = 0.$$

Now, for any $\varepsilon > 0$ and $s \in (t,T]$, by (1.7), one can find a control $u^\varepsilon(\cdot) \equiv u^{\varepsilon,s}(\cdot) \in \mathcal{U}[t,T]$, such that

$$\varepsilon(s-t) - \varphi(t,x) = \varepsilon(s-t) + V(t,x)$$
$$\ge \int_t^s f^0(r, y_{t,x}(r), u^\varepsilon(r))\, dr + V(s, y_{t,x}(s)) \tag{3.23}$$
$$\ge \int_t^s f^0(r, y_{t,x}(r), u^\varepsilon(r))\, dr - \varphi(s, y_{t,x}(s)) - g(s, y_{t,x}(s)).$$

Thus, it follows from Lemmas 3.3 and 3.4 that

$$\varepsilon \geq \frac{1}{s-t} \int_t^s f^0(r, y_{t,x}(r), u^\varepsilon(r)) dr$$
$$- \frac{\varphi(s, y_{t,x}(s)) - \varphi(t, x)}{s-t} - \frac{g(s, y_{t,x}(s))}{s-t}$$
$$= -\varphi_t(t, x) - g_t(t, x) - \langle A^* \nabla \varphi(t, x), x \rangle$$

(3.24)
$$+ \frac{1}{s-t} \int_t^s \Big\{ f^0(t, x, u^\varepsilon(r))$$
$$- \langle \nabla \varphi(t, x) + \nabla g(t, x), f(t, x, u^\varepsilon(r)) \rangle \Big\} dr + o(1)$$
$$\geq -\varphi_t(t, x) - g_t(t, x) - \langle A^* \nabla \varphi(t, x), x \rangle$$
$$+ H(t, x, -\nabla \varphi(t, x) - \nabla g(t, x)) + o(1).$$

Letting $\varepsilon \to 0$, we obtain the following inequality:

(3.25)
$$0 \geq -\varphi_t(t, x) - g_t(t, x) - \langle A^* \nabla \varphi(t, x), x \rangle$$
$$+ H(t, x, -\nabla \varphi(t, x) - \nabla g(t, x)).$$

Therefore, $V$ is also a viscosity supersolution of (3.4). This completes the proof of Theorem 3.2. □

## §4. Uniqueness of Viscosity Solutions

In this section we are going to present a uniqueness result for viscosity solutions. This result, together with those in the previous section, will give a characterization for the value function of Problem (C).

### §4.1. A perturbed optimization lemma

It is well known that if $D$ is a bounded closed subset in $\mathbb{R}^n$ and $g : D \to \mathbb{R}^n$ is lower semicontinuous and bounded from below, then there exists an $x_0 \in D$, such that $g(x_0) = \min_{x \in D} g(x)$. This is due to the fact that $D$ is compact. However, if $\mathbb{R}^n$ is replaced by some infinite dimensional Banach space $X$, we cannot expect such a result. Let us present a simple example to illustrate this point.

**Example 4.1.** Take $X = \ell^2 \equiv \{x = (x_k)_{k \geq 1} \mid \sum_{k \geq 1} x_k^2 < \infty\}$. Let $D$ be the closed unit ball in $X$ and define $h : D \to \mathbb{R}$ as follows:

(4.1)
$$h(x) = \sum_{|x_k| \geq 1/2} k(2|x_k| - 1), \qquad \forall x = (x_k)_{k \geq 1} \in X.$$

Note that the right-hand side of (4.1) is actually a finite sum. Thus, it is easy to show that $h(\cdot)$ is continuous on $D$. Let $x^m = (\delta_{mk})_{k \geq 1}$, where $\delta_{mk} = 0$ for $m \neq k$ and $\delta_{mm} = 1$. Then $x^m \in D$, and

(4.2)
$$h(x^m) = m \to \infty, \qquad \text{as } m \to \infty.$$

§4. Uniqueness of Viscosity Solutions

Thus, $h(\cdot)$ is unbounded on $D$, which is quite different from the finite dimensional case. Now, we set

(4.3) $$g(x) = (1 + h(x))^{-1}, \qquad x \in X.$$

Then $g(\cdot)$ is continuous. On the other hand, by (4.2) and (4.3),

(4.4) $$\inf_{x \in D} g(x) = 0, \qquad g(x) > 0, \quad \forall x \in D.$$

Thus, the infimum is not attained on $D$.

The above example tells us that in infinite dimensional spaces, continuous functions on some bounded closed set may be unbounded, and the infimum may be not attained (even if it is bounded from below).

On the other hand, regard $D$ as a complete metric space. By Ekeland's variational principle (Chapter 4, Lemma 2.1), we know that for any $\varepsilon > 0$, there exists an $\bar{x} \in D$, such that the function $x \mapsto g(x) + \varepsilon|x - \bar{x}|$ attains its minimum over $D$ at some point $x_0 \in D$. The term $\varepsilon|x - \bar{x}|$ is not differentiable in $x$ (at $\bar{x}$), which is not convenient. Thus, we hope to improve this result by taking into account some special properties of the underlying Banach space $X$. In the following result, we shall prove that the term $\varepsilon|x - \bar{x}|$ can be replaced by $\langle p, x - \bar{x} \rangle$, for some $p \in X^*$ with $|p|_* < \varepsilon$. This result will be very crucial later in the proof of uniqueness result for the viscosity solutions. Such a result is also very important itself in the theory of optimization in infinite dimensions. Before we state this result, let us make the following assumption:

(BP) The dual space $X^*$ of the Banach space $X$ admits a *Fréchet differentiable bump function*; namely, there exists a $\psi : X^* \to \mathbb{R}$, $0 \le \psi \le 1$, with $\{p \in X^* \mid \psi(p) > 0\}$ nonempty and bounded, in which $\psi$ is Fréchet differentiable.

Because $X^*$ is a linear space, by translation and scaling, we see that if (BP) holds, then we can assume that for any $\delta > 0$, there exists a Fréchet differentiable bump function with the support contained in the ball centered at 0 with radius $\delta$. We note that in the case where the norm $|\cdot|_*$ of $X$ is Fréchet differentiable on $X^* \setminus \{0\}$, (BP) holds. In fact, we can take any nonnegative function $\varphi$ with compact support and $\varphi \ne 0$, and set $\psi(p) = \varphi(|p|_*)$. Thus, it is clear that for any Hilbert space, (BP) holds. Also, if $X^*$ is uniformly convex, then (BP) holds. More generally, by Chapter 2, Corollary 2.22, for any reflexive Banach spaces, (BP) holds.

**Lemma 4.2.** (Ekeland–Lebourg) *Let $X$ be a Banach space with the property (BP). Let $D$ be a bounded closed subset of $X$ and $g : D \to (-\infty, +\infty]$ be a lower semicontinuous function that is proper (i.e., $g \not\equiv +\infty$) and bounded from below. Then, for any $\varepsilon > 0$, there exists a $p \in X^*$ with $|p|_* < \varepsilon$, such that the map $x \mapsto g(x) + \langle p, x \rangle$ attains its minimum over $D$ at some point $x_0 \in D$.*

*Proof.* We first define the following:

(4.5) $$G(p) = \inf_{x \in D} \{g(x) + \langle p, x \rangle\}, \qquad \forall p \in X^*.$$

As $g$ is bounded from below, the function $G(p)$ is defined for all $p \in X^*$. Also, it is clear that this function is concave; thus, it is locally Lipschitz continuous (by Lemma 2.8). Next, we define

(4.6) $$A_p(\theta) = \{x \in D \mid g(x) + \langle p, x \rangle \leq G(p) + \theta\}, \qquad \theta > 0.$$

Clearly, $A_p(\theta)$ is nonempty for any $\theta > 0$. For any $\varepsilon > 0$, set

(4.7) $$\Sigma_\varepsilon = \{p \in X^* \mid \exists \theta > 0, \text{ diam } A_p(\theta) \leq \varepsilon\},$$

(4.8) $$\begin{aligned}Y_\varepsilon = \{p \in X^* \mid \exists \eta > 0, \ x^{**} \in X^{**}, \text{ such that} \\ |q - p|_* \leq \eta \Rightarrow G(q) \geq G(p) + \langle x^{**}, q - p \rangle - \varepsilon |q - p|_*\}.\end{aligned}$$

We are going to prove several claims about $Y_\varepsilon$ and $\Sigma_\varepsilon$.

*Claim 1.* For any $\varepsilon > 0$, the set $Y_\varepsilon$ is dense in $X^*$.

To show this, we take any $p_0 \in X^*$. By assumption (BP), for any $\delta > 0$, there exists a Fréchet differentiable bump function $\psi$ with supp $\psi \subseteq \overline{\mathcal{O}_\delta(0)}$ and $0 \leq \psi(p) \leq 1$. Denote $\varphi(p) = \frac{1}{\psi(p-p_0)}$ ($\frac{1}{0} \triangleq +\infty$) and define

(4.9) $$\Gamma(p) = G(p) + \varphi(p), \qquad p \in X^*.$$

Clearly, $\Gamma$ is lower semicontinuous, proper, and bounded from below with the domain $\mathcal{D}(\Gamma) \subseteq \mathcal{O}_\delta(p_0)$. By Chapter 4, Lemma 2.1, for any $\varepsilon > 0$, we have $p_\varepsilon \in X^*$, such that

(4.10) $$\Gamma(q) \geq \Gamma(p_\varepsilon) - \frac{\varepsilon}{2}|q - p_\varepsilon|_*, \qquad \forall q \in X^*.$$

We note that (4.10) implies $p_\varepsilon \in \mathcal{D}(\Gamma)$. Thus,

(4.11) $$|p_\varepsilon - p_0|_* < \delta.$$

Consequently, $\varphi$ is Fréchet differentiable at $p_\varepsilon$. We denote $x^{**} = -D\varphi(p_\varepsilon)$. Then there exists an $\eta > 0$, such that for $|q - p_\varepsilon|_* \leq \eta$, it holds that (note (4.10) and the definition of the Fréchet differentiability of $\varphi$ at $p_\varepsilon$)

(4.12) $$\begin{aligned}G(q) &= \Gamma(q) - \varphi(q) \\ &\geq \Gamma(p_\varepsilon) - \frac{\varepsilon}{2}|q - p_\varepsilon|_* - \varphi(p_\varepsilon) - \langle x^{**}, q - p_\varepsilon \rangle - \frac{\varepsilon}{2}|q - p_\varepsilon|_* \\ &= G(p_\varepsilon) - \langle x^{**}, q - p_\varepsilon \rangle - \varepsilon|p_\varepsilon - q|_*.\end{aligned}$$

This means that $p_\varepsilon \in Y_\varepsilon$. The above proof says that for any $p_0 \in X^*$ and $\delta > 0$, there exists a $p_\varepsilon \in Y_\varepsilon \cap \mathcal{O}_\delta(p_0)$. Hence, $Y_\varepsilon$ is dense in $X^*$ for any $\varepsilon > 0$.

## §4. Uniqueness of Viscosity Solutions

*Claim 2.* For any $\varepsilon > 0$, $Y_{\varepsilon/4} \subset \Sigma_\varepsilon$. Consequently, $\Sigma_\varepsilon$ is dense in $X^*$, for any $\varepsilon > 0$.

In fact, if $p \in Y_{\varepsilon/4}$, there exist $\eta > 0$ and $x^{**} \in X^{**}$, such that

$$(4.13) \qquad |q-p|_* \leq \eta \Rightarrow G(q) \geq G(p) + \langle x^{**}, q-p \rangle - \frac{\varepsilon}{4}|q-p|_*.$$

Let $0 < \theta \leq \frac{\varepsilon\eta}{4}$. Then, for any $x \in A_p(\theta)$ and $|q-p|_* \leq \eta$, we have (note (4.5))

$$(4.14) \qquad \begin{aligned} G(q) &\leq g(x) + \langle q, x \rangle = g(x) + \langle p, x \rangle + \langle q-p, x \rangle \\ &\leq G(p) + \theta + \langle q-p, x \rangle \\ &\leq G(q) - \langle x^{**}, q-p \rangle + \frac{\varepsilon}{4}|q-p|_* + \theta + \langle q-p, x \rangle \\ &= G(q) - \langle x^{**} - x, q-p \rangle + \theta + \frac{\varepsilon}{4}|q-p|_*. \end{aligned}$$

Here, we have used the fact that $X \hookrightarrow X^{**}$. It follows from (4.14) that

$$(4.15) \qquad \langle x^{**} - x, q-p \rangle \leq \frac{\varepsilon\eta}{4} + \frac{\varepsilon}{4}|q-p|_* \leq \frac{\varepsilon\eta}{2}, \qquad \forall |q-p|_* \leq \eta.$$

Thus, $|x - x^{**}|_{**} \leq \frac{\varepsilon}{2}$. As $x \in A_p(\theta)$ is arbitrary, by the triangle inequality, we obtain that

$$(4.16) \qquad \mathrm{diam}\, A_p(\theta) \leq \varepsilon.$$

This means that $Y_{\varepsilon/4} \subset \Sigma_\varepsilon$. Thus, $\Sigma_\varepsilon$ is dense in $X^*$, proving Claim 2.

*Claim 3.* For any $\varepsilon > 0$, $\Sigma_\varepsilon$ is open.

To show this, we take $p \in \Sigma_\varepsilon$. Then there exists a $\theta > 0$, such that $\mathrm{diam}\, A_p(\theta) \leq \varepsilon$. Because $G$ is continuous, we can find $0 < \delta \leq \frac{\theta}{3C}$ ($C = \sup_{x \in D} |x|$), such that

$$(4.17) \qquad G(q) \leq G(p) + \frac{\theta}{3}, \qquad \forall |q-p|_* < \delta.$$

On the other hand, for any $x \in A_q(\frac{\theta}{3})$, we have (note $|q-p|_* < \delta$)

$$(4.18) \qquad \begin{aligned} g(x) + \langle p, x \rangle &= g(x) + \langle q, x \rangle - \langle q-p, x \rangle \\ &\leq G(q) + \frac{\theta}{3} + |q-p|_*|x| \leq G(p) + \frac{2\theta}{3} + \delta C \leq G(p) + \theta. \end{aligned}$$

This means $x \in A_p(\theta)$. Thus, $A_q(\frac{\theta}{3}) \subseteq A_p(\theta)$. Consequently,

$$(4.19) \qquad \mathrm{diam}\, A_q(\frac{\theta}{3}) \leq \varepsilon, \qquad \forall |q-p|_* < \delta.$$

Hence, by the definition of $\Sigma_\varepsilon$, we have $\mathcal{O}_\delta(p) \subset \Sigma_\varepsilon$, proving the openness of $\Sigma_\varepsilon$, which is Claim 3.

Now, we are ready to complete the proof. Let $\Sigma = \bigcap_{n\geq 1} \Sigma_{1/n}$. We claim that $\Sigma$ must be dense in $X^*$. Suppose that the contrary is true. Then we may let some open ball $\mathcal{O}$ be contained in $\Sigma^c = \bigcup_{n\geq 1} \Sigma^c_{1/n}$. Consequently, $\mathcal{O} = \bigcup_{n\geq 1}(\mathcal{O} \cap \Sigma^c_{1/n})$. But each $\Sigma^c_{1/n}$ is nowhere dense. By the Baire Category Theorem (Chapter 2, Theorem 1.7), this is not possible because $X^*$ is a Banach space. Hence, $\Sigma$ is dense in $X^*$. Now, for any $p \in \Sigma$, we have $p \in \Sigma_{1/n}$, $\forall n \geq 1$. Thus, there exists a $\theta_n > 0$, such that diam $A_p(\theta_n) \leq \frac{1}{n}$. Because $A_p(\theta)$ increases as $\theta$ increases, we may let $\theta_n$ decrease and go to 0 as $n \to \infty$. Thus,

$$(4.20) \qquad A_p(\theta_n) \supseteq A_p(\theta_{n+1}), \quad \text{diam } A_p(\theta_n) \leq \frac{1}{n}, \qquad n \geq 1.$$

Next, we pick any $x_n \in A_p(\theta_n)$. By (4.20), we see that $\{x_n\}$ is a Cauchy sequence. Thus, we may let $|x_n - \bar{x}| \to 0$, with some $\bar{x} \in D$ (as $D$ is closed). On the other hand, by the definition of $A_p(\theta_n)$, we have

$$(4.21) \qquad g(x_n) + \langle p, x_n \rangle \leq G(p) + \theta_n, \qquad \forall n \geq 1.$$

Sending $n \to \infty$ and using the lower semicontinuity of $g$, we obtain that

$$(4.22) \qquad g(\bar{x}) + \langle p, \bar{x} \rangle \leq G(p) \leq g(\bar{x}) + \langle p, \bar{x} \rangle.$$

Thus, $g(x) + \langle p, x \rangle$ attains its minimum over $D$ at $\bar{x}$. Because $p \in \Sigma$ is arbitrary, we conclude the proof. □

In the later applications, the following corollary will be used.

**Corollary 4.3.** *Let $X$ be a Banach space with the property (BP). Let $D$ be a closed and bounded subset of $X$. Let $g : D \to [-\infty, +\infty)$ be an upper semicontinuous function that is proper (i.e., $g \not\equiv -\infty$) and bounded from above. Then, for any $\varepsilon > 0$, there exists a $p \in X^*$, with $|p|_* \leq \varepsilon$, such that the function $x \mapsto g(x) - \langle p, x \rangle$ attains its maximum over $D$ at some point $\bar{x} \in D$.*

*Proof.* Apply the above Lemma 4.2 to the function $-g(x)$. □

## §4.2. The Hilbert space $X_\alpha$

In this subsection, we study a family of Hilbert spaces induced by the positive definite operator $B$. This result will be useful in the proof of uniqueness theorem for the viscosity solutions. Before stating the result, let us recall that for any self-adjoint bounded positive semidefinite operator $B$ and any $\alpha \in \mathbb{R}$, the fractional power $B^\alpha$ is a well-defined self-adjoint positive semidefinite operator. Now, let us state the following lemma.

**Lemma 4.4.** *Let $X$ be a Hilbert space and $B$ be a linear bounded operator on $X$ that is self-adjoint and positive. Then there exists a sequence of Hilbert spaces $\{X_\alpha, \alpha \in \mathbb{R}\}$, with the following properties:*

$$(4.23) \qquad X_0 = X, \quad X_\alpha \hookrightarrow X_\beta, \qquad \forall \alpha \geq \beta,$$

§4. Uniqueness of Viscosity Solutions

(4.24) $(X_{-\alpha})' \equiv$ the dual of $X_{-\alpha} = X_\alpha$, $\quad \forall \alpha \in \mathbb{R}$,

(4.25) $X_\alpha = \mathcal{R}(B^{\alpha/2})$, $\quad |x|_\alpha \equiv |x|_{X_\alpha} = |B^{-\alpha/2}x|$, $\quad \alpha \geq 0$,

(4.26) $X_{-\alpha} =$ the completion of $X$ under $|x|_{-\alpha} = |B^{\alpha/2}x|$.

*Proof.* In what follows, if $Y$ is a Banach space and $Y_0 \subset Y$, then $\overline{Y_0}^Y$ represents the closure of $Y_0$ in $Y$. Because $B$ is positive, we see that

(4.27) $$\overline{\mathcal{R}(B)}^X = \mathcal{N}(B^*)^\perp = X.$$

Thus, $B^{-1} : \mathcal{R}(B) : X \to X$ is densely defined. Clearly, it is closed. We denote

(4.28) $$|x|_{\mathcal{R}(B)} = |B^{-1}x|, \quad \forall x \in \mathcal{R}(B).$$

Then one can show that $(\mathcal{R}(B), |\cdot|_{\mathcal{R}(B)})$ is complete and thus is a Hilbert space. Next, as an operator $B \in \mathcal{L}(X, \mathcal{R}(B))$, its adjoint operator is denoted by $B' : \mathcal{R}(B)' \to X$. Here, $\mathcal{R}(B)'$ is the dual space of $\mathcal{R}(B)$ and we have identified $X$ with its dual $X'$. Now, for any $x \in X \subset \mathcal{R}(B)'$ and $y \in X$, we have

(4.29) $$\langle B'x, y \rangle = \langle x, By \rangle_{\mathcal{R}(B)', \mathcal{R}(B)} = \langle x, By \rangle = \langle Bx, y \rangle.$$

This tells us that $B'$ is an *extension* of $B$, meaning that $B'x = Bx$ for all $x \in X$. Next, we claim that $X$ is dense in $\mathcal{R}(B)'$. In fact, if we let $i : \mathcal{R}(B) \to X$ be the *embedding operator*, i.e., the operator $i$ is one-to-one and $i(x) = x$, for all $x \in \mathcal{R}(B)$, then the adjoint operator $i'$ of $i$ maps from $X$ to $\mathcal{R}(B)'$. Furthermore, for any $x \in X$ and $y \in \mathcal{R}(B)$, $\langle i'(x), y \rangle = \langle x, i(y) \rangle = \langle x, y \rangle$. Thus, by the density of $\mathcal{R}(B)$ in $X$, we have $i'(x) = x$, for all $x \in X$. Hence, it holds that (see Chapter 2, (1.29) and Proposition 1.23)

(4.30) $$\overline{X}^{\mathcal{R}(B)'} = \overline{\mathcal{R}(i')}^{\mathcal{R}(B)'} = \mathcal{N}(i)^\perp = \mathcal{R}(B)',$$

as $i$ is one-to-one. This shows that $X$ is dense in $\mathcal{R}(B)'$.

Next, for any $z \in \mathcal{R}(B)'$, and any $y \in \mathcal{R}(B)$, we have (note (4.28))

(4.31) $$\langle z, y \rangle_{\mathcal{R}(B)', \mathcal{R}(B)} = \langle B'z, B^{-1}y \rangle \leq |B'z| \, |y|_{\mathcal{R}(B)}.$$

From this, we can show that

(4.32) $$|z|_{\mathcal{R}(B)'} = |B'z|, \quad \forall z \in \mathcal{R}(B)'.$$

In particular,

(4.33) $$|x|_{\mathcal{R}(B)'} = |Bx|, \quad \forall x \in X.$$

Thus, combining the density of the embedding $X \hookrightarrow \mathcal{R}(B)'$, we see that $\mathcal{R}(B)'$ is the completion of $X$ under the norm $|Bx|$. Now, we define

(4.34)
$$\begin{cases} X_2 = \mathcal{R}(B), \quad X_{-2} = \mathcal{R}(B)', \quad X_0 = X, \\ |x|_2 = |B^{-1}x|, \quad \forall x \in X_2, \\ |x|_{-2} = |B'x|, \quad \forall x \in \mathcal{R}(B)' = X_{-2}. \end{cases}$$

Then it holds that

(4.35) $\qquad X_2 \hookrightarrow X_0 \hookrightarrow X_{-2}, \qquad X_2' = X_{-2}, \quad (X_{-2})' = X_2.$

The last equality is due to the reflexivity of $X_2$. Finally, for any $\alpha > 0$, we repeat the above argument replacing $B$ by $B^{\alpha/2}$. One can define $X_\alpha$ and $|\cdot|_\alpha$ as (4.25) so that (4.24) and (4.26) hold. For any $\alpha \geq \beta \geq 0$, because $\mathcal{R}(B^\alpha) \subset \mathcal{R}(B^\beta)$, we obtain the inclusion (4.23). $\square$

**Lemma 4.5.** *Any convex, bounded closed set $S \subset X$ is convex, bounded, and closed in $X_{-2}$.*

*Proof.* Let $S$ be convex, bounded and closed. For any $x \in X$, $|x|_{-2} \leq \|B\| |x|$; thus, we see that $S$ is bounded in $X_{-2}$. Convexity is an algebraic property. Thus, $S$ is of course convex in $X_{-2}$. Now, we prove the closeness. Let $x_k \in S$ be such that $x_k \to \bar{x}$ in $X_{-2}$ for some $\bar{x} \in X_{-2}$. By Lemma 4.4, we know that $X_{-2} = \mathcal{R}(B)'$, on which $B'$ is defined. Thus, by (4.32),

(4.36) $\qquad |x_k - \bar{x}|_{-2} \equiv |Bx_k - B'\bar{x}| \to 0, \qquad k \to \infty.$

Because $S$ is bounded, convex, and closed in $X$, which is a Hilbert space, we may assume that $x_k \xrightarrow{w} x \in S$. Then

(4.37)
$$\begin{aligned} |x_k - x|_{-1}^2 &= \langle B(x_k - x), x_k - x \rangle \\ &= \langle Bx_k, x_k \rangle - \langle Bx_k, x \rangle - \langle x_k, Bx \rangle + \langle Bx, x \rangle \\ &\to \langle B'\bar{x}, x \rangle - \langle B'\bar{x}, x \rangle - \langle x, Bx \rangle + \langle Bx, x \rangle = 0. \end{aligned}$$

This yields

(4.38) $\qquad |x_k - x|_{-2}^2 \leq \|B\| |x_k - x|_{-1}^2 \to 0.$

Hence, $\bar{x} = x \in S$. $\square$

### §4.3. A uniqueness theorem

This subsection is devoted to the proof of the following result.

**Theorem 4.6.** *Let (A1)–(A4) hold. Let $V$ and $W$ be two viscosity solutions of (3.4) that are locally bounded and weakly sequentially continuous on $[0, T) \times X$ satisfying (2.18). Then $V = W$.*

From this theorem, we can have the following characterization for the value function $V(t, x)$ of our Problem (C).

§4. Uniqueness of Viscosity Solutions

**Theorem 4.7.** *Let (A1)–(A4) hold with $B$ compact. Then the value function $V(t,x)$ is the unique viscosity solution of the corresponding HJB equation (3.4) and is locally bounded and weakly sequentially continuous on $[0,T) \times X$.*

*Proof.* By Theorems 2.2 and 2.6 together with the remark we made after Definition 2.3, we know that the value function $V(t,x)$ is locally bounded and weakly sequentially continuous in $[0,T) \times X$. By Theorem 3.2, $V(t,x)$ is a viscosity solution of (3.4). Thus, our conclusion follows from the above Theorem 4.6. □

*Proof of Theorem 4.6.* The proof of this theorem is rather long. Thus, we split it into several steps.

*Step 1.* Definition of auxiliary functions and sets.

First, we see that under (A2)–(A3), the Hamiltonian $H(t,x,p)$ defined by (3.5) is continuous and satisfies the following:

$$(4.39) \quad \begin{cases} |H(t,x,p) - H(t,x,q)| \leq L(1+|x|)|p-q|, \\ |H(t,x,p) - H(s,y,p)| \leq |p|(L|x-y| + \omega(|t-s|, |x| \vee |y|)) \\ \qquad\qquad + \omega(|t-s| + |x-y|, |x| \vee |y|), \\ \forall t, s \in [0,T], \ x, y \in X, \ p, q \in X^* \equiv X. \end{cases}$$

Let $0 < \tau < \theta < \frac{1}{L}$ and $L_0 = \frac{L}{1-\theta L}$. Define

$$(4.40) \quad \mathcal{O}(\theta, \tau) = \{(t,x) \in (T-\theta, T-\tau) \times X \mid |x| < L_0(t - T + \theta)\}.$$

Then, by (4.39), it is clear that for any $(t,x) \in \mathcal{O}(\theta, \tau)$ and $p, q \in X$,

$$(4.41) \quad \begin{aligned} |H(t,x,p) - H(t,x,q)| &\leq L(1+|x|)|p-q| \\ &\leq L(1 + L_0\theta)|p-q| = L_0|p-q|. \end{aligned}$$

Take $\varepsilon, \delta > 0$ with $\varepsilon + \delta < L_0 \theta$ and let $K > 0$, $\zeta \in C^\infty(\mathbb{R})$, such that

$$(4.42) \quad K > \sup_{\mathcal{O}(\theta, \tau)^2} |V(t,x) - W(s,y)|,$$

$$(4.43) \quad \zeta'(r) \geq 0, \quad \zeta(r) = \begin{cases} 0, & r \leq -\delta, \\ 2K, & r \geq 0, \end{cases}$$

Next, for $\alpha, \beta, \sigma > 0$, we define for any $(t,x,s,y) \in ([0,T] \times X)^2$,

$$(4.44) \quad \begin{aligned} \Psi(t,x,s,y) = &V(t,x) - W(s,y) - \frac{1}{\alpha}|x-y|_B^2 - \frac{1}{\beta}|t-s|^2 \\ &- \zeta(\langle x \rangle_\varepsilon - L_0(t - T + \theta)) \\ &- \zeta(\langle y \rangle_\varepsilon - L_0(s - T + \theta)) + \sigma(t+s), \end{aligned}$$

where $\langle x \rangle_\varepsilon = (|x|^2 + \varepsilon^2)^{1/2}$.

**Step 2.** Properties of $\Psi(t,x,s,y)$ and $\overline{\mathcal{O}(\theta,\tau)^2}$, the closure of $\mathcal{O}(\theta,\tau)^2$ in $([0,T] \times X)^2$.

By Lemma 4.5, we see that the set $\overline{\mathcal{O}(\theta,\tau)^2}$ is bounded, convex, and closed in $(\mathbb{R} \times X_{-2})^2$. Next, we claim that $\Psi(t,x,s,y)$ is upper semicontinuous on $\overline{\mathcal{O}(\theta,\tau)^2}$ with respect to the topology of $(\mathbb{R} \times X_{-2})^2$. In fact, for any $(t_k, x_k, s_k, y_k) \in \overline{\mathcal{O}(\theta,\tau)^2}$ with $(t_k, x_k, s_k, y_k) \to (t, x, s, y)$, we have $t_k \to t$, $s_k \to s$, and $|x_k - x|_B \to 0$, $|y_k - y|_B \to 0$. On the other hand, because $\overline{\mathcal{O}(\theta,\tau)^2}$ is bounded, we may assume that $x_k \xrightarrow{w} \bar{x}$ and $y_k \xrightarrow{w} \bar{y}$ in $X$. Then $Bx_k \xrightarrow{w} B\bar{x}$. Thus, $\bar{x} = x$ because $B$ is one-to-one. Similarly, $\bar{y} = y$. Therefore, by the weakly sequential continuity of $V$ and $W$, we have

(4.45) $$\lim_{k \to \infty} \{|V(t_k, x_k) - V(t, x)| + |W(s_k, y_k) - W(s, y)|\} = 0.$$

Next, it is clear that

(4.46) $$\langle x \rangle_\varepsilon \leq \lim_{k \to \infty} \langle x_k \rangle_\varepsilon, \qquad \langle y \rangle_\varepsilon \leq \lim_{k \to \infty} \langle y_k \rangle_\varepsilon.$$

Hence, it follows easily that $\Psi(t,x,s,y)$ is upper semicontinuous on $\overline{\mathcal{O}(\theta,\tau)^2}$. Also, by the local boundedness of $V(t,x)$ and $W(s,y)$, we see that $\Psi(t,x,s,y)$ is bounded. Hence, $\Psi(t,x,s,y)$ is a bounded upper semicontinuous function defined on a bounded, convex and closed subset $\overline{\mathcal{O}(\theta,\tau)^2}$ of some Hilbert space $((\mathbb{R} \times X_{-2})^2)$. Then, by Corollary 4.3, there exist $\tilde{p}, \tilde{q} \in (X_{-2})' = X_2$ and $a, b \in \mathbb{R}$ with the property $|\tilde{p}|_2, |\tilde{q}|_2, |a|, |b| < \alpha \wedge \beta$, such that the function $\Psi(t,x,s,y) - \langle \tilde{p}, x \rangle - \langle \tilde{q}, y \rangle - at - bs$ attains its maximum over $\overline{\mathcal{O}(\theta,\tau)^2}$ at some point $(t_0, x_0, s_0, y_0) \in \overline{\mathcal{O}(\theta,\tau)^2}$. By Lemma 4.4, we know that $X_2 = \mathcal{R}(B)$. Thus, there exist $p, q \in X$, such that (note (4.25))

(4.47) $$\tilde{p} = Bp, \quad \tilde{q} = Bq; \qquad |\tilde{p}|_2 = |p|, \ |\tilde{q}|_2 = |q|.$$

Hence, $|p|, |q| < \alpha \wedge \beta$ and $\Psi(t,x,s,y) - \langle Bp, x \rangle - \langle Bq, y \rangle - at - bs$ attains its maximum over $\overline{\mathcal{O}(\theta,\tau)^2}$ at point $(t_0, x_0, s_0, y_0) \in \overline{\mathcal{O}(\theta,\tau)^2}$. Finally, we should note that the point $(t_0, x_0, s_0, y_0)$ depends on $\theta, \tau, \varepsilon, \delta, \alpha, \beta, \sigma, K$.

**Step 3.** For fixed $\theta, \tau, \varepsilon, \delta, \sigma, K$, it holds that

(4.48) $$\lim_{\alpha \to 0} |x_0 - y_0|_B = 0, \qquad \lim_{\beta \to 0} |t_0 - s_0| = 0.$$

(4.49) $$\lim_{\alpha, \beta \to 0} \left( \frac{|x_0 - y_0|_B^2}{\alpha} + \frac{|t_0 - s_0|^2}{\beta} \right) = 0.$$

Let us prove the above. By the definition of $(t_0, x_0, s_0, y_0)$, we have

(4.50) $$\begin{aligned}2\Psi(t_0, x_0, s_0, y_0) &- 2\langle Bp, x_0 \rangle - 2\langle Bq, y_0 \rangle - 2at_0 - 2bs_0 \\ &\geq \Psi(t_0, x_0, t_0, x_0) + \Psi(s_0, y_0, s_0, y_0) - \langle B(p+q), x_0 \rangle \\ &\quad - \langle B(p+q), y_0 \rangle - (a+b)t_0 - (a+b)s_0.\end{aligned}$$

## §4. Uniqueness of Viscosity Solutions

This implies that

(4.51)
$$\frac{2}{\alpha}|x_0 - y_0|_B^2 + \frac{2}{\beta}|t_0 - s_0|^2 \leq V(t_0, x_0) - V(s_0, y_0)$$
$$+ W(t_0, x_0) - W(s_0, y_0) + C\alpha \wedge \beta.$$

By the local boundedness of $V$ and $W$, we see that (4.48) holds. Thus $t_0 \to \bar{t}$ and $s_0 \to \bar{t}$. On the other hand, by the boundedness of $\overline{\mathcal{O}(\theta, \tau)^2}$, we may assume that $x_0 \xrightarrow{w} \bar{x}$ and $y_0 \xrightarrow{w} \bar{y}$. Then, by the positivity of $B$ and (4.48), it is necessary that $\bar{x} = \bar{y}$. On $[0, T - \tau] \times X$, $V$ and $W$ are weakly sequentially continuous; thus, we see that the right-hand side of (4.51) tends to zero as $\alpha, \beta \to 0$. Then (4.49) follows.

We note that conditions $|x_0 - y_0|_B \to 0$ and $x_0 \xrightarrow{w} \bar{x}$ do not necessarily imply $|x_0 - \bar{x}|_B \to 0$. Thus, to get $V(t_0, x_0) - V(t_0, y_0) \to 0$, we do need the weak sequential continuity of $V$ and $W$ (for our optimal control problem, the value function has such a property as $B$ is assumed to be compact).

*Step 4.* The case $(t_0, x_0, s_0, y_0) \in \mathcal{O}(\theta, \tau)^2$ is not possible.

Suppose it is the case. Then the function

$$V(t, x) - \left[\frac{1}{\alpha}|x - y_0|_B^2 + \frac{1}{\beta}|t - s_0|^2 + \langle Bp, x \rangle + at - \sigma t\right]$$
$$- \zeta(\langle x \rangle_\varepsilon - L_0(t - T + \theta))$$

attains a local maximum at $(t_0, x_0)$. Thus, by the definition of a viscosity solution, we obtain (denote $X^0 = \langle x_0 \rangle_\varepsilon - L_0(t_0 - T + \theta)$)

(4.52)
$$\frac{2}{\beta}(t_0 - s_0) + a - \sigma - L_0\zeta'(X^0) + \langle A^*[\frac{2B(x_0 - y_0)}{\alpha} + Bp], x_0 \rangle$$
$$+ H\left(t_0, x_0, \frac{2B(x_0 - y_0)}{\alpha} + Bp + \zeta'(X^0)\frac{x_0}{\langle x_0 \rangle_\varepsilon}\right) \geq 0.$$

Also, the function

$$W(s, y) + \left[\frac{1}{\alpha}|x_0 - y|_B^2 + \frac{1}{\beta}|t_0 - s|^2 + \langle Bq, y \rangle + bs - \sigma s\right]$$
$$+ \zeta(\langle y \rangle_\varepsilon - L_0(s - T + \theta))$$

attains a local minimum at $(s_0, y_0)$. Thus, we have (denote $Y^0 = \langle y_0 \rangle_\varepsilon - L_0(s_0 - T + \theta)$)

(4.53)
$$-\frac{2}{\beta}(s_0 - t_0) - b + \sigma + L_0\zeta'(Y^0) - \langle A^*[\frac{2B(y_0 - x_0)}{\alpha} + Bq], y_0 \rangle$$
$$+ H\left(s_0, y_0, -\frac{2B(y_0 - x_0)}{\alpha} - Bq - \zeta'(Y^0)\frac{y_0}{\langle y_0 \rangle_\varepsilon}\right) \leq 0.$$

Combining (4.52) and (4.53), we obtain (note (A4), (4.41), and (4.39))

$$
\begin{aligned}
2\sigma &\leq \frac{2}{\alpha}\langle A^*B(x_0 - y_0), x_0 - y_0\rangle \\
&\quad + \|A^*B\|(|p||x_0| + |q||y_0|) + |a| + |b| \\
&\quad - L_0\zeta'(X^0) - L_0\zeta'(Y^0) + L_0[\zeta'(X^0) \\
&\quad + |Bp| + \zeta'(Y^0) + |Bq|] \\
&\quad + H(t_0, x_0, \frac{2B(x_0 - y_0)}{\alpha}) - H(s_0, y_0, \frac{2B(x_0 - y_0)}{\alpha}) \\
&\leq \frac{2}{\alpha}(c_0|x_0 - y_0|_B^2 - |x_0 - y_0|^2) + C\alpha \wedge \beta \\
&\quad + \frac{2|B(x_0 - y_0)|}{\alpha}\{L|x_0 - y_0| + \omega(|t_0 - s_0|, |x_0| \vee |y_0|)\} \\
&\quad + \omega(|t_0 - s_0| + |x_0 - y_0|, |x_0| \vee |y_0|) \\
&\leq (2c_0 + L^2\|B\|)\frac{|x_0 - y_0|_B^2}{\alpha} - \frac{|x_0 - y_0|^2}{\alpha} + C\alpha \wedge \beta \\
&\quad + \omega(|t_0 - s_0| + |x_0 - y_0|, |x_0| \vee |y_0|) \\
&\quad + \omega(|t_0 - s_0|, |x_0| \vee |y_0|)\frac{2|B(x_0 - y_0)|}{\alpha}.
\end{aligned}
$$

(4.54)

Note that for the given $\sigma > 0$, by the property of the modulus of continuity (see (2.39)), we have $C_\sigma > 0$ (depending on $|x_0| \vee |y_0|$, in general; however, because both $x_0$ and $y_0$ are in $\overline{\mathcal{O}(\theta, \tau)}$, which is a bounded set, we may choose $C_\sigma$ to be independent of $x_0$ and $y_0$) such that

$$
\begin{aligned}
\omega(|t_0 - s_0| &+ |x_0 - y_0|, |x_0| \vee |y_0|) - \frac{|x_0 - y_0|^2}{\alpha} \\
&\leq \frac{\sigma}{2} + C_\sigma(|t_0 - s_0| + |x_0 - y_0|) - \frac{|x_0 - y_0|^2}{\alpha} \\
&\leq \frac{\sigma}{2} + C_\sigma|t_0 - s_0| + \frac{\alpha C_\sigma^2}{4}.
\end{aligned}
$$

(4.55)

Thus, by (4.49), for the given $\sigma > 0$, we can find $\alpha, \beta > 0$, such that

$$
(2c_0 + L^2\|B\|)\frac{|x_0 - y_0|_B^2}{\alpha} + C\alpha \wedge \beta - \frac{|x_0 - y_0|^2}{\alpha} \\
+ \omega(|t_0 - s_0| + |x_0 - y_0|, |x_0| \vee |y_0|) < \sigma.
$$

(4.56)

Here, we note that $|x_0| \vee |y_0|$ is bounded uniformly in all the parameters because $\overline{\mathcal{O}(\theta, \tau)}$ is bounded. Next, for these fixed $\sigma, \alpha$, by (4.48), we can choose a smaller $\beta > 0$, such that

(4.57) $\qquad \omega(|t_0 - s_0|, |x_0| \vee |y_0|)\frac{2|B(x_0 - y_0)|}{\alpha} < \sigma/2.$

Combining (4.54)–(4.57), we obtain $2\sigma \leq \frac{3}{2}\sigma$, a contradiction. Thus, our claim holds.

## §4. Uniqueness of Viscosity Solutions

*Step 5.* Neither of the following two equalities holds:

(4.58) $\quad |x_0| = L_0(t_0 - T + \theta), \qquad |y_0| = L_0(t_0 - T + \theta).$

In fact, if say, $|x_0| = L_0(t_0 - T + \theta)$, then, by the definition of $(t_0, x_0, s_0, y_0)$, we have

(4.59) $\quad \begin{aligned}&\Psi(T-\tau, 0, T-\tau, 0) - a(T-\tau) - b(T-\tau) \\ &\leq \Psi(t_0, x_0, s_0, y_0) - \langle Bp, x_0 \rangle - \langle Bq, y_0 \rangle - at_0 - bs_0.\end{aligned}$

This gives

(4.60) $\quad \begin{aligned}&V(T-\tau, 0) - W(T-\tau, 0) - 2\zeta(\varepsilon - L_0(\theta - \tau)) \\ &\quad + 2\sigma(T-\tau) - a(T-\tau) - b(T-\tau) \\ &\leq V(t_0, x_0) - W(s_0, y_0) - \frac{1}{\alpha}|x_0 - y_0|_B^2 - \frac{1}{\beta}|t_0 - s_0|^2 \\ &\quad - \zeta(\langle x_0 \rangle_\varepsilon - L_0(t_0 - T + \theta)) - \zeta(\langle y_0 \rangle_\varepsilon - L_0(s_0 - T + \theta)) \\ &\quad \sigma(t_0 + s_0) - \langle Bp, x_0 \rangle - \langle Bq, y_0 \rangle - at_0 - bs_0.\end{aligned}$

We note that $\varepsilon - L_0\theta < -\delta$ (by the choice of $\varepsilon$ and $\delta$). Thus, $\varepsilon - L_0(\theta - \tau) < 0$ if $\tau > 0$ is small enough. Consequently, $\zeta(\varepsilon - L_0(\theta - \tau)) = 0$ if $\tau > 0$ is small enough. Also, in the present case, $\langle x_0 \rangle_\varepsilon - L_0(t_0 - T + \theta) = \langle x_0 \rangle_\varepsilon - |x_0| > 0$. Thus, $\zeta(\langle x_0 \rangle_\varepsilon - L_0(t_0 - T + \theta)) = 2K$. Then (4.60) implies that

(4.61) $\quad \begin{aligned}&V(T-\tau, 0) - W(T-\tau, 0) + 2\sigma(T-\tau) - a(T-\tau) - b(T-\tau) \\ &\leq -K + \sigma(t_0 + s_0) - \langle Bp, x_0 \rangle - \langle Bq, y_0 \rangle - at_0 - bs_0.\end{aligned}$

But, the above is not possible if $\tau, \alpha, \beta, \sigma$ are small enough (recall $|a|, |b| \leq \alpha \wedge \beta$ and (4.42)).

*Step 6.* Completion of the proof.

From the above, we must have either $t_0 = T - \tau$ or $s_0 = T - \tau$. Let us assume the former (the latter is similar). Then, for any $(x,t) \in \mathcal{O}(\theta, \tau)$ fixed, we have

(4.62) $\quad \begin{aligned}&V(t,x) - W(t,x) - 2\zeta(\langle x \rangle_\varepsilon - L_0(t - T + \theta)) \\ &\quad + 2\sigma t - \langle B(p+q), x \rangle - (a+b)t \\ &\leq V(T-\tau, x_0) - W(s_0, y_0) - \zeta(\langle x_0 \rangle_\varepsilon - L_0(\theta - \tau)) \\ &\quad - \zeta(\langle y_0 \rangle_\varepsilon - L_0(s_0 - T + \theta)) + \sigma(T - \tau + s_0) \\ &\quad - \langle Bp, x_0 \rangle - \langle Bq, y_0 \rangle - a(T-\tau) - bs_0.\end{aligned}$

Let $\alpha, \beta \to 0$. We must have $s_0 \to T - \tau$. By (4.48), the positivity of $B$, and the boundedness of $x_0, y_0$, we may assume that $x_0 \xrightarrow{w} \bar{x}$ and $y_0 \xrightarrow{w} \bar{x}$. Hence, by the weak sequential continuity of the functions $V$ and $W$, we have

(4.63) $\quad \begin{aligned}&V(t,x) - W(t,x) - 2\zeta(\langle x \rangle_\varepsilon - L_0(t - T + \theta)) + 2\sigma t \\ &\leq V(T-\tau, \bar{x}) - W(T-\tau, \bar{x}) \\ &\quad - 2\zeta(\langle \bar{x} \rangle_\varepsilon - L_0(\theta - \tau)) + 2\sigma(T - \tau).\end{aligned}$

For $(x,t) \in \mathcal{O}(\theta, \tau)$, we choose $\delta > 0$ so small that

(4.64) $$|x| - L_0(t - T + \theta) < -2\delta.$$

Then, for $\varepsilon > 0$ small, we have $\zeta(\langle x \rangle_\varepsilon - L_0(t - T + \theta)) = 0$. Thus, further sending $\sigma \to 0$ and $\varepsilon \to 0$, we get

(4.65) $$V(t,x) - W(t,x) \leq \sup\{V(T-\tau, \bar{x}) - W(T-\tau, \bar{x}) \mid |\bar{x}| \leq L_0\theta\}.$$

By (2.18),

(4.66) $$|V(T-\tau, \bar{x}) - h(e^{A\tau}\bar{x})| \leq \omega(\tau, |\bar{x}|) \to 0, \qquad (\tau \to 0).$$

Thus, we obtain (recall (4.40))

(4.67) $$V(t,x) \leq W(t,x), \qquad \forall (t,x) \in \overline{\mathcal{O}(\theta, 0)}.$$

Now, for any $x_0 \in X$, let

(4.68) $$\mathcal{O}_{x_0}(\theta, \tau) = \{(t,x) \in (T-\theta, T-\tau) \mid |x - x_0| < L_0(t - T + \theta)\}.$$

By applying the same argument as above, we are able to show that

(4.69) $$V(t,x) \leq W(t,x), \qquad \forall (t,x) \in \overline{\mathcal{O}_{x_0}(\theta, 0)}.$$

Hence,

(4.70) $$V(t,x) \leq W(t,x), \qquad \forall (t,x) \in [T-\theta, T] \times X, \qquad \theta < \frac{1}{L}.$$

Next, replacing $T$ by $T - \frac{1}{2L}$ and continuing the above procedure, we must have $V(t,x) \leq W(t,x)$ for all $(t,x) \in [0,T] \times X$. Finally, by symmetry, we obtain the equality $V = W$. □

## §5. Relation to Maximum Principle and Optimal Synthesis

Both the Pontryagin maximum principle and Bellman optimality principle (which leads to the HJB equation) are all necessary conditions for optimal controls. Thus, it is very natural to ask: Are there any relations between them? The first objective of this section is to answer such a question. In the second part of this section, we will find *optimal state feedback controls* via the value function, in some generalized sense. In this section, we only assume that $X$ is a Banach space.

Now, let us introduce the following notion. Suppose $\varphi : X \to \mathbb{R}$ and $x_0 \in X$. We define

(5.1) $$D^+\varphi(x_0) = \{p \in X^* \mid \varlimsup_{x \to x_0} \frac{\varphi(x) - \varphi(x_0) - \langle p, x - x_0 \rangle}{|x - x_0|} \leq 0\},$$

and

(5.2) $$D^-\varphi(x_0) = \{p \in X^* \mid \varliminf_{x \to x_0} \frac{\varphi(x) - \varphi(x_0) - \langle p, x - x_0 \rangle}{|x - x_0|} \geq 0\}.$$

## §5. Relation to Maximum Principle and Optimal Synthesis

We call $D^+\varphi(x_0)$ and $D^-\varphi(x_0)$ the *superdifferential* and *subdifferential* of $\varphi$ at $x_0$, respectively. It should be pointed out that $D^+\varphi(x_0)$ and/or $D^-\varphi(x_0)$ could be empty.

**Proposition 5.1.** *Let $\varphi : X \to \mathbb{R}$. Then*

(i) *$\varphi$ is Fréchet differentiable at $x_0 \in X$ if and only if*

(5.3) $$D^+\varphi(x_0) \cap D^-\varphi(x_0) \neq \phi.$$

*In this case, we have*

(5.4) $$D^+\varphi(x_0) = D^-\varphi(x_0) = \{\nabla\varphi(x_0)\},$$

*where $\nabla\varphi(x_0)$ is the Fréchet derivative of $\varphi$ at $x_0$.*

(ii) *If $\varphi$ is Gâteaux differentiable at $x_0$ with Gâteaux derivative $\nabla\varphi(x_0)$ and $D^+\varphi(x_0)$ is nonempty, then*

(5.5) $$D^+\varphi(x_0) = \{\nabla\varphi(x_0)\}.$$

*The same result holds for $D^-\varphi(x_0)$.*

*Proof.* (i) The "only if" part is clear. We now prove the "if" part. Thus, let (5.3) hold and $p \in D^+\varphi(x_0) \cap D^-\varphi(x_0)$. Then, by definition,

(5.6) $$0 \leq \varliminf_{x \to x_0} \frac{\varphi(x) - \varphi(x_0) - \langle p, x - x_0 \rangle}{|x - x_0|}$$
$$\leq \varlimsup_{x \to x_0} \frac{\varphi(x) - \varphi(x_0) - \langle p, x - x_0 \rangle}{|x - x_0|} \leq 0.$$

Thus, the equality holds, which means that $\varphi$ is Fréchet differentiable at $x_0$. Then (5.4) follows easily.

(ii) Let $p \in D^+\varphi(x_0)$ and let $\nabla\varphi(x_0)$ be the Gâteaux derivative of $\varphi$ at $x_0$. Then, for any $y \in X$,

(5.7) $$\langle \nabla\varphi(x_0) - p, y \rangle = \varlimsup_{\delta \to 0}\left[\frac{\varphi(x_0 + \delta y) - \varphi(x_0)}{\delta} - \langle p, y \rangle\right]$$
$$\leq \varlimsup_{x \to x_0}\left[\frac{\varphi(x) - \varphi(x_0) - \langle p, x - x_0 \rangle}{|x - x_0|}\right] \leq 0.$$

Hence, $p = \nabla\varphi(x_0)$. □

Note that (ii) of the above proposition *never* means that the Gâteaux differentiability implies the Fréchet differentiability because the sets $D^+\varphi(x_0)$ and/or $D^-\varphi(x_0)$ could be empty even though the function $\varphi$ is Gâteaux differentiable. As matter of fact, the above result tells us that if $\varphi$ is Gâteaux differentiable but not Fréchet differentiable at $x_0$, then either $D^+\varphi(x_0)$ or $D^-\varphi(x_0)$ has to be empty.

Next, for functions $\varphi : [0, T] \times X \to \mathbb{R}$, with fixed $t_0$, $\varphi(t_0, x)$ is a function of $x$. We may define its superdifferential and subdifferential in $x$, denoted by $D_x^+\varphi(t_0,x_0)$ and $D_x^-\varphi(t_0,x_0)$, respectively. On the other hand, we define $D^+\varphi(t_0,x_0)$ and $D^-\varphi(t_0,x_0)$ in the following way:

$$\tag{5.8} D^+\varphi(t_0,x_0) = \{(\alpha,p) \in \mathbb{R} \times X^* \mid \varlimsup_{t\downarrow t_0, x\to x_0} \frac{\varphi(t,x) - \varphi(t_0,x_0) - \alpha(t-t_0) - \langle p, x-x_0\rangle}{|t-t_0| + |x-x_0|} \leq 0\},$$

and

$$\tag{5.9} D^-\varphi(t_0,x_0) = \{(\alpha,p) \in \mathbb{R} \times X^* \mid \varliminf_{t\downarrow t_0, x\to x_0} \frac{\varphi(t,x) - \varphi(t_0,x_0) - \alpha(t-t_0) - \langle p, x-x_0\rangle}{|t-t_0| + |x-x_0|} \geq 0\}.$$

It is important to point out that the limit in $t$ is from the right. This fits the general irreversibility of evolution equations. We will return to this point later.

**Theorem 5.2.** *Let (A1), (A2)' and (A3)' hold. In addition, assume that $f(t,x,u)$, $f^0(t,x,u)$, and $h(x)$ are Fréchet differentiable in $x$ with $f_x(t,x,u)$, $f_x^0(t,x,u)$, and $h_x(x)$ continuous in $(x,u)$. Let $V(t,x)$ be the value function of our optimal control problem. Suppose that $(\bar{y}(\cdot), \bar{u}(\cdot))$ is an optimal pair for the problem starting from $(0,x)$. Let $\psi(\cdot)$ be the solution of the following equation:*

$$\tag{5.10} \begin{aligned}\psi(t) = -e^{A^*(T-t)}h_x(\bar{y}(T)) &+ \int_t^T e^{A^*(r-t)} f_x(r,\bar{y}(r),\bar{u}(r))^*\psi(r)\, dr \\ &- \int_t^T e^{A^*(r-t)} f_x^0(r,\bar{y}(r),\bar{u}(r))\, dr, \qquad t \in [0,T].\end{aligned}$$

*Then (recall (1.15))*

$$\tag{5.11} \begin{aligned}\langle \psi(t), f(t,\bar{y}(t),\bar{u}(t))\rangle &- f^0(t,\bar{y}(t),\bar{u}(t)) \\ &= \max_{v \in U} \big[\langle \psi(t), f(t,\bar{y}(t),v)\rangle - f^0(t,\bar{y}(t),v)\big] \\ &\equiv -H(t,\bar{y}(t),-\psi(t)), \qquad \text{a.e. } t \in [0,T].\end{aligned}$$

$$\tag{5.12} D_x^- V(t,\bar{y}(t)) \subset \{-\psi(t)\} \subset D_x^+ V(t,\bar{y}(t)), \qquad \forall t \in [0,T],$$

$$\tag{5.13} \begin{aligned}D^- V(t,\bar{y}(t)) &\subset \left\{\Big(\langle \psi(t), A\bar{y}(t)\rangle - H(t,\bar{y}(t),-\psi(t)), -\psi(t)\Big)\right\} \\ &\subset D^+ V(t,\bar{y}(t)), \qquad \forall t \in \mathcal{T}(\bar{y}(\cdot),\bar{u}(\cdot)),\end{aligned}$$

*where for any admissible pair $(y(\cdot), u(\cdot))$,*

$$\tag{5.14} \begin{aligned}\mathcal{T}(y(\cdot),u(\cdot)) \equiv \Big\{t \in [0,T] \mid{} &y(t) \in \mathcal{D}(A), \\ &\lim_{\delta \downarrow 0} \frac{1}{\delta}\int_t^{t+\delta} |f(r,y(r),u(r)) - f(t,y(t),u(t))|\, dr = 0\Big\}.\end{aligned}$$

§5. Relation to Maximum Principle and Optimal Synthesis          259

To prove this result, we need the following lemma.

**Lemma 5.3.** *Let $(y(\cdot), u(\cdot))$ be any admissible pair. Then*
(i) *For any $t \in [0, T]$,*

(5.15)
$$\sup_{s \in [t, T]} |y_{t,z}(s) - y(s) - \xi(s)| = o(|z - y(t)|), \quad \text{as } z \to y(t),$$

*where $\xi(\cdot)$ is the solution of*

(5.16)
$$\xi(s) = e^{A(s-t)}(z - y(t)) + \int_t^s e^{A(s-r)} f_x(r, y(r), u(r))\xi(r)\, dr, \quad s \in [t, T].$$

(ii) *For any $t \in \mathcal{T}(y(\cdot), u(\cdot))$,*

(5.17)
$$\sup_{s \in [\tau, T]} |y_{\tau,z}(s) - y(s) - \eta(s)| = o(\tau - t + |z - y(t)|), \quad \text{as } \tau \downarrow t, z \to y(t),$$

*where $\eta(s)$ is the solution of*

(5.18)
$$\eta(s) = e^{A(s-\tau)}\{z - y(t) - (\tau - t)[Ay(t) + f(t, y(t), u(t))]\} \\ + \int_\tau^s e^{A(s-r)} f_x(r, y(r), u(r))\eta(r)\, dr, \quad s \in [\tau, T].$$

*Proof.* (i) First of all, from (2.9), we see that (note $u(\cdot) \in \mathcal{U}[t, T]$ is fixed)

(5.19)     $|y_{t,z}(s) - y(s)| \le C|z - y(t)|, \quad \forall s \in [t, T].$

Also, by (5.16) and Gronwall's inequality,

(5.20)     $|\xi(s)| \le C|z - y(t)|, \quad \forall s \in [t, T].$

Thus, we have

$$|y_{t,z}(s) - y(s) - \xi(s)|$$
$$\le \left| \int_t^s e^{A(s-r)} \int_0^1 f_x(r, (1-\sigma)y(r) + \sigma y_{t,z}(r), u(r))\, d\sigma \right.$$
$$\left. \cdot (y_{t,z}(r) - y(r) - \xi(r))\, dr \right|$$

(5.21)
$$+ \left| \int_t^s e^{A(s-r)} \left[ \int_0^1 f_x(r, (1-\sigma)y(r) + \sigma y_{t,z}(r), u(r))\, d\sigma \right. \right.$$
$$\left. \left. - f_x(r, y(r), u(r)) \right] \xi(r)\, dr \right|$$
$$\le C \int_t^s |y_{t,z}(r) - y(r) - \xi(r)|\, dr + o(|z - y(t)|).$$

Then, by Gronwall's inequality, we obtain (5.15).

(ii) We first have the following (see (2.9)):

$$|y_{\tau,z}(s) - y(s)| \leq C|z - y(\tau)|$$
$$\leq C\{|z - y(t)| + |e^{A(\tau-t)}y(t) - y(t)|$$
$$+ |\int_t^\tau e^{A(\tau-r)}f(r, y(r), u(r))\, dr|\}$$

(5.22)
$$\leq C\{|z - y(t)| + (\tau - t)[|Ay(t)| + |f(t, y(t), u(t))|]$$
$$+ (r - t)|\frac{e^{A(\tau-t)} - I}{\tau - t}y(t) - Ay(t)|$$
$$+ \int_t^\tau |e^{A(\tau-r)}f(r, y(r), u(r)) - f(t, y(t), u(t))|\, dr\}$$
$$\leq C\{|z - y(t)| + (\tau - t)[|Ay(t)| + |f(t, y(t), u(t))|] + o(\tau - t)\}.$$

Thus, it follows that

$$y_{\tau,z}(s) - y(s) = e^{A(s-\tau)}z + \int_\tau^s e^{A(s-r)}f(r, y_{r,z}(r), u(r))\, dr$$
$$- e^{A(s-t)}y(t) - \int_t^s e^{A(s-r)}f(r, y(r), u(r))\, dr$$
$$= e^{A(s-\tau)}[z - y(t) - (\tau - t)Ay(t) - (\tau - t)f(t, y(t), u(t))]$$
$$+ \int_\tau^s e^{A(s-r)}\Big\{\int_0^1 f_x(r, (1-\sigma)y(r) + \sigma y_{\tau,z}(r), u(r))\, d\sigma\Big\}$$
$$\cdot (y_{\tau,z}(r) - y(r))\, dr$$
$$+ e^{A(s-\tau)}[(I - e^{A(\tau-t)})y(t) + (\tau - t)Ay(t)]$$
$$+ (\tau - t)e^{A(s-\tau)}f(t, y(t), u(t))$$
$$- \int_t^\tau e^{A(s-r)}f(r, y(r), u(r))\, dr$$

(5.23)
$$= \eta(s) + \int_\tau^s e^{A(s-r)}f_x(r, y(r), u(r))(y_{\tau,z}(r) - y(r) - \eta(r))\, dr$$
$$+ \int_\tau^s e^{A(s-r)}[\int_0^1 f_x(r, (1-\sigma)y(r) + \sigma y_{\tau,z}(r), u(r))\, d\sigma$$
$$- f_x(r, y(r), u(r))](y_{\tau,z}(r) - y(r))\, dr$$
$$- (\tau - t)e^{A(s-\tau)}[\frac{e^{A(\tau-t)} - I}{\tau - t}y(t) - Ay(t)]$$
$$- (\tau - t)\int_t^\tau [e^{A(s-r)}f(r, y(r), u(r))$$
$$- e^{A(s-\tau)}f(t, y(t), u(t))]\, dr$$
$$= \eta(s) + \int_\tau^s e^{A(s-r)}f_x(r, y(r), u(r))(y_{\tau,z}(r) - y(r) - \eta(r))\, dr$$
$$+ o(|\tau - t| + |z - y(t)|).$$

§5. Relation to Maximum Principle and Optimal Synthesis

Hence, by Gronwall's inequality, we obtain (5.17). □

*Proof of Theorem 5.2.* The conclusions (5.10) and (5.11) are nothing but the Pontryagin maximum principle, a proof of which is similar to that given in Chapter 4. The differences are that (i) we do not have the endpoint constraint here and thus $\psi^0$ can be taken as $-1$; and (ii) we have a Bolza type cost functional here, which leads to the terminal condition $\psi(T) = -h_x(\bar{y}(T))$. Now, let us prove (5.12) and (5.13). To this end, we let $(\bar{y}(\cdot), \bar{u}(\cdot))$ be an optimal pair. Then, for any $z \in X$ and any $t \in [0, T)$, we have (note (2.9))

(5.24)
$$V(t, z) - V(t, \bar{y}(t)) \leq \int_t^T [f^0(r, y_{t,z}(r), \bar{u}(r)) - f^0(r, \bar{y}(r), \bar{u}(r))] \, dr$$
$$+ h(y_{t,z}(T)) - h(\bar{y}(T))$$
$$= \int_t^T f_x^0(r, \bar{y}(r), \bar{u}(r)) \xi(r) \, dr + \langle h_x(\bar{y}(T)), \xi(T) \rangle + o(|z - \bar{y}(t)|).$$

By Chapter 2, Proposition 5.7, and (5.10), (5.16) of this chapter, we have

(5.25)
$$-\langle h_x(\bar{y}(T)), \xi(T) \rangle = \langle \psi(T), \xi(T) \rangle$$
$$= \langle \psi(t), z - \bar{y}(t) \rangle + \int_t^T f_x^0(r, \bar{y}(r), \bar{u}(r)) \xi(r) \, dr.$$

Thus,

(5.26)
$$\frac{V(t, z) - V(t, \bar{y}(t)) + \langle \psi(t), z - \bar{y}(t) \rangle}{|z - \bar{y}(t)|} \leq o(1).$$

This implies that

(5.27)
$$-\psi(t) \in D_x^+ V(t, \bar{y}(t)), \quad \forall t \in [0, T].$$

On the other hand, for any $p \in D_x^- V(t, \bar{y}(t))$ (if it exists), by (5.26), we have that

(5.28)
$$0 \leq \lim_{z \to \bar{y}(t)} \frac{V(t, z) - V(t, \bar{y}(t)) - \langle p, z - \bar{y}(t) \rangle}{|z - \bar{y}(t)|}$$
$$\leq \lim_{z \to \bar{y}(t)} \left\{ -\langle \psi(t) + p, \frac{z - \bar{y}(t)}{|z - \bar{y}(t)|} \rangle \right\}.$$

Hence, $p = -\psi(t)$ and (5.12) follows. Finally, let us prove (5.13). We fix a $t \in \mathcal{T}(\bar{y}(\cdot), \bar{u}(\cdot))$. For any $z \in X$ and $\tau \in (t, T]$, by (5.17) and Chapter 2, Proposition 5.7 again, we have

$$V(\tau, z) - V(t, \bar{y}(t))$$
$$\leq \int_\tau^T f^0(r, y_{\tau,z}(r), \bar{u}(r)) \, dr - \int_t^T f^0(r, \bar{y}(r), \bar{u}(r)) \, dr$$
$$+ h(y_{\tau,z}(T)) - h(\bar{y}(T))$$

$$= \int_t^T f_x^0(r,\bar{y}(r),\bar{u}(r))\eta(r)\,dr - \int_t^\tau f^0(r,\bar{y}(r),\bar{u}(r))\,dr$$
$$+ \langle h_x(\bar{y}(T)),\eta(T)\rangle + o(|\tau - t| + |z - \bar{y}(t)|)$$
$$= \langle \psi(T),\eta(T)\rangle$$
(5.29)
$$- \langle \psi(t), z - \bar{y}(t) - (\tau - t)[A\bar{y}(t) + f(t,\bar{y}(t),\bar{u}(t))]\rangle$$
$$- (\tau - t)f^0(t,\bar{y}(t),\bar{u}(t)) + \langle h_x(\bar{y}(T)),\eta(T)\rangle$$
$$+ o(|\tau - t| + |z - \bar{y}(t)|)$$
$$= -\langle \psi(t), z - \bar{y}(t)\rangle + (\tau - t)\big[\langle \psi(t), A\bar{y}(t) + f(t,\bar{y}(t),\bar{u}(t))\rangle$$
$$- f^0(t,\bar{y}(t),\bar{u}(t))\big] + o(|\tau - t| + |z - \bar{y}(t)|).$$

Thus, by (5.11), we obtain the second inclusion in (5.13). The other inclusion can be proved similar to that for (5.12) (see (5.28)). □

Next, we give a sufficient condition for an admissible pair to be optimal.

**Theorem 5.4.** *Let (A1), (A2)', and (A3)' hold. Let $(\bar{y}(\cdot),\bar{u}(\cdot))$ be an admissible pair. Suppose that*

(5.30) $\qquad \bar{y}(t) \in \mathcal{D}(A), \quad \text{a.e. } t \in [0,T]; \qquad A\bar{y}(\cdot) \in L^1(0,T;X),$

*and there exists a function $\psi(\cdot)$ such that*

(5.31)
$$\Big(\langle \psi(t), A\bar{y}(t) + f(t,\bar{y}(t),\bar{u}(t))\rangle - f^0(t,\bar{y}(t),\bar{u}(t)), -\psi(t)\Big)$$
$$\in D^+V(t,\bar{y}(t)), \qquad \text{a.e. } t \in [0,T].$$

*Then $(\bar{y}(\cdot),\bar{u}(\cdot))$ is optimal.*

*Proof.* Let $\theta(t) = V(t,\bar{y}(t))$. Then $\theta(\cdot)$ is continuous and for almost all $t \in [0,T]$, we have (see (2.20), (2.21), (2.8), and (2.10))

(5.32)
$$|\theta(s) - \theta(t)| = |V(s,\bar{y}(s)) - V(t,\bar{y}(t))|$$
$$\leq C(|s-t| + |(e^{A(s-t)} - I)\bar{y}(t)|)$$
$$\leq C(1 + |A\bar{y}(t)|)|s-t|, \qquad \forall s \in [t,T],$$

where $C$ depends on $x$, the fixed initial state. Thus, $\theta(\cdot)$ is Lipschitz continuous. On the other hand, for almost all $t \in [0,T]$, we have

(5.33) $\qquad \bar{y}(t+\delta) = \bar{y}(t) + [A\bar{y}(t) + f(t,\bar{y}(t),\bar{u}(t))]\delta + o(\delta).$

Thus, by (5.31),

(5.34)
$$\theta(t+\delta) - \theta(t) = V(t+\delta,\bar{y}(t+\delta)) - V(t,\bar{y}(t))$$
$$= V\big(t+\delta, \bar{y}(t) + [A\bar{y}(t) + f(t,\bar{y}(t),\bar{u}(t))]\delta + o(\delta)\big) - V(t,\bar{y}(t))$$
$$\leq \delta[\langle \psi(t), A\bar{y}(t) + f(t,\bar{y}(t),\bar{u}(t))\rangle - f^0(t,\bar{y}(t),\bar{u}(t))]$$
$$- \delta\langle \psi(t), A\bar{y}(t) + f(t,\bar{y}(t),\bar{u}(t))\rangle + o(\delta)$$
$$\leq -\delta f^0(t,\bar{y}(t),\bar{u}(t)) + o(\delta).$$

## §5. Relation to Maximum Principle and Optimal Synthesis

This, together with the Lipschitz continuity of $\theta(\cdot)$, implies that

$$\theta'(t) \leq -f^0(t, \bar{y}(t), \bar{u}(t)), \qquad \text{a.e. } t \in [0, T]. \tag{5.35}$$

Then, by (1.7)

$$V(t, \bar{y}(t)) \geq V(s, \bar{y}(s)) + \int_t^s f^0(r, \bar{y}(r), \bar{u}(r)) \, dr \geq V(t, \bar{y}(t)), \tag{5.36}$$
$$\forall 0 \leq t \leq s \leq T.$$

Hence, the pair $(\bar{y}(\cdot), \bar{u}(\cdot))$ is optimal. $\square$

Next, for any $(t, x) \in [0, T] \times \mathcal{D}(A)$, we define

$$G(t, x) = \Big\{ u \in U \mid \lim_{\delta \to 0} \frac{V(t + \delta, x + \delta[Ax + f(t, x, u)]) - V(t, x)}{\delta} \tag{5.37}$$
$$= -f^0(t, x, u) \Big\}.$$

We have the following result.

**Theorem 5.5.** *Let (A1), (A2)', and (A3)' hold. Let $(\bar{y}(\cdot), \bar{u}(\cdot))$ be an admissible pair, such that (5.30) holds. Then the following are equivalent:*

(i) $(\bar{y}(\cdot), \bar{u}(\cdot))$ *is optimal;*

(ii) *It holds that*

$$\bar{u}(t) \in G(t, \bar{y}(t)), \qquad \text{a.e. } t \in [0, T]. \tag{5.38}$$

*Proof.* Because $\bar{y}(t) \in \mathcal{D}(A)$, a.e. $t \in [0, T]$, by the proof of Theorem 5.4, we know that $\theta(t) = V(t, \bar{y}(t))$ is Lipschitz continuous.

(i) $\Rightarrow$ (ii). Let $(\bar{y}(\cdot), \bar{u}(\cdot))$ be optimal. By (5.33),

$$\lim_{\delta \to 0} \frac{V(t + \delta, \bar{y}(t) + \delta[A\bar{y}(t) + f(t, \bar{y}(t), \bar{u}(t))]) - V(t, \bar{y}(t))}{\delta}$$
$$= \lim_{\delta \to 0} \frac{V(t + \delta, \bar{y}(t + \delta) + o(\delta)) - V(t, \bar{y}(t))}{\delta} \tag{5.39}$$
$$= -\lim_{\delta \to 0} \frac{1}{\delta} \int_t^{t+\delta} f^0(r, \bar{y}(r), \bar{u}(r)) \, dr = -f^0(t, \bar{y}(t), \bar{u}(t)),$$
$$\text{a.e. } t \in [0, T].$$

This means that (5.38) holds.

(ii) $\Rightarrow$ (i). In this case, we have
$$\tag{5.40}$$
$$\lim_{\delta \to 0} \frac{V(t + \delta, \bar{y}(t + \delta)) - V(t, \bar{y}(t))}{\delta}$$
$$= \lim_{\delta \to 0} \frac{V(t + \delta, \bar{y}(t) + \delta[A\bar{y}(t) + f(t, \bar{y}(t), \bar{u}(t))] + o(\delta)) - V(t, \bar{y}(t))}{\delta}$$
$$= -f^0(t, \bar{y}(t), \bar{u}(t)), \qquad \text{a.e. } t \in [0, T].$$

Then, by the absolute continuity of $V(t, \bar{y}(t))$, we obtain the optimality of $\bar{u}(\cdot)$. □

We see that (5.38) gives a representation of the optimal control in terms of the corresponding optimal state trajectory. Such a form is referred to as a *state feedback control*. Formally, the optimal trajectory $\bar{y}(\cdot)$ satisfies the following:

$$(5.41) \qquad \frac{d}{dt}\bar{y}(t) \in A\bar{y}(t) + f(t, \bar{y}(t), G(t, \bar{y}(t))), \qquad t \in [0, T].$$

This is a *differential inclusion* in the unknown function $\bar{y}(\cdot)$ and the control variable does not appear explicitly. Such a system is called the *closed-loop* system for our Problem (C). Roughly speaking, in order to solve Problem (C), we first find a solution of (5.41). We then use (5.38) to determine an optimal control. In some special cases, such a procedure does give a solution to Problem (C). We refer to Theorem 5.5 as an *optimal synthesis*.

## §6. Infinite Horizon Problems

In this section we will discuss the optimal control problem in an infinite time interval. Some results and their proofs are very similar to the case of a finite horizon; thus, we will omit some proofs. Consider the following state equation:

$$(6.1) \qquad y_x(t) = e^{At}x + \int_0^t e^{A(t-r)} f(y_x(r), u(r))\, dr, \qquad t \geq 0.$$

Here, as before, $e^{At}$ is a contraction semigroup on the underlying Hilbert space $X$, and $f : X \times U \to X$ is a given map where $U$ is a metric space in which the control $u(\cdot)$ takes values. Similar to the previous sections, the subscript $x$ in $y_x(t)$ is used to emphasize the dependence of the state trajectory on the initial state $x$. The dependence of $y_x(t)$ on the control $u(\cdot) \in \mathcal{U}[0, \infty)$ should also be remembered. Next, we introduce the cost functional:

$$(6.2) \qquad J_x(u(\cdot)) = \int_0^\infty f^0(y_x(t), u(t))e^{-\lambda t}\, dt.$$

Again, we see that the dependence on the initial state $x$ is indicated. In (6.2), the constant $\lambda > 0$ is called the *discount factor*. Under some hypotheses, which will be specified below, for any $x \in X$ and $u(\cdot) \in \mathcal{U}[0, \infty)$, there exists a unique trajectory $y_x(\cdot)$ and the cost functional is well defined. Our optimal control problem is to find a control $\bar{u}(\cdot) \in \mathcal{U}[0, \infty)$, such that the cost functional (6.2) is minimized. We define the value function as follows:

$$(6.3) \qquad V(x) = \inf_{u(\cdot) \in \mathcal{U}[0,\infty)} J_x(u(\cdot)).$$

The goal of this section is to characterize the above value function in terms of viscosity solutions.

## §6. Infinite Horizon Problems

Now, let us make some assumptions: The functions $f : X \times U \to X$ and $f^0 : X \times U \to \mathbb{R}$ are continuous, satisfying one of the following:

(B1) There exist constants $L_0, L, C > 0$ and a modulus of continuity $\omega$, such that for all $x, \bar{x} \in X$ and $u \in U$,

(6.4) $$\begin{cases} |f(x,u) - f(\bar{x},u)| \leq L_0|x - \bar{x}|, & |f(x,u)| \leq L, \\ |f^0(x,u)| \leq C, & |f^0(x,u) - f^0(\bar{x},u)| \leq \omega(|x - \bar{x}|). \end{cases}$$

(B2) There exist constants $L_0, L, C, m > 0$ with $0 \leq m < \frac{\lambda}{L}$ and a local modulus of continuity $\omega(\cdot, \cdot)$ such that for all $x, \bar{x} \in X$ and $u \in U$,

(6.5) $$\begin{cases} |f(x,u) - f(\bar{x},u)| \leq L_0|x - \bar{x}|, & |f(x,u)| \leq L, \\ |f^0(x,u)| \leq Ce^{m|x|}, & |f^0(x,u) - f^0(\bar{x},u)| \leq \omega(|x - \bar{x}|, |x| \vee |\bar{x}|). \end{cases}$$

(B3) There exist constants $L_0, L, C, m > 0$ with $0 \leq m < \frac{\lambda}{L_0}$ and a local modulus of continuity $\omega(\cdot, \cdot)$, such that for all $x, \bar{x} \in X$ and $u \in U$,

(6.6) $$\begin{cases} |f(x,u) - f(\bar{x},u)| \leq L_0|x - \bar{x}|, & |f(0,u)| \leq L, \\ |f^0(x,u)| \leq C(1 + |x|)^m, \\ |f^0(x,u) - f^0(\bar{x},u)| \leq \omega(|x - \bar{x}|, |x| \vee |\bar{x}|). \end{cases}$$

**Proposition 6.1.** *Let (A1) hold and let one of (B1), (B2), and (B3) hold. Then, for any $x \in X$ and $u(\cdot) \in \mathcal{U}[0, \infty)$, the state equation (6.1) admits a unique trajectory $y_x(\cdot)$ and the cost functional (6.2) is well defined. Moreover, we have the following:*

(i) *If (B1) holds, then $V$ is bounded and uniformly continuous.*

(ii) *If (B2) holds, then $V$ is locally uniformly continuous and for some constant $K > 0$,*

(6.7) $$|V(x)| \leq Ke^{m|x|}, \qquad \forall x \in X.$$

(iii) *If (B3) holds, then $V$ is locally uniformly continuous and for some constant $K > 0$,*

(6.8) $$|V(x)| \leq K(1 + |x|)^m, \qquad \forall x \in X.$$

*Proof.* (i) First of all, under (A1) and (B1), we have

(6.9) $$|y_x(t)| \leq |x| + Lt, \qquad \forall (t, x, u(\cdot)) \in [0, \infty) \times X \times \mathcal{U}[0, \infty),$$

(6.10) $$\begin{aligned} |y_x(t) - y_{\bar{x}}(t)| &\leq e^{L_0 t}|x - \bar{x}|, \\ & \forall x, \bar{x} \in X, \ t \in [0, \infty), \ u(\cdot) \in \mathcal{U}[0, \infty). \end{aligned}$$

Thus, we see that

(6.11) $$|V(x)| \leq \int_0^\infty Ce^{-\lambda t}\,dt = \frac{C}{\lambda},$$

and for any $T > 0$,

$$
\begin{aligned}
(6.12) \quad |V(x) - V(\bar{x})| &\leq \int_0^T \omega(|x - \bar{x}|e^{L_0 t})e^{-\lambda t}\,dt + 2\int_T^\infty Ce^{-\lambda t}\,dt \\
&\leq \frac{1}{\lambda}\omega(|x - \bar{x}|e^{L_0 T}) + \frac{2C}{\lambda}e^{-\lambda T}.
\end{aligned}
$$

Thus, $V$ is bounded and uniformly continuous on $X$. This proves (i).

(ii) In this case, we still have (6.9) and (6.10). Thus, it follows that

$$
\begin{aligned}
(6.13) \quad |V(x)| &\leq C\int_0^\infty e^{m|y_x(t)|}e^{-\lambda t}\,dt \\
&\leq C\int_0^\infty e^{m|x|}e^{-(\lambda - mL)t}\,dt = \frac{C}{\lambda - mL}e^{m|x|}.
\end{aligned}
$$

This gives (6.7). Next, for any $T > 0$,

$$
\begin{aligned}
(6.14) \quad |V(x) - V(\bar{x})| &\leq \int_0^T \omega(|x - \bar{x}|e^{L_0 T}, |x| \vee |\bar{x}| + LT)e^{-\lambda t}\,dt \\
&\quad + \int_T^\infty 2Ce^{m(|x| \vee |\bar{x}|)}e^{-(\lambda - mL)t}\,dt \\
&\leq \frac{1}{\lambda}\omega(|x - \bar{x}|e^{L_0 T}, |x| \vee |\bar{x}| + LT) \\
&\quad + \frac{2C}{\lambda - mL}e^{m(|x| \vee |\bar{x}|)}e^{-(\lambda - mL)T}.
\end{aligned}
$$

This gives the local uniform continuity of the value function $V$.

(iii) In this case, we still have (6.10) and instead of (6.9), we have (by Gronwall's inequality)

$$(6.15) \quad |y_x(t)| \leq C(1 + |x|)e^{L_0 t}, \qquad \forall (t, x, u(\cdot)) \in [0, \infty) \times X \times \mathcal{U}[0, \infty),$$

for some absolute constant $C > 0$. Then, similar to the proof for case (ii), we can obtain (iii). □

Next, we state the following optimality principle whose proof is almost the same as that of Theorem 1.1.

**Proposition 6.2.** *Let (A1) hold and one of (B1)–(B3) hold. Then, for any $x \in X$ and $t > 0$,*

$$(6.16) \quad V(x) = \inf_{u(\cdot) \in \mathcal{U}[0,t]} \Big\{ \int_0^t f^0(y_x(r), u(r))e^{-\lambda r}\,dr + V(y_x(t))e^{-\lambda t} \Big\}.$$

In the case where $V \in C^1(X)$, similar to Proposition 1.2, we can show that the value function satisfies the following HJB equation:

$$(6.17) \quad \lambda V(x) - \langle \nabla V(x), Ax \rangle - H(x, \nabla V(x)) = 0, \qquad \forall x \in \mathcal{D}(A),$$

## §6. Infinite Horizon Problems

where

(6.18) $$H(x,p) = \inf_{u \in U}\{\langle p, f(x,u) \rangle + f^0(x,u)\}, \qquad x, p \in X.$$

However, in general, the value function is not necessarily $C^1$. Thus, we need to introduce the notion of viscosity solutions. In the present case, we define the set of *test functions* as follows (compare with those introduced in §3)

$$\Phi_0 = \{\varphi \in C^1(X) \mid \varphi \text{ is weakly sequentially lower semicontinuous}, \ A^*\nabla\varphi \in C(X)\},$$
$$\mathcal{G}_0 = \{g \in C^1(X) \mid \exists \rho \in C^1(\mathbb{R}), \ \rho'(r) \geq 0, \rho'(0) = 0,$$
$$g(x) = \rho(|x|), \ \forall x \in X\}.$$

We now introduce the following definition.

**Definition 6.3.** A function $v(\cdot) \in C(X)$ is called a *viscosity subsolution* (resp. *supersolution*) of (6.17) if for any $\varphi \in \Phi_0$ and $g \in \mathcal{G}_0$, whenever the function $v - \varphi - g$ attains a local maximum (resp. the function $v + \varphi + g$ attains a local minimum) at $x \in X$, we have

(6.19) $$\lambda v(x) - \langle A^*\nabla\varphi(x), x \rangle - H(x, \nabla\varphi(x) + \nabla g(x)) \leq 0,$$

(respectively,

(6.20) $$\lambda v(x) + \langle A^*\nabla\varphi(x), x \rangle - H(x, -\nabla\varphi(x) - \nabla g(x)) \geq 0.)$$

**Proposition 6.4.** *Let (A1) hold and one of (B1)–(B3) hold. Then the value function $V$ is a viscosity solution of (6.17).*

This proposition can be proved by almost the same arguments used in the proof of Theorem 3.2. We leave it to the readers. Also, we point out that assumption (A4) is not needed in the above result. In order to prove the uniqueness of viscosity solutions, we need the $B$-continuity of the value function.

**Proposition 6.5.** *Let (A1) and (A4) hold. Then, if (B1) holds, we have some modulus of continuity $\bar{\omega}(\cdot)$, such that*

(6.21) $$|V(x) - V(\bar{x})| \leq \bar{\omega}(|x - \bar{x}|_B), \qquad \forall x, \bar{x} \in X,$$

*and if (B2) or (B3) holds, we have some local modulus of continuity $\bar{\omega}(\cdot, \cdot)$, such that*

(6.22) $$|V(x) - V(\bar{x})| \leq \bar{\omega}(|x - \bar{x}|_B, |r| \vee |\bar{r}|), \qquad \forall x, \bar{x} \in X.$$

*Proof.* First of all, for any $x, \bar{x} \in X$ and $t \geq 0$, by Lemma 2.5,

(6.23) $$|y_x(t) - y_{\bar{x}}(t)|_B^2 + \int_0^t |y_x(r) - y_{\bar{x}}(r)|^2 \, dr \leq C|x - \bar{x}|_B^2.$$

Thus, in the case where (B1) holds, we have (for any $T > 0$)

$$|V(x) - V(\bar{x})| \leq \int_0^T \omega(|y_x(r) - y_{\bar{x}}(r)|)e^{-\lambda r}\,dr + Ce^{-\lambda T}$$
(6.24)
$$\leq \varepsilon + C_\varepsilon \int_0^T |y_x(r) - y_{\bar{x}}(r)|\,dr + Ce^{-\lambda T}$$
$$\leq \varepsilon + C_{\varepsilon,T}|x - \bar{x}|_B + Ce^{-\lambda T}.$$

Hence, we can find a modulus of continuity $\bar{\omega}$, such that (6.21) holds. Similarly, we can prove the other two cases. □

Next, we are going to give a characterization of the value function. It is clear that (B1) is strictly contained in (B2).

**Theorem 6.6.** *Let (A1), (A4), and (BP) hold. Let (B2) (resp. (B3)) hold. Then the value function $V$ is the unique $B$-continuous viscosity solution of (6.17) satisfying (6.7) (resp. (6.8)).*

The basic idea of the proof for the above result is the same as that for a finite horizon problem. However, there are many interesting differences. For example, we do not need the compactness of the operator $B$ here. For the readers' convenience, we present a detailed proof below.

*Proof.* We have seen that the value function is a viscosity solution of (6.17) (see Proposition 6.4). Thus, it remains to prove the uniqueness. Let $v$ and $w$ be two $B$-continuous viscosity solutions of (6.17). Let $\alpha, \varepsilon > 0$ and define

(6.25) $\quad \Psi(x,y) = v(x) - w(y) - \dfrac{1}{2\varepsilon}|x-y|_B^2 - \alpha(\mu(x) + \mu(y)), \quad x, y \in X,$

where $\mu(\cdot)$ is in $\mathcal{G}_0$ such that

(6.26) $\quad \lim\limits_{|x| \to \infty} \dfrac{|v(x)| + |w(x)|}{\mu(x)} = 0.$

For the case (B2) or (B3), we will choose $\mu(x)$ differently below. Clearly, there exists a constant $R = R_\alpha > 0$, such that

(6.27) $\quad \Psi(0,0) \geq \sup\limits_{|x|,|y| \geq R} \Psi(x,y) + 1.$

The set $S = \{(x,y) \in X \times X \mid |x|, |y| \leq R\}$ is closed and convex under $X_{-2}$ topology, and $\Psi(x,y)$ is upper semicontinuous in $X_{-2}$ topology and bounded from above. Thus, by Corollary 4.3, there exist $p, q \in X$ with $|p|, |q| < \varepsilon$, such that the function $\Psi(x,y) - \langle Bp, x\rangle - \langle Bq, y\rangle$ attains a maximum over $\overline{S}$ at some point $(\bar{x}, \bar{y})$ in the interior of $S$. We should note that this point depends on the choice of $\varepsilon, \alpha$ and $p, q$. By the maximality of $(\bar{x}, \bar{y})$, we have

(6.28) $\quad \begin{aligned} 2\Psi(\bar{x}, \bar{y}) - 2\langle Bp, \bar{x}\rangle - 2\langle Bq, \bar{y}\rangle &\geq \Psi(\bar{x}, \bar{x}) - \langle Bp, \bar{x}\rangle - \langle Bq, \bar{x}\rangle \\ &\quad + \Psi(\bar{y}, \bar{y}) - \langle Bp, \bar{y}\rangle - \langle Bq, \bar{y}\rangle. \end{aligned}$

## §6. Infinite Horizon Problems

This implies that (note Proposition 6.5)

$$(6.29) \quad \frac{1}{\varepsilon}|\bar{x}-\bar{y}|_B^2 \leq v(\bar{x}) - v(\bar{y}) + w(\bar{x}) - w(\bar{y}) + 2\|B\|^{1/2}\varepsilon|\bar{x}-\bar{y}|_B$$
$$\leq \bar{\omega}(|\bar{x}-\bar{y}|_B, R).$$

Then, we see immediately that (note the dependence of $(\bar{x},\bar{y})$ on $\varepsilon,\alpha$)

$$(6.30) \quad \lim_{\varepsilon \to 0} \frac{1}{\varepsilon}|\bar{x}-\bar{y}|_B^2 = 0, \quad \text{for any fixed } \alpha > 0.$$

On the other hand, it is seen that the function $v(x) - \frac{1}{2\varepsilon}|x-\bar{y}|_B^2 - \alpha\mu(x) - \langle Bp, x \rangle$ attains a local maximum at $\bar{x} \in X$. By the definition of a viscosity solution,

$$(6.31) \quad \lambda v(\bar{x}) - \langle \frac{1}{\varepsilon}A^*B(\bar{x}-\bar{y}) + A^*Bp, \bar{x} \rangle$$
$$- H(\bar{x}, \frac{1}{\varepsilon}B(\bar{x}-\bar{y}) + Bp + \alpha\nabla\mu(\bar{x})) \leq 0.$$

Similarly, we have

$$(6.32) \quad \lambda w(\bar{y}) + \langle \frac{1}{\varepsilon}A^*B(\bar{y}-\bar{x}) + A^*Bq, \bar{y} \rangle$$
$$- H(\bar{y}, -\frac{1}{\varepsilon}B(\bar{y}-\bar{x}) - Bq - \alpha\nabla\mu(\bar{y})) \geq 0.$$

Thus, by (A4), we have (note $|\bar{x}|, |\bar{y}| \leq R$)

$$(6.33) \quad \lambda(v(\bar{x}) - w(\bar{y})) \leq \frac{1}{\varepsilon}\langle A^*B(\bar{x}-\bar{y}), \bar{x}-\bar{y} \rangle + \langle A^*Bp, \bar{x} \rangle$$
$$+ \langle A^*Bq, \bar{y} \rangle + H(\bar{x}, \frac{1}{\varepsilon}B(\bar{x}-\bar{y}) + Bp + \alpha\nabla\mu(\bar{x}))$$
$$- H(\bar{y}, \frac{1}{\varepsilon}B(\bar{x}-\bar{y}) - Bq - \alpha\nabla\mu(\bar{y}))$$
$$\leq \frac{c_0}{\varepsilon}|\bar{x}-\bar{y}|_B^2 - \frac{1}{\varepsilon}|\bar{x}-\bar{y}|^2 + 2R\|A^*B\|\varepsilon$$
$$+ H(\bar{x}, \frac{1}{\varepsilon}B(\bar{x}-\bar{y}) + Bp + \alpha\nabla\mu(\bar{x}))$$
$$- H(\bar{y}, \frac{1}{\varepsilon}B(\bar{x}-\bar{y}) - Bq - \alpha\nabla\mu(\bar{y})).$$

Now, if (B2) holds, then

$$(6.34) \quad \begin{cases} |H(x,p) - H(x,q)| \leq L|p-q|, \\ |H(x,p) - H(y,p)| \leq L_0|p||x-y| + \omega(|x-y|, |x| \vee |y|). \end{cases}$$

Take $\mu(x) = e^{\bar{m}\langle x \rangle}$ with $m < \bar{m} < \lambda/L$. Then (6.7) implies (6.26). Also,

$$(6.35) \quad |\nabla\mu(x)| = \bar{m}\mu(x)|\frac{x}{\langle x \rangle}| \leq \bar{m}\mu(x).$$

Consequently,

$$H(\bar{x}, \frac{B(\bar{x}-\bar{y})}{\varepsilon} + Bp + \alpha\nabla\mu(\bar{x}))$$
$$- H(\bar{y}, \frac{B(\bar{x}-\bar{y})}{\varepsilon} - Bq - \alpha\nabla\mu(\bar{y}))$$
(6.36) $$\leq \left| H(\bar{x}, \frac{B(\bar{x}-\bar{y})}{\varepsilon}) - H(\bar{y}, \frac{B(\bar{x}-\bar{y})}{\varepsilon}) \right| + L(|Bp| + |Bq|)$$
$$+ \alpha L(|\nabla\mu(\bar{x})| + |\nabla\mu(\bar{y})|)$$
$$\leq L_0|\bar{x}-\bar{y}|\frac{|B(\bar{x}-\bar{y})|}{\varepsilon} + \omega(|\bar{x}-\bar{y}|, R) + 2L\|B\|\varepsilon$$
$$+ \alpha \bar{m} L\big(\mu(\bar{x}) + \mu(\bar{y})\big).$$

Hence, by (2.39) and (6.33), for any $\delta > 0$, there exists a $C_{\delta,R} > 0$, such that

$$\lambda(v(\bar{x}) - w(\bar{y})) \leq \frac{c_0}{\varepsilon}|\bar{x}-\bar{y}|_B^2 - \frac{1}{\varepsilon}|\bar{x}-\bar{y}|^2 + 2R\|A^*B\|\varepsilon$$
$$+ \frac{|\bar{x}-\bar{y}|^2}{2\varepsilon} + \frac{L_0^2\|B\||\bar{x}-\bar{y}|_B^2}{2\varepsilon} + 2L\|B\|\varepsilon$$
$$+ \delta + C_{\delta,R}|\bar{x}-\bar{y}| + \alpha\bar{m}L\big(\mu(\bar{x}) + \mu(\bar{y})\big)$$
$$\leq \Big(c_0 + \frac{L_0^2\|B\|}{2}\Big)\frac{|\bar{x}-\bar{y}|_B^2}{\varepsilon} + \delta$$
(6.37) $$+ \Big(\frac{C_{\delta,R}^2}{2} + 2L\|B\| + 2R\|A^*B\|\Big)\varepsilon + \alpha\bar{m}L\big(\mu(\bar{x}) + \mu(\bar{y})\big)$$
$$\equiv C_1\frac{|\bar{x}-\bar{y}|_B^2}{\varepsilon} + \delta + \varepsilon\widetilde{C}_{\delta,R} + \alpha\bar{m}L\big(\mu(\bar{x}) + \mu(\bar{y})\big).$$

Here, we have used the following inequality:

(6.38) $$C_{\delta,R}|\bar{x}-\bar{y}| \leq \frac{|\bar{x}-\bar{y}|^2}{2\varepsilon} + \frac{\varepsilon C_{\delta,R}^2}{2}.$$

Now, for any fixed $x \in X$, we let $\alpha > 0$ be fixed and $R \geq R_\alpha \geq |x|$. From

§6. Infinite Horizon Problems

(6.37) and the definition of $(\bar{x}, \bar{y})$, we have

(6.39)
$$\begin{aligned}
v(x) &- w(x) - 2\alpha\mu(x) - \langle Bp, x \rangle - \langle Bq, x \rangle \\
&\equiv \Psi(x,x) - \langle Bp, x \rangle - \langle Bq, x \rangle \\
&\leq \Psi(\bar{x}, \bar{y}) - \langle Bp, \bar{x} \rangle - \langle Bq, \bar{y} \rangle \\
&\leq v(\bar{x}) - w(\bar{y}) - \frac{1}{2\varepsilon}|\bar{x} - \bar{y}|_B^2 - \alpha(\mu(\bar{x}) + \mu(\bar{y})) + 2R\|B\|\varepsilon \\
&\leq \frac{C_1 |\bar{x} - \bar{y}|_B^2}{\lambda} + \frac{\delta}{\varepsilon} + \frac{\widetilde{C}_{\delta,R}}{\lambda}\varepsilon \\
&\quad - \alpha\left(1 - \frac{\bar{m}L}{\lambda}\right)(\mu(\bar{x}) + \mu(\bar{y})) + 2R\|B\|\varepsilon \\
&\leq \frac{C_1 |\bar{x} - \bar{y}|_B^2}{\lambda} + \frac{\delta}{\varepsilon} + \frac{\widetilde{C}_{\delta,R}}{\lambda}\varepsilon + 2R\|B\|\varepsilon.
\end{aligned}$$

The last inequality is due to the assumption $\frac{\bar{m}L}{\lambda} < 1$. Hence, by (6.30), we send $\varepsilon \to 0$, then send $\delta \to 0$, and finally send $\alpha \to 0$ to get $v(x) \leq w(x)$ for all $x \in X$.

Now, in the case where (B3) holds, we have (compare with (6.34))

(6.40)
$$\begin{cases} |H(x,p) - H(x,q)| \leq (L + L_0|x|)|p - q|, \\ |H(x,p) - H(y,p)| \leq L_0|p||x - y| + \omega(|x - y|, |x| \vee |y|). \end{cases}$$

Let us take $\mu(x) = \langle x \rangle^{\bar{m}}$ with $m < \bar{m} < \frac{\lambda}{L_0}$. Then, by (6.8), we see that (6.26) holds. In this case,

(6.41)
$$|\nabla\mu(x)| = \bar{m}\langle x \rangle^{\bar{m}-2}|x| \leq \bar{m}\langle x \rangle^{\bar{m}-1}.$$

Thus (compare with (6.36)),

(6.42)
$$\begin{aligned}
H(\bar{x}, &\frac{B(\bar{x} - \bar{y})}{\varepsilon} + Bp + \alpha\nabla\mu(\bar{x})) \\
&\quad - H(\bar{y}, \frac{B(\bar{x} - \bar{y})}{\varepsilon} - Bq - \alpha\nabla\mu(\bar{y})) \\
&\leq \left|H(\bar{x}, \frac{B(\bar{x} - \bar{y})}{\varepsilon}) - H(\bar{y}, \frac{B(\bar{x} - \bar{y})}{\varepsilon})\right| \\
&\quad + (L + L_0|\bar{x}|)(|Bp| + \alpha|\nabla\mu(\bar{x})|) \\
&\quad + (L + L_0|\bar{y}|)(|Bq| + \alpha|\nabla\mu(\bar{y})|) \\
&\leq L_0|\bar{x} - \bar{y}|\frac{|B(\bar{x} - \bar{y})|}{\varepsilon} + \omega(|\bar{x} - \bar{y}|, R) + 2(L + L_0 R)\|B\|\varepsilon \\
&\quad + \alpha\bar{m}L(\langle \bar{x} \rangle^{\bar{m}-1} + \langle \bar{y} \rangle^{\bar{m}-1}) + \alpha\bar{m}L_0(\langle \bar{x} \rangle^{\bar{m}} + \langle \bar{y} \rangle^{\bar{m}}).
\end{aligned}$$

Hence, by (6.33), similar to (6.37), we have

(6.43)
$$\lambda(v(\bar{x}) - w(\bar{y})) \leq \frac{c_0}{\varepsilon}|\bar{x} - \bar{y}|_B^2 - \frac{1}{\varepsilon}|\bar{x} - \bar{y}|^2 + 2R\|A^*B\|\varepsilon + \frac{|\bar{x} - \bar{y}|^2}{2\varepsilon}$$
$$+ \frac{L_0^2\|B\|\|x - \bar{y}|_B^2}{2\varepsilon} + 2(L + L_0R)\|B\|\varepsilon + \delta + C_{\delta,R}|\bar{x} - \bar{y}|$$
$$+ \alpha \bar{m} L(\langle \bar{x} \rangle^{\bar{m}-1} + \langle \bar{y} \rangle^{\bar{m}-1}) + \alpha \bar{m} L_0(\langle \bar{x} \rangle^{\bar{m}} + \langle \bar{y} \rangle^{\bar{m}})$$
$$\leq C_1 \frac{|\bar{x} - \bar{y}|_B^2}{\varepsilon} + \delta + \varepsilon \widetilde{C}_{\delta,R} + \alpha \bar{m} L(\langle \bar{x} \rangle^{\bar{m}-1} + \langle \bar{y} \rangle^{\bar{m}-1})$$
$$+ \alpha \bar{m} L_0(\langle \bar{x} \rangle^{\bar{m}} + \langle \bar{y} \rangle^{\bar{m}}).$$

Hence, for any $x \in X$, we let $\alpha > 0$ small with the corresponding $R \geq |x|$. Then, similar to (6.39), we have

(6.44)
$$v(x) - w(x) - 2\alpha\mu(x) - \langle Bp, x \rangle - \langle Bq, x \rangle$$
$$\leq \frac{C_1}{\lambda}\frac{|\bar{x} - \bar{y}|_B^2}{\varepsilon} + \frac{\delta}{\lambda} + \frac{\widetilde{C}_{\delta,R}}{\lambda}\varepsilon$$
$$- \alpha\{(1 - \frac{\bar{m}L_0}{\lambda})(\langle \bar{x} \rangle^{\bar{m}} + \langle \bar{y} \rangle^{\bar{m}}) - \frac{\bar{m}L}{\lambda}(\langle \bar{x} \rangle^{\bar{m}-1} + \langle \bar{y} \rangle^{\bar{m}-1})\}$$
$$\leq \frac{C_1}{\lambda}\frac{|\bar{x} - \bar{y}|_B^2}{\varepsilon} + \frac{\delta}{\lambda} + \frac{\widetilde{C}_{\delta,R}}{\lambda}\varepsilon + \alpha C.$$

Here, it is crucial that $\frac{\bar{m}L_0}{\lambda} < 1$ implies

(6.45) $$(1 - \frac{\bar{m}L_0}{\lambda})(\langle \bar{x} \rangle^{\bar{m}} + \langle \bar{y} \rangle^{\bar{m}}) - \frac{\bar{m}L}{\lambda}(\langle \bar{x} \rangle^{\bar{m}-1} + \langle \bar{y} \rangle^{\bar{m}-1}) \geq -C,$$

for some constant $C > 0$ independent of $\alpha$ and $R$. Hence, first sending $\varepsilon \to 0$, then sending $\delta \to 0$, and finally $\alpha \to 0$, we obtain $v(x) \leq w(x)$. □

**Remark 6.7.** It should be pointed out that for the infinite horizon problem, we do not need the compactness of the operator $B$.

As for the finite horizon problem, we can also discuss the relation between the maximum principle and the dynamic programming method. It is also possible to study the feedback optimal controls. We leave the details to the readers.

### Remarks

The dynamic programming method was initiated by Bellman [1,2] in the early 1950's. He was concerned with finite dimensional control problems. For a long time, the result was just formal (except for some special cases) due to the fact that the value function is not smooth enough to satisfy the Hamilton-Jacobi-Bellman equation in the classical sense. In the early 1980's, Crandall–Lions [1] introduced the notion of viscosity solutions for the Hamilton–Jacobi equations. This was a breakthrough in the direction of dynamic programming method. Since then, a vast number of papers have been published developing the theory of viscosity solutions. For a very nice survey, see Crandall-Ishii-Lions [1]. Soon after, the corresponding theory for infinite dimensions also appeared (Crandall–Lions [2]). A slightly different definition of viscosity solutions for infinite dimensional HJ equations was introduced by Tataru [1,2]. Besides the notion of viscosity solutions, there are other ways of studying the HJ equations in infinite dimensions. We refer the readers to the works of Barbu–Da Prato [1,2], Barbu-Da Prato-Popa [1], and Di Blasio [1].

The material of Section 1 is an infinite dimensional version of the standard result for finite dimensions. In Section 2, the notion of $B$-continuity is adopted from Crandall–Lions [2], part V, and the result concerning the semi-concavity is a modification of that given in Cannarsa-Frankowska [1]. The material of Sections 3, 4, and 6 is a proper modification of Crandall–Lions [2] parts IV and V. The major modification is the definition of viscosity solutions for evolution equations in which the test function $g(t, x)$ allows the dependence on $t$. This is necessary for the proof of the uniqueness of the viscosity solutions in Section 4. This proof is an infinite dimensional version of Ishii [1] (for finite dimensions). We feel that this proof can cover slightly more general cases than the one given in Crandall–Lions [2]. Lemma 4.2 is due to Ekeland and Lebourg [1]. This result is very closely related to the Ekeland variational principle presented in Chapter 4. We point out that a similar result to Lemma 4.2 holds true for the more general case, namely, where the space $X$ is a Banach space with the so-called Radon–Nikodým property. This is the case if $X$ is reflexive, for example. Such a result is due to Stegall [1]. In Section 5, the result concerning the relation between the maximum principle and the HJB equation is an infinite version of that of Zhou [1] (which was for finite dimensional cases); the result concerning the optimal synthesis is based on the work of Cannasa-Frankowska [1]. In Cannasa [1] and Cannasa–Frankowska [2], based on the work of Preiss [1], the semiconcavity of the value function was studied in more detail and it led to some further interesting results concerning the regularity of the value function.

# Chapter 7
# Controllability and Time Optimal Control

In the previous chapters, an existence theory and some necessary conditions for optimal controls have been studied extensively. However, the nonemptiness of the admissible pair set has not been touched. In the first part of this chapter, we will discuss this issue in some detail. Such a problem is usually referred to as the *controllablility problem*. In the second part of this chapter, we are going to study the time optimal control problem for evolutionary control systems.

## §1. Definitions of Controllability

We consider the following state equation:

$$(1.1) \qquad y(t) = e^{At}x + \int_0^t e^{A(t-s)} f(s, y(s), u(s))\, ds, \qquad t \in [0, \infty).$$

Here, as before, $e^{At}$ is a $C_0$ semigroup on some Banach space $X$, $f : [0, \infty) \times X \times U \to X$ is some given map, $U$ is a metric space; $y(\cdot)$ is the state trajectory, and $u(\cdot)$ is the control taking values in $U$. In what follows, we assume that $f$ satisfies proper conditions so that for any $x \in X$ and $u \in \mathcal{U} \equiv \{u : [0, \infty) \to U \mid u(\cdot) \text{ is measurable}\}$, (1.1) admits a unique solution $y(\cdot) \equiv y(\cdot\,; x, u) \in C([0, \infty); X)$. This is the case if, for instance, $f(t, x, u)$ is Lipschitz continuous and grows at most linearly in $x$, and uniformly in $(t, u)$.

Next, let $Q : [0, \infty) \to 2^X$, such that $Q(t) \neq \phi$, for all $t \in [0, \infty)$. Then we can pose the following general problem:

**Problem (C).** For given $x \in X$, find $u \in \mathcal{U}$, such that for some $t \geq 0$,

$$(1.2) \qquad y(t; x, u) \in Q(t).$$

Such a problem is usually referred to as the *controllability problem*. The set valued function $Q(\cdot)$ is called the *target trajectory*. We have seen that control problems with the terminal state constraint are examples of the above problem. Now, let us introduce the following sets.

$$(1.3) \qquad \mathcal{R}(t, x) = \{y(t; x, u) \mid u \in \mathcal{U}\}, \qquad (t, x) \in [0, \infty) \times X,$$

$$(1.4) \qquad \mathcal{R}([0, t], x) = \bigcup_{s \in [0, t]} \mathcal{R}(s, x), \qquad \forall (t, x) \in [0, \infty) \times X.$$

One may also define the set $\mathcal{R}([0, \infty), x)$ in a similar manner. We call $\mathcal{R}(t, x)$ and $\mathcal{R}([0, t], x)$ the *reachable set* of the system (1.1) with the initial

## §1. Definitions of Controllability

state $x$, at time $t$ and on $[0, t]$, respectively. The set $\mathcal{R}([0, \infty), x)$ is simply called the *reachable set* of the system (1.1) and simply denoted by $\mathcal{R}(x)$.

Next, we introduce definitions of various controllabilities.

**Definition 1.1.** System (1.1) is said to be

(i) *exactly (approximately) controllable on $[0, T]$* if

(1.5) $\quad \mathcal{R}([0, T], x) = X, \quad (\overline{\mathcal{R}([0, T], x)} = X, \quad \text{resp.}) \quad \forall x \in X,$

(ii) *exactly (approximately) controllable with the target $Q(\cdot)$ on $[0, T]$* if

(1.6)
$$\bigcup_{t \in [0,T]} \Big( Q(t) \bigcap \mathcal{R}(t, x) \Big) \neq \phi, \quad \forall x \in X,$$
$$\Big( \inf_{t \in [0,T]} d(Q(t), \mathcal{R}(t, x)) = 0, \quad \forall x \in X, \text{ resp.} \Big)$$

where

(1.7) $\quad d(Q(t), \mathcal{R}(t, x)) \triangleq \inf_{q \in Q(t), y \in \mathcal{R}(t, x)} |q - y|.$

(iii) *locally exactly (approximately) controllable with the target $Q(\cdot)$ on $[0, T]$* if there exists an open set $\mathcal{O} \supset Q(0)$, such that (1.6) holds for all $x \in \mathcal{O}$.

(iv) *small time locally exactly (approximately) controllable with the target $Q(\cdot)$* if for any $T > 0$, there exists a $\delta > 0$, such that for any $x \in X$ with $d(x, Q(0)) < \delta$, (1.6) holds.

(v) In the above, if $Q(t) \equiv \{0\}$, we add the word "null" in front of "controllable;" if $T > 0$ can be arbitrarily small, we add the words "infinitesimal time" in front of "controllable." For example, we say *small time locally exactly controllable with the target $Q(\cdot)$*, *infinitesimal time approximately controllable*, etc.

Similarly, we may define the various controllabilities *on $[0, \infty)$*, namely, replacing $[0, T]$ by $[0, \infty)$ in the above. In these cases, we will omit the words "on $[0, \infty)$." Thus, for example, system (1.1) is said to be *exactly controllable* if

(1.8) $\quad \mathcal{R}([0, \infty), x) = X, \quad \forall x \in X.$

From the above definition, we see immediately that

(1.9) $\quad\quad\quad\quad\quad (i) \Rightarrow (ii) \Rightarrow (iii) \Leftarrow (iv).$

Also, the notion with the words "infinitesimal time" is the strongest. Before going further, let us briefly look at the meaning of Definition 1.1. From (i), we see that system (1.1) is exactly controllable on $[0, T]$ if and only if for any $x, z \in X$, there exists a $u(\cdot) \in \mathcal{U}$ and a $t \in [0, T]$, such that

(1.10) $\quad\quad\quad\quad\quad y(t; x, u) = z.$

However, system (1.1) is approximately controllable on $[0,T]$ if and only if for any $x, z \in X$ and any $\varepsilon > 0$, there exists a $u(\cdot) \in \mathcal{U}$ and a $t \in [0,T]$, such that

$$|y(t;x,u) - z| < \varepsilon. \tag{1.11}$$

Similarly, it is not hard to give the precise meaning for the notions of various other controllabilities.

The following gives a negative result about the exact controllability.

**Proposition 1.2.** *Let $X$ be an infinite dimensional Banach space, $e^{At}$ be a compact semigroup, and $f : [0,\infty) \times X \times U \to X$ be a map such that for any $(x, u(\cdot)) \in X \times \mathcal{U}$, there exists a unique solution of (1.1). Moreover, there exists a $p > 1$, such that for any $(x, u(\cdot)) \in X \times \mathcal{U}$ and the corresponding state trajectory $y(\cdot)$, we have $f(\cdot, y(\cdot), u(\cdot)) \in L^p_{loc}(0, \infty; X)$. Then system (1.1) is not exactly controllable.*

*Proof.* For any $T > 0$, let $\mathcal{G}_T : L^p(0,T;X) \to X$ be defined by

$$\mathcal{G}_T(g(\cdot)) = \int_0^T e^{A(T-s)} g(s)\, ds, \qquad g(\cdot) \in L^p(0,T;X). \tag{1.12}$$

We claim that

$$\mathcal{R}(\mathcal{G}_{T_1}) \subseteq \mathcal{R}(\mathcal{G}_{T_2}), \qquad \forall 0 \leq T_1 \leq T_2. \tag{1.13}$$

In fact, for any $0 \leq T_1 \leq T_2$ and any $z(\cdot) \in L^p_{loc}(0, \infty; X)$, it holds that

$$\begin{aligned}\int_0^{T_1} e^{A(T_1-s)} z(s)\, ds &= \int_0^{T_1} e^{As} z(T_1 - s)\, ds \\ &= \int_0^{T_2} e^{As} \widetilde{z}(T_2 - s)\, ds = \int_0^{T_2} e^{A(T_2-s)} \widetilde{z}(s)\, ds,\end{aligned} \tag{1.14}$$

where

$$\widetilde{z}(s) = \begin{cases} 0, & s \in [0, T_2 - T_1], \\ z(s - T_2 + T_1), & s \in (T_2 - T_1, T_2]. \end{cases} \tag{1.15}$$

This shows (1.13). From our assumption,

$$\mathcal{R}([0,T], 0) \subseteq \mathcal{R}(\mathcal{G}_T). \tag{1.16}$$

On the other hand, by Chapter 3, Lemma 3.2, operator $\mathcal{G}_T$ is compact. We claim that

$$\bigcup_{T \geq 0} \mathcal{R}(\mathcal{G}_T) \neq X. \tag{1.17}$$

## §1. Definitions of Controllability

If this is not the case, then, noting (1.13), we have (let $B_n$ be the closed ball of radius $n$ centered at the origin in $L^p(0,T;X)$)

$$\tag{1.18} X = \bigcup_{m=1}^{\infty} \bigcup_{n=1}^{\infty} \mathcal{G}_m(B_n).$$

Because $X$ is of second category (see Chapter 2, Theorem 1.7), there must be $m, n \geq 1$, such that $\overline{\mathcal{G}_m(B_n)}$ contains a nonempty open set in $X$. But the compactness of $\mathcal{G}_m$ implies that $\overline{\mathcal{G}_m(B_n)}$ is compact, and it cannot contain any nonempty open set in $X$ because $\dim X = \infty$. This is a contradiction. Hence, system (1.1) must not be exactly controllable. □

We see that in the above example, the exact controllability fails because of the compactness of the semigroup $e^{At}$. Let us point out that in some cases, even without restricting the semigroup $e^{At}$ to be compact, the exact controllability may still be missing. Here is a simple example.

*Example 1.3.* Let $\varphi \in X$ be an eigenvector of the operator $A$ with the corresponding eigenvalue $\lambda \in \mathbb{R}$. Let $g : [0, \infty) \times X \times U \to \mathbb{R}$ and define

$$\tag{1.19} f(t, x, u) = g(t, x, u)\varphi, \qquad \forall (t, x, u) \in [0, \infty) \times X \times U.$$

Then, we see immediately that $\mathcal{R}(t, 0)$ is contained in the one dimensional space in $X$ spanned by $\varphi$ for any $t \geq 0$. Thus, such a system cannot even be approximately controllable. Of course, it is not exactly controllable.

Clearly, the above example can be extended to much more general situations. For example, the range of $f$ may be contained in some fixed subspace that is not equal to $X$ and spanned by a set of eigenvectors of $A$. We leave the details to the interested readers.

From the above, we see that, in general, it is very hard to expect exact controllability (even approximate controllability) in infinite dimensional spaces, especially for parabolic partial differential equations. But, when the target set is "nice," exact controllability can still be very easily obtained. Here is a simple result whose proof is left to the readers.

**Proposition 1.4.** *Let $Q$ be a set in $X$ with $0 \in \text{Int}\, Q$. Let the semigroup $e^{At}$ satisfy the following:*

$$\tag{1.20} \|e^{At}\|_{\mathcal{L}(X)} \leq M e^{-\omega t}, \qquad \forall t \geq 0,$$

*for some $M \geq 1$ and $\omega > 0$. Further, let $f : [0, \infty) \times X \times U \to X$ be measurable in $t$, continuous in $(x, u)$, and let the constants $L, \theta > 0$ exist, such that*

$$\tag{1.21} \begin{cases} |f(t, x, u) - f(t, y, u)| \leq L|x - y|, \\ \qquad\qquad\qquad\qquad \forall (t, u) \in [0, \infty) \times U, \ x, y \in X, \\ |f(t, 0, u)| \leq L, \qquad \forall (t, u) \in [0, \infty) \times U, \end{cases}$$

(1.22)  $\quad 0 \in f(t, x, U), \qquad \forall t \in [0, \infty), \ d(x, Q) \leq \theta.$

*Then system (1.1) is locally exactly controllable with the target $Q$. Moreover, if (1.22) holds for all $(t, x) \in [0, \infty) \times X$, then system (1.1) is exactly controllable with the target set $Q$.*

It is possible to give some other cases with a similar nature. In the above result, the target set is "big" in some sense. We will see later that if $Q = \{x_1\}$ is a singleton and $x_1$ is a "nice" point, say the origin, then, the exact controllability may also be obtained under some mild conditions. In some sense, the exact (approximate) controllability of the system with a given target is a sort of compatibility between the system and the target.

On the other hand, it is clear that approximate controllability is easier to obtain than exact controllability. Once the system is approximately controllable, for any target set $Q$ with a nonempty interior, the system is exactly controllable with the target set $Q$. In many applications, the target set $Q$ has a nonempty interior. For such problems, the approximate controllability is enough. Thus, the exact and approximate controllabilities are almost equally important in infinite dimensional control theory.

## §2. Controllability for Linear Systems

To obtain some deeper results, we would like to look at the linear systems first; they will be the subject of this section. Thus, our system reads

(2.1) $\quad \begin{cases} \dot{y}(t) = Ay(t) + Bu(t), \\ y(0) = x, \end{cases}$

where $A$ is the generator of the $C_0$ semigroup $e^{At}$ on some Banach space $X$ and $B \in \mathcal{L}(U, X)$. Usually, system (2.1) is abbreviated as $[A, B]$. Of course, in the present case, $U$ is a Banach space (instead of just a metric space). We let

(2.2) $\quad \mathcal{U}^p = L^p_{loc}(0, \infty; U) \triangleq \{u : [0, \infty) \to U \mid u \in L^p(0, T; U), \forall T > 0\},$

with $p \in [1, \infty]$. Clearly, $\mathcal{U}^p \subset \mathcal{U}^q$ for any $1 \leq q \leq p \leq \infty$. The state $y(\cdot)$ is represented by the following:

(2.3) $\quad y(t) = e^{At}x + \int_0^t e^{A(t-s)} Bu(s) \, ds, \qquad t \geq 0.$

Thus, any $(x, u(\cdot)) \in X \times \mathcal{U}^p$ uniquely determines the state trajectory $y(\cdot; x, u)$ in an explicit way. Next, for $T > 0$, we define operator $G_T \in \mathcal{L}(\mathcal{U}^1, X)$ as follows.

(2.4) $\quad G_T(u(\cdot)) = \int_0^T e^{A(T-s)} Bu(s) \, ds, \qquad \forall u(\cdot) \in \mathcal{U}^1.$

## §2. Controllability for Linear Systems

Clearly, the adjoint of $G_T^* \in \mathcal{L}(X^*, \mathcal{U}^\infty)$ is given by

(2.5) $$G_T^* y^* = B^* e^{A^*(T-\cdot)} y^*, \qquad \forall y^* \in X^*.$$

Because $\mathcal{U}^p \subset \mathcal{U}^1$, for any $p \in [1, \infty]$, operator $G_T$ is well defined on any $\mathcal{U}^p$ ($p \in [1, \infty]$). Similar to (1.13), we have

(2.6) $$G_{t_1}(\mathcal{U}^p) \subseteq G_{t_2}(\mathcal{U}^p), \qquad \forall 0 \leq t_1 \leq t_2, \ p \in [1, \infty].$$

This tells us that the set $G_t(\mathcal{U}^p)$ is expanding as $t$ increases.

In the present case, any set $\mathcal{U}^p$ with $p \in [1, \infty]$ can be taken as the control set. Thus, when we talk about the controllability, the control set should be specified, for example, exact controllability with the control set $\mathcal{U}^p$, etc. Clearly, the smaller the control set, the harder the controllability being obtained. In particular, say, if $[A, B]$ is exactly controllable with the control set $\mathcal{U}^\infty$, then it is exactly controllable with the control set $\mathcal{U}^p$ for any $p \in [1, \infty)$.

### §2.1. Approximate controllability

Let us first give the following basic result.

**Lemma 2.1.** *For all $p \in [1, \infty]$ and $T > 0$, it holds that*

(2.7) $$\begin{aligned} \overline{\bigcup_{t \geq 0} G_t(\mathcal{U}^p)} &= \operatorname{span}\{e^{As} Bu \mid s \geq 0, \ u \in U\} \\ &\supseteq \overline{G_T(\mathcal{U}^p)} = \operatorname{span}\{e^{As} Bu \mid s \in [0, T], \ u \in U\}, \\ &\supseteq \operatorname{span}\{A^n Bu \mid n \geq 0, \ u \in U, \ Bu \in \bigcap_{m \geq 0} \mathcal{D}(A^m)\}, \end{aligned}$$

*where $\operatorname{span}\{\cdots\}$ represents the closed subspace spanned by $\{\cdots\}$. In the case where $e^{At}$ is analytic, the first inclusion becomes an equality.*

*Proof.* We first prove the second equality in (2.7). To this end, let us take any $T > 0$ and let $s \in [0, T)$. Define

(2.8) $$\varphi_n(r) = n \chi_{[s, s+1/n]}(r), \qquad r \in [0, \infty).$$

Clearly, for any $u \in U$, $\varphi_n(\cdot) u \in L^\infty(0, \infty; U)$. Then, it follows that

(2.9) $$\begin{aligned} e^{As} Bu &= \lim_{n \to \infty} n \int_s^{s+1/n} e^{Ar} Bu \, dr \\ &= \lim_{n \to \infty} \int_0^T e^{Ar} Bu \varphi_n(r) \, dr \in \overline{G_T(\mathcal{U}^\infty)} \subseteq \overline{G_T(\mathcal{U}^p)}. \end{aligned}$$

Here $p \in [1, \infty)$ and the last inclusion follows from the inclusion $\mathcal{U}^\infty \subset \mathcal{U}^p$. Sending $s \uparrow T$, we see that

(2.10) $$e^{AT} Bu \in \overline{G_T(\mathcal{U}^\infty)} \subseteq \overline{G_T(\mathcal{U}^p)}, \qquad \forall u \in U.$$

On the other hand, for any $u \in C([0,T];U)$, we have

$$(2.11) \qquad \int_0^T e^{Ar} Bu(r)\, dr = \lim_{\max_i |\Delta r_i| \to 0} \sum_i e^{Ar_i} Bu(r_i) \Delta r_i,$$

where the right-hand side of (2.11) is the limit of the Riemann sums for the integrand $e^{Ar} Bu(r)$ on $[0,T]$. Clearly, each term in the above sum belongs to the right-hand side of the second equality in (2.7). Then, by the density of $C([0,T];U)$ in $L^p(0,T;U)$ ($p \in [1,\infty)$), we obtain that

$$(2.12) \qquad \overline{G_T(\mathcal{U}^p)} \subseteq \overline{\text{span}\,\{e^{As} Bu \mid s \in [0,T],\, u \in U\}}.$$

Hence, the second equality in (2.7) follows from (2.9), (2.10), and (2.12). The proof of the first equality in (2.7) is similar. Also, the first inclusion is clear from (2.6). Thus, it remains to prove the last inclusion in (2.7). To show this, we note that both sides of the last inclusion are closed subspaces of $X$. Now, for any $\eta \in X^*$ with

$$(2.13) \qquad \langle \eta, e^{As} Bu \rangle = 0, \qquad \forall s \in [0,T],\, u \in U,$$

we must have

$$(2.14) \qquad \langle \eta, A^n Bu \rangle = 0, \qquad \forall u \in U, \text{ with } Bu \in \mathcal{D}(A^n),\, n \geq 0.$$

This implies that

$$(2.15) \qquad \Big(\text{span}\,\{e^{As} Bu \mid s \in [0,T],\, u \in U\}\Big)^{\perp} \\ \subseteq \Big(\{A^n Bu \mid n \geq 0,\, u \in U,\, Bu \in \bigcap_{m \geq 0} \mathcal{D}(A^m)\}\Big)^{\perp}.$$

Then the last inclusion in (2.7) follows.

Finally, let $e^{At}$ be analytic and $T > 0$ be given. If $\eta \in X^*$ satisfies (2.13), then, as the left-hand side of (2.13) is analytic in a sectorial region containing the nonnegative real axis, we must have that (2.13) holds for all $s \geq 0$. This shows that

$$(2.16) \qquad \Big(\text{span}\,\{e^{As} Bu \mid s \in [0,T],\, u \in U\}\Big)^{\perp} \\ \subseteq \Big(\text{span}\,\{e^{As} Bu \mid s \geq 0,\, u \in U\}\Big)^{\perp}.$$

Then we obtain

$$(2.17) \quad \overline{\text{span}\,\{e^{As} Bu \mid s \in [0,T],\, u \in U\}} = \overline{\text{span}\,\{e^{As} Bu \mid s \geq 0,\, u \in U\}}.$$

This completes our proof. □

The following result gives some criteria for the approximate controllability of the system $[A,B]$.

§2. Controllability for Linear Systems

**Theorem 2.2.** *For any $p \in [1, \infty]$ and $T > 0$, the following are equivalent:*
  (i) $[A, B]$ *is approximately controllable on $[0, T]$ with the control set $\mathcal{U}^p$.*
  (ii) *The range of $G_T$ is dense in $X$, i.e.,*

$$\overline{G_T(\mathcal{U}^p)} = X. \tag{2.18}$$

  (iii) *Operator $G_T^*$ is injective, i.e.,*

$$\mathcal{N}(G_T^*) = \{0\}. \tag{2.19}$$

  (iv) $y^* = 0$ *whenever*

$$B^* e^{A^* t} y^* = 0, \qquad \forall t \in [0, T]. \tag{2.20}$$

*In particular* (i)–(iv) *hold if*

$$\operatorname{span} \{A^n Bu \mid n \geq 0,\ u \in U,\ Bu \in \bigcap_{m \geq 0} \mathcal{D}(A^m)\} = X. \tag{2.21}$$

*Proof.* (i) $\Rightarrow$ (ii): For any $x, z \in X$, $z + e^{AT} x$ is again some given vector. Thus, by the definition of approximate controllability, we see that for any $\varepsilon > 0$, there exists a $u(\cdot) \in \mathcal{U}^p$, such that (note (2.4))

$$\varepsilon > |z + e^{AT} x - y(T; x, u(\cdot))| = |z - G_T(u(\cdot))|. \tag{2.22}$$

This means that $z \in \overline{G_T(\mathcal{U}^p)}$ and (2.18) follows.

(ii) $\Longleftrightarrow$ (iii) is obvious because $\mathcal{N}(G_T^*) = \mathcal{R}(G_T)^\perp$ (see Chapter 2, §1.4).

(iii) $\Rightarrow$ (iv): Let $y^* \in X^*$ satisfy (2.20). Then, by (2.5),

$$y^* \in \mathcal{N}(G_T^*) = \{0\}. \tag{2.23}$$

Thus, (iv) holds.

(iv) $\Rightarrow$ (i): Suppose for some $x \in X$, $\overline{\mathcal{R}([0, T], x)} \neq X$. Consequently, there exists a

$$z \in X \setminus \overline{\mathcal{R}([0, T], x)} \subseteq X \setminus \overline{\mathcal{R}(T, x)}. \tag{2.24}$$

On the other hand, it is not hard to see that $\mathcal{R}(T, x)$ is a convex set. Thus, by the Hahn–Banach Theorem, we can find a $y^* \in X^*$, $|y^*| = 1$, such that

$$\langle y^*, z \rangle \leq \langle y^*, e^{AT} x + \int_0^T e^{A(T-s)} Bu(s)\, ds \rangle, \qquad \forall u(\cdot) \in \mathcal{U}^p. \tag{2.25}$$

The above holds if we replace $u(\cdot)$ by $ku(\cdot)$ for any $k > 0$, thus,

$$\int_0^T \langle B^* e^{A^* s} y^*, u(T - s) \rangle\, ds = 0, \qquad \forall u(\cdot) \in \mathcal{U}^p. \tag{2.26}$$

Consequently,

(2.27) $\langle B^* e^{A^*t} y^*, u \rangle = 0, \qquad \forall t \in [0,T], \ u \in U.$

Then, by (iv), we must have $y^* = 0$, which is a contradiction.

Finally, in the case where (2.21) holds, by Lemma 2.1, (ii) holds and our conclusion follows. □

**Theorem 2.3.** *Let $e^{At}$ be analytic and $p \in [1, \infty]$. Then the following are equivalent:*

(i) $[A, B]$ *is infinitesimal time approximately controllable.*

(ii) $[A, B]$ *is approximately controllable on some $[0, T]$ $(T > 0)$.*

(iii) $[A, B]$ *is approximately controllable ( on $[0, \infty)$).*

The proof of (i) ⇒ (ii) ⇒ (iii) is obvious. The proof of (iii) ⇒ (i) follows from the last assertion in Lemma 2.1.

Let us now look at the following parabolic control system:

(2.28) $\begin{cases} y_t - \Delta y = \chi_E(x) u(x,t), & \text{in } \Omega \times (0,T], \\ y|_{\partial \Omega} = 0, \quad y|_{t=0} = y_0(x). \end{cases}$

Here, $\Omega \subset \mathbb{R}^n$ is a bounded domain with a smooth boundary $\partial \Omega$ and $E \subset \Omega$ is a nonempty open set. The corresponding adjoint equation looks like the following:

(2.29) $\begin{cases} \psi_t + \Delta \psi = 0, & \text{in } \Omega \times [0,T), \\ \psi|_{\partial \Omega} = 0, \quad \psi|_{t=T} = \psi_T(x). \end{cases}$

We take the state space to be $X = L^2(\Omega)$. Then, the abstract condition (2.19) is equivalent to the following: If for some $\psi_T(\cdot)$, the solution $\psi$ of (2.29) satisfies

(2.30) $\psi(x,t) = 0, \qquad \forall (x,t) \in E \times (0,T),$

then it is necessary that $\psi_T(x) \equiv 0$ over $\Omega$ and hence $\psi(x,t) \equiv 0$ over $\Omega \times (0,T)$. Such a property is called the *unique continuation property*. By a very deep result in partial differential equations (see Mizohata [1], Saut-Scheurer [1]), we know that for equation (2.29), the above stated property holds for any $T > 0$. Hence, system (2.28) is infinitesimal time approximately controllable.

### §2.2. Exact controllability

The following lemma will be useful below.

**Lemma 2.4.** *Let $W, V$, and $Z$ be Banach spaces where $W$ is reflexive. Let $F \in \mathcal{L}(V, Z)$ and $G \in \mathcal{L}(W, Z)$. Then*

(2.31) $\begin{aligned} |G^* z^*| &\geq \delta |F^* z^*|, \quad \forall z^* \in Z^*, \text{ for some } \delta > 0 \\ &\iff \mathcal{R}(G) \supseteq \mathcal{R}(F). \end{aligned}$

## §2. Controllability for Linear Systems

*Proof.* ⇒ Fix any $v \in V$, define

$$(2.32) \qquad f(w^*) = \langle v, F^*z^* \rangle, \qquad \forall z^* \in X^*, \text{ with } w^* = G^*z^*.$$

Thus, $\mathcal{D}(f) = \mathcal{R}(G^*)$. We need to show that $f$ is well defined. In fact, for any $z_1^*, z_2^* \in Z^*$ with $G^*z_1^* = G^*z_2^*$, by our condition,

$$(2.33) \qquad |F^*(z_1^* - z_2^*)| \leq \frac{1}{\delta}|G^*(z_1^* - z_2^*)| = 0.$$

This implies that $f$ is well defined. Clearly, $f$ is linear and

$$(2.34) \qquad |f(w^*)| = |\langle v, F^*z^* \rangle| \leq |v||F^*z^*| \leq \frac{1}{\delta}|v||G^*z^*| = \frac{1}{\delta}|v||w^*|,$$

which yields that $f$ is a linear bounded functional on $\mathcal{R}(G^*)$. By the Hahn–Banach Theorem and the reflexivity of $W$, we can find $w \in W$, such that

$$(2.35) \qquad f(w^*) = \langle w^*, w \rangle, \qquad \forall w^* \in \mathcal{R}(G^*).$$

Consequently,

$$(2.36) \qquad \begin{aligned} \langle z^*, Fv \rangle &= \langle v, F^*z^* \rangle = f(w^*) = \langle w^*, w \rangle \\ &= \langle G^*z^*, w \rangle = \langle z^*, Gw \rangle, \qquad \forall z^* \in Z^*. \end{aligned}$$

Hence, $Fv = Gw$ and consequently, $\mathcal{R}(F) \subseteq \mathcal{R}(G)$.

⇐ Suppose it is not the case. Then we can find $z_n^* \in Z^*$, such that

$$(2.37) \qquad 0 \leq |G^*z_n^*| < \frac{1}{n}|F^*z_n^*|, \qquad n \geq 1.$$

From the above, we see that $F^*z_n^* \neq 0$ and we can define

$$(2.38) \qquad y_n^* = \sqrt{n}\frac{z_n^*}{|F^*z_n^*|}, \qquad n \geq 1.$$

Then

$$(2.39) \qquad \begin{cases} |G^*y_n^*| = \sqrt{n}\dfrac{|G^*z_n^*|}{|F^*z_n^*|} < \dfrac{1}{\sqrt{n}} \to 0, \\ |F^*y_n^*| = \sqrt{n} \to \infty, \end{cases} \qquad n \to \infty.$$

On the other hand, for any $v \in V$, there exists a $w \in W$, such that $Fv = Gw$. Thus,

$$(2.40) \qquad \begin{aligned} \langle F^*y_n^*, v \rangle &= \langle y_n^*, Fv \rangle = \langle y_n^*, Gw \rangle \\ &= \langle G^*y_n^*, w \rangle \to 0, \qquad n \to \infty. \end{aligned}$$

This means that $F^*y_n^* \xrightarrow{w} 0$. Thus, by the Principle of Uniform Boundedness (see Chapter 2, Theorem 1.11), the sequence $\{F^*y_n^*\}_{n \geq 1}$ must be uniformly bounded, contradicting (2.39). This completes the proof. □

Now, let us give some criteria for the exact controllability (compare with Theorem 2.2).

**Theorem 2.5.** *Let $X$ and $U$ be reflexive Banach spaces and let $p \in (1, \infty)$, $T > 0$. Then the following are equivalent:*

(i) *$[A, B]$ is exactly controllable on $[0, T]$ with the control set $\mathcal{U}^p$.*

(ii) *The range of $G_T$ coincides with $X$:*

$$G_T(\mathcal{U}^p) = X. \tag{2.41}$$

(iii) *Operator $G_T^*$ satisfies*

$$|G_T^* z^*|_{L^{p'}(0,T;U^*)} \geq \delta |z^*|_{X^*}, \qquad \forall z^* \in X^*. \tag{2.42}$$

(iv) *There exists a constant $\delta > 0$, such that $(p' = p/(p-1))$*

$$|B^* e^{A^* \cdot} z^*|_{L^{p'}(0,T;U^*)} \geq \delta |z^*|_{X^*}, \qquad \forall z^* \in X^*. \tag{2.43}$$

*Proof.* (i) $\Rightarrow$ (ii): For any $x, z \in X$, $z + e^{AT} x$ is also a given vector in $X$. Thus, there exists a $u(\cdot) \in \mathcal{U}^p$, such that

$$z + e^{AT} x = y(T; x, u(\cdot)) \equiv e^{AT} x + G_T(u(\cdot)). \tag{2.44}$$

This gives $z \in G_T(\mathcal{U}^p)$. Hence, (ii) follows.

(ii) $\Rightarrow$ (i) can be proved similarly.

(ii) $\Longleftrightarrow$ (iii): Let $W = L^p(0, T; U)$, $Z = V = X$, $F = I$, and $G = G_T$. Then, by Lemma 2.4, we obtain the equivalence between (ii) and (iii).

(iii) $\Longleftrightarrow$ (iv): By (2.5), we have

$$\begin{aligned} |G_T^* z^*|_{L^{p'}(0,T;U^*)} &= \left( \int_0^T |B^* e^{A^*(T-s)} z^*|^{p'} ds \right)^{1/p'} \\ &= \left( \int_0^T |B^* e^{A^* s} z^*|^{p'} ds \right)^{1/p'}. \end{aligned} \tag{2.45}$$

Thus, (iii) and (iv) are equivalent. □

Let us look at the following control system:

$$\begin{cases} y_{tt} - \Delta y = \chi_E(x) u(x, t), & \text{in } \Omega \times (0, T], \\ y|_{\partial \Omega} = 0, \quad y|_{t=0} = y_0(x), \quad y_t|_{t=0} = y_1(x). \end{cases} \tag{2.46}$$

Here, $\Omega \subset \mathbb{R}^n$ is a bounded domain with a smooth boundary $\partial \Omega$ and $E \subset \Omega$ is a nonempty open set. We know that the corresponding adjoint equation is the following:

$$\begin{cases} \psi_{tt} - \Delta \psi = 0, & \text{in } \Omega \times [0, T), \\ \psi|_{\partial \Omega} = 0, \quad \psi|_{t=T} = \psi_0, \quad \psi_t|_{t=T} = \psi_1. \end{cases} \tag{2.47}$$

## §2. Controllability for Linear Systems

We take the state space to be $X = L^2(\Omega) \times H^{-1}(\Omega)$. Then the abstract condition (2.42) is equivalent to the following: There exists a constant $\delta > 0$, such that

$$\text{(2.48)} \qquad \int_{E \times (0,T)} |\psi(x,t)|^2 \, dx \, dt \geq \delta \|(\psi_0, \psi_1)\|^2_{L^2(\Omega) \times H^{-1}(\Omega)},$$

where $\psi$ is the solution of (2.47). It is known that (see Lions [6]) (2.48) holds if $T > 0$ is large enough. Hence, we see that control system (2.46) is exactly controllable for some large $T > 0$.

The method that uses condition (2.42) to obtain the exact controllability in Hilbert spaces is referred to as the *Hilbert Uniqueness Method* (HUM for short). This method is applicable to many other equations and many other type of control problems (boundary control, pointwise control, etc.) In Lions [6], the readers can find an extensive study on this subject as well as a very good list of references. The parabolic case we have discussed in §2.1 has a similar basic idea.

The following result is concerned with exact null controllability.

**Theorem 2.6.** $[A, B]$ *is exactly null controllable on* $[0, T]$ *with the control set* $\mathcal{U}^p$ *if and only if*

$$\text{(2.49)} \qquad \mathcal{R}(e^{AT}) \subseteq G_T(\mathcal{U}^p).$$

*In particular, if $e^{At}$ is analytic with $A^{-1} \in \mathcal{L}(X)$ and for some $T_0 \in (0, T]$, $(e^{AT_0} - I)^{-1} \in \mathcal{L}(X)$, then $[A, I]$ is exactly null controllable.*

*Proof.* From the proof of Theorem 2.5, we can easily prove that (2.49) is a necessary and sufficient condition for the exact null controllability of $[A, B]$. Now, in the case where $e^{At}$ is analytic with $A^{-1} \in \mathcal{L}(X)$ and $1 \in \rho(e^{AT_0})$, for any $u_0 \in U$,

$$\text{(2.50)} \qquad \int_0^T e^{A(T-s)} u_0 \chi_{[T-T_0, T]}(s) \, ds$$

$$= \int_0^{T_0} e^{As} u_0 \chi_{[0, T_0]}(s) \, ds = (e^{AT_0} - I) A^{-1} u_0.$$

Hence, for any $x \in X$, by taking (note $Ae^{AT} \in \mathcal{L}(X)$)

$$\text{(2.51)} \qquad u(s) = \chi_{[T-T_0, T]}(s)(e^{AT_0} - I)^{-1} A e^{AT} x, \qquad s \in [0, T],$$

we have $e^{AT} x = G_T u$, proving (2.49). Thus, $[A, I]$ is exactly null controllable. □

Let us look at the control system (2.28) with $E = \Omega$. Let $A = \Delta$ with $\mathcal{D}(A) = H_0^1(\Omega) \cap H^2(\Omega)$. Then, it is known that the semigroup $e^{At}$ is analytic and satisfies

$$\text{(2.52)} \qquad \|e^{At}\| \leq e^{-\omega t}, \qquad \forall t \geq 0,$$

for some $\omega > 0$. Thus, for any $T > 0$, $(e^{AT} - I)^{-1} \in \mathcal{L}(X)$. By the above result, we see that system (2.28) with $E = \Omega$ is (infinitesimal time) exactly null controllable.

Similar to Theorem 2.6, we have the following result whose proof is left to the readers.

**Theorem 2.7.** *Let $\mathcal{R}(B) = X$, $A^{-1} \in \mathcal{L}(X)$, and for some $T_0 > 0$, $(e^{AT_0} - I)^{-1} \in \mathcal{L}(X)$. Then, for any $T > 0$, $x_1 \in \mathcal{D}(A)$, $[A, B]$ is exactly controllable with the target set $Q = \{x_1\}$ and with the control set $\mathcal{U}^p$.*

The above result tells us that if the target point is "nice," we may still exactly hit it in finite time.

## §3. Approximate Controllability for Semilinear Systems

In this section, we consider the following system:

$$(3.1) \qquad y(t) = e^{At}x + \int_0^t e^{A(t-s)}\big[f(y(s)) + Bu(s)\big]\,ds, \qquad \forall t \geq 0.$$

The differential form of the above is

$$(3.2) \qquad \begin{cases} \dot{y}(t) = Ay(t) + f(y(t)) + Bu(t), & t \geq 0, \\ y(0) = x. \end{cases}$$

We see that system (3.1) is less general than (1.1). However, systems like (3.1) appear in many applications. On the other hand, we may regard (3.1) as a perturbation of the linear system (2.1). In what follows, we establish the approximate controllability of (3.1). To begin with, let us make the following assumptions.

(AC1) $X$ and $U$ are Banach spaces where $X$ is separable and $X^*$ is uniformly convex; $e^{At}$ is a compact semigroup on $X$ and $B \in \mathcal{L}(U, X)$.

(AC2) $f : X \to X$ is $C^1$ with the Fréchet derivative $f'(\cdot)$, such that for some constant $L > 0$,

$$(3.3) \qquad \|f'(z)\|_{\mathcal{L}(X)} \leq L, \qquad \forall z \in X.$$

The above (AC1) means that we are essentially treating the semilinear parabolic partial differential equations. For example,

$$(3.4) \qquad \begin{cases} y_t - \Delta y = f(y) + u\chi_E, & \text{in } \Omega, \\ y\big|_{\partial\Omega} = 0, \quad y\big|_{t=0} = y_0, \end{cases}$$

where $\Omega \subset \mathbb{R}^n$ is a bounded domain with a smooth boundary $\partial\Omega$ and $E \subset \Omega$ is an open set. Some other situations such as the Neumann boundary condition, boundary control, etc., are also possible.

## §3. Approximate Ccontrollability for Semilinear Systems

It is clear that under (AC1) and (AC2), for any $x \in X$ and $u(\cdot) \in \mathcal{U}^1$, (3.1) admits a unique solution $y(\cdot) \equiv y(\cdot, x, u)$. From the previous section, we know that under certain conditions, the linear system $[A, B]$ is approximately controllable on a given (or any) $[0, T]$ ($T > 0$). We may regard (3.1) as a perturbation of system $[A, B]$, and condition (3.3) means that the perturbation $f(\cdot)$ is "small" compared with the unbounded operator $A$. Thus, it is expected that if system $[A, B]$ is approximately controllable, then, with possibly some technical conditions, system (3.1) should also be approximately controllable. We now make this more precise.

Let $x, y_T \in X$ and $T > 0$; we want to achieve the following: For any $\varepsilon > 0$, find $u(\cdot) \in \mathcal{U}^1$, such that

(3.5) $$|y(T; x, u) - y_T| \leq \varepsilon.$$

In what follows, we fix $x, y_T \in X$ and $T, \varepsilon > 0$. Let us first give a brief description of the idea with which we obtain the approximate controllability for system (3.1). To this end, we define

(3.6) $$F(z) = \int_0^1 f'(\sigma z) \, d\sigma, \qquad z \in X.$$

From our assumption,

(3.7) $$F(\cdot) \in C(X; \mathcal{L}(X)), \quad \|F(z)\|_{\mathcal{L}(X)} \leq L, \qquad \forall z \in X.$$

Then system (3.2) can be rewritten as

(3.8) $$\begin{cases} \dot{y}(t) = [A + F(y(t))] y(t) + Bu(t) + f(0), \\ y(0) = x. \end{cases}$$

Next, for any fixed $z(\cdot) \in C([0, \infty); X)$, let $w(\cdot) \equiv w(\cdot; x, z(\cdot))$ and $\xi(\cdot) \equiv \xi(\cdot; z(\cdot), u(\cdot))$ be, respectively, the mild solutions of

(3.9) $$\begin{cases} \dot{w}(t) = [A + F(z(t))] w(t) + f(0), \\ w(0) = x, \end{cases}$$

(3.10) $$\begin{cases} \dot{\xi}(t) = [A + F(z(t))] \xi(t) + Bu(t), \\ \xi(0) = 0. \end{cases}$$

Under certain conditions, we will show the following:

(1) For any $z(\cdot) \in C([0, \infty); X)$, there exists a $u \equiv u(z) \in \mathcal{U}^p$ determined by $z(\cdot)$ explicitly, such that

(3.11) $$|\xi(T; z(\cdot), u(z)) + w(T; z(\cdot)) - y_T| \leq \varepsilon.$$

(2) Define the map as follows:

(3.12) $$\Lambda(z(\cdot)) = \xi(T; z(\cdot), u(z)) + w(T; z(\cdot)), \qquad z(\cdot) \in C([0, \infty); X).$$

It turns out that this map admits a fixed point, say $\bar{y}(\cdot) \in C([0,\infty); X)$. This implies that $\bar{y}(\cdot)$ and the corresponding control, denoted by $\bar{u}(\cdot)$, consist of an admissible pair and

(3.13) $$|\bar{y}(T; x, \bar{u}) - y_T| \leq \varepsilon,$$

which gives the approximate controllability of (3.1).

Now, let us make the above rigorous. We state the following result.

**Theorem 3.1.** *Let (AC1) and (AC2) hold. Let the following hold:*

(3.14) $$\mathcal{N}(B^* G(T, \cdot; F)^*) = \{0\}, \qquad \forall F \in L^\infty(0, \infty; \mathcal{L}(X)),$$

*where $G(t, s; F)$ is the evolution operator generated by $A + F(\cdot)$. Then, for any $x, y_T \in X$ and any $\varepsilon > 0$, there exists a control $u(\cdot) \in \mathcal{U}^p$, such that (3.5) holds.*

The proof of the above theorem is long and technical. We split it into several lemmas. Note that by (AC1), $X^*$ is reflexive. Thus, the weak and weak* topologies in $X^*$ coincide.

**Lemma 3.2.** *If $F_n(\cdot), F(\cdot) \in L^\infty(0, T; \mathcal{L}(X))$ and $\eta_n, \eta \in X^*$ such that*

(3.15) $$\begin{cases} F_n(\cdot) \xrightarrow{*} F(\cdot), & \text{in } L^\infty(0, T; \mathcal{L}(X)), \\ \eta_n \xrightarrow{w} \eta, & \text{in } X^*, \end{cases}$$

*then,*

(3.16) $$G(T, \cdot; F_n)^* \eta_n \xrightarrow{s} G(T, \cdot; F)^* \eta, \qquad \text{in } C([0, T]; X^*).$$

*Proof.* Denote

(3.17) $$\psi_n(t) = G(T, t; F_n)^* \eta_n, \quad \psi(t) = G(T, t; F)^* \eta, \qquad t \in [0, T].$$

It is easily seen that

(3.18) $$\psi_n(t) = e^{A^*(T-t)} \eta_n + \int_t^T e^{A^*(T-s)} F_n(s) \psi_n(s)\, ds, \qquad t \in [0, T].$$

Because $X$ is a reflexive Banach space and $e^{At}$ is a compact semigroup, we see that $e^{A^* t}$ is a compact semigroup on $X^*$. On the other hand, from (3.15) and the Principle of Uniform Boundedness (see Chapter 2, §1.2), we have the uniform boundedness of $\{F_n(\cdot)\}$ and $\{\eta_n\}$. Then, by (3.18) and Gronwall's inequality, one obtains the uniform boundedness of $\{\psi_n(\cdot)\}$ and $\{F_n(\cdot)\psi_n(\cdot)\}$. Thus, similar to Chapter 3, §3, we can show that $\{\psi_n(\cdot)\}$ is relatively compact in $C([0, T]; X^*)$. Let $\psi(\cdot)$ be any limit point of $\psi_n(\cdot)$ in $C([0, T]; X^*)$. Passing to the limit in (3.18) along some proper subsequence, we see that $\psi(\cdot)$ satisfies

(3.19) $$\psi(t) = e^{A^*(T-t)} \eta + \int_t^T e^{A^*(T-s)} F(s) \psi(s)\, ds, \qquad t \in [0, T].$$

§3. Approximate Ccontrollability for Semilinear Systems

By the uniqueness of the solutions to (3.19), we obtain that the whole sequence $\{\psi_n(\cdot)\}$ converges to $\psi(\cdot)$ in $C([0,T];X^*)$. □

**Lemma 3.3.** *For any $\varepsilon > 0$, define*

$$(3.20) \quad J(\eta, F, \zeta) = \frac{1}{2}\int_0^T |B^*G(T,t;F)^*\eta|^2\, dt + \varepsilon|\eta| + \langle \eta, \zeta \rangle,$$

$$\forall (\eta, F(\cdot), \zeta) \in X^* \times L^\infty(0,T;\mathcal{L}(X)) \times X.$$

*Then*

  (i) *The map $\eta \mapsto J(\eta, F, \zeta)$ is continuous and strictly convex.*

  (ii) *For any bounded set $\mathcal{F} \subset L^\infty(0,T;\mathcal{L}(X))$ and compact set $Z \subset X$,*

$$(3.21) \quad \lim_{|\eta|\to\infty} \inf_{F\in\mathcal{F}, \zeta\in Z} \frac{J(\eta, F, \zeta)}{|\eta|} \geq \varepsilon.$$

*Proof.* (i) From (3.20), we see immediately that $\eta \mapsto J(\eta, F, \zeta)$ is continuous. The strict convexity follows easily from the uniform convexity of $X^*$ (see (AC1)).

(ii) Suppose it is not the case. Then there exist sequences $\{\eta_n\} \subset X^*$, $\{F_n(\cdot)\} \subset \mathcal{F}$ and $\{\zeta_n\} \subset Z$, with $|\eta_n| \to \infty$, such that

$$(3.22) \quad \lim_{n\to\infty} \frac{J(\eta_n, F_n, \zeta_n)}{|\eta_n|} < \varepsilon.$$

Without loss of generality, we may assume that

$$(3.23) \quad \begin{cases} \zeta_n \xrightarrow{s} \zeta, & \text{in } X, \\ F_n(\cdot) \xrightarrow{*} F(\cdot), & \text{in } L^\infty(0,T;\mathcal{L}(X)), \end{cases}$$

for some $\zeta \in X$ and $F(\cdot) \in L^\infty(0,T;\mathcal{L}(X))$. In fact, $Z$ is compact in $X$; thus, we may assume the first by picking a subsequence. For the second, because $X$ is separable (by (AC1)), we may let $\{\xi_m\}_{m\geq 1}$ be a countable dense set in $X$. For any $\xi_m$, $\{F_n(\cdot)\xi_m\}_{n\geq 1}$ is a bounded set in $L^\infty(0,T;X) = L^1(0,T;X^*)^*$ (because $X$ is reflexive). Thus, we may assume that $\{F_n(\cdot)\xi_m\}_{n\geq 1}$ is weakly* convergent. By the diagonal argument, we can then assume that the whole sequence $\{F_n(\cdot)\}$ is weakly* convergent to some $F(\cdot) \in L^\infty(0,T;\mathcal{L}(X))$. Next, let us set $\widetilde{\eta}_n = \eta_n/|\eta_n|$. Then $|\widetilde{\eta}_n| = 1$ and one may assume that $\widetilde{\eta}_n \xrightarrow{*} \widetilde{\eta}$, in $X^*$. Consequently, by Lemma 3.2,

$$(3.24) \quad B^*G(T,\cdot;F_n)^*\widetilde{\eta}_n \xrightarrow{s} B^*G(T,\cdot;F)^*\widetilde{\eta}, \quad \text{in } C([0,T];X^*).$$

From (3.20), it follows that

$$(3.25) \quad \frac{J(\eta_n, F_n, \zeta_n)}{|\eta_n|} = \frac{1}{2}\Big(\int_0^T |B^*G(T,t;F_n)^*\widetilde{\eta}_n|^2\, dt\Big)|\eta_n| + \varepsilon + \langle \widetilde{\eta}_n, \zeta_n \rangle.$$

Thus, noting that $|\eta_n| \to \infty$, by (3.22)–(3.25) and Fatou's Lemma,

$$(3.26) \quad \int_0^T |B^*G(T,t;F)^*\widetilde{\eta}|^2 \, dt \leq \lim_{n\to\infty} \int_0^T |B^*G(T,t;F_n)^*\widetilde{\eta}_n|^2 \, dt = 0.$$

By (3.14), we must have $\widetilde{\eta} = 0$. Hence, by (3.22), (3.23), and (3.25),

$$(3.27) \quad \varepsilon > \lim_{n\to\infty} \frac{J(\eta_n, F_n, \zeta_n)}{|\eta_n|} \geq \varepsilon + \lim_{n\to\infty} \langle \widetilde{\eta}_n, \zeta_n \rangle = \varepsilon,$$

which is a contradiction, proving (ii). □

**Lemma 3.4.** *Let $J(\eta, F, \zeta)$ be defined by (3.20). Then, for any $(F(\cdot), \zeta) \in L^\infty(0,T;\mathcal{L}(X)) \times X$, the functional $J(\cdot, F, \zeta)$ admits a unique minimum $\bar{\eta}$ that defines a map $\Phi : L^\infty(0,T;\mathcal{L}(X)) \times X \to X^*$. This $\Phi$ has the following properties:*

(i) *For any bounded set $\mathcal{F} \subset L^\infty(0,T;\mathcal{L}(X))$ and compact set $Z \subset X$, the image $\Phi(\mathcal{F} \times Z)$ is bounded in $X^*$;*

(ii) *For any $(F_n(\cdot), \zeta_n)$, $(\mathcal{F}(\cdot), \zeta)$ satisfying (3.23), it holds that*

$$(3.28) \quad \lim_{n\to\infty} |\Phi(F_n, \zeta_n) - \Phi(F, \zeta)| = 0.$$

(iii) $\Phi(F, \zeta) = 0$ *if and only if $|\zeta| \leq \varepsilon$.*

*Proof.* First of all, we note that (3.21) implies the following: For any $(F, \zeta) \in L^\infty(0,T;\mathcal{L}(X)) \times X$,

$$(3.29) \quad \lim_{|\eta|\to\infty} \frac{J(\eta, F, \zeta)}{|\eta|} \geq \varepsilon.$$

Such a property is usually referred to as the *coercivity*. Now, let $(F(\cdot), \zeta) \in L^\infty(0,T;\mathcal{L}(X)) \times X$ be given and let $\{\eta_n\}_{n\geq 1}$ be a minimizing sequence of $J(\cdot, F, \zeta)$. Because of (3.29), this sequence is bounded in $X^*$. Then we may assume that $\eta_n \xrightarrow{w} \bar{\eta}$. By Lemma 3.2 and Fatou's Lemma, we see immediately that

$$(3.30) \quad J(\bar{\eta}, F, \zeta) \leq \lim_{n\to\infty} J(\eta_n, F, \zeta) = \inf_{\eta \in X^*} J(\eta, F, \zeta).$$

Thus, $\bar{\eta}$ is a minimum. By the strict convexity of $J(\cdot, F, \zeta)$, the minimum is unique. Hence, the map $\Phi(F, \zeta) = \bar{\eta}$ is well defined.

(i) Let $\mathcal{F} \subset L^\infty(0,T;\mathcal{L}(X))$ be bounded and $Z \subset X$ be compact. From Lemma 3.3 (ii), we see that there exists a constant $R > 0$, depending on the sets $\mathcal{F}$ and $Z$, such that

$$(3.31) \quad \inf_{F \in \mathcal{F}, \zeta \in Z} \frac{J(\eta, F, \zeta)}{|\eta|} \geq \frac{\varepsilon}{2}, \quad \forall |\eta| \geq R.$$

## §3. Approximate Ccontrollability for Semilinear Systems

On the other hand, for any $(F,\zeta) \in \mathcal{F} \times Z$, by the definition of $\Phi$,

(3.32) $$J(\Phi(F,\zeta),F,\zeta) \leq J(0,F,\zeta) = 0.$$

Hence, combining (3.31) and (3.32), we have

(3.33) $$|\Phi(F,\zeta)| \leq R \equiv R(\mathcal{F},Z). \qquad \forall (F,\zeta) \in \mathcal{F} \times Z.$$

(ii) Let $(F_n,\zeta_n) \in L^\infty(0,T;\mathcal{L}(X)) \times X$ satisfy (3.23). Then the set $\{F_n\}_{n\geq 1}$ is bounded and the set $\{\zeta_n\}_{n\geq 1}$ is compact. Thus, by the above (i), we have the boundedness of $\bar\eta_n \equiv \Phi(F_n,\zeta_n)$. Consequently, we may assume that $\bar\eta_n \overset{w}{\rightharpoonup} \widetilde\eta$. From Lemma 3.2, it follows that

(3.34) $$B^*G(T,\cdot;F_n)^*\bar\eta_n \overset{s}{\to} B^*G(T,\cdot;F)^*\widetilde\eta, \qquad \text{in } C([0,T];X^*).$$

Thus, by the definition of $J$ (see (3.20)), and the optimality of both $\bar\eta \equiv \Phi(F,\zeta)$ and $\bar\eta_n \equiv \Phi(F_n,\zeta_n)$, one has

(3.35) $$\begin{aligned} J(\bar\eta,F,\zeta) \leq J(\widetilde\eta,F,\zeta) &\leq \varliminf_{n\to\infty} J(\bar\eta_n,F_n,\zeta_n) \\ &\leq \varlimsup_{n\to\infty} J(\bar\eta_n,F_n,\zeta_n) \leq \lim_{n\to\infty} J(\bar\eta,F_n,\zeta_n) = J(\bar\eta,F,\zeta). \end{aligned}$$

Hence, the equalities hold in the above. That means that $\widetilde\eta$ is also a minimum of $J(\cdot,F,\zeta)$. By the uniqueness of the minimum, it is necessary that $\widetilde\eta = \bar\eta$. Therefore,

(3.36) $$\begin{cases} \lim\limits_{n\to\infty} J(\bar\eta_n,F_n,\zeta_n) = J(\bar\eta,F,\zeta), \\ \lim\limits_{n\to\infty} \int_0^T |B^*G(T,t;F_n)^*\bar\eta_n|^2\,dt = \int_0^T |B^*G(T,t;F)^*\bar\eta|^2\,dt, \\ \lim\limits_{n\to\infty} \langle \bar\eta_n,\zeta_n\rangle = \langle \bar\eta,\zeta\rangle, \qquad |\bar\eta| \leq \varliminf\limits_{n\to\infty} |\bar\eta_n|. \end{cases}$$

These relations imply that

(3.37) $$\lim_{n\to\infty} |\bar\eta_n| = |\bar\eta|.$$

Because $X^*$ is uniformly convex, from $\bar\eta_n \overset{w}{\rightharpoonup} \bar\eta$ and (3.37), we obtain the strong convergence of $\bar\eta_n$ to $\bar\eta$.

(iii) Suppose that $|\zeta| \leq \varepsilon$. Then, for any $\eta \in X^*$,

(3.38) $$J(\eta,F,\zeta) \geq \varepsilon|\eta| + \langle\eta,\zeta\rangle \geq 0 = J(0,F,\zeta).$$

This yields $\Phi(F,\zeta) = 0$. Conversely, if $\Phi(F,\zeta) = 0$, then, for any $\eta \in X^*$ and $\delta > 0$, we have (note that $J(0,F,\zeta) = 0$)

(3.39) $$\begin{aligned} 0 \leq \frac{J(\delta\eta,F,\zeta)}{\delta} &= \frac{\delta}{2}\int_0^T |B^*G(T,t)^*\eta|^2\,dt + \varepsilon|\eta| + \langle\eta,\zeta\rangle \\ &\to \varepsilon|\eta| + \langle\eta,\zeta\rangle, \qquad (\delta \to 0). \end{aligned}$$

This implies that $|\zeta| \leq \varepsilon$. □

Now, we are ready to prove our main theorem of this section.

*Proof of Theorem 3.1.* Let $x, y_T \in X$ be any given two points and $\varepsilon > 0$ be any given accuracy. We define $F(z)$ as in (3.6). Fix any $z(\cdot) \in C([0,T]; X)$. Recall that the evolution operator generated by $A + F(z(\cdot))$ is denoted by $G(t, s; F(z))$. Let

(3.40)
$$\begin{cases} w(t; z) = G(t, 0; F(z))x + \int_0^t G(t, s; F(z))f(0) \, ds, & t \in [0, T], \\ \xi(t; z, u) = \int_0^t G(t, s; F(z))Bu(s) \, ds, & \forall t \in [0, T]. \end{cases}$$

Next, we define

(3.41)
$$\begin{cases} \zeta \stackrel{\Delta}{=} w(T; z) - y_T, \\ \bar{u}(t) \equiv \bar{u}(t; z) \stackrel{\Delta}{=} B^* G(T, t; F(z))^* \Phi(F(z), \zeta), & t \in [0, T]. \end{cases}$$

Set

(3.42) $\quad Y(z)(t) = \xi(t; z, \bar{u}(\cdot)) + w(t; z), \quad t \in [0, T].$

Then it is seen that for all $t \in [0, T]$,

(3.43) $\quad Y(z)(t) = G(t, 0; F(z))x + \int_0^t G(t, s; F(z))[f(0) + B\bar{u}(s)] \, ds,$

which is equivalent to the following:

(3.44) $\quad Y(z)(t) = e^{At}x + \int_0^t e^{A(t-s)}[F(z(s))Y(z)(s) + f(0) + B\bar{u}(s)] \, ds.$

Hence, any fixed point of the map $Y : C([0,T]; X) \to C([0,T]; X)$ coincides with the state trajectory $y(\cdot; x, \bar{u})$ of the original system (3.1).

Now, from Lemma 3.4 (iii), it follows that in the case where

(3.45) $\quad |\zeta| \equiv |w(T; z) - y_T| \leq \varepsilon,$

we have $\Phi(F(z), \zeta) = 0$, which implies $\bar{u}(\cdot) = 0$ and

(3.46) $\quad |Y(z)(T) - y_T| = |w(T; z) - y_T| \equiv |\zeta| \leq \varepsilon.$

On the other hand, if

(3.47) $\quad |\zeta| \equiv |w(T; z) - y_T| > \varepsilon,$

§3. Approximate Ccontrollability for Semilinear Systems

then $\bar{\eta} \equiv \Phi(F(z), \zeta) \neq 0$. Because $J(\cdot, F(z), \zeta)$ is Fréchet differentiable at $\bar{\eta} \neq 0$, by the optimality of $\bar{\eta}$, we must have

$$
\begin{aligned}
0 &= J_\eta(\bar{\eta}, F(z), \zeta) \\
&= \int_0^T G(T, t; F(z)) BB^* G(T, t; F(z))^* \bar{\eta} \, dt + \varepsilon \frac{\bar{\eta}}{|\bar{\eta}|} + \zeta \\
&= \xi(T; z, \bar{u}) + w(T; z) - y_T + \varepsilon \frac{\bar{\eta}}{|\bar{\eta}|} = Y(z)(T) - y_T + \varepsilon \frac{\bar{\eta}}{|\bar{\eta}|}.
\end{aligned}
\tag{3.48}
$$

Thus,

$$|Y(z)(T) - y_T| = \varepsilon. \tag{3.49}$$

Combining the above, we see that it suffices to show that the map $Y$ admits a fixed point in $C([0, T]; X)$. Let us prove this. By (3.7), (3.9), and the compactness of the semigroup $e^{At}$, one sees that the set $\{w(T; z) - y_T \mid z(\cdot) \in C([0, T]; X)\}$ is relatively compact in $X$. Then, by Lemma 3.4 (i) and (ii), we obtain that the set $\{\Phi(F(z), w(T; z) - y_T) \mid z(\cdot) \in C([0, T]; X)\}$ is bounded and relatively compact. Consequently, it follows from (3.41)–(3.44) that the map $Y : C([0, T]; X) \to C([0, T]; X)$ is continuous and compact with the image being uniformly bounded. Hence, by the Schauder fixed point theorem, we obtain a fixed point of $Y$. Let $\bar{y}(\cdot)$ be a fixed point. Then

$$|\bar{y}(T) - y_T| \leq \varepsilon, \tag{3.50}$$

and $\bar{y}(\cdot)$ is nothing but the state trajectory of (3.1) under the control

$$\bar{u}(t) = B^* G(T, t; F(\bar{y})) \Phi(F(\bar{y}), w(T, \bar{y}) - y_T). \tag{3.51}$$

This gives the approximate controllability of (3.1). □

To conclude this section, let us make a remark on the condition (3.14). By taking $F = 0$, we see that (3.14) implies (2.19). Consequently, (3.14) implies that $[A, B]$ is approximately controllable on $[0, T]$ with the control set $\mathcal{U}^p$. For system (3.4), we may write down the associated adjoint problem as follows:

$$
\begin{cases}
\psi_t + \Delta \psi + F(x, t) \psi = 0, & \text{in } \Omega \times (0, T), \\
\psi|_{\partial \Omega} = 0, \quad \psi|_{t=T} = \psi_T,
\end{cases}
\tag{3.52}
$$

where $F \in L^\infty(\Omega \times (0, T))$ is given by

$$F(x, t) = \int_0^1 f'(\sigma y(x, t)) \, d\sigma \equiv F(y(x, t)). \tag{3.53}$$

Then (3.14) is implied by the following: If $\psi(x, t)$ is the solution of (3.52) for some $\psi_T$, such that

$$\psi(x, t) = 0, \quad \forall (x, t) \in E \times (0, T), \tag{3.54}$$

then it is necessary that $\psi \equiv 0$ over $\Omega \times (0, T)$. This is again a *unique continuation property*. For system (3.52), the above property holds (provided $E$ is a nonempty open set) (see Mizohata [1], Saut–Scheurer [1]). Hence, system (3.4) is actually infinitesimal approximately controllable.

## §4. Time Optimal Control—Semilinear Systems

In this section, we consider the semilinear control system (1.1) with the target set $Q : [0, \infty) \to 2^X$.

### §4.1. Necessary conditions for time optimal pairs

Let us make some assumptions.

(T1) $X$ is a Banach space with its dual strictly convex, $U$ is a separable metric space, and $e^{At}$ is a $C_0$ semigroup on $X$.

(T2) The map $f : [0, \infty) \times X \times U \to X$ is measurable in $t$ and, Fréchet differentiable in $y$ where $f(t, y, \cdot)$ and $f_y(t, y, \cdot)$ are continuous. Moreover, there exists a constant $L > 0$, such that

(4.1) $\quad |f_y(t, y, u)|, |f(t, 0, u)| \leq L, \quad \forall (t, y, u) \in [0, \infty) \times X \times U.$

(T3) The multifunction $Q : [0, \infty) \to 2^X$ is continuous (with respect to the Hausdorff metric, see Chapter 3) and for each $t \in [0, \infty)$, $Q(t)$ is convex and closed.

Clearly, under (T1) and (T2), we see that for any $(x, u(\cdot)) \in X \times \mathcal{U}$, the state equation (1.1) admits a unique solution $y(\cdot\,; x, u) \in C([0, \infty); X)$. We let $x \in X \setminus Q(0)$, such that

(4.2) $\quad \displaystyle\bigcup_{t \in [0, \infty)} \left( \mathcal{R}(t, x) \bigcap Q(t) \right) \neq \phi.$

This means that starting from the initial state $x$, which is away from the target $Q(0)$, there is at least one control $u(\cdot) \in \mathcal{U}$, such that for some finite moment $T > 0$, we have

(4.3) $\quad y(T; x, u) \in Q(T).$

In other words, we can steer the initial state $x$ to the target $Q(\cdot)$ in finite time, or we can hit the target in finite time. We refer to (4.3) as the *exact controllability condition*. This is necessary for the following study of the time optimal control problem. Now, we let

(4.4) $\quad \mathcal{T}(x, u(\cdot)) \triangleq \{ t \geq 0 \mid y(t; x, u) \in Q(t) \}, \quad \forall u(\cdot) \in \mathcal{U}.$

The set $\mathcal{T}(x, u(\cdot))$ could be empty for some $u(\cdot) \in \mathcal{U}$. But, under our controllability condition (4.3), we have

(4.5) $\quad \displaystyle\bigcup_{u(\cdot) \in \mathcal{U}} \mathcal{T}(x, u(\cdot)) \neq \phi.$

## §4. Time Optimal Control—Semilinear Systems

Actually, it is not hard to see that (4.3) and (4.5) are equivalent. Next, let us define

(4.6) $$J(u(\cdot)) = \inf \mathcal{T}(x, u(\cdot)), \qquad \forall u(\cdot) \in \mathcal{U}.$$

Here, we take the convention that $\inf \phi = \infty$. The initial state $x$ will be fixed below; thus, we have suppressed $x$ on the left-hand side of (4.6). Now, our *time optimal control problem* is the following:

**Problem (T).** Find $\bar{u}(\cdot) \in \mathcal{U}$, such that

(4.7) $$J(\bar{u}(\cdot)) = \inf_{u(\cdot) \in \mathcal{U}} J(u(\cdot)) \stackrel{\Delta}{=} \bar{t}.$$

Any control $\bar{u}(\cdot) \in \mathcal{U}$ satisfying (4.7) is called a *time optimal control* and the optimal cost $\bar{t}$ is called the *minimum hitting time*.

The existence of time optimal controls can be obtained under certain convexity conditions. The method is very similar to that presented in Chapter 3. We leave the details as an exercise to the readers. In what follows, we would like to look at some necessary conditions for time optimal controls. Note that the cost functional of Problem (T) can also be written as

(4.8) $$J(u(\cdot)) = \int_0^{\tilde{t}(u(\cdot))} dt,$$

where $\tilde{t}(u(\cdot)) \stackrel{\Delta}{=} J(u(\cdot))$ is the *first hitting time* of the state to the target trajectory $Q(\cdot)$ under control $u(\cdot)$ (see (4.6)). Thus, in the present case, the time duration $[0, \tilde{t}(u(\cdot))]$ on which the problem is considered is not fixed. This is significantly different from the problem in Chapter 4. Hence, it is expected that the necessary conditions of optimal controls will be a little different both in the statement and in the proof. But we should say that the basic idea will be the same as that in Chapter 4.

Now, we let $(\bar{y}(\cdot), \bar{u}(\cdot))$ be an optimal pair where $\bar{t}$ is the minimum hitting time. Set

(4.9) $$\widehat{\mathcal{R}}(\bar{t}) = \{\xi(\bar{t}) \mid \xi(t) = \int_0^t e^{A(t-s)} f_y(s, \bar{y}(s), \bar{u}(s))\xi(s)\, ds$$
$$+ \int_0^t e^{A(t-s)} [f(s, \bar{y}(s), u(s)) - f(s, \bar{y}(s), \bar{u}(s))]\, ds,$$
$$\forall t \in [0, \bar{t}], \quad u(\cdot) \in \mathcal{U}\}.$$

This set is comparable with that defined in (1.32) of Chapter 4. We refer to this set as the reachable set of the variation system along the pair $(\bar{y}(\cdot), \bar{u}(\cdot))$. Our main result in this subsection is the following:

**Theorem 4.1.** (Maximum Principle). *Let (T1)–(T3) hold. Let $(\bar{y}(\cdot), \bar{u}(\cdot))$,*

$\bar t$) be optimal. Suppose that $\widehat{\mathcal{R}}(\bar t) - Q(\bar t)$ is of finite codimension in $X$. Then there exists a $\psi(\cdot) \neq 0$, such that

(4.10) $$\psi(t) = e^{A^*(\bar t - t)}\psi(\bar t) + \int_t^{\bar t} e^{A^*(s-t)} f_y(s, \bar y(s), \bar u(s))^* \psi(s)\, ds,$$
$$t \in [0, \bar t],$$

(4.11) $$\langle \psi(\bar t), z - \bar y(\bar t) \rangle \geq 0, \qquad \forall z \in Q(\bar t),$$

and

(4.12) $$\langle \psi(t), f(t, \bar x(t), \bar u(t)) \rangle = \max_{u \in U} \langle \psi(t), f(t, \bar x(t), u) \rangle,$$
$$\text{a.e. } t \in [0, \bar t].$$

Furthermore, if there exists a $\delta > 0$, such that for some fixed convex and closed set $Q$ in $X$,

(4.13) $$\begin{cases} Q(t) = Q, & \forall t \in [\bar t - \delta, \bar t], \\ \bar y(t) \in \mathcal{D}(A), & \text{a.e. } t \in [\bar t - \delta, \bar t], \quad A\bar y(\cdot) \in L^1(\bar t - \delta, \bar t; X), \\ \bar y(\bar t) \in \mathcal{D}(A), \\ \bar t \text{ is a Lebesgue point of } A\bar y(\cdot) + f(\cdot, \bar y(\cdot), \bar u(\cdot)). \end{cases}$$

then

(4.14) $$\min_{\zeta \in \partial d_Q(\bar y(\bar t))} \langle \zeta, A\bar y(\bar t) + f(\bar t, \bar y(\bar t), \bar u(\bar t)) \rangle \leq 0.$$

In particular, if the function $d_Q(\cdot)$ is $C^1$ in a neighborhood of $\bar y(\bar t)$, then

(4.15) $$\langle \psi(\bar t), A\bar y(\bar t) + f(\bar t, \bar y(\bar t), \bar u(\bar t)) \rangle \geq 0.$$

We have seen that conclusions (4.10)–(4.12) are the same as those in Chapter 4 for the fixed duration problem. However, conclusions (4.14) and (4.15) only appear here. These are an additional conditions that might be useful for determining the minimum hitting time $\bar t$.

*Proof.* Let $T > \bar t$ be a fixed constant and let $\mathcal{U}[0, T] = \{u|_{[0,T]} \mid u \in \mathcal{U}\}$. We define the distance $\bar d$ on $\mathcal{U}[0, T]$ as in Chapter 4, §3. Then $(\mathcal{U}[0, T], \bar d)$ is a complete metric space. Next, we define the following penalty functional:

(4.16) $$J_\varepsilon(u(\cdot)) = d_{Q(\bar t - \varepsilon)}\bigl(y(\bar t - \varepsilon; u(\cdot))\bigr)$$
$$\equiv \inf_{z \in Q(\bar t - \varepsilon)} |y(\bar t - \varepsilon; u(\cdot)) - z|, \qquad \forall u(\cdot) \in \mathcal{U}.$$

This penalty functional is designed for our Problem (T), and it is different from that given in Chapter 4. It is clear that $J_\varepsilon(\cdot)$ is continuous on

§4. Time Optimal Control—Semilinear Systems

$(\mathcal{U}[0,T], \bar{d})$ and positive (because $\bar{t}$ is the minimum hitting time). Also, note that $\bar{y}(\cdot) = y(\cdot\,; \bar{u}(\cdot))$, and by the continuity of $Q(\cdot)$, we have

(4.17) $\quad J_\varepsilon(\bar{u}(\cdot)) = d_{Q(\bar{t}-\varepsilon)}(\bar{y}(\bar{t}-\varepsilon)) \equiv \sigma(\varepsilon) \to 0, \qquad \varepsilon \to 0.$

Thus, by Ekeland's variational principle (Chapter 4, Lemma 2.1 and Corollary 2.2), one can find a $u^\varepsilon(\cdot) \in \mathcal{U}[0,T]$, such that

(4.18) $\begin{cases} \bar{d}(u^\varepsilon(\cdot), \bar{u}(\cdot)) \leq \sqrt{\sigma(\varepsilon)}, \\ -\sqrt{\sigma(\varepsilon)}\bar{d}(\hat{u}(\cdot), u^\varepsilon(\cdot)) \leq J_\varepsilon(\hat{u}(\cdot)) - J_\varepsilon(u^\varepsilon(\cdot)), \quad \forall \hat{u}(\cdot) \in \mathcal{U}[0,T]. \end{cases}$

Now, we let $u(\cdot) \in \mathcal{U}[0,T]$ be fixed. Then, as in Chapter 4, §4, for any $\rho \in (0,1]$, there exists a measurable set $E_\rho \subset [0,T]$, such that $|E_\rho| = \rho T$ and if one defines

(4.19) $\quad u_\rho^\varepsilon(t) = \begin{cases} u^\varepsilon(t), & t \in [0,T] \setminus E_\rho, \\ u(t), & t \in E_\rho, \end{cases}$

then $u_\rho^\varepsilon(\cdot) \in \mathcal{U}[0,T]$ and the trajectory $y_\rho^\varepsilon(\cdot)$ of (1.1) corresponding to $(x, u_\rho^\varepsilon(\cdot))$ satisfies

(4.20) $\quad y_\rho^\varepsilon(t) = y^\varepsilon(t) + \rho \xi_\varepsilon(t) + o(\rho), \qquad \text{uniformly in } t \in [0,T],$

where $\xi_\varepsilon(\cdot)$ satisfies the following variation system:

(4.21) $\begin{aligned} \xi_\varepsilon(t) = & \int_0^t e^{A(t-s)} f_y(s, y^\varepsilon(s), u^\varepsilon(s)) \xi_\varepsilon(s)\, ds \\ & + \int_0^t e^{A(t-s)} [f(s, y^\varepsilon(s), u(s)) - f(s, y^\varepsilon(s), u^\varepsilon(s))]\, ds, \\ & \hspace{6cm} \forall t \in [0,T]. \end{aligned}$

On the other hand, by the optimality of $\bar{t}$, we have

(4.22) $\quad y^\varepsilon(\bar{t}-\varepsilon) \equiv y(\bar{t}-\varepsilon; x, u^\varepsilon) \notin Q(\bar{t}-\varepsilon).$

Thus, by Chapter 4, Proposition 3.11 and the strict convexity of $X^*$,

(4.23) $\quad \partial d_{Q(\bar{t}-\varepsilon)}(y^\varepsilon(\bar{t}-\varepsilon)) \equiv \{\psi^\varepsilon\}$

is a singleton with

(4.24) $\quad |\psi^\varepsilon|_{X^*} = 1.$

Then, by (4.18) and (4.16), we obtain

(4.25) $\begin{aligned} -\sqrt{\sigma(\varepsilon)} \leq & \frac{1}{\rho}\{d_{Q(\bar{t}-\varepsilon)}(y_\rho^\varepsilon(\bar{t}-\varepsilon)) - d_{Q(\bar{t}-\varepsilon)}(y^\varepsilon(\bar{t}-\varepsilon))\} \\ & \to \langle \psi^\varepsilon, \xi_\varepsilon(\bar{t}-\varepsilon) \rangle. \end{aligned}$

Because

(4.26) $\quad \xi_\varepsilon(t) \to \xi(t) \quad$ as $\varepsilon \to 0,\quad$ uniformly in $t \in [0,T]$,

with

(4.27)
$$\begin{aligned}\xi(t) &= \int_0^t e^{A(t-s)} f_y(s,\bar{y}(s),\bar{u}(s))\xi(s)\, ds \\ &+ \int_0^t e^{A(t-s)}[f(s,\bar{y}(s),u(s)) - f(s,\bar{y}(s),\bar{u}(s))]\, ds,\end{aligned}$$
$$\forall t \in [0,T],$$

combining (4.25) with the above, we have

(4.28) $\quad \langle \psi^\varepsilon, \xi(\bar{t}) \rangle \geq -|\xi(\bar{t}) - \xi_\varepsilon(\bar{t}-\varepsilon)| - \sqrt{\sigma(\varepsilon)}.$

On the other hand, because $Q(\bar{t}-\varepsilon)$ is convex and closed, $\partial d_{Q(\bar{t}-\varepsilon)}(\cdot)$ coincides with the subdifferential of the convex function $d_{Q(\bar{t}-\varepsilon)}(\cdot)$. Thus,

(4.29) $\quad \langle \psi^\varepsilon, z - y^\varepsilon(\bar{t}-\varepsilon) \rangle \leq d_{Q(\bar{t}-\varepsilon)}(z) - d_{Q(\bar{t}-\varepsilon)}(y^\varepsilon(\bar{t}-\varepsilon)), \quad \forall z \in X.$

Consequently, for all $z \in Q(\bar{t})$, we have

(4.30)
$$\begin{aligned}\langle \psi^\varepsilon, z - \bar{y}(\bar{t}) \rangle &= \langle \psi^\varepsilon, y^\varepsilon(\bar{t}-\varepsilon) - \bar{y}(\bar{t}) \rangle + \langle \psi^\varepsilon, z - y^\varepsilon(\bar{t}-\varepsilon) \rangle \\ &\leq |y^\varepsilon(\bar{t}-\varepsilon) - \bar{y}(\bar{t})| + d_{Q(\bar{t}-\varepsilon)}(z) \\ &\leq |y^\varepsilon(\bar{t}-\varepsilon) - \bar{y}(\bar{t})| + \rho_H(Q(\bar{t}-\varepsilon), Q(\bar{t})).\end{aligned}$$

From (4.28) and (4.30), we obtain

(4.31) $\quad \langle \psi^\varepsilon, \xi - (z - \bar{y}(\bar{t})) \rangle \geq -\delta_\varepsilon, \quad \forall \xi \in \widehat{\mathcal{R}}(\bar{t}),\ z \in Q(\bar{t}),$

where $\delta_\varepsilon \to 0$ as $\varepsilon \to 0$. Because $\widehat{\mathcal{R}}(\bar{t}) - Q(\bar{t})$ is finite codimensional in $X$, as in Chapter 4, we can find a subsequence of $\{\psi^\varepsilon\}$ (still denoted by itself), such that

(4.32) $\quad \psi^\varepsilon \xrightarrow{*} \bar{\psi} \neq 0, \quad \varepsilon \to 0.$

Then, from the above, one has

(4.33) $\quad \langle \bar{\psi}, \xi(\bar{t}) \rangle \geq 0,$

(4.34) $\quad \langle \bar{\psi}, z - \bar{y}(\bar{t}) \rangle \leq 0, \quad \forall z \in Q(\bar{t}).$

Now, we let

(4.35) $\quad \psi(t) = -e^{A^*(\bar{t}-t)}\bar{\psi} + \int_t^{\bar{t}} e^{A^*(s-t)} f_y(s,\bar{y}(s),\bar{u}(s))^*\psi(s)\, ds,$
$$t \in [0,\bar{t}].$$

## §4. Time Optimal Control—Semilinear Systems

Similar to Chapter 4, §4, one can easily derive that

$$0 \geq \langle \psi(\bar{t}), \xi(\bar{t}) \rangle = \int_0^{\bar{t}} \langle \psi(s), f(s, \bar{y}(s), u(s)) - f(s, \bar{y}(s), \bar{u}(s)) \rangle \, ds,$$
(4.36)
$$\forall u(\cdot) \in \mathcal{U}[0, T].$$

Then, (4.12) follows and (4.11) follows from (4.34) (note that $\bar{\psi} = -\psi(\bar{t})$). By (4.32), we see the costate $\psi(\cdot)$ is nonzero. Finally, we let (4.13) hold. Then for any $\varepsilon \in (0, \delta)$,

(4.37)
$$\bar{y}(\bar{t} - \varepsilon) = \bar{y}(\bar{t}) - \int_{\bar{t}-\varepsilon}^{\bar{t}} \{A\bar{y}(s) + f(s, \bar{y}(s), \bar{u}(s))\} \, ds.$$

Thus,

$$0 \leq \overline{\lim_{\varepsilon \to 0}} \frac{d_Q(\bar{y}(\bar{t} - \varepsilon))}{\varepsilon}$$

(4.38)
$$= \overline{\lim_{\varepsilon \to 0}} \frac{d_Q\left(\bar{y}(\bar{t}) - \int_{\bar{t}-\varepsilon}^{\bar{t}} \{A\bar{y}(s) + f(s, \bar{y}(s), \bar{u}(s))\} \, ds\right) - d_Q(\bar{y}(\bar{t}))}{\varepsilon}$$
$$\leq \max_{\zeta \in \partial d_Q(\bar{y}(\bar{t}))} \langle \zeta, -A\bar{y}(\bar{t}) - f(\bar{t}, \bar{y}(\bar{t}), \bar{u}(\bar{t})) \rangle.$$

This proves (4.14). In the case where $d_Q(z)$ is $C^1$ near $\bar{y}(\bar{t})$, we have that

(4.39)
$$\partial d_Q(\bar{y}(\bar{t})) = \{\lambda \nabla d_Q(\bar{y}(\bar{t})) \mid \lambda \in [0, 1]\}.$$

And (4.38) is equivalent to the following:

(4.40)
$$\langle \nabla d_Q(\bar{y}(\bar{t})), A\bar{y}(\bar{t}) + f(\bar{t}, \bar{y}(\bar{t}), \bar{u}(\bar{t})) \rangle \leq 0.$$

On the other hand, from (4.23) and the weak* upper semicontinuity of the multifunction $\partial d_Q(z)$ (see Chapter 3, §3), we see that

(4.41)
$$\bar{\psi} \in \partial d_Q(\bar{y}(\bar{t})).$$

This implies that for some $\lambda \in [0, 1]$, $\bar{\psi} = \lambda \nabla d_Q(\bar{y}(\bar{t}))$. From (4.32), $\lambda \neq 0$. Hence, (4.15) follows from (4.40) and the fact that $\psi(\bar{t}) = -\bar{\psi}$. □

### §4.2. The minimum time function

In this subsection, we briefly look at the time optimal control problem from the dynamic programming point of view. For convenience, we consider the time invariant system

(4.42)
$$y_x(t) = e^{At}x + \int_0^t e^{A(t-s)} f(y_x(s), u(s)) \, ds, \qquad t \geq 0.$$

Let us make the following assumptions (compare with (T1)–(T3)).

(T1)′ $X$ is a Banach space, $U$ is a metric space, and $e^{At}$ is a $C_0$ semigroup on $X$.

(T2)′ $f : X \times U \to X$ is continuous and there exists a constant $L > 0$, such that

(4.43) $$\begin{cases} |f(x,u) - f(y,u)| \leq L|x-y|, & \forall x, y \in X, \ u \in U, \\ |f(0,u)| \leq L, & \forall u \in U. \end{cases}$$

(T3)′ $Q \subset X$ is nonempty and closed.

(T4)′ System (4.42) is for a small time locally exactly controllable with the target $Q$.

Clearly, under (T1)′ and (T2)′, for any $(x, u(\cdot)) \in X \times \mathcal{U}$, (4.42) admits a unique solution $y_x(\cdot; u)$. Next, we define (compare with (4.4))

(4.44) $$\mathcal{T}(x, u(\cdot)) \overset{\Delta}{=} \{t \geq 0 \mid y_x(t; u) \in Q\}, \qquad \forall (x, u(\cdot)) \in X \times \mathcal{U}.$$

It is easy to see that under (T3)′ and (T4)′, we have

(4.45) $$\{x \in X \mid \bigcup_{u(\cdot) \in \mathcal{U}} \mathcal{T}(x, u(\cdot)) \neq \phi\} \supset \mathcal{O} \supset Q,$$

for some open set $\mathcal{O}$. Now, we define (inf $\phi = +\infty$)

(4.46) $$T(x) = \inf_{u(\cdot) \in \mathcal{U}} \mathcal{T}(x, u(\cdot)).$$

We call $T : X \to [0, \infty]$ the *minimum time function*. It is also called the value function of our time optimal control problem. We should note that $T(x)$ could take the value $+\infty$ for some $x \in X$. We define

(4.47) $$\mathcal{C}(Q) = \{x \in X \mid T(x) < \infty\}.$$

Clearly, for any $x \in \mathcal{C}(Q)$, there exists a control $u(\cdot) \in \mathcal{U}$, such that $y_x(t; u) \in Q$ for some $t \geq 0$. Thus, we call $\mathcal{C}(Q)$ the *exact controllable set* to $Q$. Our first result is the following

**Proposition 4.2.** *Let (T1)′–(T4)′ hold. Then the set $\mathcal{C}(Q)$ is open, the minimum time function $T(\cdot)$ is continuous on $\mathcal{C}(Q)$, and*

(4.48) $$\begin{cases} T(x) = 0, & \forall x \in Q, \\ \lim_{z \in \mathcal{C}(Q), z \to x} T(z) = +\infty, & \forall x \in \partial \mathcal{C}(Q). \end{cases}$$

*Proof.* The first equality in (4.48) is clear. Now, for any $x_0 \in \mathcal{C}(Q)$, there exists a control $u_0(\cdot) \in \mathcal{U}$, such that for some $t_0 \geq T(x_0)$, we have

(4.49) $$y_{x_0}(t_0; u_0) \in Q.$$

By (T4)′, for any $\varepsilon > 0$, there exists a $\delta > 0$, such that

(4.50) $$T(z) < \varepsilon, \qquad \forall d(z, Q) < \delta.$$

## §4. Time Optimal Control—Semilinear Systems

On the other hand, by Gronwall's inequality,

$$(4.51) \qquad |y_x(t_0; u_0) - y_{x_0}(t_0; u_0)| \leq e^{Lt_0}|x - x_0|, \qquad \forall x \in X.$$

Hence, provided $|x - x_0| < \delta e^{-Lt_0}$, we have (note (4.49))

$$(4.52) \qquad d(y_x(t_0; u_0), Q) < \delta.$$

Thus, for such an $x$, we have

$$(4.53) \qquad T(x) \leq t_0 + T(y_x(t_0; u_0)) < t_0 + \varepsilon.$$

This proves the openness of $\mathcal{C}(Q)$. Further, $t_0$ in (4.53) can be chosen arbitrarily close to $T(x_0)$. Thus, (4.53) implies that

$$(4.54) \qquad T(x) \leq T(x_0) + \varepsilon, \qquad \forall |x - x_0| < \delta e^{-LT(x_0)},$$

which gives the upper semicontinuity of the function $T(\cdot)$. Next, we prove the lower semicontinuity. For given $x_0 \in \mathcal{C}(Q)$, by (4.54), we see that $T(x)$ is bounded by, say, $T(x_0) + 1$ on the open set $\mathcal{O}_\eta(x_0)$ (for some $\eta > 0$). For any $\varepsilon > 0$, we let $\delta > 0$ satisfy (4.50). Also, there exists a $u_x(\cdot) \in \mathcal{U}$ and a $t_x \geq 0$, such that

$$(4.55) \qquad \begin{cases} t_x \in [T(x), T(x) + \varepsilon) \subset [0, T(x_0) + 1], \\ y_x(t_x; u_x) \in Q. \end{cases}$$

Similar to (4.51), we have

$$(4.56) \quad |y_x(t_x; u_x) - y_{x_0}(t_x; u_x)| \leq e^{L(T(x_0)+1)}C|x - x_0|, \qquad \forall x \in \mathcal{O}_\eta(x_0).$$

Hence,

$$(4.57) \qquad d(y_{x_0}(t_x; u_x), Q) < \delta, \qquad \forall |x - x_0| < \delta e^{-L(T(x_0)+1)}.$$

Consequently, by (4.50),

$$(4.58) \qquad T(x_0) \leq t_x + T(y_{x_0}(t_x; u_x)) < T(x) + 2\varepsilon.$$

Thus, $T(\cdot)$ is continuous on $\mathcal{C}(Q)$. Finally, let $x \in \partial \mathcal{C}(Q)$ and $z \in \mathcal{C}(Q)$ and $z \to x$. If $T(z)$ is uniformly bounded, similar to (4.54), we will have $x \in \mathcal{C}(Q)$, which is not possible. $\square$

Similar to Chapter 6, we can show the following.

**Proposition 4.3.** (Principle of Optimality) *Let (T1)'–(T4)' hold. Then, for any $x \notin Q$, there exists a $t_0 > 0$, such that*

$$(4.59) \qquad T(x) = \inf_{u(\cdot) \in \mathcal{U}} \{t + T(y_x(t; u(\cdot)))\}, \qquad \forall t \in [0, t_0].$$

Consequently, we have the following HJB equation for the function $T(\cdot)$.

**Proposition 4.4.** *Suppose that $T(\cdot)$ is $C^1(\mathcal{C}(Q) \setminus Q)$. Then it holds that*

$$(4.60) \quad 1 + \inf_{u \in U} \langle DT(x), Ax + f(x, u) \rangle = 0, \qquad \forall x \in \mathcal{D}(A) \cap (\mathcal{C}(Q) \setminus Q),$$

*with the boundary condition (4.48).*

We see that problem (4.60) with boundary condition (4.48) is complicated because the boundary $\partial \mathcal{C}(Q)$ is not known *a priori*. Sometimes such a problem is referred to as a *free boundary problem*. For the present situation, we may introduce the following function

$$(4.61) \quad V(x) = \begin{cases} 1 - e^{-T(x)}, & x \in \mathcal{C}(Q), \\ 1, & x \in X \setminus \mathcal{C}(Q). \end{cases}$$

Then, under our condition, $V(\cdot) \in C(X)$, and formally, it is a solution of the following HJB equation:

$$(4.62) \quad \begin{cases} V(x) - \inf_{u \in U} \langle DV(x), Ax + f(x, u) \rangle = 0, & x \in \mathcal{D}(A) \setminus Q, \\ V\big|_Q = 0. \end{cases}$$

The advantage of (4.62) is that the boundary condition on $\partial \mathcal{C}(Q)$ disappears. Also, this is a form similar to one that we have treated in Chapter 6, §6. Hence, under proper conditions, we are able to prove the uniqueness of viscosity solutions to (4.62). This will give a characterization of the minimum time function $T(\cdot)$. We leave the details to the readers.

## §5. Time Optimal Control—Linear Systems

In the previous section, we have discussed time optimal control problems for semilinear systems. In this section, we are going to restrict ourselves to a system that is linear in the state variable. For such systems, we will have some stronger results than those presented in the previous section.

The system we are going to study here is the following:

$$(5.1) \quad \begin{cases} \dot{y}(t) = Ay(t) + b(t, u(t)), & t \geq 0, \\ y(0) = x, \end{cases}$$

where $A$ is as before and $b : [0, \infty) \times U \to X$ is a map satisfying the following conditions:

(B) $b(t, u)$ is continuous, and there exists a $\mu(\cdot) \in L^1_{loc}(0, \infty)$, such that

$$(5.2) \quad |b(t, u)| \leq \mu(t), \qquad \forall (t, u) \in [0, \infty) \times U.$$

It is by now standard that for any $(x, u(\cdot)) \in X \times \mathcal{U}$, there exists a unique mild solution $y(\cdot) \in C([0, \infty); X)$ of (5.1), which is given by

$$(5.3) \quad y(t) = e^{At}x + \int_0^t e^{A(t-s)} b(s, u(s))\, ds, \qquad \forall t \geq 0.$$

## §5. Time Optimal Control—Linear Systems

Before studying the corresponding time optimal control problem, let us first look at some properties of the reachable sets.

### §5.1. Convexity of the reachable set

In the present case, our reachable set $\mathcal{R}(t,x)$ has an explicit form:

$$(5.4) \qquad \mathcal{R}(t,x) = \{e^{At}x + \int_0^t e^{A(t-s)} b(s, u(s))\, ds \mid u(\cdot) \in \mathcal{U}\}.$$

It is clear that if $U$ is a convex set in some Banach space $Z$ and $b(t,u) \equiv Bu$ with $B \in \mathcal{L}(Z, X)$, then the set $\mathcal{R}(t,x)$ is convex. It is natural to ask that in the above general case, do we have the convexity of the reachable set $\mathcal{R}(t,x)$? The answer is negative, in general. Let us present a counterexample. To this end, we first give the following result.

**Lemma 5.1.** *The space $L^2(0,1)$ admits an orthonormal basis $\{w_n(\cdot)\}_{n\geq 0}$ satisfying the following conditions:*

$$(5.5) \qquad w_0(t) \equiv 1, \qquad w_n(t) = \pm 1, \quad \forall t \in [0,1],\ n \geq 1.$$

The orthonormal basis $\{w_n(\cdot)\}_{n\geq 0}$ is called the *Walsh system*. For the readers' conveniece, we present a proof below.

*Proof.* Define

$$(5.6) \qquad r_k(t) = \operatorname{sgn}\bigl(\sin(2^k \pi t)\bigr), \qquad k \geq 0.$$

The above $r_k(\cdot)$'s are called the *Rademacher functions*. Clearly, we have the following simple properties for $r_k(\cdot)$'s:

$$(5.7) \qquad \begin{cases} r_k(t)^2 = 1,\ r_{k-1}(t) = r_k(t/2), & \text{a.e. } t \in \mathbb{R}, \\ r_k(t+m) = r_k(t), & \forall t \in \mathbb{R},\ m \text{ is an integer}, \end{cases} \quad k \geq 1.$$

Also, for any $k_1 > k_2 > \cdots > k_\lambda \geq 0$, we have

$$(5.8) \quad \begin{aligned} & \int_0^1 r_{k_1}(t) r_{k_2}(t) \cdots r_{k_\lambda}(t)\, dt \\ &= \frac{1}{2^{k_\lambda}} \int_0^{2^{k_\lambda}} r_{k_1 - k_\lambda}(t) \cdots r_{k_{\lambda-1} - k_\lambda}(t) \operatorname{sgn}\bigl(\sin(\pi t)\bigr)\, dt \\ &= \frac{1}{2^{k_\lambda}} \sum_{j=1}^{2^{k_\lambda - 1}} \Bigl\{ \int_{2j-2}^{2j-1} r_{k_1 - k_\lambda}(t) \cdots r_{k_{\lambda-1} - k_\lambda}(t)\, dt \\ &\qquad\qquad - \int_{2j-1}^{2j} r_{k_1 - k_\lambda}(t) \cdots r_{k_{\lambda-1} - k_\lambda}(t)\, dt \Bigr\} = 0. \end{aligned}$$

In the last equality, we have used the fact that $r_k(t+m) = r_k(t)$.

Next, for any $n \geq 1$, with the decomposition (which exists uniquely)

(5.9) $$n = 2^{n_1} + 2^{n_2} + \cdots + 2^{n_\nu}, \qquad n_1 > n_2 > \cdots > n_\nu \geq 0,$$

we define

(5.10) $$w_n(t) = r_{n_1+1}(t) r_{n_2+1}(t) \cdots r_{n_\nu+1}(t), \qquad t \in \mathbb{R},$$

and set

(5.11) $$w_0(t) = \operatorname{sgn}\bigl(\sin(\pi t)\bigr), \qquad t \in \mathbb{R}.$$

We will show that the above constructed $\{w_n(\cdot)\}_{n\geq 0}$ is the Walsh system. To this end, let us first show the orthogonality of this system. Obviously,

(5.12) $$\int_0^1 w_n(t)^2 \, dt = 1, \qquad \forall n \geq 0.$$

Now, let $m \neq n$. We write $n$ as (5.9) and

(5.13) $$m = 2^{m_1} + 2^{m_2} + \cdots + 2^{m_\mu}, \qquad m_1 > m_2 > \cdots > m_\mu \geq 0.$$

Because $m \neq n$, the sequences $(m_1, \cdots, m_\mu)$ and $(n_1, \cdots, n_\nu)$ are different. Thus, noting that $r_k(t)^2 = 1$, a.e. $t \in \mathbb{R}$ and (5.8), we have some $k_1 > k_2 > \cdots > k_\lambda \geq 0$, such that

(5.14) $$\begin{aligned}
&\int_0^1 w_m(t) w_n(t) \, dt \\
&= \int_0^1 r_{m_1+1}(t) \cdots r_{m_\mu+1}(t) r_{n_1+1}(t) \cdots r_{n_\nu+1}(t) \, dt \\
&= \int_0^1 r_{k_1}(t) \cdots r_{k_\lambda}(t) \, dt = 0.
\end{aligned}$$

This proves the orthogonality of the system $\{w_n(\cdot)\}_{n\geq 0}$. To prove the completeness of this system, we first claim the following:

(5.15) $$\begin{cases} \forall h(\cdot) \in L^2(0,1), \quad \displaystyle\int_0^1 h(t) w_j(t) \, dt = 0, \qquad 0 \leq j \leq 2^k - 1, \\[4pt] \text{implies} \quad \displaystyle\int_0^{j 2^{-k}} h(t) \, dt = 0, \qquad 0 \leq j \leq 2^k - 1. \end{cases}$$

Let us prove it by induction. For $k = 0$, the claim is trivial and for $k = 1$, we have

(5.16) $$\begin{cases} 0 = \displaystyle\int_0^1 h(t) w_0(t) \, dt = \int_0^{1/2} h(t) \, dt + \int_{1/2}^1 h(t) \, dt, \\[4pt] 0 = \displaystyle\int_0^1 h(t) w_1(t) \, dt = \int_0^1 h(t) \operatorname{sgn}\bigl(\sin(2\pi t)\bigr) \, dt \\[4pt] \phantom{0} = \displaystyle\int_0^{1/2} h(t) \, dt - \int_{1/2}^1 h(t) \, dt. \end{cases}$$

## §5. Time Optimal Control—Linear Systems

Thus, it follows that $\int_0^{1/2} h(t)\, dt = 0$. This means that (5.15) holds for $k = 1$. Suppose the claim holds for $k$ and we now prove it for $k + 1$. By the definition of $w_n(t)$ and (5.7), it is easy to see that

(5.17) $\quad \begin{cases} w_{2k}(t/2) = w_k(t), \\ w_{2k+1}(t/2) = w_k(t)\operatorname{sgn}\bigl(\sin(\pi t)\bigr), \end{cases} \quad \forall t \in \mathbb{R},\ k \geq 1.$

Now, if for all $0 \leq j \leq 2^k - 1$, we have (note (5.17))

(5.18) $\quad \begin{cases} 0 = \int_0^1 h(t) w_{2j}(t)\, dt = \dfrac{1}{2} \int_0^2 h(\tfrac{t}{2}) w_j(t)\, dt, \\[4pt] 0 = \int_0^1 h(t) w_{2j+1}(t)\, dt \\[4pt] \quad = \dfrac{1}{2} \int_0^2 h(\tfrac{t}{2}) w_j(t) \operatorname{sgn}\bigl(\sin(\pi t)\bigr)\, dt \\[4pt] \quad = \dfrac{1}{2} \int_0^1 h(\tfrac{t}{2}) w_j(t)\, dt - \dfrac{1}{2} \int_1^2 h(\tfrac{t}{2}) w_j(t)\, dt, \end{cases}$

then

(5.19) $\quad \begin{cases} \int_0^1 h(\tfrac{t}{2}) w_j(t)\, dt = 0, \\[4pt] \int_0^1 h(\tfrac{t+1}{2}) w_j(t)\, dt = 0, \end{cases} \quad \forall 0 \leq j \leq 2^k - 1.$

Thus, by the induction hypothesis, we must have
(5.20)
$\begin{cases} 0 = \int_0^{j 2^{-k}} h(\tfrac{t}{2})\, dt = 2 \int_0^{j 2^{-k-1}} h(t)\, dt, \\[4pt] 0 = \int_0^{j 2^{-k}} h(\tfrac{t+1}{2})\, dt = 2 \int_{\tfrac{1}{2}}^{\tfrac{1}{2}+j 2^{-k-1}} h(t)\, dt, \end{cases} \quad \forall 0 \leq j \leq 2^k - 1.$

This implies (5.15) for $k + 1$. Now, if $h(\cdot) \in L^2(0, 1)$ is such that

(5.21) $\quad \displaystyle\int_0^1 h(t) w_n(t)\, dt = 0, \qquad \forall n \geq 0,$

then, by (5.15), we have

(5.22) $\quad \displaystyle\int_0^{j 2^{-k}} h(t)\, dt = 0, \qquad \forall 0 \leq j \leq 2^k - 1,\ k \geq 1.$

Consequently, $\int_0^t h(t)\, dt = 0$ for all $t \in [0, 1]$. Thus, $h(t) = 0$ for almost all $t \in [0, 1]$, proving the completeness of the system $\{w_n(\cdot)\}_{n \geq 0}$. $\square$

Now, we are ready to present the following example.

*Example 5.2.* Let $X = \ell^2$ and $\{w_n(\cdot)\}_{n\geq 0}$ be an orthonormal basis for $L^2(0,1)$ satisfying (5.5). We define

(5.23) $$b(t) = \left(\frac{1+w_0(t)}{2}, \frac{1+w_1(t)}{2^2}, \ldots, \frac{1+w_n(t)}{2^{n+1}}, \ldots\right).$$

Clearly,

(5.24) $$|b(t)|_{\ell^2}^2 = \sum_{n\geq 0}\left(\frac{1+w_n(t)}{2^{n+1}}\right)^2 \leq \sum_{n\geq 0}\frac{1}{4^n} = \frac{4}{3}.$$

This means that $b(\cdot) \in L^\infty(0,1;X)$. Now, we consider the following control system

(5.25) $$\dot{y}(t) = b(t)u(t), \qquad y(0) = x,$$

with the control $u(\cdot)$ taking values in the set $U = \{0,1\}$. Then, for any $x \in X$ and $u(\cdot) \in \mathcal{U}$, we have

(5.26) $$y(1;x,u) = x + \int_0^1 b(t)u(t)\,dt = x + \int_{\{u(t)=1\}} b(t)\,dt.$$

Consequently,

(5.27) $$\mathcal{R}(1,0) = \{\int_E b(t)\,dt \mid E \subset (0,1), \text{ measurable}\}.$$

We claim that the set $\mathcal{R}(1,0)$ is not convex. In fact, if it were convex, then, noting that $0, \int_0^1 b(t)\,dt \in \mathcal{R}(1,0)$, we would have some $E \subset (0,1)$, measurable, such that

(5.28) $$\frac{1}{2}\int_0^1 b(t)\,dt = \int_E b(t)\,dt.$$

Because $w_n$'s are mutually orthogonal and $w_0(t) \equiv 1$, for $t \in [0,1]$, we have

(5.29) $$\int_0^1 w_n(t)\,dt = 0, \qquad n \geq 1.$$

Hence, (5.28) implies

(5.30) $$\left(1, \frac{1}{2^2}, \frac{1}{2^3}, \ldots\right) = \int_0^1 b(t)\,dt = 2\int_E b(t)\,dt.$$

Comparing the first component on both sides, we have

(5.31) $$|E| = \frac{1}{2}.$$

Next, comparing the $(n+1)$-th components on both sides ($n \geq 1$), we have

(5.32)
$$\frac{1}{2^{n+1}} = 2 \int_E \frac{1 + w_n(t)}{2^{n+1}} \, dt$$
$$= 2 \int_{E \cap \{w_n = 1\}} \frac{1}{2^n} \, dt = \frac{1}{2^{n-1}} |E \cap \{w_n = 1\}|.$$

This yields

(5.33)
$$|E \cap \{w_n = 1\}| = \frac{1}{4}, \qquad n \geq 1.$$

On the other hand, by (5.29), we have

(5.34)
$$|\{w_n = 1\}| = |\{w_n = -1\}| = \frac{1}{2}, \qquad n \geq 1.$$

Then, combining (5.31), (5.33), and (5.34), we obtain that for all $n \geq 1$,

(5.35)
$$\begin{cases} |E \cap \{w_n = -1\}| = |E| - |E \cap \{w_n = 1\}| = \frac{1}{4}, \\ |\{w_n = 1\} \setminus E| = |\{w_n = 1\}| - |E \cap \{w_n = 1\}| = \frac{1}{4}, \\ |\{w_n = -1\} \setminus E| = |\{w_n = -1\}| - |E \cap \{w_n = -1\}| = \frac{1}{4}. \end{cases}$$

Set $h = \chi_E - \chi_{(0,1) \setminus E} \in L^2(0,1)$. By (5.31) and (5.35),

(5.36)
$$\int_0^1 h(t) w_0(t) \, dt = |E| - |(0,1) \setminus E| = \frac{1}{2} - \frac{1}{2} = 0,$$
$$\int_0^1 h(t) w_n(t) \, dt = |E \cap \{w_n = 1\}| + |E \cap \{w_n = -1\}|$$
$$\qquad - |\{w_n = 1\} \setminus E| - |\{w_n = -1\} \setminus E| = 0, \qquad n \geq 1.$$

Then, by the completeness of $\{w_n\}_{n \geq 0}$, we must have $h(\cdot) = 0$. This is a contradiction. Therefore, $\mathcal{R}(1,0)$ is not convex. □

We do, however, have the following result, which is good enough in many applications.

**Theorem 5.3.** *The closure $\overline{\mathcal{R}(t,x)}$ of the reachable set $\mathcal{R}(t,x)$ is convex.*

*Proof.* Let $y_1, y_2 \in \overline{\mathcal{R}(t,x)}$ and $\lambda \in (0,1)$. For any $\varepsilon > 0$, there exist $u_1(\cdot), u_2(\cdot) \in \mathcal{U}$, such that

(5.37)
$$\left| y_i - \int_0^t e^{A(t-s)} b(s, u_i(s)) \, ds \right| < \frac{\varepsilon}{2}, \qquad i = 1, 2.$$

By applying Chapter 4, Corollary 3.9 to the function $e^{A(t-s)} \{b(s, u_2(s)) - b(s, u_1(s))\}$, we can find a measurable set $E_\lambda \subset [0,t]$ with $|E_\lambda| = \lambda t$, such that

$$\left|\lambda\int_0^t e^{A(t-s)}\{b(s,u_2(s))-b(s,u_1(s))\}\,ds\right.$$
(5.38)
$$\left.-\int_{E_\lambda}e^{A(t-s)}\{b(s,u_2(s))-b(s,u_1(s))\}\,ds\right|<\frac{\varepsilon}{2}.$$

Now, define

(5.39)
$$\widetilde{u}(s)=\begin{cases}u_1(s),&s\in E_\lambda,\\ u_2(s),&s\in[0,t]\setminus E_\lambda.\end{cases}$$

Clearly, $\widetilde{u}(\cdot)\in\mathcal{U}$ and

$$\left|\int_0^t e^{A(t-s)}b(s,\widetilde{u}(s))\,ds-\lambda y_1-(1-\lambda)y_2\right|$$

$$=\left|-\int_{E_\lambda}e^{A(t-s)}\{b(s,u_2(s))-b(s,u_1(s))\}\,ds\right.$$

(5.40)
$$\left.+\int_0^t e^{A(t-s)}b(s,u_2(s))\,ds-\lambda y_1-(1-\lambda)y_2\right|$$

$$\leq\left|\lambda\int_0^t e^{A(t-s)}b(s,u_1(s))\,ds-\lambda y_1\right|$$

$$+\left|(1-\lambda)\int_0^t e^{A(t-s)}b(s,u_2(s))\,ds-(1-\lambda)y_2\right|+\frac{\varepsilon}{2}$$

$$<\lambda\frac{\varepsilon}{2}+(1-\lambda)\frac{\varepsilon}{2}+\frac{\varepsilon}{2}=\varepsilon.$$

Hence, $\lambda y_1+(1-\lambda)y_2\in\overline{\mathcal{R}(t,x)}$. This proves our conclusion. □

## §5.2. Encounter of moving sets

In this section, we consider the encounter of two moving sets. This will be very useful for our time optimal control problem.

Let $\Omega_1,\Omega_2:[0,\infty)\to X$ be two multifunctions. Suggestively, we call $\Omega_1(\cdot)$ and $\Omega_2(\cdot)$ *moving sets*. Suppose that

(5.41)
$$\overline{\Omega_1(0)}\bigcap\overline{\Omega_2(0)}=\phi,$$

and for some $\widetilde{t}>0$,

(5.42)
$$d(\Omega_1(\widetilde{t}),\Omega_2(\widetilde{t}))\equiv\inf_{x_1\in\Omega_1(\widetilde{t}),x_2\in\Omega_2(\widetilde{t})}|x_1-x_2|=0.$$

Clearly, the above is implied by the following:

(5.43)
$$\overline{\Omega_1(\widetilde{t})}\bigcap\overline{\Omega_2(\widetilde{t})}\neq\phi.$$

## §5. Time Optimal Control—Linear Systems

However, in general, (5.42) does not necessarily imply (5.43). We will explain this point soon. Any $\tilde{t} > 0$ satisfying (5.43) is called an *approximate encounter time*. If $\bar{t} \in (0, \tilde{t}]$ is such that

(5.44)
$$\begin{cases} \overline{\Omega_1(t)} \cap \overline{\Omega_2(t)} = \phi, & 0 \le t < \bar{t}, \\ \overline{\Omega_1(\bar{t})} \cap \overline{\Omega_2(\bar{t})} \ne \phi, \end{cases}$$

then we call $\bar{t}$ the *first approximate encounter time* of the moving sets $\Omega_1(\cdot)$ and $\Omega_2(\cdot)$.

If we drop the bars over $\Omega_1(t)$ and $\Omega_2(t)$ in (5.43) and (5.44), then we call the corresponding $\tilde{t}$ and $\bar{t}$ an *exact encounter time* and the *first exact encounter time*, respectively. It is clear that in the case where both $\Omega_1(t)$ and $\Omega_2(t)$ are closed for all $t \in [0, \infty)$, the above two sets of notions coincide. In what follows, we concentrate on the approximate case.

Our questions are the following: (i) Under what conditions does the first approximate encounter time exist? (ii) If there is a first approximate encounter time (under certain conditions), how can we determine it?

Before going further, let us give an example for which the first approximate encounter time does not exist. This will tell us that we have to impose some conditions in order to determine the existence of the first approximate encounter time for two given moving sets.

**Example 5.4.** Let $X = L^1(0, 1)$ and

(5.45)
$$\begin{cases} \Omega_1(t) = \{y(\cdot) \in L^1(0,1) \mid \int_0^1 |y(s)|\, ds \le 1\}, \\ \Omega_2(t) = \{y(\cdot) \in L^1(0,1) \mid \int_0^1 sy(s)\, ds = 2 - t\}, \end{cases} \quad t \in [0, \infty).$$

Then we see immediately that for any $t \in [0, \infty)$, both $\Omega_1(t)$ and $\Omega_2(t)$ are closed and convex. Next, we note that for any $y(\cdot) \in L^1(0, 1)$, $y(\cdot) \ne 0$,

(5.46)
$$\int_0^1 sy(s)\, ds < \int_0^1 |y(s)|\, ds.$$

This implies that

(5.47)
$$\Omega_1(t) \cap \Omega_2(t) = \phi, \quad \forall t \in [0, 1].$$

On the other hand, for any $t \in (1, 3/2]$, by setting

(5.48)
$$\bar{y}(s) = \frac{1}{2(t-1)} \chi_{[1-2(t-1),1]}(s), \quad s \in [0, 1],$$

we have

(5.49)
$$\begin{cases} \int_0^1 |\bar{y}(s)|\, ds = 1, \\ \int_0^1 s\bar{y}(s)\, ds = \dfrac{1}{2(t-1)} \int_{1-2(t-1)}^1 s\, ds \\ \qquad = \dfrac{1}{2(t-1)} \dfrac{1}{2}\left(1 - [1-2(t-1)]^2\right) = 2 - t. \end{cases}$$

This means that $\bar{y}(\cdot) \in \Omega_1(t) \cap \Omega_2(t)$ for any $t \in (1, 3/2]$. Hence, combining (5.47) and (5.49), we see that

(5.50) $$\bar{t} = 1, \qquad \Omega_1(\bar{t}) \cap \Omega_2(\bar{t}) = \phi.$$

Hence, the first approximate encounter time does not exist for $\Omega_1(\cdot)$ and $\Omega_2(\cdot)$. Next, we claim that

(5.51) $$d(\Omega_1(1), \Omega_2(1)) = 0.$$

To show this, we define, for any $n \geq 1$, that

(5.52)
$$\begin{cases} y_n(\cdot) = n\chi_{(1-1/n,1)}(\cdot), \\ z_n(\cdot) = \dfrac{2n^2}{2n-1} \chi_{(1-1/n,1)}(\cdot). \end{cases}$$

Clearly, $y_n(\cdot) \in \Omega_1(1)$ and

(5.53) $$\int_0^1 sz_n(s)\, ds = \frac{2n^2}{2n-1} \int_{1-1/n}^1 s\, ds = \frac{n^2}{2n-1}\left(1 - (1-1/n)^2\right) = 1.$$

Thus, $z_n(\cdot) \in \Omega_2(1)$. On the other hand,

(5.54) $$\int_0^1 |z_n(s) - y_n(s)|\, ds = \frac{1}{n}\left(\frac{2n^2}{2n-1} - n\right) = \frac{1}{2n-1}.$$

Consequently,

(5.55) $$d(\Omega_1(1), \Omega_2(1)) \leq \frac{1}{2n-1} \to 0.$$

Then (5.51) follows. Conclusions (5.50) and (5.51) tell us that, in general, (5.42) does not imply (5.43).

To give some positive results under certain conditions, let us first give some preliminaries. In what follows, we let $d$ represent different metrics in different metric spaces, which can be identified from the contexts. We recall the Hausdorff metric on metric space $(X, d)$ defined in Chapter 3:

(5.56) $$\rho_H(S_1, S_2) = \max\left\{ \sup_{x_1 \in S_1} d(x_1, S_2),\ \sup_{x_2 \in S_2} d(x_2, S_1)\right\},$$
$$\forall S_1, S_2 \subset X,$$

## §5. Time Optimal Control—Linear Systems

where

(5.57) $$d(x, S) = \inf_{y \in S} d(x, y), \quad \forall x \in X, \ S \subset X.$$

Also, we use the same $\rho_H$ representing the Hausdorff metric in different metric spaces. Finally, we recall that

(5.58) $$d(S_1, S_2) = \inf_{x_1 \in S_1, x_2 \in S_2} d(x_1, x_2), \quad \forall S_1, S_2 \subset X.$$

It is important to note that (5.58) is not a metric on $2^X$, as $\bar{S}_1 \cap \bar{S}_2 \neq \phi$ implies $d(S_1, S_2) = 0$. Clearly, such an $S_1$ and $S_2$ can be significantly different. The following result will be useful.

**Lemma 5.5.** *Let $S_1, S_2, S_3$ be subsets in some metric space $X$. Then*

(5.59) $$d(S_1, S_3) \leq d(S_1, S_2) + \rho_H(S_2, S_3).$$

*Proof.* For any $\varepsilon > 0$, take $x_1 \in S_1$ and $x_2 \in S_2$, such that

(5.60) $$d(x_1, x_2) \leq d(S_1, S_2) + \varepsilon.$$

Next, take $x_3 \in S_3$, such that

(5.61) $$d(x_2, S_3) \geq d(x_2, x_3) - \varepsilon.$$

Then

(5.62) $$\begin{aligned} d(S_1, S_3) &\leq d(x_1, x_3) \leq d(x_1, x_2) + d(x_2, x_3) \\ &\leq d(S_1, S_2) + d(x_2, S_3) + 2\varepsilon \\ &\leq d(S_1, S_2) + \rho_H(S_2, S_3) + 2\varepsilon. \end{aligned}$$

Hence, (5.59) follows because $\varepsilon > 0$ is arbitrary. □

The next result is not surprising, but it will be useful.

**Lemma 5.6.** *Let $X$ and $Y$ be two metric spaces. Let $K : X \to Y$ be uniformly Lipschitz continuous, i.e., there exists a constant $L > 0$, such that*

(5.63) $$d(K(x_1), K(x_2)) \leq L d(x_1, x_2), \quad \forall x_1, x_2 \in X.$$

*Then*

(5.64) $$\rho_H(K(S_1), K(S_2)) \leq L \rho_H(S_1, S_2), \quad \forall S_1, S_2 \subset X.$$

*Proof.* For any $\varepsilon > 0$, there exist $x_1 \in S_1$ and $x_2 \in S_2$, such that

(5.65) $$\begin{cases} d(K(x_1), K(S_2)) \geq \sup_{y \in K(S_1)} d(y, K(S_2)) - \varepsilon, \\ d(K(x_2), K(S_1)) \geq \sup_{y \in K(S_2)} d(y, K(S_1)) - \varepsilon. \end{cases}$$

On the other hand, by (5.63), we have

(5.66)
$$d(K(x_1), K(S_2)) = \inf_{z \in S_2} d(K(x_1), K(z))$$
$$\leq L \inf_{z \in S_2} d(x_1, z) = Ld(x_1, S_2).$$

Similarly,

(5.67)
$$d(K(x_2), K(S_1)) \leq Ld(x_2, S_1).$$

Hence, combining (5.65)–(5.67), we obtain (noting (5.56))

(5.68)
$$\rho_H(K(S_1), K(S_2))$$
$$\leq \max\{d(K(x_1), K(S_2)), d(K(x_2), K(S_1))\} + \varepsilon$$
$$\leq L\max\{d(x_1, S_2), d(x_2, S_1)\} + \varepsilon$$
$$\leq L\rho_H(S_1, S_2) + \varepsilon.$$

Then, (5.64) follows because $\varepsilon > 0$ is arbitrary. □

Now, we are ready to give the existence of the first approximate encounter time.

**Theorem 5.7.** *Let $X$ be a Banach space. Let $\Omega_1, \Omega_2 : [0, \infty) \to 2^X$ be continuous satisfying (5.41) and (5.42) for some $\tilde{t} > 0$. Then the first approximate encounter time $\bar{t}$ exists and it holds that*

(5.69)
$$\overline{\Omega_1(\bar{t})} \bigcap \overline{\Omega_2(\bar{t})} \neq \phi,$$

*provided that one of the following is true:*

*(i) $X$ is reflexive, for any $t \in [0, \infty)$, $\Omega_1(t)$ and $\Omega_2(t)$ are convex and one of them is bounded;*

*(ii) For each $t \in [0, \infty)$, $\Omega_1(t)$ is compact.*

*Proof.* By Lemma 5.5, we know that for any $s, t \in [0, \infty)$,

(5.70)
$$d(\Omega_1(t), \Omega_2(t)) \leq d(\Omega_1(t), \Omega_2(s)) + \rho_H(\Omega_2(t), \Omega_2(s))$$
$$\leq d(\Omega_1(s), \Omega_2(s)) + \rho_H(\Omega_1(t), \Omega_1(s)) + \rho_H(\Omega_2(t), \Omega_2(s)).$$

Exchanging $t$ and $s$, we see that

(5.71)
$$\big|d(\Omega_1(t), \Omega_2(t)) - d(\Omega_1(s), \Omega_2(s))\big|$$
$$\leq \rho_H(\Omega_1(t), \Omega_1(s)) + \rho_H(\Omega_2(t), \Omega_2(s)), \qquad s, t \in [0, \infty).$$

Thus, by the continuity of $\Omega_1(\cdot)$ and $\Omega_2(\cdot)$, one has the continuity of the map $t \mapsto d(\Omega_1(t), \Omega_2(t))$.

Now, from (5.42), there exists a nonincreasing sequence $t_k \in (0, \tilde{t}]$, such that

(5.72)
$$\lim_{k \to \infty} t_k = \bar{t} \equiv \inf\{t > 0 \mid d(\Omega_1(t), \Omega_2(t)) = 0\}.$$

§5. Time Optimal Control—Linear Systems

By the continuity of $d(\Omega_1(\cdot), \Omega_2(\cdot))$, we obtain

(5.73) $$d(\Omega_1(\bar{t}), \Omega_2(\bar{t})) = 0.$$

This yields that there exist $x_k^1 \in \Omega_1(\bar{t})$ and $x_k^2 \in \Omega_2(\bar{t})$, such that

(5.74) $$\lim_{k \to \infty} |x_k^1 - x_k^2| = 0.$$

Now, let us look at the two situations: In the case (i), without loss of generality, we let $\Omega_1(\bar{t})$ be bounded. Because $X$ is reflexive, we may assume that $x_k^1 \xrightarrow{w} x^1 \in \Omega_1(\bar{t})$. By the Mazur Theorem (Chapter 2, Corollary 2.8), one can find $\alpha_{jk} \geq 0$, $\sum_{j \geq k} \alpha_{jk} = 1$, such that

(5.75) $$y_k^1 \equiv \sum_{j \geq k} \alpha_{jk} x_j^1 \to x^1, \qquad \text{strongly in } X.$$

By the convexity of $\Omega_1(\bar{t})$ and $\Omega_2(\bar{t})$, $y_k^1 \in \overline{\Omega_1(\bar{t})}$ and $y_k^2 \equiv \sum_{j \geq k} \alpha_{jk} x_j^2 \in \Omega_2(\bar{t})$. Hence,

(5.76) $$|y_k^2 - x^1| \leq |y_k^1 - x^1| + \sum_{j \geq k} \alpha_{jk} |x_j^1 - x_j^2| \to 0.$$

This shows that $x^1 \in \overline{\Omega_1(\bar{t})} \cap \overline{\Omega_2(\bar{t})}$, proving our conclusion.

In the case (ii), by the compactness of $\Omega_1(\bar{t})$, we may assume that $x_k^1$ converges to $x^1 \in \Omega_1(\bar{t})$ strongly in $X$. Then

(5.77) $$|x^1 - x_k^2| \leq |x_k^1 - x^1| + |x_k^1 - x_k^2| \to 0.$$

This implies that $x^1 \in \Omega_1(\bar{t}) \cap \overline{\Omega_2(\bar{t})}$, and our conclusion follows. □

We see that in Example 5.4, the underlying Banach space $L^1(0,1)$ is not reflexive. Thus, although both $\Omega_1(t)$ and $\Omega_2(t)$ are convex and closed with $\Omega_1(t)$ being bounded, the first approximate encounter time does not exist.

Next, we want to determine the first approximate encounter time.

**Theorem 5.8.** *Let $X$ be a reflexive Banach space, and let $\Omega_1, \Omega_2 : [0, \infty) \to 2^X$ be continuous, taking bounded and convex sets. Then, the function defined by*

(5.78) $$F(t) \triangleq \max_{y^* \in X^*, |y^*| \leq 1} \min_{z \in \Omega(t)} \langle y^*, z \rangle, \qquad t \in [0, \infty),$$

*is continuous, where $\Omega(t) = \overline{\Omega_1(t) - \Omega_2(t)}$. In addition, if (5.41) and (5.42) hold for some $\bar{t} > 0$, then the first approximate encounter time $\bar{t}$ exists and is the smallest positive root of $F(t) = 0$, i.e.,*

(5.79) $$F(\bar{t}) = 0, \qquad F(t) \neq 0, \qquad \forall t \in [0, \bar{t}).$$

Furthermore, there exists a $y_0^* \in X^* \setminus \{0\}$, $|y_0^*| \leq 1$, such that

(5.80) $$F(\bar{t}) = \min_{z \in \Omega(\bar{t})} \langle y_0^*, z \rangle = 0,$$

if and only if 0 is a supporting point (see Chapter 2, §2.1) of $\Omega(\bar{t})$.

*Proof.* First of all, it is easy to see that for any $t \in [0, \infty)$, $\Omega(t)$ is convex and bounded. It is also closed. In fact, if $x_k^1 \in \Omega_1(t)$ and $x_k^2 \in \Omega_2(t)$, such that

(5.81) $$x_k^1 - x_k^2 \equiv y_k \to y, \qquad (k \to \infty),$$

then, by the reflexivity of $X$, and the boundedness and convexity of $\overline{\Omega_1(t)}$ and $\overline{\Omega_2(t)}$, we may use the Mazur Theorem (Chapter 2, Corollary 2.8) to find $\alpha_{jk} \geq 0$, $\sum_{j \geq k} \alpha_{jk} = 1$, such that

(5.82) $$z_k^1 \stackrel{\Delta}{=} \sum_{j \geq k} \alpha_{jk} x_j^1 \to x^1 \in \overline{\Omega_1(t)},$$
$$\overline{\Omega_2(t)} \ni z_k^2 \stackrel{\Delta}{=} \sum_{j \geq k} \alpha_{jk} x_j^2 = \sum_{j \geq k} \alpha_{jk}(x_j^1 - y_j) \to x^1 - y,$$

This yields $y \in \Omega(t)$, proving the closeness of $\Omega(t)$.

Next, for any $y^* \in X^*$, $|y^*| \leq 1$, we define

(5.83) $$\langle y^*, \Omega(t) \rangle \equiv \{\langle y^*, z \rangle \mid z \in \Omega(t)\} \subset \mathbb{R}.$$

By the boundedness, closeness, and convexity of $\Omega(t)$, we see that $\langle y^*, \Omega(t) \rangle$ is a finite closed interval on $\mathbb{R}$. On the other hand, the map $z \mapsto \langle y^*, z \rangle$ is Lipschitz continuous with the Lipschitz constant $|y^*| \leq 1$. Hence, by Lemma 5.6, we have

(5.84) $$\begin{aligned}|\min \langle y^*, \Omega(t) \rangle - \min \langle y^*, \Omega(s) \rangle| &\leq \rho_H(\langle y^*, \Omega(t) \rangle, \langle y^*, \Omega(s) \rangle) \\ &\leq \rho_H(\Omega(t), \Omega(s)), \qquad 0 \leq s, t < \infty, \; |y^*| \leq 1.\end{aligned}$$

Consequently,

(5.85) $$|F(t) - F(s)| \leq \rho_H(\Omega(t), \Omega(s)), \qquad \forall 0 \leq s, t < \infty.$$

This proves the continuity of $F(t)$. Applying Theorem 5.7, we see that the first approximate encounter time $\bar{t} > 0$ exists and (5.69) holds. Thus,

(5.86) $$0 \notin \Omega(t), \qquad t \in [0, \bar{t}), \qquad 0 \in \Omega(\bar{t}).$$

Hence, for any $t \in [0, \bar{t})$, there exists a $y_0^* \in X^*$, $|y_0^*| = 1$, such that

(5.87) $$F(t) \geq \min_{z \in \Omega(t)} \langle y_0^*, z \rangle > 0.$$

On the other hand, by $0 \in \Omega(\bar{t})$, it follows that

(5.88) $$\min_{z \in \Omega(\bar{t})} \langle y^*, z \rangle \leq \langle y^*, 0 \rangle = 0, \qquad \forall y^* \in X^*.$$

Hence, $F(\bar{t}) \leq 0$ and (5.79) follows from the continuity of $F(\cdot)$.

Finally, if there exists a $y_0^* \in X^* \setminus \{0\}$, $|y_0^*| \leq 1$, such that (5.80) holds, then,

(5.89) $$0 = F(\bar{t}) = \min_{z \in \Omega(\bar{t})} \langle y_0^*, z \rangle \leq \langle y_0^*, y \rangle, \qquad \forall y \in \Omega(\bar{t}).$$

This means that the hyperplane with the normal $y_0^*$ supports the set $\Omega(\bar{t})$ at 0. Conversely, if 0 is a supporting point of $\Omega(\bar{t})$, then there exists a $y_0^* \in X^*$ with $|y_0^*| = 1$, such that

(5.90) $$\langle y_0^*, z \rangle \geq 0, \qquad \forall z \in \Omega(\bar{t}).$$

Consequently,

(5.91) $$\min_{z \in \Omega(\bar{t})} \langle y_0^*, z \rangle \geq 0 = F(\bar{t}) = \max_{y^* \in X^*, |y^*| \leq 1} \min_{z \in \Omega(\bar{t})} \langle y^*, z \rangle$$
$$\geq \min_{z \in \Omega(\bar{t})} \langle y_0^*, z \rangle.$$

Hence, the equality must hold and our conclusion follows. □

### §5.3. Time optimal control

Now, we are going to consider the time optimal control problem with the given target trajectory $Q : [0, \infty) \to 2^X$ satisfying the following:

(Q) The multifunction $Q(\cdot)$ is continuous (with respect to the Hausdorff metric) and for each $t \in [0, \infty)$, $Q(t)$ is convex and bounded.

We note that in the above (Q), we have not assumed that the set $Q(t)$ is closed. This is different from (T3) stated in the previous section. Next, we assume that

(5.92) $$\begin{cases} x \notin \overline{Q(0)}, \\ \inf_{t \in [0,T]} d(\mathcal{R}(t, x), Q(t)) = 0, & \text{for some } T > 0. \end{cases}$$

In what follows, the first approximate encounter time of the moving sets $\mathcal{R}(\cdot, x)$ and $Q(\cdot)$ (if it exists) is called the *minimum approximate hitting time* of the system (5.1) starting from the initial state $x$ to the target trajectory $Q(\cdot)$. The main result of this subsection is the following.

**Theorem 5.9.** *Let (T1), (B), and (Q) hold. Let $X$ be reflexive. Let $x \in X$ satisfy (5.92). Then the minimum approximate hitting time $\bar{t}$ of the system (5.1) starting from the initial state $x$ to the target trajectory $Q(\cdot)$ exists and it is the smallest positive root of $F(t) = 0$, where for $t \in [0, \infty)$,*

(5.93) $$F(t) = \max_{y^* \in X^*, |y^*| \leq 1} \Big\{ \min_{z \in Q(t)} \langle y^*, z \rangle - \langle y^*, e^{At} x \rangle - \int_0^t \sup_{u \in U} \langle y^*, e^{A(t-s)} b(s, u) \rangle \, ds \Big\}.$$

*Proof.* First of all, we can easily prove that $\mathcal{R}(\cdot, x)$ is continuous with respect to the Hausdorff metric. Thus, by Theorem 5.3 and (5.2), we see that $\overline{\mathcal{R}(t;x)}$ is bounded, closed, convex and continuous in $t \in [0, \infty)$. Then, by (Q), applying Theorems 5.7 and 5.8, we obtain the existence of the minimum approximate hitting time $\bar{t} > 0$, and this $\bar{t}$ is the smallest positive root of $F(t) = 0$ with

(5.94)
$$\begin{aligned}
F(t) &= \max_{y^* \in X^*, |y^*| \leq 1} \min_{z \in Q(t) - \overline{\mathcal{R}(t,x)}} \langle y^*, z \rangle \\
&= \max_{y^* \in X^*, |y^*| \leq 1} \left\{ \min_{z \in Q(t)} \langle y^*, z \rangle - \sup_{y \in \mathcal{R}(t,x)} \langle y^*, y \rangle \right\} \\
&= \max_{y^* \in X^*, |y^*| \leq 1} \left\{ \min_{z \in Q(t)} \langle y^*, z \rangle - \langle y^*, e^{At} x \rangle \right. \\
&\quad \left. - \sup_{u(\cdot) \in \mathcal{U}} \int_0^t \langle y^*, e^{A(t-s)} b(s, u(s)) \rangle \, ds \right\}.
\end{aligned}$$

For any fixed $y^* \in X^*$ and $t > 0$, set

(5.95) $$M(s) = \sup_{u \in U} \langle y^*, e^{A(t-s)} b(s, u) \rangle, \qquad s \in [0, t],$$

which is measurable because $U$ is separable. We claim that

(5.96) $$\sup_{u(\cdot) \in \mathcal{U}} \int_0^t \langle y^*, e^{A(t-s)} b(s, u(s)) \rangle \, ds = \int_0^t M(s) \, ds.$$

In fact, the following is obvious:

(5.97) $$\sup_{u(\cdot) \in \mathcal{U}} \int_0^t \langle y^*, e^{A(t-s)} b(s, u(s)) \rangle \, ds \leq \int_0^t M(s) \, ds.$$

On the other hand, by the definition of $M(\cdot)$, for any $\varepsilon > 0$ and $r \in [0, t]$, there exists a $u_r \in U$, such that

(5.98) $$M(r) - \varepsilon < \langle y^*, e^{A(t-r)} b(r, u_r) \rangle.$$

Because $\langle y^*, e^{A(t-\cdot)} b(\cdot, u_r) \rangle$ is continuous, there exists a neighborhood $\mathcal{O}(r)$ of $r$, such that

(5.99) $$M(s) - \varepsilon < \langle y^*, e^{A(t-s)} b(s, u_r) \rangle, \qquad \forall s \in \mathcal{O}(r).$$

Hence, $\{\mathcal{O}(r), r \in [0, t]\}$ is an open cover of $[0, t]$. By the compactness of $[0, t]$, we can find $r_1, r_2, \cdots, r_n \in [0, t]$, such that

(5.100) $$\begin{cases} [0, t] \subset \bigcup_{k=1}^n \mathcal{O}(r_k), \\ M(s) - \varepsilon < \langle y^*, e^{A(t-s)} b(s, u_{r_k}) \rangle, \quad \forall s \in \mathcal{O}(r_k), \ 1 \leq k \leq n. \end{cases}$$

Now, we define

$$(5.101) \quad u_0(s) = u_{r_k}, \quad s \in \mathcal{O}(r_k) \setminus \bigcup_{i=1}^{k-1} \mathcal{O}(r_i), \quad 1 \leq k \leq n.$$

Clearly, $u_0(\cdot) \in \mathcal{U}$, and

$$(5.102) \quad M(s) - \varepsilon < \langle y^*, e^{A(t-s)} b(s, u_0(s)) \rangle, \quad \forall s \in [0,1].$$

Hence,

$$(5.103) \quad \begin{aligned} \int_0^t M(s)\,ds - \varepsilon t &\leq \int_0^t \langle y^*, e^{A(t-s)} b(s, u_0(s)) \rangle\,ds \\ &\leq \sup_{u(\cdot) \in \mathcal{U}} \int_0^t \langle y^*, e^{A(t-s)} b(s, u(s)) \rangle\,ds. \end{aligned}$$

Because $\varepsilon > 0$ is arbitrary, combining (5.97) and (5.103), we obtain (5.96). Then our conclusion follows. □

*Remark 5.10.* We may relax the continuity of $b(t, u)$ to the following: $b(t, u)$ is Borel measurable in $(t, u)$ and continuous in $u$. In such a case, to obtain (5.96), one needs to use some measurable selection theorems. Thus, we have to assume the completeness of the separable metric space $U$. The details are left to the interested readers.

## Remarks

The notion of controllability (for finite dimensional systems) was initiated by Kalman [2] in the early 1960s. Later, the notion was extended to infinite dimensions. See Fattorini [2,3] for approximate controllability and Russell [1] for exact controllability. Readers are also referred to the book by Curtain–Pritchard [3], and the survey papers by Russell [2] and Lions [6] for some details. Also, we would like to mention some other interesting works: Barbu [4], Barbu–Tiba [1], Kime [1], Lagnese [1], Littman [1], Naito [1], Naito–Seidman [1], Russell [3], Seidman [1,2], Triggiani [1], Wang [1], and H.X. Zhou [1–3].

The material of Sections 1 and 2 is a modification of Curtain–Pritchard [3]. Section 3 is based on the work by Fabre-Puel-Zuazua [1], in which the approximate controllability of a semilinear heat equation was discussed. Here, we present the result for abstract semilinear evolution equations with compact $C_0$ semigroup.

In 1963, Yu.V. Egorov [1,2] discussed the time optimal control problem for a one dimensional parabolic system with a boundary control and with the target set as a ball. He proved a "bang-bang" principle for such a problem. Independently, in 1964, Fattorini [1] studied the time optimal control for the linear system (on the space $X$)

(R.1) $$\begin{cases} \dot{y}(t) = Ay(t) + u(t), & u(t) \in U \equiv \{u \mid |u| \leq 1\}, \\ y(t_0) = y_0, & y(t_1) \in W. \end{cases}$$

He assumed that $X$ is a Hilbert space, $A$ generates a $C_0$ semigroup on $X$ and $W = \{y_1\}$. The existence and uniqueness of the time optimal control was established and a bang-bang principle was proved as well, namely, that the optimal control $\bar{u}(t)$ takes values on the boundary $\partial U$ of $U$ almost everywhere. A little later, Balakrishnan [1] proved that when $y_1$ is a supporting point of the reachable set of the system, the maximum principle for the optimal control holds, which implies the bang-bang principle. Friedman [3] discussed the time optimal control problem on Banach spaces where the $A$ in (R.1) was replaced by the more general $A(t)$. Under some general conditions with the target set $W$ convex and closed and Int $W \neq \phi$, the bang-bang principle was proved. Friedman also proved that in the case where $X$ is a Hilbert space with $A(t) \equiv A$ generating a $C_0$ group, the maximum principle holds. Conti [1] also proved that if the reachable set of the system or the target set $W$ has a nonempty interior, then the maximum principle holds. In 1974, Fattorini [5] proved that the maximum principle holds if $X$ is a reflexive Banach space and $y_0, y_1 \in \mathcal{D}(A)$, or in the general Banach space $X$, if $e^{At}X = X$ for some $t > 0$, or when $X$ is a Hilbert space and $e^{At}$ is an analytic semigroup.

However, there are many other papers dealing with the time optimal controls. Among them, we mention the following: Barbu [5] and Cârjă [1].

It is important to note that in the above-mentioned literature, the control domain $U$ is convex. The case with the general control domain was firstly studied by Li–Yao [1]. At the same time, Korobov–Son [1] studied the same problem with $A$ independently a bounded operator. Since then, people started to discuss the general form of the maximum principle for semilinear and nonlinear control systems. See the remarks in Chapter 4. The time optimal control problem for semilinear evolution equations was first studied by Yao [2]. He proved a maximum principle for the case where the target set has a nonempty interior. Later, Yong [5] discussed a more general case with a different approach. Also, Frankowska [2] studied a similar problem with another method.

The material of Section 4.1 is based on Yong [5] and Section 4.2 is a generalization of Peng–Yong [1]. Section 5 is based on the work of Li–Yao [1]. Example 5.2 was appeared in Li–Yao [1] for the possible nonconvexity of the reachable set. Here, we present a self-contained proof by proving Lemma 5.1. We note that this result can be found in Singer [1], but our proof seems more accessible and shorter.

# Chapter 8

# Optimal Switching and Impulse Controls

## §1. Switching and Impulse Controls

In this chapter we consider a distributed parameter system with the so-called *switching* and *impulse* controls. Let us now explain the meaning of such controls. Suppose we are given the following $m$ evolution equations (call them subsystems for the time being):

(1.1) $\qquad \dot{y}(t) = Ay(t) + f_i(y(t)), \qquad i = 1, 2, \cdots, m,$

where $A$ is the infinitesimal generator of a $C_0$ semigroup $e^{At}$ on some Banach space $X$, and $f_i : X \to X$ are $m$ given functions. Let us arbitrarily pick a sequence of times $0 = \theta_0 \leq \theta_1 \leq \theta_2 \leq \cdots$, such that $\theta_i \to \infty$ as $i \to \infty$ and a sequence of integers $\{d_i\}_{i \geq 0} \subseteq \{1, 2, \cdots, m\}$. Consider the following evolution of the system: Set $y(0-) = y(0) = x \in X$. On each time interval $[\theta_i, \theta_{i+1})$, we run the $d_i$th subsystem:

(1.2) $\qquad \begin{cases} \dot{y}(t) = Ay(t) + f_{d_i}(y_x(t)), & t \in [\theta_i, \theta_{i+1}), \\ y(\theta_i) = y(\theta_i - 0). \end{cases}$

Intuitively, what we do in the above is the following: We start the system from $x \in X$ at $t = 0$ by running the $d_0$th subsystem. At time $t = \theta_1$, the $d_0$th subsystem is turned off and at the same time, the $d_1$th subsystem is turned on. Namely, we *switch* from the $d_0$th subsystem to the $d_1$th at moment $t = \theta_1$. Then we continue the process in the similar fashion by switching from the $d_{i-1}$th subsystem to the $d_i$th at time $t = \theta_i$. We call the sequence $\{\theta_i, d_i\}_{i \geq 0}$ a *switching control* because it provides the rule of switching. Associated with the above evolution of the system, we define the following cost functional:

(1.3) $\qquad J(\{\theta_i, d_i\}_{i \geq 0}) = \sum_{i \geq 0} \int_{\theta_i}^{\theta_{i+1}} f_{d_i}^0(y(t)) e^{-\lambda t} dt + \sum_{i \geq 1} k(d_{i-1}, d_i) e^{-\lambda \theta_i}.$

Here, each integral term on the right-hand side represents the *running cost* on the time interval $[\theta_i, \theta_{i+1})$ when the $d_i$th subsystem is running. The term $k(d_{i-1}, d_i) e^{-\lambda \theta_i}$ represents the cost paid for the switch from $d_{i-1}$ to $d_i$ at time $t = \theta_i$; thus, this is called the *switching cost*. The constant $\lambda > 0$ is called the *discount factor*, which appeared in Chapter 6, §6 also, in studying optimal control problems on an infinite horizon.

Any switching control $\{\theta_i, d_i\}_{i\geq 0}$, is identified with the following piecewise constant function:

$$(1.4) \qquad d(\cdot) = \sum_{i\geq 1} d_{i-1}\chi_{[\theta_{i-1},\theta_i)}(\cdot).$$

Also, we write $f_d(y) = f(y,d)$ and $f_d^0(y) = f^0(y,d)$. Then the evolution system (1.2) can be written as

$$(1.5) \qquad \begin{cases} \dot{y}(t) = Ay(t) + f(y(t), d(t)), & t \in [0,\infty), \\ y(0) = x, \end{cases}$$

and the cost functional (1.3) can be written as

$$(1.6) \qquad J(d(\cdot)) = \int_0^\infty f^0(y(t), d(t))e^{-\lambda t}\, dt + \sum_{i\geq 1} k(d_{i-1}, d_i)e^{-\lambda \theta_i}.$$

It should be pointed out that, if, say, $\theta_1 = \theta_2$, then the term $d_1\chi_{[\theta_1,\theta_2)}(\cdot)$ in (1.4) is actually void. From the switching control point of view, at moment $t = \theta_1 = \theta_2$, we make two switches, from $d_0$ to $d_1$ and instantaneously from $d_1$ to $d_2$. Thus, as far as the trajectory $y(\cdot)$ is concerned, $(\theta_1, d_1)$ does not play any role. In other words, we may delete $(\theta_1, d_1)$ from the sequence $\{\theta_i, d_i\}_{i\geq 0}$ without changing the trajectory $y(\cdot)$. Hence, identifying $\{\theta_i, d_i\}_{i\geq 0}$ with $d(\cdot)$ given by (1.4) is convenient when we consider the trajectory $y(\cdot)$ of the system. However, it will be assumed that $k(d, \widehat{d}) > 0$ for any $d \neq \widehat{d}$; thus, we see that the sequences with or without $(\theta_1, d_1)$ are different as far as the switching cost is concerned. Thus, we should remember that in (1.4), some of the $\theta_i$'s are allowed to be the same, and we should not delete the terms $d_i\chi_{[\theta_i,\theta_{i+1})}(\cdot)$ even if $\theta_i = \theta_{i+1}$ because they represent different switching controls.

The problem of minimizing the cost functional (1.6) subject to the state equation (1.5) over some given class of switching controls is called the *optimal switching control problem*.

Next, let us explain the so-called *impulse controls*. To this end, again, we consider the following evolution equation:

$$(1.7) \qquad \begin{cases} \dot{y}(t) = Ay(t) + f(y(t)), & t \geq 0, \\ y(0) = x. \end{cases}$$

Now, we pick any sequence of times $0 \leq \tau_1 \leq \tau_2 \leq \cdots$, such that $\tau_j \to \infty$ as $j \to \infty$; and any sequence $\{\xi_j\}_{j\geq 1} \subseteq K \subseteq X$. Then, we run the system in the following way: First we run (1.7) over $[0, \tau_1)$, then, for all $j \geq 1$, we run

$$(1.8) \qquad \begin{cases} \dot{y}(t) = Ay(t) + f(y(t)), & t \in [\tau_j, \tau_{j+1}), \\ y(\tau_j) = y(\tau_j - 0) + \xi_j. \end{cases}$$

## §1. Switching and Impulse Controls

The result of the above is the following: The system (1.7) is running and at each moment $t = \tau_j$, we give the state $y(\tau_j - 0)$ a jump of size $\xi_j$. Consequently, the resulted trajectory $y(\cdot)$ is discontinuous. The jumps are referred to as the *impulses*. For this reason, the sequence $\{\tau_j, \xi_j\}_{j \geq 1}$ is called an *impulse control*. The cost associated with the above evolution of the system is defined to be the following:

$$(1.9) \qquad J(\{\tau_j, \xi_j\}_{j \geq 1}) = \int_0^\infty f^0(y(t)) e^{-\lambda t}\, dt + \sum_{j \geq 1} \ell(\xi_j) e^{-\lambda \tau_j}.$$

Here, the first term represents the *running cost* and the term $\ell(\xi_j) e^{-\lambda \tau_j}$ represents the cost paid for making the impulse $\xi_j$ at time $t = \tau_j$; thus, we call it the *impulse cost*. Similar to the switching control case, we make the following identification: Each impulse control $\{\tau_j, \xi_j\}_{j \geq 1}$ is identified with

$$(1.10) \qquad \xi(\cdot) = \sum_{j \geq 1} \xi_j \chi_{[\tau_j, \infty)}(\cdot).$$

Then the state equation can be formally written as follows:

$$(1.11) \qquad \begin{cases} \dot{y}(t) = Ay(t) + f(y(t)) + \dot{\xi}(t), & t \in [0, \infty), \\ y(0-) = x. \end{cases}$$

Here, $\dot{\xi}(\cdot)$ is a sum of a sequence of $\delta$-functions supported at $\tau_j$'s with sizes $\xi_j$'s. This will be made precise below. We write the initial condition like $y(0-) = x$ because initial jumps are allowed.

Again, similar to the switching case, if, say, $\tau_1 = \tau_2$ in (1.10), this means that *two* jumps are made at the moment $t = \tau_1 = \tau_2$. We should not replace $\xi_1 \chi_{[\tau_1, \infty)} + \xi_2 \chi_{[\tau_2, \infty)}$ by $(\xi_1 + \xi_2)\chi_{[\tau_1, \infty)}$ because, in general, they correspond to different impulse costs ($\ell(\xi_1 + \xi_2) < \ell(\xi_1) + \ell(\xi_2)$, in general, see below), although the corresponding trajectories are the same. Finally, the "zero"-impulse, say, $\xi_1 = 0$ in (1.10), is allowed. Again, we should keep the "zero"-impulse term because the impulse cost will exist ($\ell(0) > 0$, in general), even though the trajectory is unchanged.

The problem of minimizing the cost functional (1.9) subject to the state equation (1.11) over some given class of impulse controls is called the *optimal impulse control problem*.

We note that the above described switching and impulse control problems are quite different from the one we studied in the previous chapters. The major differences are: (i) The cost functional contains a term of summation representing either switching or impulse costs; and (ii) for the switching control problem, the initial value of the control is relevant due to the positive switching cost, and for the impulse control problem, the trajectory is discontinuous.

The purpose of this chapter is to give a unified method of treating a problem where both switching and impulse control appear in the system. We will use the dynamic programming method together with viscosity solutions to characterize the value function of the problem and then to construct an optimal control via the value function. Here, we take an approach different from the one presented in the previous chapter. Basically, we approximate our problem by a sequence of "nicer" problems to which a good theory of viscosity solutions applies, etc. Then we prove that the value function of the approximate problem converges to that of the original problem. The advantage of this approach is that we do not have to make many restrictive assumptions on the space $X$ and the operator $A$. Also, the approximating procedure gives us some hint for approaching the problem numerically.

To conclude this section, we would like to point out that it is possible to develop a theory of optimal switching and impulse controls under the framework of Chapter 6.

## §2. Preliminary Results

Let us begin with some notation and assumptions.

(S1) $\Lambda = \{1, 2, \cdots, m\}$, $X$ is a Banach space, and $K$ is a closed subset of $X$ such that for any $r > 0$, the set $K \cap B_r(0)$ is compact, where $B_r(0)$ is the ball centered at 0 with radius $r$. Moreover, the following holds:

(2.1) $$\xi_1, \xi_2 \in K \Rightarrow \xi_1 + \xi_2 \in K.$$

(S2) $A : \mathcal{D}(A) \subset X \to X$ generates a $C_0$ semigroup on $X$. Moreover, there exists a constant $\widehat{L} \geq 1$, such that

(2.2) $$\|e^{At}\| \leq \widehat{L}, \qquad t \geq 0.$$

(S3) $f : X \times \Lambda \to X$, $f^0 : X \times \Lambda \to \mathbb{R}$ are continuous. There exist constants $L > 0$ and $0 < \delta \leq 1$, such that for all $x, \widehat{x} \in X$, $d \in \Lambda$,

(2.3) $$|f(x, d) - f(\widehat{x}, d)| \leq L|x - \widehat{x}|, \qquad |f(0, d)| \leq L,$$

(2.4) $$|f^0(x, d) - f^0(\widehat{x}, d)| \leq L|x - \widehat{x}|^\delta, \qquad |f^0(x, d)| \leq L.$$

(S4) $k : \Lambda \times \Lambda \to \mathbb{R}^+ \equiv [0, \infty)$ is continuous, such that

(2.5) $$k(d, \widetilde{d}) < k(d, \widehat{d}) + k(\widehat{d}, \widetilde{d}), \qquad \forall d \neq \widehat{d} \neq \widetilde{d},$$

(2.6) $$k(d, d) = 0, \qquad \min_{d \neq \widetilde{d}} k(d, \widetilde{d}) \equiv k_0 > 0.$$

(S5) $\ell : X \to \mathbb{R}^+$ is continuous, such that

(2.7) $$\ell(\xi + \widehat{\xi}) < \ell(\xi) + \ell(\widehat{\xi}), \qquad \forall \xi, \widehat{\xi} \in K,$$

## §2. Preliminary Results

$$(2.8) \quad \inf_{\xi \in K} \ell(\xi) \equiv \ell_0 > 0, \quad \lim_{\xi \in K, |\xi| \to \infty} \frac{|\xi|^\delta}{\ell(\xi)} = 0.$$

Let us make some remarks on the above assumptions. First of all, from Chapter 2, Proposition 4.7, we know that for the $C_0$ semigroup $e^{At}$, there always exist constants $\omega \in \mathbb{R}$, $\widehat{L} \geq 1$, such that

$$(2.9) \quad \|e^{At}\| \leq \widehat{L} e^{\omega t}, \quad t \geq 0.$$

On the other hand, by replacing $A$ by $A - \omega I$ and $f(y,d)$ by $f(y,d) + \omega y$, if necessary, we may assume that $\omega = 0$. Thus, if we change the constant $L$ in (2.3) properly, assumption (2.2) covers the general case.

Second, we should note that in (2.3) and (2.4), the constant $L$ may be different in different places. Moreover, the growth condition in (2.3) for $f$ can be replaced by

$$(2.10) \quad |f(x,d)| \leq L_1 + L_2|x|, \quad \forall (x,d) \in X \times \Lambda.$$

We take all these constants as the same $L$ just for simplicity. It follows from (2.8) that there exists a constant $\alpha_0 > 0$, such that

$$(2.11) \quad \ell(\xi) \geq \alpha_0 |\xi|^\delta, \quad \forall \xi \in K.$$

Next, let us introduce the following control sets: Let $\lambda > 0$ be given. For any $d \in \Lambda$, we define

$$\mathcal{A}^d = \Big\{ d(\cdot) = \sum_{i \geq 1} d_{i-1} \chi_{[\theta_{i-1}, \theta_i)}(\cdot) : [0, +\infty) \to \Lambda \,\Big|\, d_0 = d,\; \theta_0 = 0;$$
$$\theta_i \in [0, +\infty],\; \forall i \geq 1;\; \theta_i \uparrow +\infty;\; d_{i+1} \neq d_i,\; \text{if } \theta_{i+1} < \infty;$$
$$\sum_{i \geq 1} k(d_{i-1}, d_i) e^{-\lambda \theta_i} < \infty \Big\}.$$

Also, we define

$$\mathcal{X} = \Big\{ \xi(\cdot) = \sum_{j \geq 1} \xi_j \chi_{[\tau_j, \infty)}(\cdot) : [0, +\infty) \to X \,\Big|$$
$$\tau_j \in [0, +\infty],\; \forall j \geq 1;\; \tau_j \uparrow +\infty \Big\},$$

$$\mathcal{K} = \Big\{ \xi(\cdot) = \sum_{j \geq 1} \xi_j \chi_{[\tau_j, \infty)}(\cdot) \in \mathcal{X} \,\Big|\, \xi_j \in K,\; \forall j \geq 1;$$
$$\sum_{j \geq 1} \ell(\xi_j) e^{-\lambda \tau_j} < \infty \Big\}.$$

We call any $d(\cdot) \in \mathcal{A}^d$ an *admissible switching control*, any $\xi(\cdot) \in \mathcal{X}$ an *impulse control*, and any $\xi(\cdot) \in \mathcal{K}$ an *admissible impulse control*, respectively. Here, we should note that the switching control $d(\cdot)$ takes initial value $d$. This notation is suggestive and will be helpful in sequel. The meaning of $d(\cdot)$ and $d$ can be distinguished from the context.

For any pair $(d(\cdot), \xi(\cdot)) \in \mathcal{A}^d \times \mathcal{X}$, we consider the following system

(2.12)
$$\begin{cases} \dot{y}_x(t) = Ay_x(t) + f(y_x(t), d(t)) + \dot{\xi}(t), & t \geq 0, \\ y_x(0^-) = x. \end{cases}$$

Let us define a (mild) solution of (2.12) to be any solution $y_x(\cdot)$ of the following equation:

(2.13)
$$y_x(t) = e^{At}x + \int_0^t e^{A(t-s)} f(y_x(s), d(s))\, ds + \sum_{\tau_j \leq t} e^{A(t-\tau_j)} \xi_j$$
$$= e^{At}x + \int_0^t e^{A(t-s)} f(y_x(s), d(s))\, ds + \sum_{j \geq 1} e^{A(t-\tau_j)} \xi_j \chi_{[\tau_j, \infty)}(t).$$

It is clear that under (S1)–(S3), for any $(x,d) \in X \times \Lambda$ and $(d(\cdot), \xi(\cdot)) \in \mathcal{A}^d \times \mathcal{X}$, there exists a unique (mild) solution $y_x(\cdot)$ of (2.12). We call $y_x(\cdot)$ the trajectory of the system corresponding to the control $(d(\cdot), \xi(\cdot))$. Here, as in the previous chapter, the subscript $x$ in $y_x(\cdot)$ is used to emphasize the dependence of the trajectory $y_x(\cdot)$ on the initial state $x$. The dependence on the control $(d(\cdot), \xi(\cdot))$ should be kept in mind. From (2.13), we see that $y_x(\cdot)$ is right continuous. That is why we write $y_x(0^-) = x$ in (2.12). From now on, we will not distinguish (2.12) and (2.13).

The following Gronwall type of inequality will be frequently used.

**Lemma 2.1.** *Let $\varphi(t), \psi(t), \zeta(t) \geq 0$ for all $t \geq 0$. Suppose that*

(2.14)
$$\varphi(t) \leq \psi(t) + \int_0^t \zeta(s)\varphi(s)\, ds, \quad t \geq 0.$$

*Then*

(2.15)
$$\varphi(t) \leq \psi(t) + \int_0^t \zeta(s)\psi(s) e^{\int_s^t \zeta(r)\, dr}\, ds, \quad t \geq 0.$$

The proof is obvious and we leave it to the readers. Next, let us give some basic results about the trajectories $y_x(\cdot)$ of our system (2.13).

**Lemma 2.2.** *Let (S1)–(S5) hold. Let $x, \bar{x} \in X$ and*

$$d(\cdot) = \sum_{i \geq 1} d_{i-1} \chi_{[\theta_{i-1}, \theta_i)}(\cdot), \quad \bar{d}(\cdot) = \sum_{i \geq 1} \bar{d}_{i-1} \chi_{[\bar{\theta}_{i-1}, \bar{\theta}_i)}(\cdot) \in \mathcal{A}^d,$$

$$\xi(\cdot) = \sum_{j \geq 1} \xi_j \chi_{[\tau_j, \infty)}(\cdot), \quad \bar{\xi}(\cdot) = \sum_{j \geq 1} \bar{\xi}_j \chi_{[\bar{\tau}_j, \infty)}(\cdot) \in \mathcal{X},$$

*with $\theta_i \leq \bar{\theta}_i$ and $\tau_j \leq \bar{\tau}_j$. Let $y(\cdot)$ and $\bar{y}(\cdot)$ be the trajectories of the system corresponding to $(d(\cdot), \xi(\cdot), x)$ and $(\bar{d}(\cdot), \bar{\xi}(\cdot), \bar{x})$, respectively. Then, for all $t \geq 0$,*

(2.16)
$$|y(t)| \leq e^{\widehat{L}Lt}\Big\{1 + \widehat{L}|x| + \sum_{j \geq 1} \widehat{L} e^{-\widehat{L}L\tau_j} |\xi_j| \chi_{[\tau_j, \infty)}(t)\Big\}.$$

## §2. Preliminary Results

$$
\begin{aligned}
|y(t) - \bar{y}(t)| \leq \widehat{L} e^{\widehat{L}Lt} \Big\{ &|x - \bar{x}| + \sum_{j\geq 1} |\xi_j| \chi_{[\tau_j, \bar{\tau}_j)}(t) \\
&+ \sum_{j\geq 1} \Big( \{|(e^{A(\bar{\tau}_j - \tau_j)} - I)\bar{\xi}_j| + |\xi_j - \bar{\xi}_j|\} e^{-\widehat{L}L\tau_j} \\
&\quad + \widehat{L}L|\xi_j||\tau_j - \bar{\tau}_j| \Big) \chi_{[\tau_j, \infty)}(t) \\
&+ 2L\Big(2 + \widehat{L}|x| + \sum_{j\geq 1} \widehat{L} e^{-\widehat{L}L\tau_j} |\xi_j| \chi_{[\tau_j, \infty)}(t)\Big) \\
&\quad \cdot \sum_{i\geq 1} |\theta_i - \bar{\theta}_i| \chi_{[\theta_i, \infty)}(t) \Big\}.
\end{aligned}
\tag{2.17}
$$

*Proof.* By (2.2) and (2.13) we have

$$
\begin{aligned}
|y(t)| &\leq \widehat{L}|x| + \int_0^t \widehat{L}L(1 + |y(s)|) \, ds + \sum_{j\geq 1} \widehat{L}|\xi_j| \chi_{[\tau_j, \infty)}(t) \\
&= \widehat{L}|x| + \widehat{L}Lt + \sum_{j\geq 1} \widehat{L}|\xi_j| \chi_{[\tau_j, \infty)}(t) + \widehat{L}L \int_0^t |y(s)| \, ds.
\end{aligned}
\tag{2.18}
$$

Then, by Lemma 2.1, we obtain

$$
\begin{aligned}
|y(t)| &\leq \widehat{L}|x| + \widehat{L}Lt + \sum_{j\geq 1} \widehat{L}|\xi_j| \chi_{[\tau_j, \infty)}(t) \\
&\quad + \widehat{L}L \int_0^t \Big(\widehat{L}|x| + \widehat{L}Ls + \sum_{j\geq 1} \widehat{L}|\xi_j| \chi_{[\tau_j, \infty)}(s)\Big) e^{\widehat{L}L(t-s)} \, ds \\
&= \widehat{L} e^{\widehat{L}Lt} |x| + e^{\widehat{L}Lt} - 1 + \sum_{j\geq 1} \widehat{L}|\xi_j| e^{\widehat{L}L(t-\tau_j)} \chi_{[\tau_j, \infty)}(t) \\
&\leq e^{\widehat{L}Lt} \Big\{1 + \widehat{L}|x| + \sum_{j\geq 1} \widehat{L} e^{-\widehat{L}L\tau_j} |\xi_j| \chi_{[\tau_j, \infty)}(t)\Big\}.
\end{aligned}
\tag{2.19}
$$

This proves (2.16). Now, we prove (2.17). By (2.2) and (2.13) again,

$$
\begin{aligned}
|y(t) - \bar{y}(t)| &\leq \widehat{L}|x - \bar{x}| + \int_0^t \widehat{L}L|y(s) - \bar{y}(s)| \, ds \\
&\quad + \Big\{\int_0^t \widehat{L}|f(s, y(s), d(s)) - f(s, y(s), \bar{d}(s))| \, ds\Big\} \\
&\quad + \Big\{\sum_{j\geq 1} |e^{A(t-\tau_j)} \xi_j \chi_{[\tau_j, \infty)}(t) - e^{A(t-\bar{\tau}_j)} \bar{\xi}_j \chi_{[\bar{\tau}_j, \infty)}(t)|\Big\} \\
&\equiv \widehat{L}|x - \bar{x}| + \int_0^t \widehat{L}L|y(s) - \bar{y}(s)| \, ds + \psi_1(t) + \psi_2(t).
\end{aligned}
\tag{2.20}
$$

By Lemma 2.1,

$$|y(t) - \bar{y}(t)| \leq \left\{ \widehat{L}|x - \bar{x}| + \widehat{L}L \int_0^t \widehat{L}|x - \bar{x}| e^{\widehat{L}L(t-s)} \, ds \right\}$$

(2.21)
$$+ \left\{ \psi_1(t) + \widehat{L}L \int_0^t \psi_1(s) e^{\widehat{L}L(t-s)} \, ds \right\}$$

$$+ \left\{ \psi_2(t) + \widehat{L}L \int_0^t \psi_2(s) e^{\widehat{L}L(t-s)} \, ds \right\}$$

$$\equiv I_0 + I_1 + I_2.$$

Now, let us estimate each term. First of all,

(2.22) $\quad I_0 = \widehat{L}|x - \bar{x}| + \widehat{L}|x - \bar{x}|(e^{\widehat{L}Lt} - 1) = \widehat{L}e^{\widehat{L}Lt}|x - \bar{x}|.$

Next, by (2.3) and (2.16),

(2.23)
$$I_1 = \psi_1(t) - \int_0^t \psi_1(s) d(e^{\widehat{L}L(t-s)})$$
$$= e^{\widehat{L}Lt}\psi_1(0) + \int_0^t \psi_1'(s) e^{\widehat{L}L(t-s)} \, ds$$
$$= \int_0^t e^{\widehat{L}L(t-s)} \widehat{L}|f(s, y(s), d(s)) - f(s, y(s), \bar{d}(s))| \, ds$$
$$\leq 2\widehat{L}L \int_0^t e^{\widehat{L}L(t-s)} (1 + |y(s)|) \sum_{i \geq 1} \chi_{[\theta_i, \bar{\theta}_i)}(s) \, ds$$
$$\leq 2\widehat{L}L \int_0^t e^{\widehat{L}L(t-s)} \Big( 1 + e^{\widehat{L}Ls}\{1 + \widehat{L}|x|$$
$$+ \sum_{j \geq 1} \widehat{L} e^{-\widehat{L}L\tau_j} |\xi_j| \chi_{[\tau_j, \infty)}(s)\} \Big) \sum_{i \geq 1} \chi_{[\theta_i, \bar{\theta}_i)}(s) \, ds$$
$$\leq 2\widehat{L}L e^{\widehat{L}Lt} \Big\{ 2 + \widehat{L}|x| + \widehat{L} \sum_{j \geq 1} e^{-\widehat{L}L\tau_j} |\xi_j| \chi_{[\tau_j, \infty)}(t) \Big\}$$
$$\cdot \sum_{i \geq 1} |\theta_i - \bar{\theta}_i| \chi_{[\theta_i, \infty)}(t).$$

To estimate $I_2$, we first look at the following:

(2.24)
$$\psi_2(t) \equiv \sum_{j \geq 1} |e^{A(t - \tau_j)} \xi_j \chi_{[\tau_j, \infty)}(t) - e^{A(t - \bar{\tau}_j)} \bar{\xi}_j \chi_{[\bar{\tau}_j, \infty)}(t)|$$
$$\leq \sum_{j \geq 1} \widehat{L} \Big\{ [|(e^{A(\bar{\tau}_j - \tau_j)} - I)\bar{\xi}_j| + |\xi_j - \bar{\xi}_j|] \chi_{[\bar{\tau}_j, \infty)}(t)$$
$$+ |\xi_j| \chi_{[\tau_j, \bar{\tau}_j)}(t) \Big\}.$$

## §2. Preliminary Results

Thus, by Lemma 2.1, it follows that

(2.25)
$$\begin{aligned}
I_2 &\equiv \psi_2(t) + \widehat{LL}\int_0^t \psi_2(s)e^{\widehat{LL}(t-s)}\,ds \\
&\leq \widehat{L}\sum_{j\geq 1}\Big(|(e^{A(\bar{\tau}_j-\tau_j)}-I)\bar{\xi}_j| + |\xi_j - \bar{\xi}_j|\Big)\{\chi_{[\bar{\tau}_j,\infty)}(t) \\
&\quad + \widehat{LL}\int_0^t \chi_{[\bar{\tau}_j,\infty)}(s)e^{\widehat{LL}(t-s)}\,ds\} \\
&\quad + \widehat{L}\sum_{j\geq 1}|\xi_j|\{\chi_{[\tau_j,\bar{\tau}_j)}(t) + \widehat{LL}\int_0^t \chi_{[\tau_j,\bar{\tau}_j)}(s)e^{\widehat{LL}(t-s)}\,ds\} \\
&= \widehat{L}\sum_{j\geq 1}\Big(|(e^{A(\bar{\tau}_j-\tau_j)}-I)\bar{\xi}_j| + |\xi_j - \bar{\xi}_j|\Big)e^{\widehat{LL}(t-\bar{\tau}_j)}\chi_{[\bar{\tau}_j,\infty)}(t) \\
&\quad + \widehat{L}\sum_{j\geq 1}|\xi_j|\Big\{\chi_{[\tau_j,\bar{\tau}_j)}(t) + \widehat{L}e^{\widehat{LL}t}(\bar{\tau}_j-\tau_j)\chi_{[\tau_j,\infty)}(t)\Big\}.
\end{aligned}$$

Combining (2.21)–(2.25), we obtain (2.17). □

It is clear that in the case where $d(\cdot) = \bar{d}(\cdot)$, $\xi(\cdot) = \bar{\xi}(\cdot)$, (2.17) becomes the following familiar expression:

(2.26) $$|y_x(t) - y_{\bar{x}}(t)| \leq \widehat{L}e^{\widehat{LL}t}|x-\bar{x}|, \qquad \forall t \geq 0.$$

For any $(d(\cdot),\xi(\cdot)) \in \mathcal{A}^d \times \mathcal{K}$ and the corresponding trajectory $y_x(\cdot)$, we consider the following cost functional:

(2.27)
$$\begin{aligned}
J_x^d(d(\cdot),\xi(\cdot)) &= \int_0^\infty f^0(y_x(s),d(s))e^{-\lambda s}\,ds \\
&\quad + \sum_{i\geq 1} k(d_{i-1},d_i)e^{-\lambda\theta_i} + \sum_{j\geq 1}\ell(\xi_j)e^{-\lambda\tau_j}.
\end{aligned}$$

The right-hand side of (2.27) represents the sum of *running*, *switching*, and *impulse* costs. The constant $\lambda > 0$, which is the same as that in the definition of $\mathcal{A}^d$ and $\mathcal{K}$, is called the *discount factor*.

It is clear that under (S1)–(S5), for any $(x,d) \in X \times \Lambda$, $(d(\cdot),\xi(\cdot)) \in \mathcal{A}^d \times \mathcal{K}$, the cost functional (2.27) is well defined. Now, we can state our optimal control problem.

**Problem (SI).** Given $(x,d) \in X \times \Lambda$, find $(d^*(\cdot),\xi^*(\cdot)) \in \mathcal{A}^d \times \mathcal{K}$, such that

(2.28) $$J_x^d(d^*(\cdot),\xi^*(\cdot)) = \inf_{\mathcal{A}^d \times \mathcal{K}} J_x^d(d(\cdot),\xi(\cdot)).$$

We define

(2.29) $$\begin{cases} V^d(x) = \inf_{\mathcal{A}^d \times \mathcal{K}} J_x^d(d(\cdot),\xi(\cdot)), & (x,d) \in X \times \Lambda, \\ V(x) = (V^1(x), V^2(x), \cdots, V^m(x)), & x \in X. \end{cases}$$

Function $V(\cdot)$ is called the *value function* of Problem (SI). We see that this value function is vector valued.

Let us make some further remarks on the switching and impulse controls. If $d(\cdot) \equiv \sum_{i\geq 1} d_{i-1}\chi_{[\theta_{i-1},\theta_i)}(\cdot)$ with $\theta_1 = +\infty$, then $d(t) \equiv d$, which means that by applying it, we do not switch the system over $[0,\infty)$. Similarly, if $\xi(\cdot) \equiv \sum_{j\geq 1} \xi_j \chi_{[\tau_j,\infty)}(\cdot)$ with $\tau_1 = +\infty$, then by applying it, we do not make any impulse. This is different from applying $\xi^0(\cdot) \equiv \sum_{j\geq 1} 0 \cdot \chi_{[\tau_j,\infty)}(\cdot)$ with $\tau_1 < \infty$, $\tau_j \uparrow +\infty$, because the impulse cost (see (2.27)) will be different. Any impulse control $\sum_{j\geq 1} \xi_j \chi_{[\tau_j,\infty)} \in \mathcal{X}$ with some $\tau_j < \infty$ and $\xi_j = 0$ can never be optimal; thus, we exclude such controls from $\mathcal{K}$—the set of admissible impulse controls. Hereafter, $\xi(\cdot) \equiv 0$ always means that $\tau_1 = +\infty$.

## §3. Properties of the Value Function

In this section, we present some properties of the value function. The first result is concerned with the boundedness and the continuity of the value function.

**Proposition 3.1.** *Let (S1)–(S5) hold. Then the value function $V(\cdot)$ satisfies the following:*

$$(3.1) \qquad |V^d(x)| \leq \frac{L}{\lambda}, \qquad \forall (x,d) \in X \times \Lambda,$$

$$(3.2) \qquad |V^d(x) - V^d(\widehat{x})| \leq C_0 |x - \bar{x}|^\sigma, \qquad \forall x, \bar{x} \in X, \ d \in \Lambda,$$

where $\sigma \in (0, \frac{\lambda}{\widehat{L}L})$, $\sigma \leq \delta$, and $C_0 = \frac{2\widehat{L}^\sigma L}{\lambda - \sigma \widehat{L}L}$.

*Proof.* By taking $d(t) \equiv d$, $\xi(t) \equiv 0$, we have (recall (2.4))

$$(3.3) \qquad V^d(x) \leq J^d_x(d(\cdot), \xi(\cdot)) = \int_0^\infty f^0(y_x(s), d) e^{-\lambda s} ds \leq \frac{L}{\lambda}.$$

On the other hand, for any $(d(\cdot), \xi(\cdot)) \in \mathcal{A}^d \times \mathcal{K}$, by (2.4), (2.6) and (2.8), it holds that

$$(3.4) \qquad J^d_x(d(\cdot), \xi(\cdot)) \geq \int_0^\infty f^0(y_x(s), d(s)) e^{-\lambda s} ds \geq -\frac{L}{\lambda}.$$

Hence, (3.1) follows. To prove (3.2), we first observe that by (2.4),

$$(3.5) \qquad |f^0(x,d) - f^0(\bar{x},d)| \leq 2L|x - \bar{x}|^\sigma, \qquad \forall x, \bar{x} \in X, \ d \in \Lambda, \ \sigma \leq \delta.$$

In fact, in the case where $|x - \bar{x}| \leq 1$,

$$|f^0(x,d) - f^0(\bar{x},d)| \leq L|x - \bar{x}|^\delta \leq L|x - \bar{x}|^\sigma,$$

and in the case where $|x - \bar{x}| \geq 1$,

$$|f^0(x,d) - f^0(\bar{x},d)| \leq 2L \leq 2L|x - \bar{x}|^\sigma.$$

§3. Properties of the Value Function

Thus, (3.5) follows. Now, we take $\sigma \in (0, \frac{\lambda}{LL})$ with $\sigma \leq \delta$. Then, for any $x, \bar{x} \in X$, $(d(\cdot), \xi(\cdot)) \in \mathcal{A}^d \times \mathcal{K}$, by (2.26), we have

(3.6)
$$|J_x^d(d(\cdot), \xi(\cdot)) - J_{\bar{x}}^d(d(\cdot), \xi(\cdot))| \leq 2L \int_0^\infty |y_x(s) - y_{\bar{x}}(s)|^\sigma e^{-\lambda s}\, ds$$
$$\leq \frac{2\widehat{L}^\sigma L}{\lambda - \sigma \widehat{L} L} |x - \widehat{x}|^\sigma.$$

Thus, (3.2) follows. □

Next, let us make some further observations. For given $(x, d) \in X \times \Lambda$, suppose that $(d^*(\cdot), \xi^*(\cdot)) \in \mathcal{A}^d \times \mathcal{K}$ is an optimal control. Then, by (3.1) and (2.4), we have

(3.7)
$$\frac{L}{\lambda} \geq V^d(x) = J_x^d(d^*(\cdot), \xi^*(\cdot))$$
$$\geq -\frac{L}{\lambda} + \sum_{i \geq 1} k(d_{i-1}^*, d_i^*) e^{-\lambda \theta_i^*} + \sum_{j \geq 1} \ell(\xi_j^*) e^{-\lambda \tau_j^*}.$$

Hence,

(3.8)
$$\sum_{i \geq 1} k(d_{i-1}^*, d_i^*) e^{-\lambda \theta_i^*} + \sum_{j \geq 1} \ell(\xi_j^*) e^{-\lambda \tau_j^*} \leq \frac{2L}{\lambda}.$$

This suggests that we need only to consider the controls in a much smaller class. More precisely, we introduce the following sets:

$$\mathcal{A}_0^d = \left\{ d(\cdot) = \sum_{i \geq 1} d_{i-1} \chi_{[\theta_{i-1}, \theta_i)}(\cdot) \in \mathcal{A}^d \,\Big|\, \sum_{i \geq 1} k(d_{i-1}, d_i) e^{-\lambda \theta_i} \leq \frac{2L}{\lambda} \right\},$$

$$\mathcal{K}_0 = \left\{ \xi(\cdot) = \sum_{j \geq 1} \xi_j \chi_{[\tau_j, \infty)}(\cdot) \in \mathcal{K} \,\Big|\, \sum_{j \geq 1} \ell(\xi_j) e^{-\lambda \tau_j} \leq \frac{2L}{\lambda} \right\}.$$

It is clear that $\mathcal{A}_0^d \times \mathcal{K}_0 \neq \phi$. We have the following result.

**Proposition 3.2.** *Let (S1)–(S5) hold. Then*

(3.9)
$$V^d(x) = \inf_{\mathcal{A}_0^d \times \mathcal{K}_0} J_x^d(d(\cdot), \xi(\cdot)), \qquad \forall (x, d) \in X \times \Lambda.$$

*Proof.* Let $(d(\cdot), \xi(\cdot)) \in \mathcal{A}^d \times \mathcal{K} \setminus (\mathcal{A}_0^d \times \mathcal{K}_0)$. Suppose, say, $d(\cdot) \in \mathcal{A}^d \setminus \mathcal{A}_0^d$. Then

(3.10)
$$\frac{2L}{\lambda} < \sum_{i \geq 1} k(d_{i-1}, d_i) e^{-\lambda \theta_i}$$
$$\leq J_x^d(d(\cdot), \xi(\cdot)) - \int_0^\infty f^0(y_x(s), d(s)) e^{-\lambda s}\, ds$$
$$\leq J_x^d(d(\cdot), \xi(\cdot)) + \frac{L}{\lambda}.$$

Therefore, by (3.1),

$$J_x^d(d(\cdot), \xi(\cdot)) > \frac{L}{\lambda} \geq V^d(x). \tag{3.11}$$

This means that such a pair can never be optimal. We can prove the same thing if $\xi(\cdot) \in \mathcal{K} \setminus \mathcal{K}_0$. Hence, (3.9) follows. $\square$

The following lemma will be useful in §6.

**Lemma 3.3.** *There exists an increasing function $C(\cdot)$, such that for any $d(\cdot) = \sum_{i \geq 1} d_{i-1} \chi_{[\theta_{i-1}, \theta_i)}(\cdot) \in \mathcal{A}_0^d$ and $\xi(\cdot) = \sum_{j \geq 1} \xi_j \chi_{[\tau_j, \infty)}(\cdot) \in \mathcal{K}_0$, it holds that:*

$$\sum_{i \geq 1} \chi_{[\theta_i, \infty)}(t) \leq C(t), \qquad t \geq 0, \tag{3.12}$$

$$\sum_{j \geq 1} (1 + |\xi_j|) \chi_{[\tau_j, \infty)}(t) \leq C(t), \qquad t \geq 0. \tag{3.13}$$

*Proof.* For any given $d(\cdot) \in \mathcal{A}_0^d$ and $\xi(\cdot) \in \mathcal{K}_0$ as in the lemma, we denote

$$\widehat{i}_t(d(\cdot)) = \max\{i \geq 0 \mid \theta_i \leq t\} \equiv \sum_{i \geq 1} \chi_{[\theta_i, \infty)}(t), \tag{3.14}$$

$$\widehat{j}_t(\xi(\cdot)) = \max\{j \geq 1 \mid \tau_j \leq t\} \equiv \sum_{j \geq 1} \chi_{[\tau_j, \infty)}(t). \tag{3.15}$$

(Here, we take the convention that $\max \phi = 0$.) By (2.6), for any $t \geq 0$,

$$\frac{2L}{\lambda} \geq \sum_{i \geq 1} k(d_{i-1}, d_i) e^{-\lambda \theta_i} \geq k_0 \widehat{i}_t(d(\cdot)) e^{-\lambda t}. \tag{3.16}$$

Thus,

$$\widehat{i}_t(d(\cdot)) \leq \frac{2L}{\lambda k_0} e^{\lambda t}, \qquad \forall t \geq 0. \tag{3.17}$$

Similarly, by (2.8),

$$\widehat{j}_t(\xi(\cdot)) \leq \frac{2L}{\lambda \ell_0} e^{\lambda t}, \qquad \forall t \geq 0. \tag{3.18}$$

Also, we have

$$\frac{2L}{\lambda} \geq \sum_{j \geq 1} \ell(\xi_j) e^{-\lambda \tau_j} \geq \ell(\xi_j) e^{-\lambda \tau_j}, \qquad \forall j \geq 1. \tag{3.19}$$

Hence, it follows from (2.11) that

$$\alpha_0 |\xi_j|^\delta \leq \ell(\xi_j) \leq \frac{2L}{\lambda} e^{\lambda \tau_j}, \qquad \forall j \geq 1. \tag{3.20}$$

Thus, noting (3.15), we obtain

(3.21) $$|\xi_j| \leq [\frac{2L}{\alpha_0 \lambda}]^{1/\delta} e^{(\lambda/\delta)t}, \quad \forall j \leq \widehat{j}_t(\xi(\cdot)).$$

Then, by (3.18) and (3.21),

(3.22) $$\sum_{j\geq 1} |\xi_j| \chi_{[\tau_j, \infty)}(t) \leq \widehat{j}_t(\xi(\cdot)) \max_{1\leq j \leq \widehat{j}_t(\xi(\cdot))} |\xi_j|$$
$$\leq \frac{1}{\ell_0 \alpha_0^{1/\delta}} [\frac{2L}{\lambda}]^{(1+\delta)/\delta} e^{[\lambda(1+\delta)/\delta]t}, \quad \forall t \geq 0.$$

Combining (3.17), (3.18), and (3.22), we can construct a function $C(t)$ independent of $d(\cdot) \in \mathcal{A}_0^d$ and $\xi(\cdot) \in \mathcal{K}_0$, such that (3.12) and (3.13) hold. $\square$

The above lemma tells us that any control $(d(\cdot), \xi(\cdot)) \in \mathcal{A}_0^d \times \mathcal{K}_0$ has some kind of uniformity property in the following sense: For any given $t > 0$, the number of switches of $d(\cdot)$ within $[0, t]$ is uniformly bounded; and the number as well as the sizes of impulses of $\xi(\cdot)$ within $[0, t]$ are uniformly bounded. On the other hand, by Proposition 3.2, controls in $\mathcal{A}_0^d \times \mathcal{K}_0$ are enough to determine the value function $V(\cdot)$. This observation will be very useful in sequel.

## §4. Optimality Principle and the HJB Equation

It is seen that Problem (SI) is quite different from the one we studied in Chapter 6. Thus, to find the form of optimality principle and the corresponding Hamilton-Jacobi-Bellman equation is very interesting. This section is devoted to the derivation of the optimality principle and the HJB equation for Problem (SI).

Let us first introduce some notation. For any $v(\cdot) = (v^1(\cdot), v^2(\cdot), \cdots, v^m(\cdot)) : X \to \mathbb{R}^m$, we define

(4.1) $$M^d[v](x) = \min_{\bar{d} \neq d} \{v^{\bar{d}}(x) + k(d, \bar{d})\},$$

(4.2) $$N[v^d](x) = \inf_{\xi \in K} \{v^d(x+\xi) + \ell(\xi)\}.$$

We call $M^d$ and $N$ the *switching* and *impulse obstacles*, respectively. It is seen that the value of $M^d[v](x)$ only depends on the value of some other components of $v$ at the same location $x$. Thus, it is a "local" operator. However, the value of $N[v^d](x)$ possibly depends on the values of $v^d$ at some other points (besides $x$). Thus, it is a "nonlocal" operator. The following result is the Bellman optimality principle for our Problem (SI).

**Theorem 4.1.** *The value function $V(\cdot)$ satisfies the following: For any $(x, d) \in X \times \Lambda$,*

(4.3) $$V^d(x) \leq \min\{M^d[V](x), N[V^d](x)\},$$

(4.4) $$V^d(x) \leq \int_0^t f^0(y_x(s), d)e^{-\lambda s}\, ds + V^d(y_x(t))e^{-\lambda t}, \quad \forall t \geq 0,$$

where $y_x(\cdot)$ is the solution of

(4.5) $$\begin{cases} \dot{y}_x(s) = Ay_x(s) + f(y_x(s), d), & 0 < s \leq t, \\ y_x(0) = x. \end{cases}$$

Moreover, if the strict inequality holds at some point $x$ in (4.3), then there exists a $\delta_0 > 0$, such that

(4.6) $$V^d(x) = \int_0^t f^0(y_x(s), d)e^{-\lambda s}\, ds + V^d(y_x(t))e^{-\lambda t}, \quad 0 \leq t \leq \delta_0,$$

where $y_x(\cdot)$ satisfies (4.5) with $t = \delta_0$.

*Proof.* For any $\bar{d}(\cdot) = \sum_{i \geq 1} \bar{d}_{i-1} \chi_{[\bar{\theta}_{i-1}, \bar{\theta}_i)}(\cdot) \in \mathcal{A}^{\bar{d}}$, $\xi(\cdot) \in \mathcal{K}$, and $d \in \Lambda \setminus \{\bar{d}\}$, we define

(4.7) $$\begin{cases} d(\cdot) = \sum_{i \geq 1} d_{i-1} \chi_{[\theta_{i-1}, \theta_i)}(\cdot), \\ d_0 = d; \quad (d_i, \theta_i) = (\bar{d}_{i-1}, \bar{\theta}_{i-1}), \quad i \geq 1. \end{cases}$$

Then

(4.8) $$V^d(x) \leq J_x^d(d(\cdot), \xi(\cdot)) = J_x^{\bar{d}}(\bar{d}(\cdot), \xi(\cdot)) + k(d, \bar{d}).$$

Taking the infimum over $(\bar{d}(\cdot), \xi(\cdot)) \in \mathcal{A}^{\bar{d}} \times \mathcal{K}$, we obtain

(4.9) $$V^d(x) \leq V^{\bar{d}}(x) + k(d, \bar{d}), \quad \forall \bar{d} \neq d.$$

Similarly, we can show that

(4.10) $$V^d(x) \leq V^d(x + \xi) + \ell(\xi), \quad \forall \xi \in \mathcal{K}.$$

This proves (4.3).

Next, for any $d(\cdot) \in \mathcal{A}^d$, $\xi(\cdot) = \sum_{j \geq 1} \xi_j \chi_{[\tau_j, \infty)} \in \mathcal{K}$, and $t \geq 0$, we define

(4.11) $$\widetilde{d}(s) = \begin{cases} d, & s < t, \\ d(s-t), & s \geq t, \end{cases} \qquad \widetilde{\xi}(s) = \sum_{j \geq 1} \xi_j \chi_{[\tau_j + t, \infty)}(s).$$

Then

(4.12) $$V^d(x) \leq J_x^d(\widetilde{d}(\cdot), \widetilde{\xi}(\cdot)) \\ = \int_0^t f^0(y_x(s), d)e^{-\lambda s}\, ds + J_{y_x(t)}^d(d(\cdot), \xi(\cdot))e^{-\lambda t}.$$

Hence, by taking the infimum on the right-hand side, we obtain (4.4).

## §4. Optimality Principle and the HJB Equation

Now, we assume that the strict inequality in (4.3) holds. For any $\varepsilon > 0$, there exists a pair of controls $d^\varepsilon(\cdot) = \sum_{i\geq 1} d^\varepsilon_{i-1}\chi_{[\theta^\varepsilon_{i-1},\theta^\varepsilon_i)} \in \mathcal{A}^d$ and $\xi^\varepsilon(\cdot) = \sum_{j\geq 1} \xi^\varepsilon_j \chi_{[\tau^\varepsilon_j,\infty)}(\cdot) \in \mathcal{K}$, such that

$$V^d(x) + \varepsilon \geq J^d_x(d^\varepsilon(\cdot), \xi^\varepsilon(\cdot)). \tag{4.13}$$

We first claim that $\theta^\varepsilon_1, \tau^\varepsilon_j > 0$. In fact, if, say, $\theta^\varepsilon_1 = 0$, then we set

$$\widetilde{d}^\varepsilon(\cdot) = \sum_{i\geq 2} d^\varepsilon_i \chi_{[\theta^\varepsilon_{i-1},\theta^\varepsilon_i)}(\cdot), \tag{4.14}$$

which yields

$$\begin{aligned}V^d(x) + \varepsilon &\geq k(d, d^\varepsilon_1) + J^{d^\varepsilon_1}_x(\widetilde{d}^\varepsilon(\cdot), \xi^\varepsilon(\cdot)) \\ &\geq k(d, d^\varepsilon_1) + V^{d_1}(x) \geq M^d[V](x).\end{aligned} \tag{4.15}$$

This contradicts our assumption. Similarly, we can prove that $\tau^\varepsilon_1 > 0$. Next, we claim that there exists a $\delta > 0$, such that

$$\min\{\theta^\varepsilon_1, \tau^\varepsilon_1\} \geq \delta, \qquad \forall \varepsilon > 0. \tag{4.16}$$

Suppose the contrary. Then we may assume, say, $\theta^\varepsilon_1 \leq \tau^\varepsilon_1$ and $\theta^\varepsilon_1 \to 0$. Clearly, on the interval $[0, \theta^\varepsilon_1)$, there are no switches and impulses. Thus, the state $y_x(\cdot)$, which coincides with the solution of (4.5), satisfies $y_x(t) \to x$ as $t \to 0$. Hence, by (4.13),

$$\begin{aligned}V^d(x) + \varepsilon &\geq k(d, d^\varepsilon_1)e^{-\lambda\theta^\varepsilon_1} + \int_0^{\theta^\varepsilon_1} f^0(y_x(s), d)e^{-\lambda s}\, ds \\ &\quad + V^{d^\varepsilon_1}(y_x(\theta^\varepsilon_1 - 0))e^{-\lambda\theta^\varepsilon_1} \\ &\geq k(d, d^\varepsilon_1) + V^{d^\varepsilon_1}(x) + o(1) \geq M^d[V](x) + o(1).\end{aligned} \tag{4.17}$$

This contradicts our assumption again. The proof for the case where $\tau^\varepsilon_1 \leq \theta^\varepsilon_1$ and $\tau^\varepsilon_1 \to 0$ is similar. Therefore, (4.16) holds. Now, for any $t \in [0, \delta)$,

$$V^d(x) + \varepsilon \geq \int_0^t f^0(y_x(s), d)e^{-\lambda s}\, ds + V(y_x(t))e^{-\lambda t}. \tag{4.18}$$

Let $\varepsilon \to 0$; we obtain (4.6) for some $\delta > 0$. □

Similar to Chapter 6, we have the following result.

**Proposition 4.2.** *Suppose that the value function $V(x)$ is $C^1$. Then it is a solution of the following Hamilton-Jacobi-Bellman equation:*

$$\begin{cases}\max\{\lambda V^d(x) - \langle DV^d(x), Ax + f(x,d)\rangle - f^0(x,d), \\ \qquad V^d(x) - M^d[V](x), \; V^d(x) - N[V^d](x)\} = 0, \\ \qquad\qquad \forall x \in \mathcal{D}(A),\; d \in \Lambda,\end{cases} \tag{4.19}$$

The proof is very similar to that given in Chapter 6 for the usual optimal control problem. Here, we need to use Theorem 4.1. The details are left to the readers. It is very interesting that the above HJB equation is quite different from the one we have in Chapter 6. Roughly speaking, the above says that the value function $V(\cdot)$ satisfies the following:

(4.20) $\begin{cases} V^d(x) \leq \min\{M^d[V](x), N[V^d](x)\}, & \forall (x,d) \in X \times \Lambda; \\ \lambda V^d(x) - \langle DV^d(x), Ax + f(x,d) \rangle - f^0(x,d) = 0, \\ \text{on } \{x \in \mathcal{D}(A) \mid V^d(x) < \min\{M^d[V](x), N[V^d](x)\}, \ d \in \Lambda. \end{cases}$

Such a system is referred to as a system of *quasivariational inequalities*.

Again, because the value function is not necessarily $C^1$, the above proposition is only formal. To make it rigorous, we need to work more. This leads to sections that follow.

## §5. Construction of an Optimal Control

The basic motivation of studying the value function is that, formally, we can use it to construct optimal controls from the value function. In this section, we rigorously construct (not just formally) an optimal control via the value function for Problem (SI). We assume that the value function $V(\cdot)$ satisfies Proposition 3.1. The construction is as follows.

Let $(x,d) \in X \times \Lambda$ be given. Set

(5.1) $$\theta_0^* = \tau_0^* = 0, \qquad d_0^* = d.$$

Then, we solve (4.5) to get $y_x(\cdot)$, which is continuous. Let

(5.2) $$s_1^* = \inf\left\{t \geq 0 \,\bigg|\, V^{d_0^*}(y_x(t)) = \min\{M^{d_0^*}[V](y_x(t)), N[V^{d_0^*}](y_x(t))\}\right\}.$$

Here, we take the convention that $\inf \phi = \infty$. All the functions involved in (5.2) are continuous; thus, $s_1^*$ is well defined. Suppose $s_1^* < \infty$. If

(5.3) $$V^{d_0^*}(y_x(s_1^*)) = M^{d_0^*}[V](y_x(s_1^*)),$$

then we define

(5.4) $$\theta_1^* = s_1^*,$$

(5.5) $d_1^* = \min\{\widetilde{d} \in \Lambda \setminus \{d_0^*\} \mid M^{d_0^*}[V](y_x(\theta_1^*)) = V^{\widetilde{d}}(y_x(\theta_1^*)) + k(d_0^*, \widetilde{d})\}.$

If

(5.6) $$V^{d_0^*}(y_x(s_1^*)) = N[V^{d_0^*}](y_x(s_1^*)) < M^{d_0^*}[V](y_x(s_1^*)),$$

§5. Construction of an Optimal Control

then we define

(5.7) $$\tau_1^* = s_1^*.$$

By the definition of $N$ and the compactness of $K \cap B_r(0)$, for any $r > 0$, there exists a $\zeta_1 \in K$, such that

(5.8) $$\begin{aligned}V^{d_0^*}(y_x(\tau_1^* - 0)) &= V^{d_0^*}(y_x(\tau_1^* - 0) + \zeta_1) + \ell(\zeta_1) \\ &= \inf_{\xi \in K} \{V^{d_0^*}(y_x(\tau_1^* - 0) + \xi) + \ell(\xi)\}.\end{aligned}$$

In this case, we define

(5.9) $$\xi_1^* = \zeta_1.$$

Note that there may exist more than one $\zeta_1$ satisfying (5.8). In that case, we take any one of them. In general, suppose that we have determined $d(\cdot)$ up to $(\theta_i^*, d_i^*)$ and $\xi(\cdot)$ up to $(\tau_j^*, \xi_j^*)$, and

(5.10) $$s_{i+j}^* = \theta_i^* \vee \tau_j^* < \infty,$$

where $a \vee b = \max\{a, b\}$, for all $a, b \in \mathbb{R}$. Then we can solve (4.5) on $[s_{i+j}^*, \infty)$ where $d$ is $d_i^*$ and the initial state is $y_x(s_{i+j}^* + 0)$. Let

(5.11) $$s_{i+j+1}^* = \inf\left\{t \geq s_{i+j}^* \,\Big|\, V^{d_i^*}(y_x(t)) = \min\{M^{d_i^*}[V](y_x(t)), N[V^{d_i^*}](y_x(t))\}\right\}.$$

If $s_{i+j+1}^* = \infty$, we are done. Suppose that $s_{i+j+1}^* < \infty$. Again, we have two cases. If

(5.12) $$V^{d_i^*}(y_x(s_{i+j+1}^*)) = M^{d_i^*}[V](y_x(s_{i+j+1}^*)),$$

then we define

(5.13) $$\theta_{i+1}^* = s_{i+j+1}^*,$$

(5.14) $$\begin{aligned}d_{i+1}^* = \min\{\widetilde{d} \in \Lambda \setminus \{d_i^*\} \,\Big|\, &M^{d_i^*}[V](y_x(\theta_{i+1}^*)) \\ &= V^{\widetilde{d}}(y_x(\theta_{i+1}^*)) + k(d_i^*, \widetilde{d})\}.\end{aligned}$$

For the other case, i.e.,

(5.15) $$V^{d_i^*}(y_x(s_{i+j+1}^*)) = N[V^{d_i^*}](y_x(s_{i+j+1}^*)) < M^{d_i^*}[V](y_x(s_{i+j+1}^*)),$$

we define

(5.16) $$\tau_{j+1}^* = s_{i+j+1}^*,$$

and let $\xi_{j+1}^* \in K$ be such that

$$
\begin{aligned}
(5.17)\quad V^{d_i^*}(y_x(\tau_{j+1}^* - 0)) &= V^{d_i^*}(y_x(\tau_{j+1}^* - 0) + \xi_{j+1}^*) + \ell(\xi_{j+1}^*) \\
&= N[V^{d_i^*}](y_x(\tau_{j+1}^* - 0)).
\end{aligned}
$$

By induction, we complete the construction. Now, we state the following result.

**Proposition 5.1.** *The control* $(d^*(\cdot), \xi^*(\cdot))$ *constructed above is an optimal control.*

*Proof.* First of all, by (2.5), (4.1), and the definitions of $d_i^*$ and $\theta_i^*$, we see that

$$(5.18)\quad V^{d_i^*}(y_x(\theta_i^*)) < M^{d_i^*}[V](y_x(\theta_i^*)).$$

Thus, it follows that

$$(5.19)\quad \theta_{i+1}^* > \theta_i^*, \quad \text{if} \quad \theta_i^* < \infty.$$

Similarly, by (2.7), we can obtain

$$(5.20)\quad \tau_{j+1}^* > \tau_j^*, \quad \text{if} \quad \tau_j^* < \infty,$$

and then,

$$(5.21)\quad V^{d_i^*}(y_x(\tau_j^* + 0)) < N[V^{d_i^*}](y_x(\tau_j^* + 0)).$$

We should note that the resulting trajectory $y_x(\cdot)$ under $(d^*(\cdot), \xi^*(\cdot))$ only jumps at $\tau_j^*$'s. Thus, we only put $\pm 0$ at $t = \tau_j^*$ to emphasize the discontinuity of $y_x(\cdot)$ at these points.

From the above analysis, we see that at $y_x(s_{i+j}^* + 0)$, we have

$$
\begin{aligned}
(5.22)\quad V^{d_i^*}(y_x(s_{i+j}^* + 0)) < \min \{ &M^{d_i^*}[V](y_x(s_{i+j}^* + 0), \\
&N[V^{d_i^*}](y_x(s_{i+j}^* + 0)) \}.
\end{aligned}
$$

Then, by Theorem 4.1 and the definition of $s_{i+j+1}^*$, we have

$$
\begin{aligned}
(5.23)\quad V^{d_i^*}(y_x(s_{i+j}^* + 0)) &= \int_0^{s_{i+j+1}^* - s_{i+j}^*} f^0(y_{y_x(s_{i+j}^* + 0)}(s), d_i^*) e^{-\lambda s}\, ds \\
&\quad + V^{d_i^*}(y_{y_x(s_{i+j}^* + 0)}(s_{i+j+1}^* - s_{i+j}^* - 0)) e^{-\lambda(s_{i+j+1}^* - s_{i+j}^*)} \\
&= \Big\{ \int_{s_{i+j}^*}^{s_{i+j+1}^*} f^0(y_x(s), d_i^*) e^{-\lambda s}\, ds \\
&\qquad + V^{d_i^*}(y_x(s_{i+j+1}^* - 0)) e^{-\lambda s_{i+j+1}^*} \Big\} e^{\lambda s_{i+j}^*}.
\end{aligned}
$$

§6. Approximation of the Control Problem

Now, combining the above with (5.14) and (5.17), we have

(5.24)
$$\begin{aligned}
e^{-\lambda s^*_{i+j}} V^{d^*_i}(y_x(s^*_{i+j}+0)) &= \int_{s^*_{i+j}}^{s^*_{i+j+1}} f^0(y_x(s), d^*_i) e^{-\lambda s}\, ds \\
&+ e^{-\lambda s^*_{i+j+1}} \Big\{ \{V^{d^*_{i+1}}(y_x(s^*_{i+j+1}+0)) + k(d^*_i, d^*_{i+1})\} \bar{\eta}(\theta^*_{i+1}, \tau^*_{j+1}) \\
&+ \{V^{d^*_i}(y_x(s^*_{i+j+1}+0)) + \ell(\xi^*_{j+1})\}\{1 - \bar{\eta}(\theta^*_{i+1}, \tau^*_{j+1})\}\Big\},
\end{aligned}$$

where, $\bar{\eta}(\theta, \tau) = \chi_{[0,\infty)}(\theta - \tau)$. Hence, if $s_{i_0+j_0} < \infty$, then (note (3.1))

(5.25)
$$\begin{aligned}
V^{d^*_0}(x) &= \int_0^{s^*_{i_0+j_0}} f^0(y_x(s), d^*(s)) e^{-\lambda s}\, ds + \sum_{i=1}^{i_0-1} k(d^*_{i-1}, d^*_i) e^{-\lambda \theta^*_i} \\
&+ \sum_{j=1}^{j_0-1} \ell(\xi^*_j) e^{-\lambda \tau^*_j} + e^{-\lambda s^*_{i_0+j_0}} \Big\{ V^{d^*_{i_0}}(y_x(s^*_{i_0+j_0+1}+0)) \bar{\eta}(\theta^*_{i_0}, \tau^*_{j_0}) \\
&+ V^{d^*_{i_0-1}}(y_x(s^*_{i_0+j_0+1}+0))\{1 - \bar{\eta}(\theta^*_{i_0}, \tau^*_{j_0})\} \\
&+ k(d^*_{i_0-1}, d^*_{i_0}) \bar{\eta}(\theta^*_{i_0}, \tau^*_{j_0}) + \ell(\xi^*_{j_0})\{1 - \bar{\eta}(\theta^*_{i_0}, \tau^*_{j_0})\}\Big\} \\
&\geq \int_0^{s^*_{i_0+j_0}} f^0(y_x(s), d^*(s)) e^{-\lambda s}\, ds + \sum_{i=1}^{i_0-1} k(d^*_{i-1}, d^*_i) e^{-\lambda \theta^*_i} \\
&+ \sum_{j=1}^{j_0-1} \ell(\xi^*_j) e^{-\lambda \tau^*_j} - e^{-\lambda s^*_{i_0+j_0}} \frac{L}{\lambda}.
\end{aligned}$$

Then, by (2.4) and (3.1), we have

(5.26)
$$\begin{aligned}
&\sum_{i=1}^{i_0-1} k(d^*_{i-1}, d^*_i) e^{-\lambda \theta^*_i} + \sum_{j=1}^{j_0-1} \ell(\xi^*_j) e^{-\lambda \tau^*_j} \\
&\leq V^{d^*_0}(x) - \int_0^{s^*_{i_0+j_0}} f^0(y_x(s), d^*(s)) e^{-\lambda s}\, ds + e^{-\lambda s^*_{i_0+j_0}} \frac{L}{\lambda} \leq \frac{3L}{\lambda}.
\end{aligned}$$

From this, we see that $\theta^*_i \uparrow \infty$ and $\tau^*_j \uparrow \infty$. This proves that $(d^*(\cdot), \xi^*(\cdot)) \in \mathcal{A}^d \times \mathcal{K}$. Then, letting $i_0 + j_0 \to \infty$ in (5.25), we get

(5.27)
$$V^{d^*_0}(x) = J^{d^*_0}_x(d^*(\cdot), \xi^*(\cdot)).$$

Therefore, $(d^*(\cdot), \xi^*(\cdot))$ is optimal. □

The above construction shows that, provided the value function is given, we can explicitly construct an optimal control that is of *feedback form*. Hence, the problem of finding or identifying the value function becomes very important.

## §6. Approximation of the Control Problem

As we pointed out in §1, it is possible to adopt the theory we presented in Chapter 6 to characterize the value function via the HJB equation (4.19). In that case, we have to impose some conditions on the operator $A$ and restrict $X$ to be a Hilbert space, etc. Here, we present another approach. We know that the difficult term in (4.19) is $\langle DV^d(x), Ax \rangle$. Thus, it is natural for us to approximate this term in some sense. This leads to the following approximation approach.

We start with an approximation of $A$. Let $\mu_0 \geq 0$ be given and $\{A_\mu \mid \mu \in [\mu_0, \infty)\}$ be a family of bounded linear operators on $X$ satisfying the following:

$$\text{(6.1)} \qquad \lim_{\mu \to \infty} A_\mu x = Ax, \qquad \forall x \in \mathcal{D}(A),$$

$$\text{(6.2)} \qquad \lim_{\mu \to \infty} e^{A_\mu t} x = e^{At} x, \qquad \forall x \in X, \text{ uniformly in } t \in [0, T], \quad \forall T > 0,$$

$$\text{(6.3)} \qquad \|e^{A_\mu t}\| \leq \widehat{L}, \qquad \forall t \geq 0, \mu \geq \mu_0.$$

Here, we assume (6.3) for the semigroup $e^{A_\mu t}$ just for the simplicity of presentation. In fact, by Chapter 2, §4, we can always choose $A_\mu$ to be the Yosida approximation of $A$, i.e., $A_\mu = \mu A(\mu - A)^{-1}$. Then (6.1) and (6.2) hold and for any $\eta > 0$, provided $\mu$ large enough, we have

$$\text{(6.4)} \qquad \|e^{A_\mu t}\| \leq \widetilde{L} e^{\eta t}, \qquad \forall t \geq 0,$$

for some $\widetilde{L} \geq 1$. This is possible due to the fact that we have (2.2). Then we may replace the original $A$ by $A - \eta I$ and $f(y, d)$ by $f(y, d) + \eta y$. In doing so, we have (6.1)–(6.3) with $A_\mu$ being replaced by $A_\mu - \eta I$. From this analysis, we see that by assuming (6.1)–(6.3), we have put no additional restrictions on the problem. In what follows, we will keep (6.1)–(6.3).

**Remark 6.1.** In the following, the operator $A_\mu$ does not have to be the Yosida approximation of $A$. Any approximation satisfying (6.1)–(6.3) works (e.g., it could be the finite dimensional approximation for $A$, like the Galerkin method, or the finite element method for elliptic operators). Thus, this approach seems to lead to some numerical consideration.

Next, we consider the following system:

$$\text{(6.5)} \qquad \begin{cases} \dot{y}_x^\mu(t) = A_\mu y_x^\mu(t) + f(y_x^\mu(t), d(t)) + \dot{\xi}(t), \\ y_x^\mu(0^-) = x. \end{cases}$$

Here $(d(\cdot), \xi(\cdot)) \in \mathcal{A}^d \times \mathcal{K}$. Similar to (2.13), we interpret (6.5) in the

## §6. Approximation of the Control Problem

following way:

(6.6)
$$y_x^\mu(t) = e^{A_\mu t}x + \int_0^t e^{A_\mu(t-s)} f(y_x^\mu(s), d(s))\, ds \\ + \sum_{j\geq 1} e^{A_\mu(t-\tau_j)} \xi_j \chi_{[\tau_j,\infty)}(t).$$

Hereafter, we will not distinguish (6.5) and (6.6). Our approximate optimal control problem can be stated as follows:

**Problem (SI)$_\mu$.** Given $(x,d) \in X \times \Lambda$, find $(d^*(\cdot), \xi^*(\cdot)) \in \mathcal{A}^d \times \mathcal{K}$, such that

(6.7)
$$J_x^{\mu,d}(d^*(\cdot), \xi^*(\cdot)) = \inf_{\mathcal{A}^d \times \mathcal{K}} J_x^{\mu,d}(d(\cdot), \xi(\cdot)),$$

where

(6.8)
$$J_x^{\mu,d}(d(\cdot), \xi(\cdot)) = \int_0^\infty f^0(y_x^\mu(s), d(s)) e^{-\lambda s}\, ds \\ + \sum_{i\geq 1} k(d_{i-1}, d_i) e^{-\lambda \theta_i} + \sum_{j\geq 1} \ell(\xi_j) e^{-\lambda \tau_j}.$$

Here, we should point out that the cost functional is the same as (2.27). We put the superscript $\mu$ in $J_x^{\mu,d}(d(\cdot), \xi(\cdot))$ to indicate that the state is subject to (6.6). Similar to §2, we define the value function $V_\mu(\cdot)$ of Problem (SI)$_\mu$ as follows:

(6.9) $\quad V_\mu^d(x) \equiv \inf_{\mathcal{A}^d \times \mathcal{K}} J_x^{\mu,d}(d(\cdot), \xi(\cdot)), \qquad \forall (x,d) \in X \times \Lambda,$

(6.10) $\quad V_\mu(x) \equiv (V_\mu^1(x), V_\mu^2(x), \cdots, V_\mu^m(x)), \qquad \forall x \in X.$

Clearly, all the results presented in §§2–5 hold for Problem (SI)$_\mu$. In particular, (3.1), (3.2), and (3.9) hold for $V_\mu(x)$.

We expect that $V_\mu^d(x)$ converges to $V^d(x)$ as $\mu \to \infty$. If this is the case, then Problem (SI)$_\mu$ will be a good approximation of our original Problem (SI). This is actually the case. We state the main result of this section as follows.

**Theorem 6.2.** Let (S1)–(S3) hold. Then, for any $(x,d) \in X \times \Lambda$,

(6.11) $\quad \lim_{\mu \to \infty} V_\mu^d(x) = V^d(x).$

To prove this result, we need some lemmas.

**Lemma 6.3.** Let (S2) and (S3) hold. Then, for any $x \in X$, $T > 0$, and $(d(\cdot), \xi(\cdot)) \in \mathcal{A}^d \times \mathcal{K}$,

(6.12) $\quad \lim_{\mu \to \infty} |y_x^\mu(t) - y_x(t)| = 0, \qquad$ uniformly in $t \in [0,T]$, where $y_x^\mu(\cdot)$ and $y_x(\cdot)$ are the trajectories of (6.6) and (2.13) corresponding to $(d(\cdot), \xi(\cdot))$, respectively.

*Proof.* For any $(x,d) \in X \times \Lambda$, $(d(\cdot), \xi(\cdot)) \in \mathcal{A}^d \times \mathcal{K}$, and $T > 0$, we have ($t \in [0,T]$)

$$
\begin{aligned}
(6.13) \quad |y_x^\mu(t) - y_x(t)| &\leq |(e^{A_\mu t} - e^{At})x| + \Big| \int_0^t e^{A_\mu(t-s)} f(y_x^\mu(s), d(s))\, ds \\
&\quad - \int_0^t e^{A(t-s)} f(y_x(s), d(s))\, ds \Big| \\
&\quad + \sum_{j \geq 1} |(e^{A_\mu(t-\tau)} - e^{A(t-\tau)})\xi_j| \chi_{[\tau_j, \infty)}(t) \\
&\leq \sup_{0 \leq r \leq T} |(e^{A_\mu r} - e^{Ar})x| + \widehat{L}L \int_0^t |y_x^\mu(s) - y_x(s)|\, ds \\
&\quad + \int_0^T \sup_{0 \leq r \leq T} |(e^{A_\mu r} - e^{Ar}) f(y_x(s), d(s))|\, ds \\
&\quad + \sum_{\tau_j \leq T} \sup_{0 \leq r \leq T} |(e^{A_\mu r} - e^{Ar})\xi_j|.
\end{aligned}
$$

By (6.1), (6.3), and the Dominated Convergence Theorem,

$$
\begin{aligned}
(6.14) \quad \psi_T^\mu &\equiv \sup_{0 \leq r \leq T} |(e^{A_\mu r} - e^{Ar})x| \\
&\quad + \int_0^T \sup_{0 \leq r \leq T} |(e^{A_\mu r} - e^{Ar}) f(y_x(s), d(s))|\, ds \\
&\quad + \sum_{\tau_j \leq T} \sup_{0 \leq r \leq T} |(e^{A_\mu r} - e^{Ar})\xi_j| \to 0, \quad \mu \to \infty.
\end{aligned}
$$

Thus, by Lemma 2.1, we have

$$
(6.15) \quad \sup_{0 \leq t \leq T} |y_x^\mu(t) - y_x(t)| \leq e^{\widehat{L}LT} \psi_T^\mu \to 0, \quad (\mu \to 0).
$$

Then our conclusion follows. $\square$

**Lemma 6.4.** *Let (S1) hold. Then, for any $\eta > 0$, there exists a countable set $K(\eta) \subset \mathcal{D}(A)$, such that for any $r > 0$, $K(\eta) \cap B_r(0)$ is a finite set and*

$$
(6.16) \quad \{\xi \in K \mid \ell(\xi) \leq r\} \subset \bigcup_{\bar{\xi} \in K(\eta), \ell(\bar{\xi}) \leq r + \eta} B_\eta(\bar{\xi}).
$$

*Proof.* Let $\eta > 0$ be fixed. For any $i \geq 0$, the set $\{\xi \in K \mid i\eta \leq \ell(\xi) \leq (i+1)\eta\}$ is compact (assuming it is nonempty, otherwise the conclusion is trivial). Thus, there exist finitely many points, denoted by $\xi_j \in K$,

§6. Approximation of the Control Problem

$m_i + 1 \leq j \leq m_{i+1}$, such that

(6.17) $$\begin{cases} i\eta \leq \ell(\xi_j) \leq (i+1)\eta, & m_i + 1 \leq j \leq m_{i+1}, \\ \{\xi \in K \mid i\eta \leq \ell(\xi) \leq (i+1)\eta\} \subset \bigcup_{j=m_i+1}^{m_{i+1}} B_\eta(\xi_j). \end{cases}$$

On the other hand, because $\mathcal{D}(A)$ is dense in $X$ and $\ell(\cdot)$ is continuous, for each $\xi_j$, we can find a $\bar{\xi}_j \in \mathcal{D}(A)$, such that

(6.18) $$|\xi_j - \bar{\xi}_j| < \eta, \qquad \ell(\bar{\xi}_j) \leq \ell(\xi_j) + \eta.$$

Now, let us define $K(2\eta) = \{\bar{\xi}_j, j \geq 1\}$. We claim that this set satisfies our requirements (with $\eta$ being replaced by $2\eta$). In fact, by the definition of $\xi_j$, one has (see (2.11))

(6.19) $$\lim_{j \to \infty} |\xi_j| = \infty.$$

Thus, by (6.18), the same is true for $\bar{\xi}_j$. This means that the set $K(2\eta) \cap B_r(0)$ is finite for each $r > 0$. Now, for any $r > 0$, we can find an $i$ such that $i\eta \leq r \leq (i+1)\eta$. Then it follows that

(6.20) $$\{\xi \in K \mid \ell(\xi) \leq r\} \subset \{\xi \in K \mid \ell(\xi) \leq (i+1)\eta\} \\ \subset \bigcup_{j=1}^{m_{i+1}} B_\eta(\xi_j) \subset \bigcup_{j=1}^{m_{i+1}} B_{2\eta}(\bar{\xi}_j).$$

Here, for any $\bar{\xi}_j$ with $j \leq m_{i+1}$, we have

(6.21) $$\ell(\bar{\xi}_j) \leq \ell(\xi_j) + \eta \leq (i+2)\eta \leq r + 2\eta.$$

Because $\eta > 0$ is arbitrary, replacing $\eta$ by $\eta/2$, we obtain (6.16). □

*Proof of Theorem 6.2.* For any $\varepsilon, \eta > 0$, we choose $K(\eta)$ as in Lemma 6.4, and define

$$\mathcal{A}^d(\varepsilon) = \left\{ \bar{d}(\cdot) = \sum_{i \geq 1} \bar{d}_{i-1} \chi_{[\bar{\theta}_{i-1}, \bar{\theta}_i)}(\cdot) \in \mathcal{A}^d \mid \bar{\theta}_i \in \{0, \varepsilon, 2\varepsilon, \cdots\} \right\},$$

$$\mathcal{K}(\varepsilon, \eta) = \left\{ \bar{\xi}(\cdot) = \sum_{j \geq 1} \bar{\xi}_j \chi_{[\bar{\tau}_j, \infty)}(\cdot) \in \mathcal{X} \mid \bar{\xi}_j \in K(\eta), \bar{\tau}_j \in \{0, \varepsilon, 2\varepsilon, \cdots\} \right\}.$$

In what follows, for any

$$d(\cdot) = \sum_{i \geq 1} d_{i-1} \chi_{[\theta_{i-1}, \theta_i)}(\cdot) \in \mathcal{A}_0^d, \qquad \xi(\cdot) = \sum_{j \geq 1} \xi_j \chi_{[\tau_j, \infty)}(\cdot) \in \mathcal{K}_0,$$

we always let

$$\bar{d}(\cdot) = \sum_{i \geq 1} \bar{d}_{i-1} \chi_{[\bar{\theta}_{i-1}, \bar{\theta}_i)}(\cdot) \in \mathcal{A}^d(\varepsilon), \qquad \bar{\xi}(\cdot) = \sum_{j \geq 1} \bar{\xi}_j \chi_{[\bar{\tau}_j, \infty)}(\cdot) \in \mathcal{K}(\varepsilon, \eta),$$

with

(6.22) $$\begin{cases} 0 \leq \bar{\theta}_i - \theta_i < \varepsilon, & \bar{d}_i = d_i, \quad i \geq 1, \\ 0 \leq \bar{\tau}_j - \tau_j < \varepsilon, & |\bar{\xi}_j - \xi_j| < \eta, \quad j \geq 1. \end{cases}$$

This is possible due to Lemma 6.4. Next, we let $y(\cdot)$ and $\bar{y}(\cdot)$ be the trajectories of (2.13) corresponding to $(d(\cdot), \xi(\cdot))$ and $(\bar{d}(\cdot), \bar{\xi}(\cdot))$. Then, by Lemmas 2.2 and 3.3, we obtain (note that $y(\cdot)$ and $\bar{y}(\cdot)$ have the same initial states)

$$|y(t)-\bar{y}(t)| \leq \widehat{L}e^{\widehat{L}Lt}\Big\{\sum_{j\geq 1}|\xi_j|\chi_{[\tau_j,\bar{\tau}_j)}(t) + \sum_{j\geq 1}\Big(\widehat{L}L|\xi_j||\tau_j - \bar{\tau}_j|$$
$$\{|(e^{A(\bar{\tau}_j-\tau_j)} - I)\bar{\xi}_j| + |\bar{\xi}_j - \xi_j|\}e^{-\widehat{L}L\tau_j}\Big)\chi_{[\tau_j,\infty)}(t)$$
$$+ 2L\Big(2 + \widehat{L}|x| + \sum_{j\geq 1}\widehat{L}e^{-\widehat{L}L\tau_j}|\xi_j|\chi_{[\tau_j,\infty)}(t)\Big)$$
$$\cdot \sum_{i\geq 1}|\theta_i - \bar{\theta}_i|\chi_{[\theta_i,\infty)}(t)\Big\}$$

(6.23)
$$\leq \widehat{L}e^{\widehat{L}Lt}\Big\{\sum_{j\geq 1}|\xi_j|\chi_{[\tau_j,\bar{\tau}_j)}(t)$$
$$+ \sum_{j\geq 1}\Big(\varepsilon\Big|\frac{e^{A(\bar{\tau}_j-\tau_j)} - I}{\bar{\tau}_j - \tau_j}\bar{\xi}_j\Big| + \eta + \widehat{L}L|\xi_j|\varepsilon\Big)\chi_{[\tau_j,\infty)}(t)$$
$$+ 2L\varepsilon\Big(2 + \widehat{L}|x| + \sum_{j\geq 1}\widehat{L}|\xi_j|\chi_{[\tau_j,\infty)}(t)\Big)\sum_{i\geq 1}\chi_{[\theta_i,\infty)}(t)\Big\}$$
$$\leq C(t)\Big\{\sum_{j\geq 1}\chi_{[\tau_j,\bar{\tau}_j)}(t) + \eta + \varepsilon$$
$$+ \varepsilon\Big(\sup_{0\leq r\leq 1, \bar{\xi}\in K(\eta), |\bar{\xi}|\leq C(t)}\Big|\frac{e^{Ar}-I}{r}\bar{\xi}\Big|\Big)\Big\},$$

where $C(t)$ is some increasing function that is independent of $(d(\cdot), \xi(\cdot)) \in \mathcal{A}_0^d \times \mathcal{K}_0$. Here, we should note that by Lemma 6.4, the set $K(\eta) \cap B_{C(t)}(0)$ is finite for each $\eta > 0$ and $t \geq 0$; thus,

(6.24) $$\sup_{0\leq r\leq 1, \bar{\xi}\in K(\eta), |\bar{\xi}|\leq C(t)} \Big|\frac{e^{Ar}-I}{r}\bar{\xi}\Big| \leq C(\eta, t), \quad t \geq 0, \eta > 0,$$

where $C(\eta, t)$ is a function, independent of $(d(\cdot), \xi(\cdot)) \in \mathcal{A}_0^d \times \mathcal{K}_0$ that might go to infinity as $\eta \to 0$ or $t \to \infty$. Thus, we can rewrite (6.23) as follows:

(6.25) $$|y(t) - \bar{y}(t)| \leq C(t)\Big\{\sum_{j\geq 1}\chi_{[\tau_j,\bar{\tau}_j)}(t) + \eta + \varepsilon\Big\} + \varepsilon C(\eta, t),$$
$$\forall t \geq 0,\, \eta > 0,\, (d(\cdot),\, \xi(\cdot)) \in \mathcal{A}_0^d \times \mathcal{K}(\varepsilon, \eta).$$

## §6. Approximation of the Control Problem

In exactly the same manner, for any $\mu \geq \mu_0$ and any $(d(\cdot), \xi(\cdot)) \in \mathcal{A}_0^d \times \mathcal{K}_0$, by constructing $(\bar{d}(\cdot), \bar{\xi}(\cdot)) \in \mathcal{A}^d(\eta) \times \mathcal{K}(\varepsilon, \eta)$ as above, the trajectories $y^\mu(\cdot)$ and $\bar{y}^\mu(\cdot)$ of (6.6) corresponding to $(d(\cdot), \xi(\cdot))$ and $(\bar{d}(\cdot), \bar{\xi}(\cdot))$ also satisfy (6.25), where the constants $C(t)$ and $C(\eta, t)$ are independent of $\mu$ and $(d(\cdot), \xi(\cdot)) \in \mathcal{A}_0^d \times \mathcal{K}_0$. Next, letting $T > 0$, we consider the following:

$$
\begin{aligned}
|J_x^{\mu,d}(d(\cdot), \xi(\cdot)) - J_x^d(d(\cdot), \xi(\cdot))| &\leq \frac{2L}{\lambda} e^{-\lambda T} \\
&\quad + \int_0^T |f^0(y^\mu(t), d(t)) - f^0(y(t), d(t))| e^{-\lambda t}\, dt \\
&\leq \frac{2L}{\lambda} e^{-\lambda T} + L \int_0^T |\bar{y}^\mu(t) - \bar{y}(t)|^\delta\, dt \\
&\quad + L \int_0^T \left\{ |y^\mu(t) - \bar{y}^\mu(t)|^\delta + |y(t) - \bar{y}(t)|^\delta \right\} dt \\
&\leq \frac{2L}{\lambda} e^{-\lambda T} + C(T)\eta + \varepsilon C(\eta, T) + L \int_0^T |\bar{y}^\mu(t) - \bar{y}(t)|^\delta\, dt.
\end{aligned}
\tag{6.26}
$$

Here, $C(T)$ and $C(\eta, T)$ are not necessarily the same as those in (6.25), but they are still independent of $(d(\cdot), \xi(\cdot)) \in \mathcal{A}_0^d \times \mathcal{K}_0$ and $\varepsilon > 0$. Thus, for any $\bar{\varepsilon} > 0$, we first take $T > 0$ large enough so that

$$
\frac{2L}{\lambda} e^{-\lambda T} < \bar{\varepsilon}.
\tag{6.27}
$$

Second, for this fixed $T > 0$, we take $\eta > 0$ sufficiently small so that

$$
C(T)\eta < \bar{\varepsilon}.
\tag{6.28}
$$

Third, we take $\varepsilon > 0$ small enough so that

$$
\varepsilon C(\eta, T) < \bar{\varepsilon}.
\tag{6.29}
$$

Finally, for the fixed $T, \varepsilon, \eta > 0$, from Lemma 3.3 and (6.22), we see that

$$
|\bar{\xi}(t)| \leq R(T), \quad \forall 0 \leq t \leq T,\ \xi(\cdot) \in \mathcal{K}_0,
\tag{6.30}
$$

for some $R(T)$ independent of $\xi(\cdot) \in \mathcal{K}_0$. By the definition of $\mathcal{A}^d(\varepsilon)$ and $\mathcal{K}(\varepsilon, \eta)$, we know that the set

$$
\{(\bar{d}(\cdot), \bar{\xi}(\cdot))|_{[0,T]} \mid (\bar{d}(\cdot), \bar{\xi}(\cdot)) \in \mathcal{A}^d(\varepsilon) \times \mathcal{K}(\varepsilon, \eta), \\ |\bar{\xi}(t)| \leq R(T),\ t \in [0, T]\}
$$

consists of finitely many elements. Thus, by Lemma 6.3,

$$
\lim_{\mu \to \infty} \int_0^T |\bar{y}^\mu(t) - \bar{y}(t)|^\delta\, dt = 0,
\tag{6.31}
$$

uniformly in $(\bar{d}(\cdot), \bar{\xi}(\cdot)) \in \mathcal{A}^d(\varepsilon) \times \mathcal{K}(\varepsilon, \eta)$ with $|\bar{\xi}(t)| \leq R(T)$, $t \in [0, T]$. Combining the above, we obtain

$$\text{(6.32)} \qquad \lim_{\mu \to \infty} |J_x^{\mu,d}(d(\cdot), \xi(\cdot)) - J_x^d(d(\cdot), \xi(\cdot))| = 0,$$

uniformly in $(d(\cdot), \xi(\cdot)) \in \mathcal{A}_0^d \times \mathcal{K}_0$. Then, by (3.9) and the same for $V_\mu^d(x)$, we obtain the convergence (6.11). □

## §7. Viscosity Solutions

Let $\mu \geq \mu_0$ be given. We consider the following Hamilton-Jacobi-Bellman system:

(7.1)
$$\begin{cases} \max\{\lambda v^d(x) - \langle Dv^d(x), A_\mu x + f(x, d) \rangle - f^0(x, d), \\ \quad v^d(x) - M^d[v](x), \; v^d(x) - N[v^d](x)\} = 0, \quad \forall x \in X, \; d \in \Lambda, \end{cases}$$

where $\langle \cdot, \cdot \rangle$ is the duality between $X$ and $X^*$. Here, we should note that (7.1) is the same as (4.19) except that $A$ is replaced by $A_\mu$. Consequently, $\mathcal{D}(A)$ is replaced by $X$. As we pointed out in the previous section, all the results of §§2–5 hold for Problem $(SI)_\mu$. Thus, the above (7.1) is the corresponding Hamilton-Jacobi-Bellman equation for the value function $V_\mu(x)$ of Problem $(SI)_\mu$. To make the theory rigorous, we need to introduce the notion of viscosity solution for the present case. In what follows, we let $C(X)^m$ be the set of all $\mathbb{R}^m$-valued continuous functions and $BUC(X)^m$ be the set of all $\mathbb{R}^m$-valued bounded uniformly continuous functions.

**Definition 7.1.** A function $v(\cdot) \in C(X)^m$ is called a *viscosity subsolution* (resp. *supersolution*) of (7.1), if for any $d \in \Lambda$ and any continuous function $\varphi(\cdot)$ with $v^d(\cdot) - \varphi(\cdot)$ attaining a local maximum (resp. minimum) at $x_0 \in X$, where $\varphi(\cdot)$ is Fréchet differentiable at $x_0$, it holds that

$$\text{(7.2)} \quad \begin{aligned} &\max\{\lambda v^d(x_0) - \langle D\varphi(x_0), A_\mu x_0 + f(x_0, d) \rangle - f^0(x_0, d), \\ &\quad v^d(x_0) - M^d[v](x_0), v^d(x_0) - N[v^d](x_0)\} \leq 0 \quad (\geq 0 \text{ resp.}). \end{aligned}$$

If a function $v(\cdot) \in C(X)^m$ is both a viscosity subsolution and supersolution of (7.1), we call it a *viscosity solution* of (7.1).

We see that the above definition is different from the one given in Chapter 6 in two aspects: (i) Here, we have a system of equations, and the definition is given accordingly; and (ii) the above definition is simpler than that given in Chapter 6 in the sense that we do not need the function $g$ and certain special properties of the function $\varphi$. The fact that $A$ is replaced by $A_\mu$ plays a very important role in the simplification of the above definition.

Our first result is the following.

**Proposition 7.2.** *The value function $V_\mu(\cdot)$ is a viscosity solution of (7.1).*

## §7. Viscosity Solutions

*Proof.* Let $\varphi(\cdot)$ be a function such that for some $d \in \Lambda$, the function $V_\mu^d - \varphi$ attains a local maximum at $x \in X$ and $\varphi$ is Fréchet differentiable at this point. Then, for $t > 0$ small enough, we have

$$V_\mu^d(y_x^\mu(t)) - \varphi(y_x^\mu(t)) \le V_\mu^d(x) - \varphi(x), \tag{7.3}$$

where $y_x^\mu(\cdot)$ is the solution of the following:

$$\begin{cases} \dot{y}_x^\mu(t) = A_\mu y_x^\mu(t) + f(y_x^\mu(t), d), \\ y_x^\mu(0) = x. \end{cases} \tag{7.4}$$

We recall that Theorem 4.1 holds for $V_\mu(\cdot)$. Thus, it follows that

$$\begin{aligned}
0 &\le V_\mu^d(y_x^\mu(t))e^{-\lambda t} - V_\mu^d(x) + \int_0^t f^0(y_x^\mu(s), d)e^{-\lambda s}\, ds \\
&\le (e^{-\lambda t} - 1)V_\mu^d(x) + e^{-\lambda t}(\varphi(y_x^\mu(t)) - \varphi(x)) \\
&\quad + \int_0^t f^0(y_x^\mu(s), d)e^{-\lambda s}\, ds.
\end{aligned} \tag{7.5}$$

Dividing by $t$ and sending $t \to 0$, we obtain

$$0 \le -\lambda V_\mu^d(x) + \langle D\varphi(x), A_\mu x + f(x,d) \rangle + f^0(x,d). \tag{7.6}$$

Combining with (4.3), we have

$$\max\{\lambda V_\mu^d(x) - \langle D\varphi(x), A_\mu x + f(x,d) \rangle - f^0(x,d), \\ V_\mu^d(x) - M^d[V_\mu](x), V_\mu^d(x) - N[V_\mu^d](x)\} \le 0. \tag{7.7}$$

This means that $V_\mu(\cdot)$ is a viscosity subsolution of (7.1). Next, we let $V_\mu^d - \varphi$ attain a local minimum at $x$. We need to show that

$$\max\{\lambda V_\mu^d(x) - \langle D\varphi(x), A_\mu x + f(x,d) \rangle - f^0(x,d), \\ V_\mu^d(x) - M^d[V_\mu](x), V_\mu^d(x) - N[V_\mu^d](x)\} \ge 0. \tag{7.8}$$

Clearly, if

$$V_\mu^d(x) = \min\{M^d[V_\mu](x), N[V_\mu^d](x)\}, \tag{7.9}$$

then (7.8) holds. Otherwise, let

$$V_\mu^d(x) < \min\{M^d[V_\mu](x), N[V_\mu^d](x)\}. \tag{7.10}$$

In this case, by Theorem 4.1, we know that there exists a $\delta_0 > 0$, such that

$$0 = V_\mu^d(x) - V_\mu^d(y_x^\mu(t))e^{-\lambda t} + \int_0^t f^0(y_x^\mu(s), d)e^{-\lambda s}\, ds, \quad 0 \le t \le \delta_0, \tag{7.11}$$

where $y_x^\mu(t)$ is the solution of (7.4). Then, dividing by $t$ and sending $t \to 0$, we obtain

$$\lambda V_\mu^d(x) - \langle D\varphi(x), A_\mu x + f(x,d) \rangle - f^0(x,d) = 0. \tag{7.12}$$

Hence, (7.8) holds. This proves our conclusion. □

In the rest of this section, we want to prove the uniqueness of the viscosity solutions of (7.1). To this end, let us make some assumptions on the space $X$.

(S6) Both the norms of $X$ and $X^*$ are Fréchet differentiable.

*Remark 7.3.* From Chapter 2, Proposition 2.20, we know that if $X$ is reflexive, we can always assume that (S6) holds. In particular, if $X$ is a Hilbert space, (S6) holds. Thus, (S6) is very general. Also, it is clear that (S6) implies the assumption (BP) stated in Chapter 6, §4.1 (the existence of a Fréchet differentiable bump function).

In what follows, we denote $\nu(x) = |x|^2$. Then, by (S6), $\nu(x)$ is Fréchet differentiable everywhere (including $x = 0$) and $|D\nu(x)| \leq 2|x|$. The following lemma will be used in the proof of our main uniqueness result.

**Lemma 7.4.** *Let $v$ be a viscosity subsolution of (7.1). Then*

$$v^d(x) \leq \min\{M^d[v](x), N[v^d](x)\}, \qquad \forall (x,d) \in X \times \Lambda. \tag{7.13}$$

*Proof.* We prove (7.13) by contradiction. Suppose for some $d \in \Lambda$ and $x_0 \in X$, we have

$$v^d(x_0) > \min\{M^d[v](x_0), N[v^d](x_0)\}. \tag{7.14}$$

Then, by the continuity of the function $v$ and the definition of the operators $M$ and $N$, one can find an $\varepsilon > 0$, such that

$$v^d(x) > \min\{M^d[v](x), N[v^d](x)\}, \qquad \forall x \in B_\varepsilon(x_0). \tag{7.15}$$

Now, let us define

$$\Psi(x) = v^d(x) - C\nu(x - x_0) \equiv v^d(x) - C|x - x_0|^2, \qquad x \in X, \tag{7.16}$$

with

$$C > \frac{2}{\varepsilon^2}\big(\sup_{x \in B_\varepsilon(x_0)} |v^d(x)| + |x_0| + \varepsilon\big). \tag{7.17}$$

Because $\Psi(x)$ is continuous and bounded on the closed ball $B_\varepsilon(x_0)$, by Chapter 6, Corollary 4.3, we can find a $p \in X^*$, with $|p| \leq 1$, such that $\Psi(x) - \langle p, x \rangle$ attains its maximum over $B_\varepsilon(x_0)$ at some point $\bar{x}_0 \in B_\varepsilon(x_0)$. We claim that $|\bar{x}_0 - x_0| < \varepsilon$. In fact, if $|\bar{x}_0 - x_0| = \varepsilon$, then

$$\begin{aligned}
\Psi(\bar{x}_0) - \langle p, \bar{x}_0 \rangle &= v^d(\bar{x}_0) - C\varepsilon^2 - \langle p, \bar{x}_0 \rangle \\
&< v^d(\bar{x}_0) - 2\big(\sup_{x \in B_\varepsilon(x_0)} |v^d(x)| + |x_0| + \varepsilon\big) + |\bar{x}_0| \\
&\leq -\sup_{x \in B_\varepsilon(x_0)} |v^d(x)| - 2|x_0| - 2\varepsilon + |x_0| + \varepsilon \\
&< v^d(x_0) - \langle p, x_0 \rangle = \Psi(x_0) - \langle p, x_0 \rangle.
\end{aligned} \tag{7.18}$$

## §7. Viscosity Solutions

This contradicts the definition of $\bar{x}_0$. Hence, $\bar{x}_0$ is a local maximum of the function $\Psi(x) - \langle p, x \rangle$. Then, by the definition of viscosity subsolution, we obtain

$$\begin{aligned}(7.19) \quad \max \{\lambda v^d(\bar{x}_0) - \langle p + CD\nu(\bar{x}_0 - x_0), A_\mu \bar{x}_0 + f(\bar{x}_0, d) \rangle \\ - f^0(\bar{x}_0, d), v^d(\bar{x}_0) - M^d[v](\bar{x}_0), \ v^d(\bar{x}_0) - N[v^d](\bar{x}_0)\} \leq 0.\end{aligned}$$

Thus,

$$(7.20) \qquad v^d(\bar{x}_0) \leq \min \{M^d[v](\bar{x}_0), N[v^d](\bar{x}_0)\}.$$

This contradicts (7.15) as $\bar{x}_0 \in B_\varepsilon(x_0)$. $\square$

The main result of this section is the following theorem.

**Theorem 7.5.** *Let (S1) and (S3)–(S6) hold. Let $v(\cdot), \bar{v}(\cdot) \in BUC(X)^m$ be two viscosity solutions of (7.1). Then $v(\cdot) = \bar{v}(\cdot)$.*

*Proof.* Let $v(\cdot)$ and $\bar{v}(\cdot)$ be two bounded uniformly continuous viscosity solutions of (7.1). We may let

$$(7.21) \quad \begin{cases} |v^d(x)|, |\bar{v}^d(x)| \leq C_0, & \forall (x, d) \in X \times \Lambda, \\ |v^d(x) - v^d(y)| + |\bar{v}^d(x) - \bar{v}^d(y)| \leq \bar{\omega}(|x - y|), \\ & \forall x, y \in X, \ d \in \Lambda, \end{cases}$$

with some constant $C_0$ and modulus of continuity $\bar{\omega}(\cdot)$. Next, let us recall the constants $\ell_0, \alpha_0, \lambda$, and $\delta$ defined in §2. For any $\alpha \in (0, 1)$ with

$$(7.22) \qquad C_1 \stackrel{\Delta}{=} \left(\frac{2C_0 + 1}{\alpha_0}\right)^{1/\delta} < \frac{\ell_0}{2\alpha},$$

we define $\beta, \bar{m}, R > 0$ as follows:

$$(7.23) \quad \begin{cases} 0 < \beta < 1 - \dfrac{2\alpha C_1}{\ell_0}, \quad 0 < \bar{m} < \min\left\{\dfrac{\lambda}{\|A_\mu\| + L}, 1\right\}, \\ R > C_1 + \left(\dfrac{4(C_0 + 1)}{\alpha}\right)^{1/\bar{m}}. \end{cases}$$

Then there exists an $\varepsilon_0 = \varepsilon_0(\alpha, C_1, \beta, \bar{m}, R) \in (0, 1)$, such that for all $0 < \varepsilon < \varepsilon_0$, it holds that

$$(7.24) \qquad (1 - \beta)\ell_0 > 2\alpha C_1 + \varepsilon(1 + C_1),$$

$$(7.25) \quad \begin{aligned}\max_{d \in \Lambda} (\beta v^d(0) - \bar{v}^d(0)) - 2\alpha - 1 &\geq -(\beta + 1)C_0 - 2\alpha - 1 \\ &> (\beta + 1)C_0 - \alpha(R - C_1)^{\bar{m}} + \varepsilon R.\end{aligned}$$

Let $\langle x \rangle = \sqrt{1 + \nu(x)}$ (recall $\nu(x) = |x|^2$). Then

$$(7.26) \qquad D\langle x \rangle^{\bar{m}} = \frac{\bar{m} D\nu(x)}{2\langle x \rangle^{2-\bar{m}}}, \quad |D\langle x \rangle^{\bar{m}}| \leq \bar{m}\langle x \rangle^{\bar{m}-1} \leq 1.$$

Consequently, the map $x \mapsto \langle x \rangle^{\overline{m}}$ is Lipschitz continuous with a Lipschitz constant of 1. Now, for given $\alpha \in (0,1)$, $\overline{m}$ as above, and $\varepsilon \in (0, \varepsilon_0)$, we define

$$(7.27) \quad \Psi^d(x,y) = \beta v^d(x) - \bar{v}^d(y) - \frac{1}{\varepsilon}\nu(x-y) - \alpha(\langle x \rangle^{\overline{m}} + \langle y \rangle^{\overline{m}}), \quad x, y \in X.$$

Clearly,

$$(7.28) \quad \lim_{|x|+|y|\to\infty} \max_{d\in\Lambda} \Psi^d(x,y) = -\infty, \quad \text{uniformly in } \varepsilon > 0.$$

Thus, the function $\max_{d\in\Lambda} \Psi^d(x,y)$ is continuous and bounded on the convex closed set $G_R \equiv \{(x,y) \in X \times X \mid |x|, |y| \leq R\}$. Now, by Chapter 6, Corollary 4.3, we can find $p_\varepsilon, q_\varepsilon \in X^*$, with $|p_\varepsilon| + |q_\varepsilon| < \varepsilon$, such that the function $\max_{d\in\Lambda} \Psi^d(x,y) - \langle p_\varepsilon, x \rangle - \langle q_\varepsilon, y \rangle$ attains its maximum over the set $G_R$ at some point $(x_0, y_0) \in G_R$.

*Claim 1.* For any $|\xi_0| \leq C_1$ (see (7.22)), $(x_0 + \xi_0, y_0 + \xi_0) \in G_R$; consequently, $(x_0, y_0) \in \text{Int } G_R$.

Suppose this is not the case. We may assume, say, $|x_0 + \xi_0| > R$. Then

$$(7.29) \quad |x_0| \geq R - C_1.$$

Thus, by (7.21) and (7.25),

$$(7.30) \quad \begin{aligned} &\max_{d\in\Lambda} \Psi^d(x_0, y_0) - \langle p_\varepsilon, x_0 \rangle - \langle q_\varepsilon, y_0 \rangle \\ &\leq (\beta+1)C_0 - \alpha(R-C_1)^{\overline{m}} + \varepsilon R \\ &\leq \max_{d\in\Lambda}\left(\beta v^d(0) - \bar{v}^d(0)\right) - 2\alpha - 1 = \max_{d\in\Lambda} \Psi^d(0,0) - 1. \end{aligned}$$

This contradicts the definition of $(x_0, y_0)$. Hence, Claim 1 is true.

Now, let $d_0 \in \Lambda$ be such that

$$(7.31) \quad \max_{d\in\Lambda}\left[\beta v^d(x_0) - \bar{v}^d(y_0)\right] = \beta v^{d_0}(x_0) - \bar{v}^{d_0}(y_0).$$

*Claim 2.* It holds that

$$(7.32) \quad \bar{v}^{d_0}(y_0) < \min\left\{M^{d_0}[\bar{v}](y_0), N[\bar{v}^{d_0}](y_0)\right\}.$$

In fact, if for some $d_1 \neq d_0$,

$$(7.33) \quad \bar{v}^{d_0}(y_0) = M^{d_0}[\bar{v}](y_0) = \bar{v}^{d_1}(y_0) + k(d_0, d_1),$$

then note that by Lemma 7.4,

$$(7.34) \quad v^{d_0}(x_0) \leq M^{d_0}[v](x_0) \leq v^{d_1}(x_0) + k(d_0, d_1),$$

we obtain (note (2.6), $d_0 \neq d_1$, and $\beta < 1$)

$$(7.35) \quad \begin{aligned} \beta v^{d_0}(x_0) - \bar{v}^{d_0}(y_0) &\leq \beta\left(v^{d_1}(x_0) + k(d_0, d_1)\right) - \bar{v}^{d_1}(y_0) - k(d_0, d_1) \\ &= \beta v^{d_1}(x_0) - \bar{v}^{d_1}(y_0) - (1-\beta)k(d_0, d_1) < \beta v^{d_1}(x_0) - \bar{v}^{d_1}(y_0). \end{aligned}$$

§7. Viscosity Solutions

This contradicts the definition of $d_0$ (see (7.31)). Thus, (7.33) is impossible. Now, suppose

(7.36)
$$\bar{v}^{d_0}(y_0) = N[\bar{v}^{d_0}](y_0) = \inf_{\xi \in K}[\bar{v}^{d_0}(y_0 + \xi) + \ell(\xi)]$$
$$\geq \bar{v}^{d_0}(y_0 + \xi_0) + \ell(\xi_0) - \varepsilon,$$

for some $\xi_0 \in K$. By (2.11) and (7.21), we see that (note $0 < \varepsilon < 1$)

(7.37) $\quad 1 + 2C_0 \geq \varepsilon + \bar{v}^{d_0}(y_0) - \bar{v}^{d_0}(y_0 + \xi_0) \geq \ell(\xi_0) \geq \alpha_0|\xi_0|^\delta.$

This gives $|\xi_0| \leq C_1$ (recall that $C_1$ is defined in (7.22)). By Claim 1 above, we have $(x_0 + \xi_0, y_0 + \xi_0) \in G_R$. Thus, noting Lemma 7.4, (7.22)–(7.24), (7.26), and (7.36), we have

(7.38)
$$\max_{d \in \Lambda} \Psi^d(x_0, y_0) - \langle p_\varepsilon, x_0 \rangle - \langle q_\varepsilon, y_0 \rangle$$
$$\leq \beta v^{d_0}(x_0) - \bar{v}^{d_0}(y_0) - \alpha(\langle x_0 \rangle^{\bar{m}} + \langle y_0 \rangle^{\bar{m}}) - \langle p_\varepsilon, x_0 \rangle - \langle q_\varepsilon, y_0 \rangle$$
$$\leq \beta\{v^{d_0}(x_0 + \xi_0) + \ell(\xi_0)\} - \bar{v}^{d_0}(y_0 + \xi_0) - \ell(\xi_0) + \varepsilon$$
$$\quad - \alpha(\langle x_0 + \xi_0 \rangle^{\bar{m}} + \langle y_0 + \xi_0 \rangle^{\bar{m}}) - \langle p_\varepsilon, x_0 + \xi_0 \rangle$$
$$\quad - \langle q_\varepsilon, y_0 + \xi_0 \rangle + 2\alpha|\xi_0| + \varepsilon|\xi_0|$$
$$\leq \beta v^{d_0}(x_0 + \xi_0) - \bar{v}^{d_0}(y_0 + \xi_0) - \alpha(\langle x_0 + \xi_0 \rangle^{\bar{m}} + \langle y_0 + \xi_0 \rangle^{\bar{m}})$$
$$\quad - \langle p_\varepsilon, x_0 + \xi_0 \rangle - \langle q_\varepsilon, y_0 + \xi_0 \rangle - (1 - \beta)\ell_0 + \varepsilon(1 + C_1) + 2\alpha C_1$$
$$< \max_{d \in \Lambda} \Psi^d(x_0 + \xi_0, y_0 + \xi_0) - \langle p_\varepsilon, x_0 + \xi_0 \rangle - \langle q_\varepsilon, y_0 + \xi_0 \rangle.$$

This again contradicts the definition of $(x_0, y_0)$. Hence, (7.36) is also impossible. Thus, we must have (7.32).

We should note that for fixed $\alpha$ and $R$, $(x_0, y_0)$ and $d_0$ depend on the parameter $\varepsilon$.

*Claim 3.* For fixed $0 < \alpha < 1$, it holds that

(7.39) $\quad \dfrac{1}{\varepsilon}|x_0 - y_0|^2 = o(1), \qquad \text{as } \varepsilon \to 0.$

In fact, by the definitions of $(x_0, y_0)$ and $d_0$, we know that

(7.40)
$$2[\Psi^{d_0}(x_0, y_0) - \langle p_\varepsilon, x_0 \rangle - \langle q_\varepsilon, y_0 \rangle]$$
$$\geq [\Psi^{d_0}(x_0, x_0) - \langle p_\varepsilon + q_\varepsilon, x_0 \rangle] + [\Psi^{d_0}(y_0, y_0) - \langle p_\varepsilon + q_\varepsilon, y_0 \rangle].$$

This implies (see (7.21) and (7.27))

(7.41)
$$\dfrac{2}{\varepsilon}|x_0 - y_0|^2$$
$$\leq \beta|v^{d_0}(x_0) - v^{d_0}(y_0)| + |\bar{v}^{d_0}(x_0) - \bar{v}^{d_0}(y_0)| + \varepsilon|x_0 - y_0|$$
$$\leq \bar{\omega}(|x_0 - y_0|) + \varepsilon|x_0 - y_0|.$$

Then, we see that (7.39) holds.

Now, we want to use the definition of viscosity solutions to derive the comparison estimate. From the above, we see that the function $v^{do}(x) - \frac{1}{\varepsilon\beta}\nu(x-y_0) - \frac{\alpha}{\beta}\langle x \rangle^{\overline{m}} - \frac{1}{\beta}\langle p_\varepsilon, x\rangle$ attains a local maximum at point $x = x_0$. Thus, by the definition of viscosity solutions, we have

(7.42)
$$\max\Big\{\lambda v^{do}(x_0) - \frac{1}{\beta}\langle \frac{D\nu(x_0-y_0)}{\varepsilon} + \frac{\alpha\overline{m}D\nu(x_0)}{2\langle x_0\rangle^{2-\overline{m}}} + p_\varepsilon,$$
$$A_\mu x_0 + f(x_0, d_0)\rangle - f^0(x_0, d_0),$$
$$v^{do}(x_0) - M^{do}[v](x_0),\ v^{do}(x_0) - N[v^{do}](x_0)\Big\} \le 0.$$

Also, the function $\bar{v}^{do}(y) + \frac{1}{\varepsilon}\nu(x_0-y) + \alpha\langle y\rangle^{\overline{m}} + \langle q_\varepsilon, y\rangle$ attains a local minimum at the point $y = y_0$. Thus,

(7.43)
$$\max\Big\{\lambda\bar{v}^{do}(y_0) - \langle \frac{D\nu(x_0-y_0)}{\varepsilon} - \frac{\alpha\overline{m}D\nu(y_0)}{2\langle y_0\rangle^{2-\overline{m}}} - q_\varepsilon,$$
$$A_\mu y_0 + f(y_0, d_0)\rangle - f^0(y_0, d_0),$$
$$\bar{v}^{do}(y_0) - M^{do}[\bar{v}](y_0),\ \bar{v}^{do}(y_0) - N[\bar{v}^{do}](y_0)\Big\} \ge 0.$$

From (7.42), we have

(7.44)
$$\lambda\beta v^{do}(x_0) - \langle \frac{D\nu(x_0-y_0)}{\varepsilon} + \frac{\alpha\overline{m}D\nu(x_0)}{2\langle x_0\rangle^{2-\overline{m}}} + p_\varepsilon,$$
$$A_\mu x_0 + f(x_0, d_0)\rangle - \beta f^0(x_0, d_0) \le 0,$$

and from (7.43) and (7.32), we have

(7.45)
$$\lambda\bar{v}^{do}(y_0) - \langle \frac{D\nu(x_0-y_0)}{\varepsilon} - \frac{\alpha\overline{m}D\nu(y_0)}{2\langle y_0\rangle^{2-\overline{m}}} - q_\varepsilon,$$
$$A_\mu y_0 + f(y_0, d_0)\rangle - f^0(y_0, d_0) \ge 0.$$

Hence, by (2.3), (2.4), (7.23), $|x_0|, |y_0| \le R$, and $|D\nu(x)| \le 2|x|$, we have

(7.46)
$$\lambda(\beta v^{do}(x_0) - \bar{v}^{do}(y_0))$$
$$\le \langle \frac{D\nu(x_0-y_0)}{\varepsilon}, A_\mu(x_0-y_0) + f(x_0,d_0) - f(y_0,d_0)\rangle$$
$$+ \frac{\alpha}{2}\Big\{\langle \frac{\overline{m}D\nu(x_0)}{\langle x_0\rangle^{2-\overline{m}}}, A_\mu x_0 + f(x_0,d_0)\rangle$$
$$+ \langle \frac{\overline{m}D\nu(y_0)}{\langle y_0\rangle^{2-\overline{m}}}, A_\mu y_0 + f(y_0,d_0)\rangle\Big\}$$
$$+ \beta f^0(x_0,d_0) - f^0(y_0,d_0)$$
$$+ \langle p_\varepsilon, A_\mu x_0 + f(x_0,d_0)\rangle + \langle q_\varepsilon, A_\mu y_0 + f(y_0,d_0)\rangle$$

## §7. Viscosity Solutions

$$\leq \frac{2|x_0-y_0|^2}{\varepsilon}(\|A_\mu\|+L) + L|x_0-y_0|^\delta + (1-\beta)L$$
$$+ \varepsilon\{\|A_\mu\|R + L(1+R)\}$$
$$+ \alpha\overline{m}\langle x_0\rangle^{\overline{m}-1}\{1+(\|A_\mu\|+L)|x_0|\}$$
$$+ \alpha\overline{m}\langle y_0\rangle^{\overline{m}-1}\{1+(\|A_\mu\|+L)|y_0|\}.$$

Now, for any $x \in X$ and any $0 < \alpha < 1$, we may take $R > 0$ large enough satisfying (7.23) so that $|x| \leq R$. Then

$$\beta v^d(x) - \bar{v}^d(x) - 2\alpha\langle x\rangle^{\overline{m}} - \langle p_\varepsilon + q_\varepsilon, x\rangle$$
$$\equiv \Psi^d(x,x) - \langle p_\varepsilon + q_\varepsilon, x\rangle$$
$$\leq \Psi^{d_0}(x_0, y_0) - \langle p_\varepsilon, x_0\rangle - \langle q_\varepsilon, y_0\rangle$$
$$\equiv \beta v^{d_0}(x_0) - \bar{v}^{d_0}(y_0) - \alpha\{\langle x_0\rangle^{\overline{m}} + \langle y_0\rangle^{\overline{m}}\}$$
$$-\frac{1}{\varepsilon}|x_0-y_0|^2 - \langle p_\varepsilon, x_0\rangle - \langle q_\varepsilon, y_0\rangle$$

(7.47)
$$\leq \frac{2(\|A_\mu\|+L)}{\lambda}\frac{|x_0-y_0|^2}{\varepsilon} + \frac{L}{\lambda}|x_0-y_0|^\delta + \frac{L}{\lambda}(1-\beta)$$
$$+ \frac{\varepsilon}{\lambda}\{\|A_\mu\|R + L(1+R)\} + \varepsilon R$$
$$- \alpha\Big\{\Big(1 - \frac{\overline{m}(\|A_\mu\|+L)}{\lambda}\Big)(\langle x_0\rangle^{\overline{m}} + \langle y_0\rangle^{\overline{m}})$$
$$- \overline{m}(\langle x_0\rangle^{\overline{m}-1} + \langle y_0\rangle^{\overline{m}-1})\Big\}$$
$$\leq C\Big\{\frac{|x_0-y_0|^2}{\varepsilon} + |x_0-y_0|^\delta + (1-\beta) + \varepsilon(1+R) + \alpha\Big\},$$

where $C$ is an absolute constant. Keep in mind that $(x_0, y_0, d_0)$ depends on $\varepsilon$. Because $\Lambda$ is a finite set, we may let $d_0$ be identical along some sequence $\varepsilon \to 0$. Thus, by sending $\varepsilon \to 0$ in (7.47) along a proper sequence, and taking (7.39) into account, we obtain

(7.48) $$\beta v^d(x) - \bar{v}^d(x) - 2\alpha\langle x\rangle^{\overline{m}} \leq C(1-\beta+\alpha).$$

Next, letting $\beta \to 1 - \frac{2\alpha C_1}{\ell_0}$ (see (7.23)), we have

(7.49) $$(1 - \frac{\alpha C_1}{\ell_0})v^d(x) - \bar{v}^d(x) - 2\alpha\langle x\rangle^{\overline{m}} \leq \alpha C(1 + \frac{2C_1}{\ell_0}),$$

where $C$ is an absolute constant. Then, sending $\alpha \to 0$, we get

(7.50) $$v^d(x) - \bar{v}^d(x) \leq 0.$$

This is true for all $(x,d) \in X \times \Lambda$. Hence, the theorem is proved. $\square$

In the above proof, we see that the main clue is very similar to that given in Chapter 6. The appearance of the switching and impulse obstacles brings a few difficulties that are overcome by Claims 1 and 2 above. From the above, we obtain immediately the following characterization of the value function $V_\mu(\cdot)$.

**Corollary 7.6.** *Let (S1) and (S3)–(S6) hold. Then the value function $V_\mu(\cdot)$ of the problem $(SI)_\mu$ is the unique bounded uniformly continuous viscosity solution of (7.1).*

*Remark 7.7.* Note that in the above proof, we do not need the compactness of $K$, and also, we only need the inequalities (2.5) and (2.7) without strictness. The compactness of $K$ was used in two places, the construction of an optimal control and the approximation of the control problem, whereas the strictness of (2.5) and (2.7) were only used in the construction of an optimal control (see (5.18)–(5.21)).

## §8. Problem in Finite Horizon

In this section we consider the optimal switching and impulse control problems in finite horizons for time varying systems. From Chapter 6, we expect that, for this case, the value function also depends on the time variable $t$. On the other hand, we see from §2 of this chapter that because of the appearance of the impulse control, the trajectory is *not necessarily* continuous in the time variable. Consequently, it is not obvious whether the value function is continuous in the time variable. However, from Chapter 6 and the previous sections, the continuity of the value function seems very important in characterizing the value function by the viscosity solution theory. In this section we will prove that under certain conditions, the value function of our optimal control problem is continuous on $[0, T] \times X$. Once we obtain such a continuity, the rest of the theory will be very similar to that for an infinite horizon problem.

The system we are going to consider is the following:

$$(8.1) \quad \begin{cases} \dot{y}_{t,x}(s) = Ay_{t,x}(s) + f(s, y_{t,x}(s), d(s)) + \dot{\xi}(s), & s \in (t, T], \\ y_{t,x}(t-0) = x. \end{cases}$$

Here, as in Chapter 6, $y_{t,x}(\cdot)$ represents the trajectory corresponding to the initial data $(t, x) \in [0, T) \times X$. Remember that $y_{t,x}(\cdot)$ also depends on the controls $d(\cdot)$ and $\xi(\cdot)$. Similar to §2, we let

$$\mathcal{A}^d[t, T] = \Big\{ d(\cdot) = \sum_{i \geq 1} d_{i-1} \chi_{[\theta_{i-1}, \theta_i)}(\cdot) : [t, T] \to \Lambda \,\Big|\, d_0 = d, \ \theta_0 = t;$$
$$\theta_i \in [t, T], \ \forall i \geq 1; \ \theta_i \uparrow T; \ d_{i+1} \neq d_i, \text{ if } \theta_{i+1} < T;$$
$$\sum_{i \geq 1} k(\theta_i, d_{i-1}, d_i) < \infty \Big\},$$

$$\mathcal{K}[t, T] = \Big\{ \xi(\cdot) = \sum_{j \geq 1} \xi_j \chi_{[\tau_j, T]}(\cdot) \in \mathcal{X} \,\Big|\, \xi_j \in K, \ \forall j \geq 1;$$
$$\sum_{j \geq 1} \ell(\tau_j, \xi_j) < \infty \Big\}.$$

## §8. Problem in Finite Horizon

We call any $d(\cdot) \in \mathcal{A}^d[t,T]$ an *admissible switching control* on $[t,T]$ and any $\xi(\cdot) \in \mathcal{K}[t,T]$ an *admissible impulse control* on $[t,T]$, respectively. Here, we note that the switching control $d(\cdot)$ takes initial value $d$ at the initial moment $t$. Also, we recall the identifications between $\sum_{i\geq 1} d_{i-1}\chi_{[\theta_{i-1},\theta_i)}(\cdot)$ and $\{\theta_i, d_i\}_{i\geq 0}$, as well as $\sum_{j\geq 1}\xi_j\chi_{[\tau_j,T]}(\cdot)$ and $\{\tau_j,\xi_j\}_{j\geq 1}$ (see §1).

For any $(t,x) \in [0,T] \times X$ and $(d(\cdot),\xi(\cdot)) \in \mathcal{A}^d[t,T] \times \mathcal{X}[t,T]$, the state equation (8.1) is understood as the following integral equation: (compare (2.13)):

$$\begin{aligned}
(8.2)\quad y_{t,x}(s) &= e^{A(s-t)}x + \int_t^s e^{A(s-r)}f(r,y_{t,x}(r),d(r))\,dr \\
&\quad + \sum_{\tau_j \leq s} e^{A(s-\tau_j)}\xi_j \\
&= e^{A(s-t)}x + \int_t^s e^{A(s-r)}f(r,y_{t,x}(r),d(r))\,dr \\
&\quad + \sum_{j\geq 1} e^{A(s-\tau_j)}\xi_j \chi_{[\tau_j,T]}(t).
\end{aligned}$$

The cost functional that we are going to minimize is the following:

$$\begin{aligned}
(8.3)\quad J_{t,x}^d(d(\cdot),\xi(\cdot)) &= \int_t^T f^0(s,y_{t,x}(s),d(s))\,ds + h(y_{t,x}(T)) \\
&\quad + \sum_{i\geq 1} k(\theta_i,d_{i-1},d_i) + \sum_{j\geq 1}\ell(\tau_j,\xi_j).
\end{aligned}$$

The meaning of the right-hand side of (8.3) is similar to that of (2.27). The differences are the following: First, we have the term $h(y_{t,x}(T))$, which represents the cost (or penalty) for the final state. Second, the dependence of the switching and impulse costs on the time variable is more general than in (2.27).

Now, let us make some assumptions. We keep (S1) and (S2) stated in §2, calling them (S1)' and (S2)' hereafter (for the simplicity of the statements of our results). We replace (S3)–(S5) by the following:

(S3)' $f : [0,T] \times X \times \Lambda \to X$, $f^0 : [0,T] \times X \times \Lambda \to \mathbb{R}$ and $h : X \to \mathbb{R}$ are continuous. There exists a constant $L > 0$ and a modulus of continuity $\omega_0$, such that for all $(t,d) \in [0,T] \times \Lambda$, $x, \hat{x} \in X$,

(8.4) $\quad |f(t,x,d) - f(t,\hat{x},d)| \leq L|x - \hat{x}|, \qquad |f(t,x,d)| \leq L(1+|x|),$

(8.5) $\quad |f^0(t,x,d) - f^0(t,\hat{x},d)| \leq \omega_0(|x-\hat{x}|), \qquad |f^0(t,x,d)| \leq L,$

(8.6) $\quad |h(x) - h(\hat{x})| \leq \omega_0(|x-\hat{x}|), \qquad |h(x)| \leq L.$

(S4)' $k : [0,T] \times \Lambda \times \Lambda \to \mathbb{R}^+ \equiv [0,\infty)$ is continuous, and there exists a modulus of continuity $\omega_1$, such that

(8.7) $\quad k(t,d,\widetilde{d}) < k(t,d,\widehat{d}) + k(t,\widehat{d},\widetilde{d}), \qquad \forall d \neq \widehat{d} \neq \widetilde{d}, \ t \in [0,T],$

(8.8) $\quad k(t,d,d) = 0, \qquad \min_{\widetilde{d} \neq d} k(t,d,\widetilde{d}) \equiv k_0 > 0, \qquad \forall t \in [0,T],$

(8.9) $\quad \begin{aligned} k(s,d,\widetilde{d}) - k(t,d,\widetilde{d}) &\leq \omega_1(s-t) k(t,d,\widetilde{d}), \\ & \forall 0 \leq t \leq s \leq T, \ d,\widetilde{d} \in \Lambda. \end{aligned}$

(S5)' $\ell : [0,T] \times X \to \mathbb{R}^+$ is continuous, and there exists a modulus of continuity $\omega_2$ and a constant $0 < \delta \leq 1$, such that

(8.10) $\quad \ell(t, \xi + \widehat{\xi}) < \ell(t, \xi) + \ell(t, \widehat{\xi}), \qquad \forall \xi, \widehat{\xi} \in K, \ t \in [0,T],$

(8.11) $\quad \inf_{\xi \in K, t \in [0,T]} \ell(t,\xi) \equiv \ell_0 > 0, \qquad \lim_{\xi \in K, |\xi| \to \infty} \frac{|\xi|^\delta}{\inf_{t \in [0,T]} \ell(t,\xi)} = 0.$

(8.12) $\quad \ell(s,\xi) - \ell(t,\xi) \leq \omega_2(s-t) \ell(t,\xi), \qquad \forall 0 \leq t \leq s \leq T, \ \xi \in K.$

Moreover, we assume the following.

(S6)' The functions $h$, $\ell$, and the set $K$ are compatible in the following sense:

(8.13) $\quad h(x) \leq \inf_{\xi \in K} \{h(x + \xi) + \ell(T, \xi)\}, \qquad \forall x \in X.$

Comparing (S3) with (S3)', we see that the continuity of $f^0$ in $x$ is relaxed. Comparing (S4), (S5) with (S4)', (S5)', we find that conditions (8.9) and (8.12) concerning the dependence on $t$ are imposed. These conditions hold if $k(t,d,\widetilde{d})$ and $\ell(t,\xi)$ are nonincreasing in $t$.

Similar to §2, we see that under (S1)'–(S5)', for any $(t,x,d) \in [0,T] \times X \times \Lambda$, $(d(\cdot), \xi(\cdot)) \in \mathcal{A}^d[t,T] \times \mathcal{K}[t,T]$, there exists a unique state trajectory $y_{t,x}(\cdot)$ to the state equation (8.2) and the cost functional (8.3) is well defined. Hence, we may state our optimal control problem as in §2:

**Problem (SI)'.** *Given* $(t,x,d) \in [0,T] \times X \times \Lambda$, *find* $(d^*(\cdot), \xi^*(\cdot)) \in \mathcal{A}^d[t,T] \times \mathcal{K}[t,T]$, *such that*

(8.14) $\quad J^d_{t,x}(d^*(\cdot), \xi^*(\cdot)) = \inf_{\mathcal{A}^d[t,T] \times \mathcal{K}[t,T]} J^d_{t,x}(d(\cdot), \xi(\cdot)).$

## §8. Problem in Finite Horizon

Next, we define the value function as follows:

(8.15)
$$\begin{cases} V^d(t,x) = \inf_{\mathcal{A}^d[t,T] \times \mathcal{K}[t,T]} J^d_{t,x}(d(\cdot), \xi(\cdot)), \\ \qquad\qquad\qquad\qquad (t,x,d) \in [0,T) \times X \times \Lambda, \\ V(t,x) = (V^1(t,x), V^2(t,x), \cdots, V^m(t,x)), \\ \qquad\qquad\qquad\qquad (t,x) \in [0,T) \times X. \\ V^d(T,x) = h(x), \qquad (x,d) \in X \times \Lambda. \end{cases}$$

Our main result of this section is the following.

**Theorem 8.1.** *Let (S1)'–(S6)' hold. Then, the value function $V(\cdot,\cdot)$ is bounded and continuous on $[0,T] \times X$.*

*Proof.* Let us first prove the boundedness of the value function. To this end, for any $(t,x,d) \in [0,T] \times X \times \Lambda$, we define $(d_0(\cdot), \xi_0(\cdot)) \in \mathcal{A}^d[t,T] \times \mathcal{K}[t,T]$ by $d_0(s) \equiv d$ and $\xi_0(\cdot) \equiv 0$ (i.e., no switchings and no impulses). Then, by the boundedness of $f^0$ and $h$, we have

(8.16) $$V^d(t,x) \leq J^d_{t,x}(d_0(\cdot), \xi_0(\cdot)) \leq (T+1)L.$$

On the other hand, for any $(d(\cdot), \xi(\cdot)) \in \mathcal{A}^d[t,T] \times \mathcal{K}[t,T]$, by the nonnegativity of the switching and impulse costs, and the boundedness of $f^0$ and $h$ again, we have

(8.17) $$J^d_{t,x}(d(\cdot), \xi(\cdot)) \geq -L(T+1).$$

Hence, by taking the infimum over $(d(\cdot), \xi(\cdot)) \in \mathcal{A}^d[t,T] \times \mathcal{K}[t,T]$ and then combining with (8.16), we obtain the boundedness of the value function:

(8.18) $$|V^d(t,x)| \leq (T+1)L, \qquad \forall (t,x,d) \in [0,T] \times X \times \Lambda.$$

Next, let us prove the continuity of the value function $V(t,x)$ in $x$. To this end, we first note that under (S1)'–(S3)', similar to Lemma 2.2, for any $(t,d) \in [0,T] \times \Lambda$, $x, \bar{x} \in X$, and any $(d(\cdot), \xi(\cdot)) \in \mathcal{A}^d[t,T] \times \mathcal{K}[t,T]$, we have

(8.19) $$|y_{t,x}(s) - y_{t,\bar{x}}(s)| \leq \widehat{L}e^{\widehat{LL}(s-t)}|x - \bar{x}| \leq C|x - \bar{x}|, \quad s \in [t,T].$$

with $C = \widehat{L}e^{\widehat{L}LT}$. Thus, it follows that

(8.20)
$$\begin{aligned}
|J^d_{t,x}(d(\cdot),\xi(\cdot)) &- J^d_{t,\bar{x}}(d(\cdot),\xi(\cdot))| \\
&\leq \int_t^T |f^0(s,y_{t,x}(s),d(s)) - f^0(s,y_{t,\bar{x}}(s),d(s))|\,ds \\
&\quad + |h(y_{t,x}(T)) - h(y_{t,\bar{x}}(T))| \\
&\leq \int_t^T \omega_0(|y_{t,x}(s) - y_{t,\bar{x}}(s)|)\,ds + \omega_0(|y_{t,x}(T) - y_{t,\bar{x}}(T)|) \\
&\leq \int_t^T \omega_0(C|x - \bar{x}|)\,ds + \omega_0(C|x - \bar{x}|) \\
&\leq (T+1)\omega_0(C|x - \bar{x}|).
\end{aligned}$$

Taking the infimum with respect to $(d(\cdot),\xi(\cdot)) \in \mathcal{A}^d[t,T] \times \mathcal{K}[t,T]$, we obtain the continuity of the value function $V(t,x)$ with respect to $x$. Actually, this continuity is uniform in $(t,x) \in [0,T] \times X$.

Finally, we need to prove the continuity in the time variable, which is much harder. In order to obtain such a continuity, we first define the following subsets of the control sets:

(8.21)
$$\begin{aligned}
\mathcal{A}^d_0[t,T] &\equiv \{d(\cdot) \equiv \{\theta_i, d_i\}_{i\geq 0} \in \mathcal{A}^d[t,T] \mid \\
&\qquad\qquad \sum_{i\geq 1} k(\theta_i, d_{i-1}, d_i) \leq 3(T+1)L\,\}, \\
\mathcal{K}_0[t,T] &\equiv \{\xi(\cdot) \equiv \{\tau_j, \xi_j\}_{j\geq 1} \in \mathcal{K}[t,T] \mid \\
&\qquad\qquad \sum_{j\geq 1} \ell(\tau_j, \xi_j) \leq 3(T+1)L\,\}.
\end{aligned}$$

We claim that

(8.22)
$$V^d(t,x) = \inf_{\mathcal{A}^d_0[t,T] \times \mathcal{K}_0[t,T]} J^d_{t,x}(d(\cdot),\xi(\cdot)),$$
$$\forall (t,x,d) \in [0,T] \times X \times \Lambda.$$

In fact, for any $(d(\cdot),\xi(\cdot)) \notin \mathcal{A}^d_0[t,T] \times \mathcal{K}_0[t,T]$, we have that either $d(\cdot) \notin \mathcal{A}^d_0[t,T]$ or $\xi(\cdot) \notin \mathcal{K}_0[t,T]$. Let us assume the former (the latter case can be proved similarly). Then (note (8.18))

(8.23)
$$\begin{aligned}
J^d_{t,x}(d(\cdot),\xi(\cdot)) &\geq -(T+1)L + \sum_{i\geq 1} k(\theta_i, d_{i-1}, d_i) \\
&> 2(T+1)L \geq V^d(t,x) + (T+1)L.
\end{aligned}$$

Hence, we must have (8.22). Now, let us fix $(x,d) \in X \times \Lambda$ and take $0 \leq t \leq \bar{t} \leq T$. For any $\bar{d}(\cdot) \equiv \{\bar{\theta}_i, \bar{d}_i\}_{i\geq 0} \in \mathcal{A}^d[\bar{t},T]$ and $\bar{\xi}(\cdot) \equiv \{\bar{\tau}_j, \bar{\xi}_j\}_{j\geq 1} \in \mathcal{K}[\bar{t},T]$, we define $d(\cdot) \equiv \{\theta_i, d_i\}_{i\geq 0} \in \mathcal{A}^d[t,T]$ and $\xi(\cdot) \equiv \{\tau_j, \xi_j\}_{j\geq 1} \in \mathcal{K}[t,T]$ as

§8. Problem in Finite Horizon

follows:

(8.24)
$$\begin{cases} d_i = \bar{d}_i, & \theta_i = \bar{\theta}_i, & i \geq 1, \\ \xi_j = \bar{\xi}_j, & \tau_j = \bar{\tau}_j, & j \geq 1. \end{cases}$$

The above definition means that under $(d(\cdot), \xi(\cdot))$, there is neither switchings nor impulses on $[t, \bar{t})$, and they act the same as $(\bar{d}(\cdot), \bar{\xi}(\cdot))$ on $[\bar{t}, T]$. Hence, similar to Chapter 6, Lemma 2.1, we have

(8.25) $$|y_{t,x}(\bar{t}) - x| \leq C|(e^{A(\bar{t}-t)} - I)x| + C(1+|x|)(\bar{t}-t).$$

Then, noting (8.20), we obtain

(8.26)
$$\begin{aligned} V^d(t,x) &\leq J^d_{t,x}(d(\cdot), \xi(\cdot)) \\ &= \int_t^{\bar{t}} f^0(s, y_{t,x}(s), d(s))\, ds + J^d_{\bar{t}, y_{t,x}(\bar{t})}(\bar{d}(\cdot), \bar{\xi}(\cdot)) \\ &\leq L(\bar{t}-t) + J^d_{\bar{t},x}(\bar{d}(\cdot), \bar{\xi}(\cdot)) + (T+1)\omega_0(C|y_{t,x}(\bar{t})-x|) \\ &\leq \omega(|\bar{t}-t|, x) + J^d_{\bar{t},x}(\bar{d}(\cdot), \bar{\xi}(\cdot)), \end{aligned}$$

with $\omega(r; x) = Lr + (T+1)\omega_0(C|(e^{Ar} - I)x| + C(1+|x|)r)$. Taking the infimum in $(\bar{d}(\cdot), \bar{\xi}(\cdot)) \in \mathcal{A}^d[\bar{t}, T] \times \mathcal{K}[\bar{t}, T]$, we obtain from the above that

(8.27) $$V^d(t,x) \leq \omega(|\bar{t}-t|; x) + V^d(\bar{t}, x).$$

Next, we need to establish the inequality in the other direction. To this end, let us consider the cases $\bar{t} < T$ and $\bar{t} = T$ separately. First, let $\bar{t} < T$. Fix $(x, d) \in X \times \Lambda$. For any $\bar{d}(\cdot) \equiv \{\bar{\theta}_i, \bar{d}_i\}_{i \geq 0} \in \mathcal{A}^d_0[\bar{t}, T]$ and $\bar{\xi}(\cdot) \equiv \{\bar{\tau}_j, \bar{\xi}_j\}_{j \geq 1} \in \mathcal{K}_0[\bar{t}, T]$, we define $d(\cdot) \equiv \{\theta_i, d_i\}_{i \geq 0} \in \mathcal{A}^d[\bar{t}, T]$ and $\xi(\cdot) \equiv \{\tau_j, \xi_j\}_{j \geq 1} \in \mathcal{K}[\bar{t}, T]$ as follows:

(8.28)
$$\begin{cases} d_i = \bar{d}_i, & i \geq 1, \\ \theta_i = \bar{t}, & \text{if } \bar{\theta}_i \leq \bar{t}, \\ \theta_i = \bar{\theta}_i, & \text{if } \bar{\theta}_i > \bar{t}. \end{cases}$$

(8.29)
$$\begin{cases} \xi_j = \bar{\xi}_j, & j \geq 1, \\ \tau_j = \bar{t}, & \text{if } \bar{\tau}_j \leq \bar{t}, \\ \tau_j = \bar{\tau}_j, & \text{if } \bar{\tau}_j > \bar{t}. \end{cases}$$

From the above definition, it is seen that when constructing $d(\cdot)$ from $\bar{d}(\cdot)$, we move all the switchings on $[t, \bar{t})$ to $\bar{t}$. Likewise, we move all the impulses of $\bar{\xi}(\cdot)$ on $[t, \bar{t})$ to $\bar{t}$ when we construct $\xi(\cdot)$ from $\bar{\xi}(\cdot)$. Let

$\bar{y}_{t,x}(\cdot) = y_{\bar{t},x}(\cdot\,;\bar{d}(\cdot),\bar{\xi}(\cdot))$. Clearly,

(8.30)
$$\begin{aligned}
J^d_{t,x}(d(\cdot),\xi(\cdot)) - V^d(\bar{t},x) &\geq \int_t^{\bar{t}} f^0(s, y_{t,x}(s), d(s))\, ds \\
&\quad + \int_{\bar{t}}^T \left[ f^0(s, y_{t,x}(s), d(s)) - f^0(s, \bar{y}_{\bar{t},x}(s), d(s)) \right] ds \\
&\quad + h(y_{t,x}(T)) - h(\bar{y}_{\bar{t},x}(T)) \\
&\quad + \sum_{\theta_i < \bar{t}} \left[ k(\theta_i, d_{i-1}, d_i) - k(\bar{t}, d_{i-1}, d_i) \right] \\
&\quad + \sum_{\tau_j < \bar{t}} \left[ \ell(\tau_j, \xi_j) - \ell(\bar{t}, \xi_j) \right] \\
&\geq -L(\bar{t}-t) - \int_{\bar{t}}^T \omega_0(|y_{t,x}(s) - \bar{y}_{\bar{t},x}(s)|)\, ds \\
&\quad - \omega_0(|y_{t,x}(T) - \bar{y}_{\bar{t},x}(T)|) \\
&\quad - \omega_1(\bar{t}-t)\sum_{i\geq 1} k(\theta_i, d_{i-1}, d_i) - \omega_2(\bar{t}-t)\sum_{j\geq 1}\ell(\tau_j,\xi_j) \\
&\geq -L(\bar{t}-t) - 3(T+1)L\left[\omega_1(\bar{t}-t) + \omega_2(\bar{t}-t)\right] \\
&\quad - (T+1)\omega_0(C|y_{t,x}(\bar{t}) - x - \xi(\bar{t})|).
\end{aligned}$$

Here we should note that

(8.31) $$\bar{y}_{\bar{t},x}(\bar{t}) = \bar{y}_{\bar{t},x}(\bar{t}+0) = x + \xi(\bar{t}).$$

On the other hand, by (8.11), we have some constant $\alpha_0 > 0$, such that

(8.32) $$\ell(t,\xi) \geq \alpha_0 |\xi|^\delta, \qquad \forall (t,\xi) \in [0,T] \times K.$$

Thus, for any $\xi(\cdot) \in \mathcal{K}_0[t,T]$, it holds that

(8.33) $$\begin{cases} \max_{j\geq 1}|\xi_j| \leq \left(\dfrac{3(T+1)L}{\alpha_0}\right)^{1/\delta}, \\ \widehat{j}(\xi(\cdot)) \equiv \text{number of jumps of } \xi(\cdot) \leq \dfrac{3(T+1)L}{\ell_0}. \end{cases}$$

We also note that $K \cap B_R(0)$ is compact for each $R > 0$. Thus, there exists a modulus of continuity $\widetilde{\omega}$, such that for all $\xi(\cdot) \in \mathcal{K}_0[t,T]$,

(8.34) $$\sum_{j\geq 1} |(e^{A(\tau_j - t)} - I)\xi_j| \chi_{[\tau_j, T]}(s) \leq \widetilde{\omega}(s-t), \qquad \forall s \in [t,T].$$

Then, from (8.2), (8.33), and (8.34), we have

$$|y_{t,x}(s) - x - \xi(s)| \leq |(e^{A(s-t)} - I)x| + C\int_t^s (1 + |y_{t,x}(r)|)\,dr$$

(8.35)
$$+ \sum_{j\geq 1} |(e^{A(\tau_j-t)} - I)\xi_j|\chi_{[\tau_j,T]}(s)$$

$$\leq |(e^{A(s-t)} - I)x| + C(s-t) + \widetilde{\omega}(s-t)$$

$$+ C\int_t^s |y_{t,x}(r) - x - \xi(r)|\,dr.$$

By Gronwall's inequality,

(8.36) $\quad |y_{t,x}(s) - x - \xi(s)| \leq \widehat{\omega}(s-t, x), \qquad s \in [t, T],$

for some local modulus of continuity $\widehat{\omega}(\cdot; x)$. Therefore, it follows from (8.30) that

(8.37) $\quad J_{t,x}^d(d(\cdot), \xi(\cdot)) - V^d(\bar{t}, x) \geq -\omega(\bar{t} - t, x),$

where $\omega(\cdot, \cdot)$ is a local modulus of continuity that is independent of $(d(\cdot), \xi(\cdot)) \in \mathcal{A}_0^d \times \mathcal{K}_0$. Taking the infimum for $(d(\cdot), \xi(\cdot)) \in \mathcal{A}_0^d[t,T] \times \mathcal{K}_0[t,T]$, by (8.22), we obtain

(8.38) $\quad V^d(t,x) - V^d(\bar{t},x) \geq -\omega(\bar{t} - t, x).$

This proves the continuity of the value function in $t$ for $t < T$. Finally, let us consider $\bar{t} = T$. By (8.10), (8.12), (8.13), and (8.36), we have

(8.39)
$$J_{t,x}^d(d(\cdot), \xi(\cdot)) - h(x) = \int_t^T f^0(s, y_{t,x}(s), d(s))\,ds$$
$$+ h(y_{t,x}(T)) - h(x) + \sum_{i\geq 1} k(\theta_i, d_{i-1}, d_i) + \sum_{j\geq 1} \ell(\tau_j, \xi_j)$$
$$\geq -L(T-t) + h(y_{t,x}(T)) - h(x + \xi(T))$$
$$+ \big[h(x + \xi(T)) + \ell(T, \xi(T)) - h(x)\big] - 3(T+1)L\omega_2(T-t)$$
$$\geq -\omega(T-t; x).$$

Hence, we obtain the continuity of the value function. □

As we said at the beginning of this section, one may follow a similar argument given in the previous sections to give an approximation of the value function, construct optimal controls, and characterize the value function for the approximate control problem via viscosity solutions, etc. We do not repeat these details here and suggest that the readers give a complete presentation as an exercise.

### Remarks

The optimal switching problem for ordinary differential equations was first studied by Capuzzo–Dolcetta and Evans [1]. For stochastic differential equations, this problem is a generalization of the optimal stopping problem and thus it can be traced back much earlier; see Wald [1], Shiryaev [1], and Evans–Friedman [1]. The optimal switching problem for partial differential equations (or evolution equations) was first studied by Stojanovic and Yong [1–2]. Later, Yong [6] investigated the two-person zero-sum differential games with both players using switching controls. It was proved that such a game admits an Elliott–Kalton type value.

The optimal impulse control problem for ordinary differential equations in an infinite horizon was first discussed by Barles [1]. For the stochastic case, readers are referred to the book by Bensoussan–Lions [1] for extensive details and many relevant references. The material of this chapter is based on the work of Yong [2]. The basic idea was initiated in Stojanovic–Yong [1–2]. Some other related works are Yong [1,9,12], and Yong-Zhang [1].

# Chapter 9

# Linear Quadratic Optimal Control Problems

## §1. Formulation of the Problem

In this chapter we consider the optimal control problem with a linear state equation and a quadratic cost functional. Such problems are referred to as *linear-quadratic optimal control problems*, or LQ problems for short.

### §1.1. Examples of unbounded control problems

Let us first present some interesting examples of unbounded control problems. To avoid some technical details that are not quite relevant to our discussion below, we are going to quote some standard and technical results from the literature.

Let $\Omega \subset \mathbb{R}^n$ be a bounded domain with a smooth boundary $\partial\Omega$. Let $a_{ij}(\cdot) \in C^1(\bar{\Omega})$, $a_0(\cdot) \in C(\bar{\Omega})$, satisfying the following:

$$(1.1) \quad \begin{cases} \sum_{i,j=1}^m a_{ij}(x)\xi_i\xi_j \geq \lambda|\xi|^2, \\ a_{ij}(x) = a_{ji}(x), \quad a_0(x) \geq \lambda, \end{cases} \quad \text{a.e. } x \in \Omega, \xi \in \mathbb{R}^n,$$

where $\lambda > 0$ is a constant. Define the differential operator $\mathcal{A}$ as follows:

$$(1.2) \quad \mathcal{A}y(x) = -\sum_{i,j=1}^n \left(a_{ij}(x)y_{x_j}(x)\right)_{x_i} + a_0(x)y(x).$$

In what follows, we consider several kinds of control problems.

1. *Dirichlet boundary control for parabolic equations.*

Consider the following control system:

$$(1.3) \quad \begin{cases} y_t + \mathcal{A}y = 0, & \text{in } \Omega \times (0,T), \\ y\big|_{\partial\Omega} = u, \\ y\big|_{t=0} = y_0. \end{cases}$$

Here, $y$ is the state and $u$ is called the *Dirichlet boundary control*. Set $X = L^2(\Omega)$, $U = L^2(\partial\Omega)$, and $\mathcal{U} = L^2(0,T;U)$. Define

$$(1.4) \quad A_D y(x) = \mathcal{A}y(x), \quad \forall y \in \mathcal{D}(A_D) \triangleq H_0^1(\Omega) \cap H^2(\Omega).$$

Then we know that $-A_D$ generates an analytic semigroup $e^{-A_D t}$ on $X$ and the fractional power $A_D^\alpha$ is well defined for all $\alpha \in \mathbb{R}$ (see Fujiwara [1], Lasiecka [1]). Moreover, it holds that

(1.5)
$$\mathcal{D}(A_D^\alpha) \triangleq \{y(\cdot) \in L^2(\Omega) \mid A_D^\alpha y(\cdot) \in L^2(\Omega)\} = \begin{cases} H^{2\alpha}(\Omega), & 0 \leq \alpha < 1/4, \\ H_0^{2\alpha}(\Omega), & 1/4 < \alpha < 3/4, \end{cases}$$

and

(1.6)
$$\|A_D^\alpha e^{-A_D t}\| \leq C t^{-\alpha}, \qquad \forall t > 0.$$

Next, we consider the following problem:

(1.7)
$$\begin{cases} \mathcal{A}\xi(x) = 0, & \text{in } \Omega, \\ \xi(x) = h(x), & \text{on } \partial\Omega. \end{cases}$$

From Lions–Magenes [1], we know that for any $s \in \mathbb{R}$, $h \in H^{s-\frac{1}{2}}(\partial\Omega)$, there exists a unique (weak) solution $\xi(\cdot) \in H^s(\Omega)$, and there exists some absolute constant $C > 0$ (depending on $s$ and independent of $h$), such that

(1.8)
$$\|\xi\|_{H^s(\Omega)} \leq C \|h\|_{H^{s-\frac{1}{2}}(\partial\Omega)}, \qquad \forall h \in H^{s-\frac{1}{2}}(\partial\Omega).$$

Thus, $h \mapsto \xi$ is a linear bounded operator from $H^{s-\frac{1}{2}}(\partial\Omega)$ to $H^s(\Omega)$. This map is called the *Dirichlet map*, denoted by $G_D$. From (1.8) and the continuity of the embedding $L^2(\partial\Omega) \hookrightarrow H^{-2\varepsilon}(\partial\Omega)$ ($\varepsilon > 0$), it follows that

(1.9) $\quad \|G_D h\|_{H^{\frac{1}{2}-2\varepsilon}(\Omega)} \leq C \|h\|_{H^{-2\varepsilon}(\partial\Omega)} \leq C \|h\|_{L^2(\partial\Omega)}, \quad \forall h \in L^2(\partial\Omega).$

Combining (1.5) and (1.9), we obtain

(1.10)
$$G_D(L^2(\partial\Omega)) \subseteq H^{\frac{1}{2}-2\varepsilon}(\Omega) = \mathcal{D}(A^{\frac{1}{4}-\varepsilon}).$$

Consequently, $A^{\frac{1}{4}-\varepsilon} G_D : L^2(\partial\Omega) \to L^2(\Omega)$ is closed and everywhere defined. Thus, by the Closed Graph Theorem (see Chapter 2, §1.2), this operator is bounded. Next, for a fixed $\varepsilon \in (0, \frac{1}{4})$, we denote

(1.11)
$$B = A_D^{\frac{1}{4}-\varepsilon} G_D \in \mathcal{L}(L^2(\partial\Omega); L^2(\Omega)).$$

We note that in (1.5), $\alpha = \frac{1}{4}$ is excluded. Thus, it is necessary to take $\varepsilon \in (0, \frac{1}{4})$ in the above. Now, take $u \in C^\infty(\partial\Omega \times [0,T])$. It is clear that $G_D u$, as a solution of (1.7) with $h$ being replaced by $u$, is smooth in the parameter $t$ and $(G_D u)_t = G_D u_t$. Consider the following equation:

(1.12)
$$\begin{cases} z_t + \mathcal{A}z = -G_D u_t, & \text{in } \Omega \times (0,T), \\ z\big|_{\partial\Omega} = 0, \\ z\big|_{t=0} = y_0(x) - (G_D u)(x, 0). \end{cases}$$

## §1. Formulation of the Problem

Clearly, if $z$ is a solution of (1.12), then $y = z + G_D u$ is a solution of (1.3). On the other hand, using the abstract evolution equation formulation, we can write down the mild solution of (1.12) as follows (set $A = A_D$):

$$
\begin{aligned}
z(t) &= e^{-At}\bigl(y_0 - (G_D u)(0)\bigr) - \int_0^t e^{-A(t-s)} (G_D u)_s\, ds \\
&= e^{-At}\bigl(y_0 - (G_D u)(0)\bigr) - (G_D u)(t) + e^{-At}(G_D u)(0) \\
&\quad + \int_0^t A e^{-A(t-s)} (G_D u)(s)\, ds \\
&= e^{-At} y_0 + \int_0^t A^{\frac{3}{4}+\varepsilon} e^{-A(t-s)} B u(s)\, ds - (G_D u)(t).
\end{aligned}
\tag{1.13}
$$

Consequently, the solution $y(\cdot\,; y_0, u)$ of (1.3) (for smooth $u$) is given by

$$
y(t; y_0, u) = e^{-At} y_0 + \int_0^t A^{\frac{3}{4}+\varepsilon} e^{-A(t-s)} B u(s)\, ds.
\tag{1.14}
$$

On the other hand, we have the following lemma.

**Lemma 1.1.** *Let $e^{-At}$ be an analytic semigroup on a Hilbert space $X$. Let $\alpha \in (0,1)$ and (1.6) holds with $A_D$ is replaced by $A$. Then, with the same constant $C > 0$ in (1.6), for any $0 \le a < b$, $z(\cdot) \in L^2(a, b; X)$, it holds that*

$$
\left| \int_a^\cdot A^\alpha e^{-A(\cdot - s)} z(s)\, ds \right|_{L^2(a,b;X)} \le \frac{C(b-a)^{1-\alpha}}{1-\alpha} |z(\cdot)|_{L^2(a,b;X)}.
\tag{1.15}
$$

*Proof.* For any $z(\cdot) \in L^2(a,b;X)$, by (1.6),

$$
\begin{aligned}
\int_a^b \left| \int_a^t A^\alpha e^{-A(t-s)} z(s)\, ds \right|^2 dt &\le C^2 \int_a^b \left( \int_a^t \frac{|z(s)|}{(t-s)^\alpha}\, ds \right)^2 dt \\
&\le C^2 \int_a^b \left( \int_a^t \frac{ds}{(t-s)^\alpha} \right)\left( \int_a^t \frac{|z(s)|^2}{(t-s)^\alpha}\, ds \right) dt \\
&\le C^2 \frac{(b-a)^{1-\alpha}}{1-\alpha} \int_a^b \int_s^b \frac{|z(s)|^2}{(t-s)^\alpha}\, dt\, ds \\
&\le C^2 \frac{(b-a)^{2-2\alpha}}{(1-\alpha)^2} \int_a^b |z(s)|^2\, ds.
\end{aligned}
\tag{1.16}
$$

Thus, (1.15) follows. □

Applying the above lemma to our case with $\alpha = \frac{3}{4} + \varepsilon < 1$ and $[a, b] = [0, T]$, we have that the solution $y(\cdot)$ of (1.14) satisfies the following *a priori* estimate ($\mathcal{U} \triangleq L^2(0, T; L^2(\partial\Omega))$):

$$
\|y(\cdot\,; y_0, u)\|_{L^2(0,T;X)} \le C_\alpha (|y_0|_X + \|u\|_{\mathcal{U}}), \quad \forall (y_0, u) \in X \times \mathcal{U}.
\tag{1.17}
$$

Hence, by the density of smooth functions in $\mathcal{U}$, we see that (1.14) holds for all $(y_0, u) \in X \times \mathcal{U}$.

**2. Neumann boundary control for parabolic equations.**

Consider the following control system:

$$(1.18) \quad \begin{cases} y_t + \mathcal{A}y = 0, & (x,t) \in \Omega \times (0,T), \\ y_{\nu_{\mathcal{A}}} = u(x,t), & (x,t) \in \partial\Omega \times (0,T), \\ y(x,0) = y_0(x), & x \in \Omega, \end{cases}$$

where $\mathcal{A}$ is defined by (1.2) and $\nu_{\mathcal{A}}$ is the conormal derivative associated with $\mathcal{A}$ on the boundary $\partial\Omega$ (see Chapter 2, §6). Here, $u(x,t)$ is called the *Neumann boundary control*. Similar to the Dirichlet boundary control problem, we may define

$$(1.19) \quad A_N y = \mathcal{A}y, \quad \forall y \in \mathcal{D}(A_N) \stackrel{\Delta}{=} \{y \in H^2(\Omega) | y_{\nu_{\mathcal{A}}} = 0\}.$$

It is known that $-A_N$ generates an analytic semigroup $e^{-A_N t}$ on $X \equiv L^2(\Omega)$, and the fractional power $A_N^\alpha$ is defined for all $\alpha \in \mathbb{R}$. Moreover, it is known that (see Fujiwara [1], Lasiecka [1]) for all $\alpha \in [0, \frac{3}{4})$

$$(1.20) \quad \mathcal{D}(A_N^\alpha) \equiv \{y(\cdot) \in L^2(\Omega) \mid A_N^\alpha y(\cdot) \in L^2(\Omega)\} = H^{2\alpha}(\Omega),$$

and (1.6) holds with $A_D$ being replaced by $A_N$. Now, we define the *Neumann map* $G_N$ as follows: For any $h \in H^{s-\frac{3}{2}}(\partial\Omega)$, $\xi = G_N h$ is the solution of the following:

$$(1.21) \quad \begin{cases} \mathcal{A}\xi = 0, & \text{in } \Omega, \\ \xi_{\nu_{\mathcal{A}}}|_{\partial\Omega} = h, & \text{on } \partial\Omega. \end{cases}$$

Then it is known that (see Lions–Magenes [1])

$$(1.22) \quad \|G_N h\|_{H^s(\Omega)} \leq C\|h\|_{H^{s-3/2}(\partial\Omega)}, \quad \forall h \in H^{s-3/2}(\partial\Omega).$$

Thus, for a fixed $\varepsilon \in (0, 1/4)$, similar as before (compare (1.11)),

$$(1.23) \quad B \stackrel{\Delta}{=} A^{\frac{3}{4}-\varepsilon} G_N : L^2(\partial\Omega) \to L^2(\Omega)$$

is a bounded linear operator. Again, $\varepsilon > 0$ is necessary because $\alpha = \frac{3}{4}$ is excluded in (1.20). Similar to the Dirichlet boundary control system, we have

$$(1.24) \quad y(t; y_0, u) = e^{-At} y_0 + \int_0^t A^{\frac{1}{4}+\varepsilon} e^{-A(t-s)} Bu(s)\, ds.$$

## §1. Formulation of the Problem

### 3. Pointwise control problem.

Let $n \leq 3$, $x_0 \in \Omega$, and $\delta(\cdot - x_0)$ be the Dirac $\delta$-function concentrated at $x_0$. Consider the following control system:

$$
(1.25) \quad \begin{cases} y_t + \mathcal{A}y = u(t)\delta(x - x_0), & \text{in } \Omega \times (0,T), \\ y|_{\partial \Omega} = 0, \\ y|_{t=0} = y_0. \end{cases}
$$

We note that for any $\alpha > \frac{n}{4}$, $H_0^{2\alpha}(\Omega) \hookrightarrow C(\overline{\Omega})$ (see Chapter 2, §6). Consequently, $\delta(\cdot - x_0) \in C(\overline{\Omega})^* \hookrightarrow H^{-2\alpha}(\Omega)$. Let $A = A_D$ and $\frac{n}{4} < \alpha < 1$ (which is possible as $n \leq 3$). Because $A^{-\alpha} \in \mathcal{L}(H^{-2\alpha}(\Omega); L^2(\Omega))$, we obtain that

$$
(1.26) \quad b \stackrel{\Delta}{=} A^{-\alpha}\delta(\cdot - x_0) \in L^2(\Omega).
$$

Now, let $h_k(\cdot) \in C^\infty(\Omega)$ be a sequence such that $h_k(\cdot) \stackrel{s}{\to} \delta(\cdot - x_0)$ in $H^{-2\alpha}(\Omega)$ and let $y_k(\cdot; y_0, u)$ be the mild solution of (1.25) with $\delta(\cdot - x_0)$ replaced by $h_k$. Then, using abstract formulation, one has

$$
(1.27) \quad \begin{aligned} y_k(t; y_0, u) &= e^{-At}y_0 + \int_0^t e^{-A(t-s)} h_k u(s)\, ds \\ &= e^{-At}y_0 + \int_0^t A^\alpha e^{-A(t-s)} A^{-\alpha} h_k u(s)\, ds. \end{aligned}
$$

Note that $A^{-\alpha} h_k \stackrel{s}{\to} b$ in $L^2(\Omega)$. Thus, letting $k \to \infty$ in (1.27), we obtain

$$
(1.28) \quad y(t; y_0, u) = e^{-At}y_0 + \int_0^t A^\alpha e^{-A(t-s)} bu(s)\, ds.
$$

Clearly, the above analysis remains true if $u(t)\delta(x - x_0)$ is replaced by $\sum_{k=1}^m u_k(t)\delta(x - x_k)$ with $x_k \in \Omega$, $1 \leq k \leq m$.

From the above examples, it is seen that for parabolic systems with Dirichlet, or Neumann boundary controls or pointwise control, we can represent the system as the following form:

$$
(1.29) \quad y(t) = e^{-At}x + \int_0^t A^\alpha e^{A(t-s)} Bu(s)\, ds, \qquad t \in [0,T],
$$

with $B \in \mathcal{L}(U, X)$ and $\alpha \in [0,1)$. Some more careful analysis will give a similar result for the so-called *patch control problem*. By this, we mean the following control system:

$$
(1.30) \quad \begin{cases} y_t + \mathcal{A}y = 0, & \text{in } \Omega \times (0,T), \\ y|_{\Gamma_0} = 0, & y_{\nu_A}|_{\partial\Omega \setminus \Gamma_0} = u, \\ y|_{t=0} = y_0. \end{cases}
$$

Here, $\Gamma_0$ is a submanifold of $\partial\Omega$. In (1.30), the Neumann boundary control is applied on the part of the boundary $\partial\Omega\setminus\Gamma_0$ and a fixed Dirichlet boundary condition is imposed on the rest of the boundary.

We note that the *distributed control problems* for evolution equations (not necessarily the parabolic type) correspond to the case $\alpha = 0$. Of course, in this case, we do not need the analyticity of $e^{-At}$. On the other hand, we may consider parabolic equations with mixed distributed, boundary, and pointwise controls. The same representation (1.29) holds for many of these systems. Hence, system (1.29) is very general. Because the operator $A^\alpha B$ is an unbounded operator if $\alpha > 0$, we refer to (1.29) as an *unbounded control system*.

**Remark 1.2.** If we let $z = A^{-\alpha}y$, then

$$(1.31) \qquad z(t) = e^{-At}A^{-\alpha}x + \int_0^t e^{-A(t-s)}Bu(s)\,ds, \qquad t \in [0,T].$$

This is a usual (bounded) control system. Of course, by doing this, we have changed the state space from $X$ to $A^{-\alpha}X$, which is smaller than the original one. Although we have a bounded control system (1.31) in this setting, the unboundedness will be brought into the cost functional. Consequently, the essential difficulty still remains. Hence, in what follows, we will not use this setting.

### §1.2. The LQ problem

In this subsection we are going to give a formulation of the linear quadratic optimal control problem, which will be studied below.

Let $X$ and $U$ be two real Hilbert spaces, $-A$ generate an analytic semigroup on $X$ with $0 \in \rho(A)$, and the fractional power $A^\beta$ be defined for all $\beta \in \mathbb{R}$, $B \in \mathcal{L}(U, X)$. Then, for any $\gamma > 0$, there exists a $C_\gamma > 0$, such that (see Chapter 2, §4.3)

$$(1.32) \qquad \begin{cases} \|A^\gamma e^{-At}\| \le C_\gamma t^{-\gamma}, & t > 0, \\ |(I - e^{-At})x| \le C_\gamma t^\gamma |A^\gamma x|, & x \in \mathcal{D}(A^\gamma),\ t > 0. \end{cases}$$

We consider state equation (1.29) with $0 < \alpha < 1$. The case $\alpha = 0$ is much easier and we leave the corresponding details to the readers. (For the case $\alpha = 0$, we do not need the analyticity of $e^{-At}$; the fractional powers are not needed either.) Let $\mathcal{U} = L^2(0, T; U)$. We have the following simple result.

**Proposition 1.3.** *Under our setting, the following hold:*

(i) *Let $\alpha \in (0, 1)$. Then the state trajectory $y(\,\cdot\,; x, u) \in L^2(0, T; X)$, for any $(x, u(\cdot)) \in X \times \mathcal{U}$.*

(ii) *If $\alpha \in (0, \frac{1}{2})$, the state trajectory $y(\,\cdot\,; x, u) \in C([0, T]; X)$ for any $(x, u(\cdot)) \in X \times \mathcal{U}$.*

§1. Formulation of the Problem

(iii) If $\alpha \in [\frac{1}{2}, 1)$, the state trajectory $y(\,\cdot\,;x,u)$ is not necessarily in $C([0,T];X)$.

*Proof.* (i) follows from Lemma 1.1.

(ii) Let $\alpha \in (0, \frac{1}{2})$. Then $s \mapsto s^{-\alpha}$ is in $L^2(0,T)$. Thus, for any $u(\cdot) \in \mathcal{U}$, $A^\alpha e^{-A(t-\cdot)}Bu(\cdot) \in L^1(0,t;X)$. Consequently, we can easily prove that $y(\,\cdot\,;x,u) \in C([0,T];X)$.

(iii) We construct a counterexample. Let $\{\varphi_n\}_{n\geq 1}$ be an orthonormal basis of the Hilbert space $X = U$, $B = I$, and let $A$ be given by

$$(1.33) \qquad A\varphi_n = n^2 \varphi_n, \qquad n \geq 1.$$

Then $e^{-At}$ is an analytic semigroup. (This can be proved by using Proposition 4.19 of Chapter 2). Now, let us take

$$(1.34) \qquad u(t) = \sum_{n\geq 1} e^{-n^2(T-t)} \varphi_n, \qquad t \in [0,T].$$

It is seen that

$$(1.35) \qquad \begin{aligned} \int_0^T |u(t)|^2\,dt &= \sum_{n\geq 1} \int_0^T e^{-2n^2(T-t)}\,dt \\ &= \sum_{n\geq 1} \frac{1-e^{-2n^2 T}}{2n^2} \leq \frac{1}{2}\sum_{n\geq 1} \frac{1}{n^2} < \infty. \end{aligned}$$

This implies that $u(\cdot) \in \mathcal{U}$. On the other hand,

$$(1.36) \qquad \begin{aligned} \left| \int_0^t A^\alpha e^{-A(t-s)} Bu(s)\,ds \right|^2 &= \sum_{n\geq 1} \left( \int_0^t n^{2\alpha} e^{-n^2(t-s)-n^2(T-s)}\,ds \right)^2 \\ &= \sum_{n\geq 1} n^{4\alpha} e^{-n^2(T+t)} \frac{e^{2n^2 t}-1}{2n^2} \\ &\geq \frac{1}{2} \sum_{n\geq 1} n^{4\alpha-2} e^{-n^2(T-t)} - \frac{1}{2} \sum_{n\geq 1} n^{4\alpha-2} e^{-n^2 T}. \end{aligned}$$

Because $\alpha \geq \frac{1}{2}$, the second term is finite (independent of $t$) and the first term goes to infinity as $t \to T$. Thus, $y(\,\cdot\,;x,u)$ cannot be continuous at $t = T$, proving our conclusion. $\square$

From the above proposition, we see that if $\alpha \geq \frac{1}{2}$, which is the case for parabolic equations with the Dirichlet boundary control, the state trajectory $y(\,\cdot\,;x,u)$ is not necessarily continuous. Hence, in general, the map $(x,u(\cdot)) \mapsto y(T;x,u)$ is not necessarily everywhere defined. However, we need to consider the following type of cost functional:

$$(1.37) \qquad \begin{aligned} J(x;u(\cdot)) &= \langle Q_1 y(T), y(T) \rangle \\ &\quad + \int_0^T \{ \langle Qy(t), y(t) \rangle + \langle Ru(t), u(t) \rangle \}\,dt, \end{aligned}$$

where $y(\cdot) = y(\cdot\,;x,u)$ is the mild solution of (1.29), and $Q_1^* = Q_1 \in \mathcal{L}(X)$, $Q^* = Q \in \mathcal{L}(X)$, $R^* = R \in \mathcal{L}(U)$. Because of the possible discontinuity of $y(\cdot\,;x,u)$ at $T$, we see that $J(x;u(\cdot))$ is not necessarily everywhere defined.

On the other hand, if a pair $(x,u(\cdot))$ ensures the state trajectory $y(\cdot\,;x,u)$ to be continuous at $t = T$, $y(T;x,u)$ is well defined and so is the cost functional $J(x;u(\cdot))$. We now discuss this issue more carefully. To this end, let us denote $\mathcal{X} = L^2(0,T;X)$ and recall $\mathcal{U} = L^2(0,T;U)$. Clearly, $\mathcal{X}$ and $\mathcal{U}$ are Hilbert spaces. Define the operators $L : \mathcal{U} \to \mathcal{X}$, $G_t : \mathcal{U} \to X$, and $L_0 : \mathcal{D}(L_0) \subseteq \mathcal{U} \to X$ as follows:

(1.38)
$$\begin{cases} (Lu(\cdot))(t) = \int_0^t A^\alpha e^{-A(t-s)} Bu(s)\, ds, \quad t \in [0,T],\ u(\cdot) \in \mathcal{U}, \\[4pt] G_t u(\cdot) = \int_0^t e^{-A(t-s)} Bu(s)\, ds, \quad t \in [0,T],\ u(\cdot) \in \mathcal{U}, \\[4pt] \mathcal{D}(L_0) \equiv \Big\{ u(\cdot) \in \mathcal{U} \;\Big|\; \lim_{t \to T} \int_0^t A^\alpha e^{-A(t-s)} Bu(s)\, ds \text{ exists}\Big\}, \\[4pt] L_0 u(\cdot) = \lim_{t \to T} \int_0^t A^\alpha e^{-A(t-s)} Bu(s)\, ds, \quad \forall u(\cdot) \in \mathcal{D}(L_0). \end{cases}$$

Clearly, $G_t \in \mathcal{L}(\mathcal{U};X)$. From Proposition 1.3, we have that $L \in \mathcal{L}(\mathcal{U};\mathcal{X})$ and $L_0 \in \mathcal{L}(\mathcal{U};X)$ if $\alpha \in [0, \frac{1}{2})$. For the case where $\alpha \in [\frac{1}{2}, 1)$, the following result is valid.

**Proposition 1.4.** *Let $\alpha \in [\frac{1}{2}, 1)$. Then the operator $L_0$ is densely defined, closable, and*

(1.39) $\quad \mathcal{D}(L_0) \supseteq \{u(\cdot)\chi_{[0,T-\delta)}(\cdot) + u_0 \chi_{[T-\delta,T]}(\cdot) \mid u(\cdot) \in \mathcal{U},\ u_0 \in U,\ \delta > 0\}$

*Moreover, $A^\alpha G_T$ is closed and is an extension of $L_0$.*

*Proof.* It is clear that the set on the right-hand side of (1.39) is dense in $\mathcal{U}$. Thus, (1.39) implies that $L_0$ is densely defined. Now, let us prove (1.39). For any $u(\cdot) \in \mathcal{U}$, $u_0 \in U$, and $\delta > 0$, we define

(1.40) $\quad u_\delta(s) = u(s)\chi_{[0,T-\delta)}(s) + u_0 \chi_{[T-\delta,T]}(s), \qquad s \in [0,T].$

Then, $u_\delta(\cdot) \in \mathcal{U}$ and for any $t \in (T - \frac{\delta}{2}, T]$, we have

(1.41)
$$\begin{aligned} g(t) &\triangleq \int_0^t A^\alpha e^{-A(t-s)} Bu_\delta(s)\, ds \\ &= \int_0^{T-\delta} A^\alpha e^{-A(t-s)} Bu(s)\, ds + \int_0^{t-T+\delta} A^\alpha e^{-As} Bu_0\, ds \\ &= A^\alpha e^{-A\delta/2} \int_0^{T-\delta} e^{-A(t-\delta/2-s)} Bu(s)\, ds \\ &\quad + A^{\alpha-1}\big(I - e^{-A(t-T+\delta)}\big) Bu_0. \end{aligned}$$

## §1. Formulation of the Problem

By (1.32), we see that $A^\alpha e^{-A\delta/2} \in \mathcal{L}(X)$. Thus, $g(\cdot)$ is continuous at $t = T$, which implies that $u_\delta(\cdot) \in \mathcal{D}(L_0)$. This proves (1.39).

Now, for any $u(\cdot) \in \mathcal{D}(L_0)$ and $z \in \mathcal{D}(A^*)$, by the definition of $L_0$, we have

$$\begin{aligned}
\langle L_0 u(\cdot), z \rangle &= \langle \lim_{t \to T} \int_0^t A^\alpha e^{-A(t-s)} B u(s)\, ds, z \rangle \\
&= \lim_{t \to T} \int_0^t \langle u(s), B^*(A^*)^\alpha e^{-A^*(t-s)} z \rangle\, ds \\
&= \int_0^T \langle u(s), B^*(A^*)^\alpha e^{-A^*(T-s)} z \rangle\, ds.
\end{aligned} \tag{1.42}$$

Next, let $u_k(\cdot) \in \mathcal{D}(L_0)$, such that

$$\begin{cases} u_k(\cdot) \xrightarrow{s} 0, & \text{in } \mathcal{U}, \\ L_0 u_k(\cdot) \xrightarrow{s} y, & \text{in } X, \end{cases} \tag{1.43}$$

for some $y \in X$. Then, by (1.42), for any $z \in \mathcal{D}(A^*)$, we have

$$\begin{aligned}
\langle y, z \rangle &= \lim_{k \to \infty} \langle L_0 u_k(\cdot), z \rangle \\
&= \lim_{k \to \infty} \int_0^T \langle u_k(s), B^*(A^*)^\alpha e^{-A^*(T-s)} z \rangle\, ds = 0.
\end{aligned} \tag{1.44}$$

By the density of $\mathcal{D}(A^*)$, we have $y = 0$. Thus, $L_0$ is closable (see Chapter 2, §4.1).

$A^\alpha$ is closed and $G_T \in \mathcal{L}(\mathcal{U}; X)$; thus, we have the closeness of $A^\alpha G_T$. We now prove that $A^\alpha G_T$ is an extension of $L_0$. For any $u(\cdot) \in \mathcal{D}(L_0)$, we have the existence of the limit

$$\lim_{t \to T} |A^\alpha G_t u - L_0 u| = 0. \tag{1.45}$$

Because $G_t u \xrightarrow{s} G_T u$ in $X$ and $A^\alpha$ is closed, we have that

$$G_T u \in \mathcal{D}(A^\alpha), \quad A^\alpha G_T u = L_0 u. \tag{1.46}$$

This shows that $A^\alpha G_T$ is an extension of $L_0$. □

In what follows, we denote $L_1 = A^\alpha G_T$. Because $L_1$ is closed and $\mathcal{D}(L_1) \supseteq \mathcal{D}(L_0)$, $L_1$ looks easier to treat than $L_0$. Thus, we will take the convention that

$$y(T; x, u) = e^{-AT} x + L_1 u, \quad \forall (x, u(\cdot)) \in X \times \mathcal{D}(L_1). \tag{1.47}$$

From the above result, we see that the cost functional (1.37) is defined on $X \times \mathcal{D}(L_1)$, which is dense in $X \times \mathcal{U}$.

Now, our optimal control problem can be stated as follows.

**Problem (LQ).** For given $x \in X$, find $u(\cdot) \in \mathcal{D}(L_1)$, so that the cost functional (1.37) is minimized.

As the state equation is linear and the cost functional is quadratic (in the state and the control), the above problem is referred to as the *linear-quadratic* optimal control problem (LQ problem, for short). If for given $x \in X$, it holds that

$$\inf_{u(\cdot) \in \mathcal{D}(L_1)} J(x; u(\cdot)) > -\infty, \tag{1.48}$$

we say that the Problem (LQ) is *well posed* at $x$. In addition, if there exists a $\bar{u}(\cdot) \in \mathcal{D}(L_1)$ (depending on $x$), such that

$$J(x; \bar{u}(\cdot)) = \inf_{u(\cdot) \in \mathcal{D}(L_1)} J(x; u(\cdot)), \tag{1.49}$$

we say that Problem (LQ) is *solvable* at $x$. In this case, we call $\bar{u}(\cdot)$ an *optimal control* of Problem (LQ); also the corresponding trajectory $\bar{y}(\cdot)$ and the pair $(\bar{y}(\cdot), \bar{u}(\cdot))$ are referred to as an *optimal trajectory* and an *optimal pair*, respectively. If Problem (LQ) is well posed (resp. solvable) at all $x \in X$, we simply say that Problem (LQ) is well posed (resp. solvable) (on $X$).

We note that in the case $\alpha \in [0, 1/2)$, in particular, for $\alpha = 0$, $\mathcal{D}(L_1) = \mathcal{U}$. The Problem (LQ) with $R \gg 0$ (meaning that $R - \delta I \geq 0$ for some $\delta > 0$, said to be *uniformly positive definite*) is usually referred to as a *regular* LQ problem. If, moreover, $Q \geq 0$, $Q_1 \geq 0$, and $\alpha = 0$, we call the problem a *standard* or *classical* LQ problem.

**Remark 1.5.** It should be pointed out that we allow the operators $Q$ and $Q_1$ to be indefinite. Also, we have not yet assumed any conditions on $R$. Suitable conditions will be imposed a little later.

It is straightforward that the adjoint operators $L^* \in \mathcal{L}(\mathcal{X}; \mathcal{U})$ and $L_1^* : \mathcal{D}(L_1^*) \subseteq X \to \mathcal{U}$ of $L$ and $L_1$ are given by

$$\begin{cases} (L^* y)(t) = \int_t^T B^* (A^*)^\alpha e^{-A^*(s-t)} y(s)\, ds, & t \in [0, T],\ \forall y \in \mathcal{X}, \\ L_1^* x = B^* (A^*)^\alpha e^{-A^*(T-t)} x, & t \in (0, T],\ \forall x \in \mathcal{D}(L_1^*). \end{cases} \tag{1.50}$$

We note that $B^* (A^*)^\alpha e^{-A^*(T-t)} x$ is defined for all $x \in X$, $0 < t < T$. However, for general $x$, such an expression does not necessarily belong to $\mathcal{U} \equiv L^2(0, T; U)$.

Next, we regard $Q$ and $R$ as operators on $\mathcal{X}$ and $\mathcal{U}$, respectively:

$$\begin{cases} (Qy(\cdot))(t) = Qy(t), & \forall y(\cdot) \in \mathcal{X}, \\ (Ru(\cdot))(t) = Ru(t), & \forall u(\cdot) \in \mathcal{U}, \end{cases} \quad \forall t \in [0, T]. \tag{1.51}$$

Then, for any $(x, u(\cdot)) \in X \times \mathcal{D}(L_1)$, we have

(1.52)
$$\begin{aligned}J(x; u(\cdot)) &= \langle (R + L^*QL)u(\cdot), u(\cdot) \rangle + \langle Q_1 L_1 u(\cdot), L_1 u(\cdot) \rangle \\ &\quad + 2\langle L^*Q e^{-A\cdot} x, u(\cdot) \rangle + 2\langle Q_1 e^{-AT} x, L_1 u(\cdot) \rangle \\ &\quad + \langle \big( \int_0^T e^{-A^*t} Q e^{-At}\, dt + A^{-A^*T} Q_1 e^{-AT} \big) x, x \rangle.\end{aligned}$$

Let us introduce the following notation:

(1.53)
$$\begin{cases} \Phi_0 = R + L^*QL, \quad \Theta_0 = L^*Q e^{-A\cdot}, \\ \Gamma = \int_0^T e^{-A^*t} Q e^{-At}\, dt + e^{-A^*T} Q_1 e^{-AT}. \end{cases}$$

Then we can rewrite (1.52) as follows:

(1.54)
$$\begin{aligned}J(x; u(\cdot)) &= \langle \Phi_0 u(\cdot), u(\cdot) \rangle + \langle Q_1 L_1 u(\cdot), L_1 u(\cdot) \rangle \\ &\quad + 2\langle \Theta_0 x, u(\cdot) \rangle + 2\langle Q_1 e^{-AT} x, L_1 u(\cdot) \rangle + \langle \Gamma x, x \rangle.\end{aligned}$$

Clearly, for any $x \in X$, $J(x; \cdot)$ is a densely defined functional on the Hilbert space $\mathcal{U}$. Hence, our original Problem (LQ) is transformed to a minimization problem for such a quadratic functional with $x \in X$ as a parameter.

## §2. Well Posedness and Solvability

Our first result is the following:

**Proposition 2.1.** *Let Problem (LQ) be well posed at some $x \in X$. Then*

(2.1) $\quad J(0; u(\cdot)) \equiv \langle \Phi_0 u, u \rangle + \langle Q_1 L_1 u, L_1 u \rangle \geq 0, \quad \forall u \in \mathcal{D}(L_1).$

*In particular, (2.1) holds if Problem (LQ) is well posed at $x = 0$.*

*Proof.* Suppose (2.1) fails. Then, for some $u_0 \in \mathcal{D}(L_1)$, we have

(2.2) $\quad \langle \Phi_0 u_0, u_0 \rangle + \langle Q_1 L_1 u_0, L_1 u_0 \rangle < 0.$

Thus, by (1.54), for any $k > 0$,

(2.3)
$$\begin{aligned}J(x; ku_0(\cdot)) &= k^2 \Big\{ \langle \Phi_0 u_0, u_0 \rangle + \langle Q_1 L_1 u_0, L_1 u_0 \rangle \\ &\quad + \frac{2}{k} \big( \langle \Theta_0 x, u_0 \rangle + \langle Q_1 e^{-AT} x, L_1 u_0 \rangle \big) + \frac{\langle \Gamma x, x \rangle}{k^2} \Big\} \\ &\leq \frac{k^2}{2} \Big\{ \langle \Phi_0 u_0, u_0 \rangle + \langle Q_1 L_1 u_0, L_1 u_0 \rangle \Big\},\end{aligned}$$

provided $k$ is large enough. The above clearly implies that

(2.4) $\quad \inf_{u(\cdot) \in \mathcal{D}(L_1)} J(x; u(\cdot)) = -\infty,$

which contradicts the well posedness of Problem (LQ). $\square$

The following result gives an equivalent condition for the solvability of Problem (LQ).

**Theorem 2.2.** *Let (2.1) hold. Then $(x, \bar{u}) \in X \times \mathcal{D}(L_1)$ satisfies (1.49) if and only if*

(2.5)
$$\begin{cases} Q_1(L_1\bar{u} + e^{-AT}x) \in \mathcal{D}(L_1^*), \\ L_1^*Q_1(L_1\bar{u} + e^{-AT}x) + \Phi_0\bar{u} + \Theta_0 x = 0. \end{cases}$$

*Proof.* First, let $(x, \bar{u}) \in X \times \mathcal{D}(L_1)$ such that (1.49) holds. Then, for any $u \in \mathcal{D}(L_1)$, we have

(2.6)
$$0 \leq \lim_{\lambda \to 0} \frac{J(x; \bar{u} + \lambda u) - J(x; \bar{u})}{\lambda}$$
$$= 2\langle \Phi_0\bar{u} + \Theta_0 x, u \rangle + 2\langle Q_1(L_1\bar{u} + e^{-AT}x), L_1 u \rangle.$$

Because $X \times \mathcal{D}(L_1)$ is a subspace, we must have the equality in the above. Hence, it follows that for all $u \in \mathcal{D}(L_1)$,

(2.7)
$$|\langle Q_1(L_1\bar{u} + e^{-AT}x), L_1 u \rangle| \leq |\langle \Phi_0\bar{u} + \Theta_0 x, u \rangle| \leq C|u|_\mathcal{U}.$$

This yields (2.5).

Conversely, let $(x, \bar{u}(\cdot)) \in X \times \mathcal{D}(L_1)$ satisfy (2.5). Then, for any $u(\cdot) \in \mathcal{D}(L_1)$, we have

(2.8)
$$J(x; u) - J(x; \bar{u}) = J(x; \bar{u} + u - \bar{u}) - J(x; \bar{u})$$
$$= 2\langle \Phi_0\bar{u} + \Theta_0 x + L_1^*Q_1(L_1\bar{u} + e^{-AT}x), u - \bar{u} \rangle$$
$$+ \langle \Phi_0(u - \bar{u}), u - \bar{u} \rangle + \langle Q_1 L_1(u - \bar{u}), L_1(u - \bar{u}) \rangle \geq 0.$$

Thus, (1.49) holds. $\square$

**Corollary 2.3.** *Suppose the operator*

(2.9) $$\Phi \triangleq \Phi_0 + L_1^*Q_1 L_1 \equiv R + L^*QL + L_1^*Q_1 L_1 : \mathcal{D}(L_1^*Q_1 L_1) \subseteq \mathcal{U} \to \mathcal{U}$$

*is boundedly invertible, i.e., $\Phi^{-1} \in \mathcal{L}(\mathcal{U})$. Then, for any $x \in \mathcal{D}(L_1^*Q_1 e^{-AT})$, $J(x; \cdot)$ admits a unique minimizer $\bar{u}(\cdot) \in \mathcal{D}(L_1^*Q_1 L_1)$, which is given by*

(2.10)
$$\bar{u}(\cdot) = -\Phi^{-1}(\Theta_0 + L_1^*Q_1 e^{-AT})x.$$

*Proof.* Define $\bar{u}(\cdot) \in \mathcal{D}(L_1^*Q_1 L_1)$ by (2.10). Then it holds that

(2.11)
$$\Phi\bar{u}(\cdot) + \Theta_0 x + L_1^*Q_1 e^{-AT}x = 0.$$

This implies (2.5). By Theorem 2.2, $(x, \bar{u}(\cdot))$ satisfies (1.49). $\square$

From Proposition 2.1, we see that in order for Problem (LQ) to be solvable, the operator $\Phi$ defined by (2.9) has to be nonnegative on $\mathcal{D}(\Phi)$. Thus, one expects that the following condition

(2.12) $$\langle \Phi_0 u, u \rangle + \langle Q_1 L_1 u, L_1 u \rangle \geq \delta|u|^2, \qquad \forall u \in \mathcal{D}(L_1),$$

## §2. Well Posedness and Solvability

with some $\delta > 0$, might ensure the bounded invertibility of $\Phi$, which guarantees the solvability of Problem (LQ) for $x \in \mathcal{D}(L_1^* Q_1 e^{-AT})$ by Corollary 2.3. But, we point out that, in general, (2.12) does not necessarily imply the bounded invertibility of the operator $\Phi$. Here is an example.

*Example 2.4.* Let $X = U$. Let $Q = 0$, $R = B = I$, and $\frac{1}{2} \leq \alpha < 1$. Thus, $L_1$ is unbounded, closed, and densely defined; and so is $L_1^*$. Hence, we can find a $q \notin \mathcal{D}(L_1^*)$, $q \neq 0$. Define $Q_1 \in \mathcal{L}(X)$ as follows:

$$(2.13) \qquad Q_1 z = \langle z, q \rangle q, \qquad \forall z \in X.$$

Then $\Phi = I + L_1^* Q_1 L_1$ has the domain

$$(2.14) \qquad \mathcal{D}(\Phi) = \{ u \in \mathcal{D}(L_1) \mid \langle L_1 u, q \rangle = 0 \}.$$

Consequently,

$$(2.15) \qquad \Phi u = u, \qquad \forall u \in \mathcal{D}(\Phi).$$

On the other hand, for any $u \in \mathcal{D}(L_1)$ (note that $\Phi_0 = I$ in the present case),

$$(2.16) \qquad \langle \Phi_0 u, u \rangle + \langle Q_1 L_1 u, L_1 u \rangle = |u|^2 + \big|\langle L_1 u, q \rangle\big|^2 \geq |u|^2.$$

Thus, (2.12) holds. From (2.15), we see that $\mathcal{R}(\Phi) = \mathcal{D}(\Phi) \subseteq \mathcal{D}(L_1)$, which does not coincide with $\mathcal{U}$. Thus, $\Phi$ is not boundedly invertible.

On the other hand, we have the following simple result.

**Lemma 2.5.** *Let $\Phi : \mathcal{D}(\Phi) \subset \mathcal{U} \to \mathcal{U}$ be self-adjoint such that $\Phi \gg 0$, i.e., for some $\delta > 0$,*

$$(2.17) \qquad \langle \Phi u, u \rangle \geq \delta |u|^2, \qquad \forall u \in \mathcal{D}(\Phi).$$

*Then $\Phi$ is boundedly invertible with*

$$(2.18) \qquad \|\Phi^{-1}\| \leq \frac{1}{\delta}.$$

*Proof.* From (2.17), we see that $\Phi$ is injective. Thus, $\Phi^{-1} : \mathcal{R}(\Phi) \to \mathcal{U}$ is well defined. On the other hand, we claim that $\mathcal{R}(\Phi)$ is closed. In fact, if $\Phi u_n \xrightarrow{s} w$ in $\mathcal{U}$, then it follows from (2.18) that

$$(2.19) \qquad |u_m - u_n| \leq \frac{1}{\delta} |\Phi u_m - \Phi u_n| \to 0, \qquad m, n \to \infty.$$

Thus, $u_n \xrightarrow{s} u$ in $\mathcal{U}$ for some $u$. By the closeness of $\Phi$, we see that $u \in \mathcal{D}(\Phi)$ and $w = \Phi u \in \mathcal{R}(\Phi)$. This proves the closeness of $\mathcal{R}(\Phi)$. Next, as both $\Phi^* = \Phi$ and $\Phi^{-1}$ are well defined, then

$$(2.20) \qquad (\Phi^{-1})^* = (\Phi^*)^{-1} = \Phi^{-1},$$

is well defined. This implies that $\mathcal{D}(\Phi^{-1}) \equiv \mathcal{R}(\Phi)$ is dense in $\mathcal{U}$. By the closeness of $\mathcal{R}(\Phi)$, we must have $\mathcal{R}(\Phi) = \mathcal{U}$. Hence, $\Phi^{-1} \in \mathcal{L}(\mathcal{U})$ and (2.18) follows easily from (2.17). □

From Lemma 2.5, we see that the operator constructed in Example 2.4 is symmetric and uniformly positive. But it is not self-adjoint.

The above analysis suggests that we should introduce the following technical assumptions:

(H1) The operator $L_1^* Q_1 L_1 : \mathcal{D}(L_1^* Q_1 L_1) \subseteq \mathcal{U} \to \mathcal{U}$ is self-adjoint.

Let us state some other technical hypotheses.

(H2) The operator $L_1^* Q_1 e^{-AT} \in \mathcal{L}(X, \mathcal{U})$.

(H3) There exists a $\beta > \alpha - \frac{1}{2}$, such that

$$\text{(2.21)} \qquad \mathcal{R}(Q_1) \subseteq \mathcal{D}((A^*)^\beta).$$

(H4) There exists a $\beta > \alpha - \frac{1}{2}$, such that

$$\text{(2.22)} \qquad \mathcal{R}(Q) \subseteq \mathcal{D}((A^*)^\beta).$$

Assumptions (H1) and (H2) are basic, and we will see that (H2) is implied by (H3), which is easier to check. Assumption (H4) will be used in §3.4 and §4. We will make some further remarks on (H1)–(H4) a little later. Now, let us set

$$\text{(2.23)} \qquad \begin{cases} \Phi = \Phi_0 + L_1^* Q_1 L_1 \equiv R + L^* Q L + L_1^* Q_1 L_1, \\ \Theta = \Theta_0 + L_1^* Q_1 e^{-AT} \equiv L^* Q e^{-A \cdot} + L_1^* Q_1 e^{-AT}. \end{cases}$$

Then

$$\text{(2.24)} \qquad \mathcal{D}(\Phi) = \mathcal{D}(L_1^* Q_1 L_1), \qquad \mathcal{D}(\Theta) = \mathcal{D}(L_1^* Q_1 e^{-AT}).$$

Under (H1), $\Phi : \mathcal{D}(\Phi) \subseteq \mathcal{U} \to \mathcal{U}$ is self-adjoint. We can rewrite the cost functional $J(x; u(\cdot))$ as follows:

$$\text{(2.25)} \qquad \begin{aligned} J(x; u(\cdot)) &= \langle \Phi u, u \rangle + 2 \langle \Theta x, u \rangle + \langle \Gamma x, x \rangle, \\ &\forall (x, u) \in \mathcal{D}(\Theta) \times \mathcal{D}(\Phi). \end{aligned}$$

The next result summarizes the above analysis and tells us something more.

**Theorem 2.6.** (i) *If Problem (LQ) is well posed at some $x \in X$, then $\Phi \geq 0$.*

(ii) *If $\Phi \geq 0$ and $x \in \mathcal{D}(\Theta)$, then $\bar{u}(\cdot)$ is a minimizer of $J(x; \cdot)$ if and only if $\bar{u}(\cdot) \in \mathcal{D}(\Phi)$ and*

$$\text{(2.26)} \qquad \Phi \bar{u}(\cdot) + \Theta x = 0.$$

§2. Well Posedness and Solvability

(iii) *If (H1) holds and $\Phi \gg 0$, then $\Phi^{-1} \in \mathcal{L}(\mathcal{U})$ and for any $x \in \mathcal{D}(\Theta)$, $J(x;\cdot)$ admits a unique minimizer $\bar{u}(\cdot)$ given by*

(2.27) $$\bar{u}(\cdot) = -\Phi^{-1}\Theta x.$$

*In this case, it holds that*

(2.28) $$V(x) \triangleq \inf_{u(\cdot) \in \mathcal{D}(L_1)} J(x; u(\cdot)) = J(x; \bar{u}(\cdot))$$
$$= \langle \Gamma x, x \rangle - \langle \Phi^{-1}\Theta x, \Theta x \rangle, \qquad \forall x \in \mathcal{D}(\Theta).$$

*In addition, if (H2) holds, then $P \equiv \Gamma - \Theta^*\Phi^{-1}\Theta \in \mathcal{L}(X)$ is self-adjoint and $V(\cdot)$ is a bilinear form defined on $X$.*

The proof is immediate and we leave it to the readers. Note that in the case without (H2), the value function $V(\cdot)$ is not necessarily everywhere defined (an exceptional case will be indicated in Theorem 2.9). This will sometimes cause certain difficulties.

Now, let us make some remarks on (H1)–(H4). First of all, we point out that (H3) implies, by the Closed Graph Theorem, that $(A^*)^\beta Q_1 \in \mathcal{L}(X)$ and $Q_1 A^\beta$ admits a continuous extension $((A^*)^\beta Q_1)^*$. A similar conclusion holds for (H4). Next, we should note that when $\alpha \in [0, \frac{1}{2})$, (H1)–(H4) are automatically true because in this case, $L_1 \in \mathcal{L}(\mathcal{U}, X)$ and we may take $\beta = 0$ in (H3) and (H4). For the case where $\alpha \in [\frac{1}{2}, 1)$, the following result gives a sufficient condition that ensures (H1).

**Proposition 2.7.** *Assumption (H1) holds if $Q_1 \gg 0$ or $-Q_1 \gg 0$.*

*Proof.* We recall Chapter 2, Proposition 4.3 (iii): If $K$ is a densely defined closed operator, then $K^*K$ is self-adjoint. Now, in the case $Q_1 \gg 0$, we have that $Q_1^{1/2}L_1$ is densely defined and closed. In fact, the domain of $Q_1^{1/2}L_1$ coincides with $\mathcal{D}(L_1)$, which is dense in $\mathcal{U}$. On the other hand, if $u_k(\cdot) \xrightarrow{s} u(\cdot)$ in $\mathcal{U}$ and $Q_1^{1/2} L_1 u_k(\cdot) \xrightarrow{s} z$ in $X$, then, by the invertibility of $Q_1^{1/2}$, we see that $L_1 u_k(\cdot) \xrightarrow{s} y \equiv Q_1^{-1/2} z$ in $X$. Consequently, by the closeness of $L_1$, one has $u(\cdot) \in \mathcal{D}(L_1)$ and $y = L_1 u(\cdot)$, which implies $z = Q_1^{1/2} L_1 u(\cdot)$. This proves the closeness of $Q_1^{1/2} L_1$. Thus, the operator

(2.29) $$L_1^* Q_1 L_1 = (Q_1^{\frac{1}{2}} L_1)^* (Q_1^{\frac{1}{2}} L_1)$$

is self-adjoint. Here, we have used the fact that $(Q_1^{1/2} L_1)^* = L_1^* Q_1^{1/2}$. This is true because $Q_1^{1/2}$ is bounded (and self-adjoint) (see Chapter 2, §4.1). The case $-Q_1 \gg 0$ can be proved similarly. □

We should note that if $Q_1$ is not invertible, then $Q_1^{1/2} L_1$ is not necessarily closed. To see this, we let $L_1$ be unbounded and closed. Take any $u(\cdot) \notin \mathcal{D}(L_1)$ and $u_k(\cdot) \xrightarrow{s} u(\cdot)$ in $\mathcal{U}$. Clearly, $0 L_1 u_k(\cdot) = 0$ and $u(\cdot) \notin \mathcal{D}(0 L_1)$. This shows that $0 L_1$ is not closed. But $0 L_1$ admits an obvious continuous extension, the zero operator. On the other hand, we point out that in the case where $Q_1 = 0$, the term $\langle Q_1 y(T;x,u), y(T;x,u) \rangle$ in the cost functional disappears. Of course in discussing such a case, the above assumption (H1) is not needed. For convenience, we say that (H1) trivially holds for such a case.

Now, we look at (H2). First of all, $\mathcal{D}(L_1^* Q_1 e^{-AT})$ is a subspace and it could be very "small." For example, in the case of Example 2.4, we have that

(2.30) $\quad \mathcal{D}(L_1^* Q_1 e^{-AT}) = \{x \in X \mid \langle e^{-AT} x, q \rangle = 0\} = \{e^{-A^*T} q\}^\perp.$

Thus, $\mathcal{D}(L_1^* Q_1 e^{-AT})$ is not dense in $X$. The following result collects some cases for which (H2) holds.

**Proposition 2.8.** *Assumption (H2) holds in the following cases:*

(i) $L_1^* Q_1 \in \mathcal{L}(X, \mathcal{U})$. *This is the case if (H3) holds, in particular, if $Q_1 = 0$.*

(ii) $Q_1$ *is given by*

(2.31) $\quad \begin{cases} Q_1 x = \sum_{n \geq 1} a_n \langle q_n, x \rangle q_n, & \forall x \in X, \\ q_n \in \mathcal{D}(A^*), & \sum_{n \geq 1} |a_n| \, |q_n| \, |A^* q_n| < \infty. \end{cases}$

(iii) $A^* = A$ *and* $AQ_1 = Q_1 A$, *in particular, $Q_1 = \lambda I$ for some $\lambda \in \mathbb{R}$.*

*Proof.* (i) It is clear that $L_1^* Q_1 \in \mathcal{L}(X, \mathcal{U})$ implies (H2). Now, if (H3) holds, then $(A^*)^\beta Q_1 \in \mathcal{L}(X)$. Also, we note that $(A^*)^{\alpha-\beta} e^{-A^*(T-\cdot)} \in \mathcal{L}(X, \mathcal{U})$. Consequently,

(2.32) $\quad \begin{aligned} L_1^* Q_1 &= B^* (A^*)^\alpha e^{-A^*(T-\cdot)} Q_1 \\ &= B^* (A^*)^{\alpha-\beta} e^{-A(T-\cdot)} (A^*)^\beta Q_1 \in \mathcal{L}(X, \mathcal{U}). \end{aligned}$

(ii) It is easy to see that (2.31) implies $\mathcal{R}(Q_1) \subseteq \mathcal{D}(A^*)$. Thus, (H3) holds and so does (H2).

(iii) We note that in the present case, $e^{-AT} Q_1 = Q_1 e^{-AT}$. Thus, for any $x \in X$ and $v \in \mathcal{U}$, we have

(2.33) $\quad \begin{aligned} |\langle Q_1 e^{-AT} x, L_1 v \rangle| &= \left| \left\langle Q_1 x, A^\alpha e^{-AT} \int_0^T e^{-A(T-s)} Bv(s) \, ds \right\rangle \right| \\ &\leq C T^{-\alpha} \left| \int_0^T e^{-A(T-s)} Bv(s) \, ds \right| \leq C |v|_\mathcal{U}. \end{aligned}$

This implies that $\mathcal{R}(Q_1 e^{-AT}) \subseteq \mathcal{D}(L_1^*)$. Hence, by the Closed Graph Theorem, $L_1^* Q_1 e^{-AT}$ is bounded, which is (H2). $\square$

The above tells us that (H1) and (H2) are very general. Actually, they are satisfied by many interesting problems. We give a partial list of such problems.

§2. Well Posedness and Solvability

(i) Distributed control problems (for any evolution equations, parabolic equations, wave equations, beam equations);

(ii) Neumann boundary control problems for parabolic equations;

(iii) $Q_1 = 0$, Dirichlet boundary control problems, pointwise control problems for parabolic equations;

(iv) Various combinations of (i)–(iii).

From Theorem 2.6 we see that under (H1) and (H2), if $\Phi \gg 0$, then Problem (LQ) is uniquely solvable (for any initial state $x \in X$) and the value function $V(x)$ is a bounded bilinear form on $X$. However, we point out that (H2) does not include the following important case: $\alpha \in [\frac{1}{2}, 1)$ (which is the case for the Dirichlet boundary control problem and the pointwise control with $n = 2, 3$) and $Q_1 \gg 0$. In such a case $L_1^*$ is unbounded and $Q_1$ is invertible. Thus, $L_1^* Q_1$ is unbounded and (H2) is not necessarily satisfied. The following result gives a compensation.

**Theorem 2.9.** *Let (H1) hold, $\Phi \gg 0$, $\Phi_0 \gg 0$, and $Q_1 \gg 0$ or $-Q_1 \gg 0$. Then, for any $x \in X$, $J(x; \cdot)$ admits a unique minimizer $\bar{u}(\cdot)$ given by*

$$(2.34) \qquad \bar{u}(\cdot) = -\Phi^{-1} \Theta_0 x - (L_1 \Phi^{-1})^* Q_1 e^{-AT} x,$$

*and the value function is a bounded bilinear form on $X$.*

*Proof.* We split the proof into several steps.

Step 1. $L_1 \Phi^{-1} \in \mathcal{L}(\mathcal{U}, X)$.

$\mathcal{R}(\Phi^{-1}) = \mathcal{D}(L_1^* Q_1 L_1) \subseteq \mathcal{D}(L_1)$; thus, we see that $L_1 \Phi^{-1}$ is closed with the domain $\mathcal{U}$. Thus, by the Closed Graph Theorem, $L_1 \Phi^{-1} \in \mathcal{L}(\mathcal{U}, X)$.

Step 2. $L_1 (L_1 \Phi^{-1})^* \in \mathcal{L}(X)$.

We let $\lambda = \pm 1$ such that $Q_0 \equiv \lambda Q_1 \gg 0$. Set $M = Q_0^{1/2} L_1 \Phi_0^{-1/2}$. Then $M : \mathcal{D}(M) \subset \mathcal{U} \to X$ is a densely defined closed operator. Thus, by polar decomposition (see Chapter 2, §4.1), we have a partial isometry $K \in \mathcal{L}(\mathcal{U}, X)$, such that

$$(2.35) \qquad M = K M_0, \quad M_0 \triangleq (M^* M)^{1/2}.$$

Consequently,

$$\begin{aligned}
L_1 \Phi^{-1} L_1^* &= L_1 (\Phi_0 + L_1 Q_1^* L_1)^{-1} L_1^* \\
&= L_1 \Phi_0^{-\frac{1}{2}} \left( I + \lambda \Phi_0^{-\frac{1}{2}} L_1^* Q_0 L_1 \Phi_0^{-\frac{1}{2}} \right)^{-1} \Phi_0^{-\frac{1}{2}} L_1^* \\
(2.36) \qquad &= Q_0^{-\frac{1}{2}} M (I + \lambda M^* M)^{-1} M^* Q_0^{-\frac{1}{2}} \\
&= Q_0^{-\frac{1}{2}} K M_0 (I + \lambda M_0^2)^{-1} M_0 K^* Q_0^{-\frac{1}{2}} \\
&\subseteq Q_0^{-\frac{1}{2}} K M_0^2 (I + \lambda M_0^2)^{-1} K^* Q_0^{-\frac{1}{2}} \equiv \Psi.
\end{aligned}$$

Clearly, $\Psi \in \mathcal{L}(X)$. Now, for any $z \in X$, by the density of $\mathcal{D}(L_1^*)$ in $X$, we can find $z_n \in \mathcal{D}(L_1^*)$ such that $z_n \xrightarrow{s} z$ in $X$. Clearly, $(L_1 \Phi^{-1})^* z_n \in \mathcal{D}(L_1)$

and $(L_1\Phi^{-1})^*z_n \xrightarrow{s} (L_1\Phi^{-1})^*z$ in $\mathcal{U}$. From (2.36), we have

$$(2.37) \qquad L_1(L_1\Phi^{-1})^*z_n = L_1\Phi^{-1}L_1^*z_n = \Psi z_n,$$

which is bounded uniformly. Then we may assume that $L_1(L_1\Phi^{-1})^*z_n$ converges weakly to some $y$. Because the graph $\mathcal{G}(L_1)$ is a closed subspace in $\mathcal{U} \times X$, it is weakly closed by Mazur's Theorem. Thus,

$$(2.38) \qquad (L_1\Phi^{-1})^*z \in \mathcal{D}(L_1), \qquad L_1(L_1\Phi^{-1})^*z = y.$$

This shows that $\mathcal{R}((L_1\Phi^{-1})^*) \subset \mathcal{D}(L_1)$. Thus, $L_1(L_1\Phi^{-1})^* \in \mathcal{L}(X)$.

*Step 3.* It holds that

$$(2.39) \qquad \begin{cases} \mathcal{R}(I - Q_1L_1(L_1\Phi^{-1})^*) \subseteq \mathcal{D}(L_1^*), \\ L_1^*(I - Q_1L_1(L_1\Phi^{-1})^*) = \Phi_0(L_1\Phi^{-1})^*. \end{cases}$$

To show this, we pick any $z \in X$. There exist $z_n \in \mathcal{D}(L_1^*)$, such that $z_n \xrightarrow{s} z$ in $X$. Let us observe the following: For any $v \in \mathcal{D}(L_1)$,

$$(2.40) \qquad \begin{aligned} \langle (I &- Q_1L_1(L_1\Phi^{-1})^*)z_n, L_1v \rangle \\ &= \langle z_n, L_1v \rangle - \langle L_1^*Q_1L_1\Phi^{-1}L_1^*z_n, v \rangle \\ &= \langle z_n, L_1v \rangle - \langle (I - \Phi_0\Phi^{-1})L_1^*z_n, v \rangle \\ &= \langle \Phi_0\Phi^{-1}L_1^*z_n, v \rangle = \langle \Phi_0(L_1\Phi^{-1})^*z_n, v \rangle. \end{aligned}$$

Sending $n \to \infty$, we see that

$$(2.41) \quad \langle (I - Q_1L_1(L_1\Phi^{-1})^*)z, L_1v \rangle = \langle \Phi_0(L_1\Phi^{-1})^*z, v \rangle, \quad \forall v \in \mathcal{D}(L_1^*).$$

This implies that for any $z \in X$,

$$(2.42) \qquad \begin{cases} (I - Q_1L_1(L_1\Phi^{-1})^*)z \in \mathcal{D}(L_1^*), \\ L_1^*(I - Q_1L_1(L_1\Phi^{-1})^*)z = \Phi_0(L_1\Phi^{-1})^*z. \end{cases}$$

This proves (2.39).

*Step 4.* The control $\bar{u}$ defined in (2.34) is the unique minimizer of $J(x;\cdot)$.

For any $v \in \mathcal{D}(L_1)$, we have (note (2.39))

$$(2.43) \qquad \begin{aligned} \langle Q_1L_1\bar{u} &+ Q_1e^{-AT}x, L_1v \rangle \\ &= \langle -Q_1L_1[(\Phi^{-1}\Theta_0 x + (L_1\Phi^{-1})^*Q_1e^{-AT}x] + Q_1e^{-AT}x, L_1v \rangle \\ &= \langle [I - Q_1L_1(L_1\Phi^{-1})^*]Q_1e^{-AT}x - Q_1L_1\Phi^{-1}\Theta_0 x, L_1v \rangle \\ &= \langle \Phi_0(L_1\Phi^{-1})^*Q_1e^{-AT}x - L_1^*Q_1L_1\Phi^{-1}\Theta_0 x, v \rangle \\ &= \langle -\Phi_0\bar{u} - \Phi_0\Phi^{-1}\Theta_0 x - L_1^*Q_1L_1\Phi^{-1}\Theta_0 x, v \rangle \\ &= \langle -\Phi_0\bar{u} - \Theta_0 x, v \rangle. \end{aligned}$$

This implies that $(x,\bar{u})$ satisfies (2.5). Hence, by Theorem 2.2, $\bar{u}$ is a minimizer of $J(x;\cdot)$. Now, if $\bar{v}$ is another minimizer of $J(x;\cdot)$, then, by Theorem 2.2, it is necessary that $Q_1 L_1(\bar{u}-\bar{v}) \in \mathcal{D}(L_1^*)$ and

$$(2.44) \qquad 0 = L_1^* Q_1 L_1(\bar{u}-\bar{v}) + \Phi_0(\bar{u}-\bar{v}) = \Phi(\bar{u}-\bar{v}).$$

As $\Phi \gg 0$, we must have $\bar{v} = \bar{u}$. This gives the uniqueness.

The rest of the conclusions are clear. □

To conclude this section, we give the following result for the definite LQ problems. The proof is left to the readers.

**Corollary 2.10.** *Let $R \gg 0$, $Q \geq 0$, and either $Q_1 \gg 0$, or $Q_1 \geq 0$ with $\alpha \in [0, \frac{1}{2})$. Then, for any $x \in X$, Problem (LQ) admits a unique optimal control $\bar{u}(\cdot) \in \mathcal{U}$ and there exists a self-adjoint operator $P \in \mathcal{L}(X)$, such that*

$$(2.45) \qquad V(x) \equiv \inf_{u(\cdot) \in \mathcal{U}} J(x; u(\cdot)) = \langle Px, x \rangle, \qquad \forall x \in X.$$

## §3. State Feedback Control

We note that the optimal control $\bar{u}(\cdot)$ determined by (2.10) (or (2.27), (2.34)) is not easy to compute as $\Phi^{-1}$ is actually very complicated. In this section, under certain conditions, we will find a simpler form of optimal control, in which the control is a linear function of the corresponding state. Such a form of control is very useful in engineering.

### §3.1. Two-point boundary value problem

Let us first give a further necessary condition for the well posedness of Problem (LQ).

**Proposition 3.1.** *Let Problem (LQ) be well posed at some $x \in X$. Then*

$$(3.1) \qquad R \geq 0.$$

*Proof.* Suppose that (3.1) is not the case. Then there exists some $u_0 \in U$ such that $\langle Ru_0, u_0 \rangle < 0$. Now, for any small $\varepsilon > 0$, we define

$$(3.2) \qquad u_\varepsilon(t) = u_0 \chi_{[0,\varepsilon]}(t), \qquad t \in [0,T].$$

Clearly,

$$(3.3) \qquad \int_0^T \langle Ru_\varepsilon(t), u_\varepsilon(t) \rangle \, dt = \varepsilon \langle Ru_0, u_0 \rangle.$$

Let $y_\varepsilon(\cdot) = y(\cdot\,;0,u_\varepsilon)$. Then

$$\int_0^T \langle Qy_\varepsilon(t), y_\varepsilon(t)\rangle \, dt \le \|Q\| \Big\{ \int_0^\varepsilon \Big| \int_0^t A^\alpha e^{-A(t-s)} Bu_0 \, ds \Big|^2 dt$$

$$+ \int_\varepsilon^T \Big| \int_0^\varepsilon A^\alpha e^{-A(t-s)} Bu_0 ds \Big|^2 dt \Big\}$$

(3.4) $\le C \Big\{ \int_0^\varepsilon \Big( \int_0^t s^{-\alpha} ds \Big)^2 dt$

$$+ \int_\varepsilon^T \Big| A^{\frac{\alpha}{2}} e^{-A(t-\varepsilon)} \int_0^\varepsilon A^{\frac{\alpha}{2}} e^{-A(\varepsilon-s)} Bu_0 \, ds \Big|^2 dt \Big\}$$

$$\le C \Big\{ \int_0^\varepsilon t^{2-2\alpha} dt + \int_0^{T-\varepsilon} \|A^{\frac{\alpha}{2}} e^{-At}\|^2 dt \Big| \int_0^\varepsilon A^{\frac{\alpha}{2}} e^{-As} Bu_0 \, ds \Big|^2 \Big\}$$

$$\le C \Big\{ \varepsilon^{3-2\alpha} + \int_0^{T-\varepsilon} t^{-\alpha} dt \Big( \int_0^\varepsilon s^{-\frac{\alpha}{2}} ds \Big)^2 \Big\} \le C\{\varepsilon^{3-2\alpha} + \varepsilon^{2-\alpha}\}.$$

Also, it holds that

(3.5)
$$\langle Q_1 y_\varepsilon(T), y_\varepsilon(T)\rangle \le C \Big| \int_0^\varepsilon A^\alpha e^{-A(T-s)} Bu_0 \, ds \Big|^2$$
$$= C \Big| A^\alpha e^{-A(T-\varepsilon)} \int_0^\varepsilon e^{-A(\varepsilon-s)} Bu_0 \, ds \Big|^2 \le C\varepsilon^2.$$

Hence,

(3.6) $\quad J(0;u_\varepsilon(\cdot)) \le \varepsilon\{ \langle Ru_0, u_0\rangle + C(\varepsilon^{2-2\alpha} + \varepsilon^{1-\alpha} + \varepsilon)\} < 0,$

provided $\varepsilon > 0$ is small enough. This contradicts the well posedness of the Problem (LQ) at some $x \in X$ (see Proposition 2.1). Hence, (3.1) holds. □

Recall that the well posedness of Problem (LQ) implies the well posedness of the problem at some $x \in X$. Thus, the above result also gives a necessary condition for the well posedness of Problem (LQ).

It is reasonable that we require Problem (LQ) to be well posed, at least, at some $x \in X$. To ensure this, from the above proposition, we should assume (3.1). In what follows, we will assume a little more, namely, that

(3.7) $\qquad\qquad\qquad\qquad R \gg 0.$

The case where (3.1) holds instead of (3.7) is usually referred to as the *singular* LQ problem. We will not consider such a problem in this book.

In what follows, we will assume (H1) and (H3). The case covered by Theorem 2.9, namely $Q_1 \gg 0$ or $-Q_1 \gg 0$ and $\alpha \in [\frac{1}{2}, 1)$, for which the (H3) does not hold, will not be discussed. A parallel theory seems possible

## §3. State Feedback Control

to establish. We suggest that the interested readers study such a case using the following ideas and find some differences.

The following result gives a relation between Problem (LQ) and a two-point boundary value problem.

**Theorem 3.2.** *Let (H1) and (H3) hold. Let $\Phi \geq 0$ and $R \gg 0$. Then Problem (LQ) is solvable at $x \in X$ if and only if the following two-point boundary value problem*

(3.8)
$$\begin{cases} y(t) = e^{-At}x + \int_0^t A^\alpha e^{-A(t-s)} BR^{-1} B^* \psi(s) \, ds, \\ \psi(t) = -(A^*)^\alpha e^{-A^*(T-t)} Q_1 y(T) - \int_t^T (A^*)^\alpha e^{-A^*(\sigma-t)} Q y(\sigma) \, d\sigma, \\ \qquad\qquad\qquad\qquad\qquad\qquad\qquad\qquad\qquad\qquad t \in [0, T], \end{cases}$$

*admits a solution $(\bar{y}(\cdot), \psi(\cdot)) \in \mathcal{X} \times \mathcal{X}$ ($\mathcal{X} \equiv L^2(0, T; X)$) with*

(3.9)
$$\int_0^T e^{-A(T-s)} BR^{-1} B^* \psi(s) \, ds \in \mathcal{D}(A^\alpha).$$

*In this case,*

(3.10)
$$\bar{u}(t) = R^{-1} B^* \psi(t), \qquad t \in [0, T]$$

*gives an optimal control and the function $y(\cdot)$ obtained in solving (3.8) is the corresponding optimal state trajectory. In addition, if $\Phi \gg 0$, then (3.8) admits a unique solution and Problem (LQ) is uniquely solvable (on $X$) with the optimal control given by (3.10).*

*Proof.* From Theorem 2.2, we know that under our assumptions, Problem (LQ) is solvable at $x \in X$ with an optimal control $\bar{u}$ if and only if $\bar{u} \in \mathcal{D}(L_1^* Q_1 L_1)$ and

(3.11)
$$\begin{aligned} 0 &= \Phi \bar{u} + \Theta x \\ &= R\bar{u} + L^* QL\bar{u} + L_1^* Q_1 L_1 \bar{u} + L^* Q e^{-A \cdot} x + L_1^* Q_1 e^{-AT} x. \end{aligned}$$

Let $\bar{y} = e^{-A \cdot} x + L\bar{u}$ be the corresponding optimal trajectory. Because $\bar{u} \in \mathcal{D}(L_1)$, $\bar{y}(T) = e^{-AT} x + L_1 \bar{u}$ is well defined. Set

(3.12)
$$\begin{aligned} \psi(t) = &-(A^*)^\alpha e^{-A^*(T-t)} Q_1 \bar{y}(T) \\ &- \int_t^T (A^*)^\alpha e^{-A^*(\sigma-t)} Q \bar{y}(\sigma) \, d\sigma, \qquad t \in [0, T]. \end{aligned}$$

Here, we note that by (H3), $(A^*)^\beta Q_1 \in \mathcal{L}(X)$ and $(A^*)^{\alpha-\beta} e^{-A^*(T-\cdot)} \in \mathcal{L}(X, \mathcal{X})$. Consequently,

(3.13) $\qquad (A^*)^\alpha e^{-A^*(T-\cdot)} Q_1 = (A^*)^{\alpha-\beta} e^{-A^*(T-\cdot)} (A^*)^\beta Q_1 \in \mathcal{L}(X, \mathcal{X}).$

Thus, the first term on the right-hand side of (3.12) is in $\mathcal{X}$. By Lemma 1.1, we can show that the second term on the right-hand side of (3.12) is also in $\mathcal{X}$ (as $\bar{y}(\cdot) \in \mathcal{X}$). Hence, the function $\psi(\cdot)$ defined by (3.12) is in $\mathcal{X}$. Next, by (3.11), we have

$$
\begin{aligned}
B^*\psi &= -L_1^* Q_1 \bar{y}(T) - L^* Q \bar{y} \\
&= -L_1^* Q_1 (e^{-AT} x + L_1 \bar{u}) - L^* Q(e^{-A\cdot} x + L\bar{u}) \\
&= -L_1^* Q_1 e^{-AT} x - L^* Q e^{-A\cdot} x - (L_1^* Q_1 L_1 + L^* Q L) \bar{u} = R\bar{u}.
\end{aligned}
$$
(3.14)

This yields (3.10). Clearly, the pair $(\bar{y}(\cdot), \psi(\cdot)) \in \mathcal{X}$ is a solution of the two-point boundary value problem (3.8). Finally, by (3.10), we have

(3.15) $$R^{-1} B^* \psi(\cdot) \in \mathcal{D}(L_1),$$

which is equivalent to (3.9).

Conversely, let $(y(\cdot), \psi(\cdot)) \in \mathcal{X} \times \mathcal{X}$ be a solution of (3.8) with the property (3.9). Then, we define $\bar{u}$ by (3.10). Clearly, $\bar{u} \in \mathcal{D}(L_1)$. Under (H3), we know that $L_1^* Q_1 \in \mathcal{L}(\mathcal{U}, X)$. Thus, $Q_1(L_1 \bar{u} + e^{-AT} x) \in \mathcal{D}(L_1^*)$. Also, it follows from (3.8) and (3.10) that

$$
\begin{aligned}
R\bar{u} &= B^* \psi \\
&= -L_1^* Q_1 e^{-AT} x - L^* Q e^{-A\cdot} x - L_1^* Q_1 L_1 \bar{u} - L^* Q L \bar{u}.
\end{aligned}
$$
(3.16)

Hence, by Theorem 2.2, $\bar{u}$ is an optimal control for the Problem (LQ) with the initial state $x$.

The rest of the conclusions are clear. $\square$

It is seen that the function $\psi(\cdot)$ is an auxiliary function that indirectly relates optimal control and the corresponding optimal state. Our next goal is to eliminate this auxiliary function and link the optimal control and the state directly. This will be carried out in the next two subsections by using the idea of dynamic programming, which was discussed intensively in the previous chapters.

## §3.2. The Problem $(LQ)_t$

Let $t \in [0, T)$. We consider the LQ Problem on the time interval $[t, T]$. Thus, the state equation is the following:

(3.17) $$y_{t,x}(s) = e^{-A(s-t)} x + \int_t^s A^\alpha e^{-A(s-r)} Bu(r)\, dr, \quad s \in [t, T],$$

and the cost functional takes the form

$$
\begin{aligned}
J_{t,x}(u(\cdot)) &= \langle Q_1 y_{t,x}(T), y_{t,x}(T) \rangle \\
&\quad + \int_t^T \{ \langle Q y_{t,x}(s), y_{t,x}(s) \rangle + \langle R u(s), u(s) \rangle \}\, ds.
\end{aligned}
$$
(3.18)

Here, $y_{t,x}(\cdot) \equiv y(\cdot\,; t, x)$ indicates the dependence of the trajectory on $(t, x)$.

## §3. State Feedback Control

For this problem, we denote $\mathcal{U}[t,T] = L^2(t,T;U)$ and $\mathcal{X}[t,T] = L^2(t,T;X)$. As before, let us define $L(t) \in \mathcal{L}(\mathcal{U}[t,T];\mathcal{X}[t,T])$ and $L_1(t) : \mathcal{D}(L_1(t)) \subset \mathcal{U}[t,T] \to X$ as follows:

$$(3.19) \quad \begin{cases} (L(t)u)(s) = \int_t^s A^\alpha e^{-A(s-r)} Bu(r)\, dr, & s \in [t,T],\ u \in \mathcal{U}[t,T], \\ L_1(t)u = \int_t^T A^\alpha e^{-A(T-r)} Bu(r)\, dr, & u \in \mathcal{D}(L_1(t)). \end{cases}$$

Then it holds that

$$(3.20) \quad \begin{cases} (L(t)^*y)(s) = \int_s^T B^*(A^*)^\alpha e^{-A^*(\sigma-s)} y(\sigma)\, d\sigma, \\ \qquad\qquad\qquad\qquad\qquad s \in [t,T],\ y \in \mathcal{X}[t,T]; \\ (L_1(t)^*x)(s) = B^*(A^*)^\alpha e^{-A^*(T-s)} x, \quad s \in [t,T],\ x \in \mathcal{D}(L_1(t)^*). \end{cases}$$

Let us further define

$$(3.21) \quad \begin{cases} \Phi(t) = R + L(t)^* Q L(t) + L_1(t)^* Q_1 L_1(t), \\ \Theta(t) = L(t)^* Q e^{-A(\cdot - t)} + L_1(t)^* Q_1 e^{-A(T-t)}, \\ \Gamma(t) = \int_t^T e^{-A^* s} Q e^{-As}\, ds + e^{-A^*(T-t)} Q_1 e^{-A(T-t)}. \end{cases}$$

Here, we regard $R \in \mathcal{L}(\mathcal{U}[t,T])$, $Q \in \mathcal{L}(\mathcal{X}[t,T])$. As before, we may rewrite the cost functional $J_{t,x}(u(\cdot))$ as follows:

$$(3.22) \quad \begin{aligned} J_{t,x}(u(\cdot)) = \langle \Phi(t)u, u \rangle + 2\langle \Theta(t)x, u \rangle + \langle \Gamma(t)x, x \rangle, \\ \forall (x, u(\cdot)) \in \mathcal{D}(\Theta(t)) \times \mathcal{D}(\Phi(t)). \end{aligned}$$

Then we can state the following LQ problem parameterized by $t \in [0,T)$.

**Problem (LQ)$_t$.** Find $\bar{u}(\cdot) \in \mathcal{D}(\Phi(t))$, such that the cost functional $J_{t,x}(u(\cdot))$ is minimized.

This family of problems will eventually lead to a direct relation between the optimal control and the optimal state trajectory. Similar to the result for Problem (LQ), we know that Problem (LQ)$_t$ is solvable if the following hold:

(H1)$_t$ The operator $L_1(t)^* Q L_1(t) : \mathcal{D}(L_1(t)^* Q_1 L_1(t)) \subset \mathcal{U}[t,T] \to \mathcal{U}[t,T]$ is self-adjoint;

(H2)$_t$ The operator $L_1(t)^* Q_1 e^{-A(T-t)} \in \mathcal{L}(X, \mathcal{U}[t,T])$;

and $\Phi(t) \gg 0$. But this amounts to saying that in order for the Problem (LQ)$_t$ to be solvable for all $t \in [0,T)$, infinitely many conditions (parameterized by $t \in [0,T]$) need to be imposed. Of course, this is not expected. Our next goal is to show that (H1) and $L_1^* Q_1 \in \mathcal{L}(X, \mathcal{U})$ imply (H1)$_t$ and

(H2)$_t$ for all $t \in [0,T]$. To this end, we first introduce the restriction operator: $K_t \in \mathcal{L}(\mathcal{U}[0,T];\mathcal{U}[t,T])$ as follows:

$$(3.23) \qquad (K_t u)(s) = u\big|_{[t,T]}(s), \qquad u \in \mathcal{U}[0,T].$$

It is easy to see that the adjoint operator $K_t^*$ of $K_t$ is given by

$$(3.24) \qquad (K_t^* v)(s) = \begin{cases} 0, & s \in [0,t), \\ v(s), & s \in [t,T], \end{cases} \qquad v \in \mathcal{U}[t,T].$$

The following result will be useful.

**Proposition 3.3.** *It holds that*

$$(3.25) \qquad L(t)^* Q L(t) = K_t L^* Q L K_t^*,$$

$$(3.26) \qquad L_1(t) = L_1 K_t^*, \qquad L_1(t)^* = K_t L_1^*.$$

*Proof.* For any $u(\cdot) \in \mathcal{U}[t,T]$, by the definition of $K_t^*$, we see that

$$(3.27) \quad \begin{aligned} (L(t)^* Q L(t) u)(s) &= \int_s^T B^*(A^*)^\alpha e^{-A^*(\sigma-s)} Q \int_t^\sigma A^\alpha e^{-A(\sigma-r)} B u(r) \, dr \, d\sigma \\ &= \int_s^T B^*(A^*)^\alpha e^{-A^*(\sigma-s)} Q \int_0^\sigma A^\alpha e^{-A(\sigma-r)} (K_t^* u)(r) \, dr \, d\sigma \\ &= \int_s^T B^*(A^*)^\alpha e^{-A^*(\sigma-s)} (Q L K_t^* u)(\sigma) \, d\sigma = (K_t L^* Q L K_t^* u)(s). \end{aligned}$$

This proves (3.25).

Next, for any $u(\cdot) \in \mathcal{D}(L_1(t))$, we have

$$(3.28) \qquad \mathcal{D}(A^\alpha) \ni \int_t^T e^{-A(T-r)} B u(r) \, dr = \int_0^T e^{-A(T-r)} B(K_t^* u)(r) \, dr.$$

This yields $K_t^* u \in \mathcal{D}(L_1)$ or $u \in \mathcal{D}(L_1 K_t^*)$ and

$$(3.29) \qquad L_1(t) u = L_1 K_t^* u.$$

The above tells us that $L_1 K_t^*$ is an extension of $L_1(t)$. On the other hand, for any $u \in \mathcal{D}(L_1 K_t^*)$, we have $K_t^* u \in \mathcal{D}(L_1)$, which implies

$$(3.30) \qquad \int_t^T e^{-A(T-r)} B u(r) \, dr = \int_0^T e^{-A(T-r)} B(K_t^* u)(r) \, dr \in \mathcal{D}(A^\alpha),$$

that is, $u \in \mathcal{D}(L_1(t))$. Hence, $\mathcal{D}(L_1 K_t^*) \subseteq \mathcal{D}(L_1(t))$, proving $L_1(t) = L_1 K_t^*$.

§3. State Feedback Control

Finally, for any $z \in \mathcal{D}(L_1(t)^*)$, we have $B^*(A^*)^\alpha e^{-A^*(T-\cdot)}z \in \mathcal{U}[t,T]$. Thus,

(3.31)
$$\int_0^T |B^*(A^*)^\alpha e^{-A^*(T-r)}z|^2 \, dr \leq \int_t^T |B^*(A^*)^\alpha e^{-A^*(T-r)}z|^2 \, dr$$
$$+ \|B^*\|^2 \frac{C}{(T-t)^{2\alpha}} \int_0^t |e^{-A^*(t-r)}z|^2 \, dr < \infty.$$

This tells us that $z \in \mathcal{D}(L_1^*)$ and

(3.32) $\qquad L_1(t)^*z = K_t L_1^* z, \qquad \forall z \in \mathcal{D}(L_1(t)^*).$

On the other hand, it is clear that $\mathcal{D}(L_1^*) \subseteq \mathcal{D}(L_1(t)^*)$. Hence, $L_1(t)^* = K_t L_1^*$. □

We note that in general, one only has $(L_1 K)^* \supseteq K^* L_1^*$ if $K$ is a bounded operator (see Chapter 2, §4). The last equality in (3.26) holds because of the specialty of the operator $K_t$. Now, we can prove the following result.

**Proposition 3.4.** *Let (H1) hold and $L_1^* Q_1 \in \mathcal{L}(X, \mathcal{U})$. Then $(H1)_t$ and $(H2)_t$ hold for all $t \in [0, T)$. In particular, this is the case if (H1) and (H3) hold.*

*Proof.* First of all, by (3.26), we have

(3.33) $\qquad L_1(t)^* Q_1 = K_t L_1^* Q_1 \in \mathcal{L}(X, \mathcal{U}[t,T]).$

Thus, $(H2)_t$ holds. Next, noting that $K_t$ and $L_1^* Q_1$ are bounded, and $L_1^* Q_1 L_1$ is self-adjoint, we have (note Chapter 2, Proposition 4.4(i))

(3.34)
$$\begin{aligned}(L_1(t)^* Q_1 L_1(t))^* &= (K_t L_1^* Q_1 L_1 K_t^*)^* \\ &= (L_1 K_t^*)^* (L_1^* Q_1)^* K_t^* = L_1(t)^* (L_1^* Q_1)^* K_t^* \\ &= K_t L_1^* (L_1^* Q_1)^* K_t^* = K_t (L_1^* Q_1 L_1)^* K_t^* \\ &= K_t L_1^* Q_1 L_1 K_t^* = L_1(t)^* Q_1 L_1(t).\end{aligned}$$

This means that $(H1)_t$ holds. Finally, if (H1) and (H3) hold, then by Proposition 2.8, $L_1^* Q_1 \in \mathcal{L}(X, \mathcal{U})$. Thus, $(H1)_t$ and $(H2)_t$ hold as well. □

Next, regarding $R \in \mathcal{L}(\mathcal{U}[t,T])$, we have $R = K_t R K_t^*$. Thus, by the definition of $\Phi(t)$ and $\Phi$, and Proposition 3.3, we have

(3.35)
$$\begin{aligned}\Phi(t) &= R + L(t)^* Q L(t) + L_1(t)^* Q_1 L_1(t) \\ &= R + K_t L^* Q L K_t^* + K_t L_1^* Q_1 L_1 K_t^* = K_t \Phi K_t^*.\end{aligned}$$

Hence, we see that

(3.36) $\qquad \begin{cases} \Phi \geq 0 & \Rightarrow \quad \Phi(t) \geq 0, \quad \forall t \in [0,T), \\ \Phi \gg 0 & \Rightarrow \quad \Phi(t) \gg 0, \quad \forall t \in [0,T). \end{cases}$

Similar to the Problem (LQ), and noting the above proposition, we have the following result (see Theorem 3.2).

**Theorem 3.5.** *Let (H1) and (H3) hold. Let $\Phi \geq 0$ and $R \gg 0$. Then Problem $(LQ)_t$ is solvable at $x \in X$ if and only if there exists a pair $(\bar{y}_{t,x}(\cdot), \psi_{t,x}(\cdot)) \in \mathcal{X}[t,T] \times \mathcal{X}[t,T]$ $(\mathcal{X}[t,T] \equiv L^2(t,T;X))$ with*

$$(3.37) \qquad \int_t^T e^{-A(T-r)} BR^{-1} B^* \psi_{t,x}(r)\, dr \in \mathcal{D}(A^\alpha),$$

*solving the following two-point boundary value problem:*

$$(3.38) \quad \begin{cases} \bar{y}_{t,x}(s) = e^{-A(s-t)} x + \displaystyle\int_t^s A^\alpha e^{-A(s-r)} BR^{-1} B^* \psi_{t,x}(r)\, dr, \\ \psi_{t,x}(s) = -(A^*)^\alpha e^{-A^*(T-s)} Q_1 \bar{y}_{t,x}(T) \\ \qquad\qquad - \displaystyle\int_s^T (A^*)^\alpha e^{-A^*(\sigma-s)} Q \bar{y}_{t,x}(\sigma)\, d\sigma, \quad s \in [t,T]. \end{cases}$$

*In this case,*

$$(3.39) \qquad \bar{u}_{t,x}(s) = R^{-1} B^* \psi_{t,x}(s), \qquad s \in [t,T]$$

*gives an optimal control and the function $\bar{y}_{t,x}(\cdot)$ is the corresponding optimal state trajectory. In addition, if $\Phi \gg 0$, then (3.38) admits a unique solution and Problem $(LQ)_t$ is uniquely solvable (on $X$) with the optimal control given by (3.39).*

### §3.3. A Fredholm integral equation

In this subsection, we would like to show that, under certain conditions, the function $\psi_{t,x}(\cdot)$ determined by (3.38) can be written in terms of $\bar{y}_{t,x}(\cdot)$. Consequently, the optimal control $\bar{u}_{t,x}(\cdot)$ will be expressed by the optimal trajectory $\bar{y}_{t,x}(\cdot)$. We now give a heuristic derivation of representing $\psi(\cdot)$ in terms of $\bar{y}(\cdot)$. Let (H1) and (H3) hold, $\Phi \gg 0$, and $R \gg 0$. Then let $t \in [0,T)$ and consider Problem $(LQ)_t$. By Theorem 3.5, the optimal pair $(\bar{y}_{t,x}(\cdot), \bar{u}(\cdot))$ and the auxiliary function $\psi_{t,x}(\cdot)$ are related by (3.37)–(3.39). Substituting the first equation in (3.38) into the second one, we obtain that (the subscripts $t, x$ are suppressed)

$$(3.40) \quad \begin{aligned} \psi(s) &= -(A^*)^\alpha e^{-A^*(T-s)} Q_1 \big\{ e^{-A(T-t)} x \\ &\qquad + \int_t^T A^\alpha e^{-A(T-r)} BR^{-1} B^* \psi(r)\, dr \big\} \\ &\quad - \int_s^T (A^*)^\alpha e^{-A^*(\sigma-s)} Q \big\{ e^{-A(\sigma-t)} x \\ &\qquad + \int_t^\sigma A^\alpha e^{-A(\sigma-r)} BR^{-1} B^* \psi(r)\, dr \big\}\, d\sigma \\ &= -S_0(s,t) x - \int_t^T [S_1(s,r) + S_2(s,r)] BR^{-1} B^* \psi(r)\, dr, \end{aligned}$$

## §3. State Feedback Control

where we have changed the order of the double integral in the last term, and

(3.41)
$$\begin{cases} S_0(s,t)x = (A^*)^{\alpha-\beta}e^{-A^*(T-s)}(A^*)^\beta Q_1 e^{-A(T-t)}x \\ \qquad\qquad + \int_s^T (A^*)^\alpha e^{-A^*(\sigma-s)} Q e^{-A(\sigma-t)} x \, d\sigma, \\ S_1(s,r)x = \int_{s\vee r}^T (A^*)^\alpha e^{-A^*(\sigma-s)} Q A^\alpha e^{-A(\sigma-r)} x \, d\sigma, \\ S_2(s,r)x = (A^*)^{\alpha-\beta} e^{-A^*(T-s)} (A^*)^\beta Q_1 A^\alpha e^{-A(T-r)} x. \end{cases}$$

Clearly, (3.40) is a linear equation in $\psi(\cdot)$. Suppose that for each $x \in X$, this equation admits a unique solution. Then the solution should be linear in $x$. Thus, $\psi_{t,x}(\cdot) = H(\cdot,t)x$ is expected, and the operator $H(\cdot,t)$ should be a solution (in some sense) of the following equation:

(3.42)
$$H(s,t) = -S_0(s,t) - \int_t^T [S_1(s,r) + S_2(s,r)] BR^{-1} B^* H(r,t) \, dr,$$
$$s \in [t,T].$$

We refer to (3.42) as the *Fredholm integral equation* for our Problem (LQ)$_t$. To study equation (3.42), the following preliminary result is necessary.

**Lemma 3.6.** *Let (H1) and (H3) hold. Let $S_i$ ($i = 0,1,2$) be defined as above. Then*

(3.43) $$|S_0(\cdot,t)x|_{\mathcal{X}[t,T]} \leq C|x|, \qquad \forall x \in X,$$

(3.44) $$\left|\int_t^T S_1(\cdot,r)\varphi(r)\,dr\right|_{\mathcal{X}[t,T]} \leq C|\varphi|_{\mathcal{X}[t,T]}, \qquad \forall \varphi \in \mathcal{X}[t,T].$$

*Moreover, the operator $\Psi : \mathcal{D}(\Psi) \subseteq \mathcal{U}[t,T] \to \mathcal{X}[t,T]$ defined by*

(3.45) $$(\Psi u)(s) = \int_t^T S_2(s,r) B u(r) \, dr, \qquad \forall u \in \mathcal{D}(\Psi),$$

*is closed, densely defined with $\mathcal{D}(\Psi) = \mathcal{D}(L_1(t))$, and*

(3.46) $$|(\Psi u)(\cdot)|_{\mathcal{X}[t,T]} \leq C|L_1(t) u|_X, \qquad \forall u \in \mathcal{D}(L_1(t)).$$

*Proof.* For any $x \in X$, it follows from (H3) that (note a similar result to Lemma 1.1)

(3.47)
$$\int_t^T |S_0(s,t)x|^2 \, ds \leq C\Big\{ \int_t^T \frac{|x|^2}{(T-s)^{2(\alpha-\beta)}} \, ds$$
$$+ \int_t^T \Big|\int_s^T (A^*)^\alpha e^{-A^*(\sigma-s)} Q e^{-A(\sigma-t)} x \, d\sigma\Big|^2 \, ds\Big\}$$
$$\leq C\Big\{|x|^2 + \int_t^T |Q e^{-A(\sigma-t)} x|^2 \, d\sigma\Big\} \leq C|x|^2.$$

This gives (3.43). Next, for any $\varphi \in \mathcal{X}[t,T]$,

$$\int_t^T \Big| \int_t^T S_1(s,r)\varphi(r)\,dr \Big|^2 ds$$

(3.48)
$$\leq \int_t^T \Big| \int_t^T \Big\{ \int_{s\vee r}^T (A^*)^\alpha e^{-A^*(\sigma-s)} Q A^\alpha e^{-A(\sigma-r)}\,d\sigma \Big\} \varphi(r)\,dr \Big|^2 ds$$

$$\leq C \int_t^T \Big\{ \int_t^T \int_{s\vee r}^T (\sigma-s)^{-\alpha}(\sigma-r)^{-\alpha} |\varphi(r)|\,d\sigma\,dr \Big\}^2 ds.$$

Denote

(3.49)
$$g(s,r) = \int_{s\vee r}^T (\sigma-s)^{-\alpha}(\sigma-r)^{-\alpha}\,d\sigma, \qquad r,s \in [t,T],\ r \neq s.$$

Then, it is clear that $g(s,r) = g(r,s)$ and

(3.50)
$$\int_t^T g(s,r)\,ds = \int_t^T g(s,r)\,dr$$
$$= \int_t^T \int_{s\vee r}^T (\sigma-s)^{-\alpha}(\sigma-r)^{-\alpha}\,d\sigma\,dr$$
$$= \int_s^T \int_t^\sigma (\sigma-s)^{-\alpha}(\sigma-r)^{-\alpha}\,dr\,d\sigma$$
$$= \frac{1}{1-\alpha} \int_s^T (\sigma-s)^{-\alpha}(\sigma-t)^{1-\alpha}\,d\sigma$$
$$\leq \frac{(T-t)^{1-\alpha}(T-s)^{1-\alpha}}{(1-\alpha)^2} \leq \frac{T^{2(1-\alpha)}}{(1-\alpha)^2}.$$

Hence, we have

(3.51)
$$\int_t^T \Big| \int_t^T S_1(s,r)\varphi(r)\,dr \Big|^2 ds$$
$$\leq C \int_t^T \Big\{ \int_t^T g(s,r)|\varphi(r)|\,dr \Big\}^2 ds$$
$$\leq C \int_t^T \Big\{ \int_t^T g(s,r)\,dr \Big\}\Big\{ \int_t^T g(s,r)|\varphi(r)|^2\,dr \Big\}\,ds$$
$$\leq C \int_t^T |\varphi(r)|^2 \Big\{ \int_t^T g(s,r)\,ds \Big\}\,dr \leq C \int_t^T |\varphi(r)|^2 dr.$$

Hence, (3.44) holds. Now, by the definition of $\Psi$, we see that

(3.52)
$$\Psi = (A^*)^{\alpha-\beta} e^{-A^*(T-\cdot)}(A^*)^\beta Q_1 L_1(t),$$

and under (H3), $(A^*)^{\alpha-\beta} e^{-A^*(T-\cdot)}(A^*)^\beta Q_1 \in \mathcal{L}(X, \mathcal{X}[t,T])$. Thus, (3.46) follows and $\mathcal{D}(\Psi) = \mathcal{D}(L_1(t))$. Finally, we show that $\Psi$ is closed. To this end, we first claim that $\Psi^*$ is densely defined. In fact, for any $\varepsilon \in (0, T-t)$,

## §3. State Feedback Control

let $z(\cdot) \in L^2(t,T; \mathcal{D}(A^\alpha))$ and $z_\varepsilon(\cdot) = z(\cdot)\chi_{[t,T-\varepsilon]}(\cdot)$. Then, for any $u(\cdot) \in \mathcal{L}(L_1(t)) = \mathcal{D}(\Psi)$, (note $Q_1 A^\beta \subseteq ((A^*)^\beta Q_1)^* \in \mathcal{L}(X)$),

$$
\begin{aligned}
(3.53) \quad & |\langle (\Psi u)(\cdot), z_\varepsilon(\cdot) \rangle| \\
&= \left| \int_t^T \langle (A^*)^\alpha e^{-A^*(T-s)} Q_1 \int_t^T A^\alpha e^{-A(T-r)} Bu(r)\, dr, z_\varepsilon(s) \rangle\, ds \right| \\
&\leq C \left| \int_t^T A^{\alpha-\beta} e^{-A(T-r)} Bu(r)\, dr \right| \left| \int_t^{T-\varepsilon} A^\alpha e^{-A(T-s)} z(s)\, ds \right| \\
&\leq C_\varepsilon |u|_{\mathcal{U}[t,T]} |z|_{\mathcal{X}[t,T]}.
\end{aligned}
$$

Because the set of all functions $z_\varepsilon(\cdot)$ $(\varepsilon > 0)$ constructed above is dense in $\mathcal{X}[t,T]$, we obtain that $\Psi^*$ is densely defined. Now, let us take $u_n \in \mathcal{D}(\Psi)$, such that

$$(3.54) \quad u_n \xrightarrow{s} u, \qquad \Psi u_n \xrightarrow{s} z.$$

This implies that (note (3.52) and the definition of $L_1(t)^*$)

$$(3.55) \quad \lim_{n\to\infty} |B^* z - L_1^*(t) Q_1 L_1(t) u_n|_{\mathcal{U}[t,T]} = \lim_{n\to\infty} |B^* z - B^* \Psi u_n|_{\mathcal{U}[t,T]} = 0.$$

By (H1) and (H3) (note Proposition 3.4), we know that $L_1(t)^* Q_1 L_1(t)$ is self-adjoint. Thus, (3.54) and (3.55) yield that $u \in \mathcal{D}(L_1(t)) = \mathcal{D}(\Psi)$. Then, for any $\xi(\cdot) \in \mathcal{D}(\Psi^*)$,

$$
\begin{aligned}
(3.56) \quad \langle \Psi u - z, \xi \rangle &= \langle u, \Psi^* \xi \rangle - \langle z, \xi \rangle \\
&= \lim_{n\to\infty} \left\{ \langle u_n, \Psi^* \xi \rangle - \langle \Psi u_n, \xi \rangle \right\} = 0.
\end{aligned}
$$

Hence, by the density of $\mathcal{D}(\Psi^*)$, we see that $\Psi u = z$. This proves the closeness of $\Psi$. □

Based on the above lemma, we may now introduce the following definition.

**Definition 3.7.** Let $t \in [0,T)$. An operator valued function $H(\cdot, t) \in \mathcal{L}(X, \mathcal{X}[t,T])$ is called a *strong solution* of (3.42) if the following hold:

(i) For any $x \in X$, $R^{-1} B^* H(\cdot, t) x \in \mathcal{D}(L_1(t))$, i.e.,

$$(3.57) \quad \int_t^T e^{-A(T-r)} B R^{-1} B^* H(r,t) x\, dr \in \mathcal{D}(A^\alpha).$$

(ii) For any $x \in X$, it holds that

$$
\begin{aligned}
(3.58) \quad H(s,t)x = &- S_0(s,t)x \\
&- \int_t^T \{S_1(s,r) + S_2(s,r)\} BR^{-1} B^* H(r,t) x\, dr,
\end{aligned}
$$

a.e. $s \in [t,T]$.

Now, we give a relation between the Fredholm equation (3.42) and the two-point boundary value problem (3.38).

**Theorem 3.8.** *Let $t \in [0, T)$ be given.*

*(i) Suppose that (3.42) admits a strong solution $H(\cdot, t)$. Then, for any given $x \in X$, (3.38) admits a solution $(\bar{y}_{t,x}(\cdot), \psi_{t,x}(\cdot)) \in \mathcal{X}[t,T] \times \mathcal{X}[t,T]$, such that (3.37) holds and*

$$\psi_{t,x}(s) = H(s,t)x, \qquad a.e. \ s \in [t, T]. \tag{3.59}$$

*(ii) For any $x \in X$, (3.38) admits a unique solution $(\bar{y}_{t,x}(\cdot), \psi_{t,x}(\cdot)) \in \mathcal{X}[t,T] \times \mathcal{X}[t,T]$ with property (3.37) if and only if (3.42) admits a unique strong solution $H(\cdot, t)$.*

*Proof.* (i) Let $H(\cdot, t)$ be a strong solution of (3.42). For any $x \in X$, we define $\psi_{t,x}(\cdot)$ by (3.59) and define $\bar{y}_{t,x}(\cdot)$ through the first equation in (3.38). Using condition (3.57), we see that (3.37) holds and $\bar{y}_{t,x}(T)$ is well defined. Then, reversing the procedure in (3.40), we obtain

$$\begin{aligned}\psi_{t,x}(s) &= H(s,t)x \\ &= -S_0(s,t)x - \int_t^T \{S_1(s,r) + S_2(s,r)\} BR^{-1}B^* \psi_{t,x}(r)\, dr \\ &= -(A^*)^\alpha e^{A^*(T-s)} Q_1 \bar{y}_{t,x}(T) - \int_s^T (A^*)^\alpha e^{A^*(\sigma-s)} Q \bar{y}_{t,x}(\sigma)\, d\sigma.\end{aligned} \tag{3.60}$$

This shows that the two-point boundary value problem (3.38) admits a solution with the property (3.37). We should note that in (3.60), the derivation is rigorous because of Lemma 3.6.

(ii) For any $x \in X$, by the uniqueness of the solution $(\bar{y}_{t,x}(\cdot), \psi_{t,x}(\cdot))$ to (3.38), we have the uniqueness of the solution $\psi_{t,x}(\cdot)$ to (3.40). Thus, the map $x \mapsto \psi_{t,x}(\cdot)$ is linear. Let us define the operator valued function $H(\cdot, t) : X \to \mathcal{X}[t,T]$ by (3.59). From (3.37), it follows that (3.57) holds. Next, we claim that $H(\cdot, t)$ is closed. In fact, if $x_n \xrightarrow{s} x$ in $X$ and $\psi_n(\cdot,,t)x_n \xrightarrow{s} \psi(\cdot)$ in $\mathcal{X}[t,T]$, then by Lemma 3.6, we see that

$$\begin{cases} S_0(\cdot,t)x_n \xrightarrow{s} S_0(\cdot,t)x, \\ \displaystyle\int_t^T S_1(\cdot,r)BR^{-1}B^*\psi_n(r)\, dr \xrightarrow{s} \int_t^T S_1(\cdot,r)BR^{-1}B^*\psi(r)\, dr. \end{cases} \tag{3.61}$$

Consequently (note (3.45) and (3.40)),

$$\Psi R^{-1} B^* \psi_n(\cdot) \xrightarrow{s} -\psi(\cdot) - S_0(\cdot,t)x - \int_t^T S_1(\cdot,r) BR^{-1} B^* \psi(r)\, dr. \tag{3.62}$$

Hence, by the closeness of $\Psi$, we obtain that $\psi(\cdot) \in \mathcal{D}(\Psi)$ and it is a solution of (3.60). By the uniqueness, it is necessary that $H(\cdot, t)x = \psi(\cdot)$.

This proves the closeness of $H(\cdot,t)$. On the other hand, as this operator is defined everywhere on $X$, by the Closed Graph Theorem again, we have $H(\cdot,t) \in \mathcal{L}(X, \mathcal{X}[t,T])$. Finally, as for (3.40), we obtain (3.58). Thus, $H(\cdot,t)$ is a strong solution of (3.42).

Now, suppose that $\widetilde{H}(\cdot,t)$ is another solution of (3.42). Then, by (i), for each $t \in [0,T]$ and $x \in X$, $\widetilde{H}(\cdot,t)$ determines a solution $(\widetilde{y}_{t,x}(\cdot), \widetilde{\psi}_{t,x}(\cdot))$ of (3.38). Hence, by the uniqueness, we must have

$$(3.63) \qquad H(s,t)x = \psi_{t,x}(s) = \widetilde{\psi}_{t,x}(s) = \widetilde{H}(s,t)x,$$
$$\text{a.e. } s \in [t,T], \ x \in X.$$

This gives $\widetilde{H}(\cdot,t) = H(\cdot,t)$, proving the uniqueness.

Conversely, let $H(\cdot,t)$ be the unique strong solution of (3.42). Then, by (i), we know that for any $x \in X$, (3.38) admits a solution $(\bar{y}_{t,x}(\cdot), \psi_{t,x}(\cdot))$ with property (3.37). Suppose that for some $\bar{x} \in X$, (3.38) admits another solution $(\widetilde{y}(s), \widetilde{\psi}(s))$ with the property (3.37), which is different from $(\bar{y}_{t,\bar{x}}(\cdot), \psi_{t,\bar{x}}(\cdot))$ obtained by $H(\cdot,t)$. Then, by (3.38), we must have

$$(3.64) \qquad \widehat{\psi}(\cdot) \triangleq \psi_{t,\bar{x}}(\cdot) - \widetilde{\psi}(\cdot) \neq 0,$$

and by (3.40), $\widehat{\psi}(\cdot)$ satisfies the following homogeneous Fredholm integral equation:

$$(3.65) \qquad \widehat{\psi}(s) = -\int_t^T \{S_1(s,r) + S_2(s,r)\} BR^{-1}B^* \widehat{\psi}(r)\, dr, \quad s \in [t,T].$$

Now, let us fix an $x_0 \in X \setminus \{0\}$. For any $x \in X$ having the unique decomposition $x = \lambda x_0 + x_1$, with $\langle x_0, x_1 \rangle = 0$ and $\lambda \in \mathbb{R}$, we define

$$(3.66) \qquad \widehat{H}(s)x = \lambda \widehat{\psi}(s), \quad s \in [t,T].$$

Then $\widehat{H}(s) \in \mathcal{L}(X, \mathcal{X}[t,T])$, $R^{-1}B^*\widehat{H}(\cdot)x \in \mathcal{D}(L_1(t))$, and

$$(3.67) \ \widehat{H}(s)x = -\int_t^T \{S_1(s,r) + S_2(s,r)\} BR^{-1}B^* \widehat{H}(r)x\, dr, \quad s \in [t,T].$$

This implies that (3.42) admits more than one solution, a contradiction. □

### §3.4. State feedback representation of optimal controls

Note that under (H1) and (H3), if $\Phi \gg 0$ and $R \gg 0$, then, for any $(t,x) \in [0,T) \times X$, Problem (LQ)$_t$ admits a unique optimal control $\bar{u}_{t,x}(\cdot)$, the two-point boundary value problem (3.38) admits a unique solution $(\bar{y}_{t,x}(\cdot), \psi_{t,x}(\cdot))$, in which $\bar{y}_{t,x}(\cdot)$ is the optimal state trajectory, and the Fredholm integral equation (3.42) admits a unique strong solution $H(\cdot,t)$.

Now, if $(\bar{y}_{t,x}(s), \psi_{t,x}(s))$ and $H(s,t)$ are continuous in $s \in [t,T)$, then, by the uniqueness of the solutions to (3.38), we must have

(3.68) $$\begin{cases} \bar{y}_{t,x}(s) = y_{r,\bar{y}_{t,x}(r)}(s), \\ \psi_{t,x}(s) = \psi_{r,\bar{y}_{t,x}(r)}(s), \end{cases} \quad \forall t \leq r \leq s \leq T.$$

Hence, by (3.59), we have

(3.69) $$\psi_{t,x}(s) = \psi_{r,\bar{y}_{t,x}(r)}(s) = H(s,r)\bar{y}_{t,x}(r).$$

Sending $s \to r$, we obtain the relation

(3.70) $$\bar{u}_{t,x}(r) = R^{-1}B^*\psi_{t,x}(r) = R^{-1}B^*H(r,r)\bar{y}_{t,x}(r), \quad r \in [t,T).$$

In particular, the optimal control $\bar{u}(\cdot) \equiv \bar{u}_{0,x}(\cdot)$ of the Problem (LQ) is of the form

(3.71) $$\bar{u}(s) = R^{-1}B^*H(s,s)\bar{y}(s), \quad s \in [0,T).$$

Such a control is called a *linear state feedback control*. However, we note that, in general, it is not clear whether we have the above required continuity. To obtain such a continuity, we adopt assumption (H4). Some other assumptions are possible, see Lasiecka-Triggiani [1]. Let us now give the main result of this subsection.

**Theorem 3.9.** *Let (H1), (H3), and (H4) hold. Let $\Phi \gg 0$ and $R \gg 0$. Then, for any $(t,x) \in [0,T) \times X$, Problem $(LQ)_t$ is uniquely solvable with the optimal pair $(\bar{y}_{t,x}(\cdot), \bar{u}_{t,x}(\cdot))$ and the Fredholm integral equation (3.42) admits a unique strong solution $H(\cdot,t)$, such that for some absolute constant $C > 0$ (independent of $(t,x) \in [0,T) \times X$), it holds that*

(3.72) $$|\bar{u}_{t,x}(\cdot)|_{\mathcal{U}[t,T]}, \ |\bar{y}_{t,x}(\cdot)|_{\mathcal{X}[t,T]}, \ |Q_1\bar{y}_{t,x}(T)|_X \leq C|x|.$$

*Moreover, $(\bar{y}_{t,x}(\cdot), \bar{u}_{t,x}(\cdot))$ and $H(\cdot,t)x$ are continuous on $[t,T)$ and with an absolute constant $C > 0$,*

(3.73) $$|\bar{u}_{t,x}(s)|, \ |\bar{y}_{t,x}(s)|, \ |H(s,t)x| \leq \frac{C|x|}{(T-s)^\alpha}, \quad \forall s \in [t,T).$$

*Consequently, (3.70) and (3.71) hold.*

*Proof.* First of all, by our assumption and Theorems 3.5 and 3.8, we know that Problem $(LQ)_t$ is uniquely solvable for all $t \in [0,T)$ with the optimal pair $(\bar{y}_{t,x}(\cdot), \bar{u}_{t,x}(\cdot))$ and both the equations (3.38) and (3.42) are uniquely solvable with the solutions $(\bar{y}_{t,x}(\cdot), \psi_{t,x}(\cdot))$ and $H(\cdot,t)$, respectively. Moreover, relations (3.37), (3.39), and (3.59) hold. On the other hand, similar to Theorem 2.6 (see (2.27) in particular), we have that

(3.74) $$\bar{u}_{t,x}(\cdot) = -\Phi(t)^{-1}\Theta(t)x.$$

§3. State Feedback Control

$\Phi(t)$ is self-adjoint and for any $v \in \mathcal{U}[t,T]$, we have (note (3.35))

(3.75) $\quad \langle \Phi(t)v, v \rangle = \langle K_t \Phi K_t^* v, v \rangle \geq \delta |K_t^* v|_{\mathcal{U}[0,T]}^2 = \delta |v|_{\mathcal{U}[t,T]}^2 .$

Hence, by Lemma 2.5, $\Phi(t)$ is boundedly invertible and

(3.76) $\quad \|\Phi(t)^{-1}\|_{\mathcal{L}(\mathcal{U}[t,T])} \leq \dfrac{1}{\delta}, \qquad \forall t \in [0, T).$

On the other hand, by the definition of $\Theta(t)$ (see (3.21)) and (H3), we have

(3.77)
$$\begin{aligned}|(\Theta(t)x)(s)| &\leq \left| \int_s^T (A^*)^\alpha e^{-A^*(\sigma-s)} Q e^{-A(\sigma-t)} x \, d\sigma \right| \\&\quad + |(A^*)^\alpha e^{-A^*(T-s)} Q_1 e^{-A(T-t)} x| \\&\leq \int_s^T \dfrac{C|x|}{(\sigma - s)^\alpha} \, d\sigma + \dfrac{C|x|}{(T-s)^{\alpha-\beta}} \leq \dfrac{C|x|}{(T-s)^{\alpha-\beta}}.\end{aligned}$$

Clearly, in the above, the constant $C$ is independent of $t \in [0,T)$. Because $\alpha - \beta < 1/2$, (3.77) implies that

(3.78) $\quad |(\Theta(t)x)(\cdot)|_{\mathcal{U}[t,T]} \leq C|x|, \qquad \forall (t,x) \in [0,T) \times X.$

Combining (3.74), (3.76), and (3.78), the estimate for $|\bar{u}_{t,x}(\cdot)|_{\mathcal{U}[t,T]}$ in (3.72) follows. Consequently, one can easily obtain the estimate for $|\bar{y}_{t,x}(\cdot)|_{\mathcal{X}[t,T]}$ in (3.72). Next, it follows from (H3) that

(3.79)
$$\begin{aligned}|Q_1 \bar{y}_{t,x}(T)| &\leq |Q_1 e^{-A(T-t)} x| \\&\quad + \left| \int_t^T Q_1 A^\alpha e^{-A(T-r)} B \bar{u}_{t,x}(r) \, dr \right| \\&\leq C|x| + \int_t^T |((A^*)^\beta Q_1)^* A^{\alpha-\beta} e^{-A(T-r)} B \bar{u}_{t,x}(r)| \, dr \\&\leq C|x| + C \int_t^T \dfrac{|\bar{u}_{t,x}(r)|}{(T-r)^{\alpha-\beta}} \, dr \\&\leq C|x| + C|(T-\cdot)^{-(\alpha-\beta)}|_{L^2(t,T)} |\bar{u}_{t,x}(\cdot)|_{\mathcal{U}[t,T]} \leq C|x|.\end{aligned}$$

Here, we have used the estimate for $|\bar{u}_{t,x}(\cdot)|_{\mathcal{U}[t,T]}$. Thus, we have proved the last estimate in (3.72).

Now, by the definition of $H(\cdot, t)$, (3.38), and (3.72), (also note (H3) and (H4)) we have that for any $s \in [t, T)$,

(3.80)
$$\begin{aligned}|H(s,t)x| \equiv |\psi_{t,x}(s)| &\leq |(A^*)^\alpha e^{-A^*(T-s)} Q_1 \bar{y}_{t,x}(T)| \\&\quad + \left| \int_t^s (A^*)^\alpha e^{-A^*(s-r)} Q \bar{y}_{t,x}(r) \, dr \right| \\&\leq \dfrac{C|Q_1 \bar{y}_{t,x}(T)|}{(T-s)^\alpha} + \int_t^s \dfrac{C|\bar{y}_{t,x}(r)|}{(s-r)^{\alpha-\beta}} \, dr \\&\leq \dfrac{C|x|}{(T-s)^\alpha} + C|\bar{y}_{t,x}(\cdot)|_{\mathcal{X}[t,T]} \leq \dfrac{C|x|}{(T-s)^\alpha}.\end{aligned}$$

This gives the third estimate in (3.73). Consequently, the first estimate in (3.73) follows from (3.39) and (3.80). Then, by the state equation and the estimate for $|\bar{u}_{t,x}(s)|$,

(3.81)
$$\begin{aligned}|\bar{y}_{t,x}(s)| &\leq \left|e^{-A(s-t)}x + \int_t^s A^\alpha e^{-A(s-r)}B\bar{u}_{t,x}(r)\,dr\right| \\ &\leq C|x| + \int_t^s \frac{C|x|\,dr}{(s-r)^\alpha(T-s)^\alpha} \leq \frac{C|x|}{(T-s)^\alpha}.\end{aligned}$$

Hence, (3.73) holds.

Next, let $\varepsilon \in (0, \beta - \alpha + \frac{1}{2})$ and let $t \in [0, T)$ be fixed. We now prove the continuity of $\psi_{t,x}(\cdot)$ on $[t, T)$. To show this, we take $t \leq s \leq \hat{s} < T$ and consider the following:

(3.82)
$$\begin{aligned}|\psi_{t,x}(\hat{s}) &- \psi_{t,x}(s)| \\ &\leq |(A^*)^\alpha e^{-A^*(T-\hat{s})}(I - e^{-A^*(\hat{s}-s)})Q_1\bar{y}_{t,x}(T)| \\ &+ \int_s^{\hat{s}} |(A^*)^\alpha e^{-A^*(\sigma-s)}Q\bar{y}_{t,x}(\sigma)|\,d\sigma \\ &+ \int_{\hat{s}}^T |(A^*)^\alpha e^{-A^*(\sigma-\hat{s})}(I - e^{-A^*(\hat{s}-s)})Q\bar{y}_{t,x}(\sigma)|\,d\sigma \\ &\equiv I_1 + I_2 + I_3.\end{aligned}$$

We estimate each term in the above separately. By (H3), (1.32), and (3.72), we have

(3.83)
$$\begin{aligned}I_1 &= |(A^*)^{\alpha+\varepsilon}e^{-A^*(T-\hat{s})}(I - e^{-A^*(\hat{s}-s)})(A^*)^{-\varepsilon}Q_1\bar{y}_{t,x}(T)| \\ &\leq \frac{C_\varepsilon}{(T-\hat{s})^{\alpha+\varepsilon}}|\hat{s}-s|^\varepsilon |Q_1\bar{y}_{t,x}(T)| \leq \frac{C_\varepsilon|x|}{(T-\hat{s})^{\alpha+\varepsilon}}|\hat{s}-s|^\varepsilon.\end{aligned}$$

By (H4), (3.72) and the fact that $r^{1/2-\alpha+\beta} \leq Cr^\varepsilon$, for $r \in [0, T]$,

(3.84)
$$\begin{aligned}I_2 &= \int_s^{\hat{s}} |(A^*)^{\alpha-\beta}e^{-A^*(\sigma-s)}(A^*)^\beta Q\bar{y}_{t,x}(\sigma)|\,d\sigma \\ &\leq C\int_s^{\hat{s}} \frac{|\bar{y}_{t,x}(\sigma)|}{(\sigma-s)^{\alpha-\beta}}\,d\sigma \\ &\leq C\Big(\int_s^{\hat{s}} \frac{d\sigma}{(\sigma-s)^{2(\alpha-\beta)}}\Big)^{1/2} |\bar{y}_{t,x}(\cdot)|_{\mathcal{X}[s,\hat{s}]} \\ &\leq C|x|\,|\hat{s}-s|^\varepsilon.\end{aligned}$$

Finally, by (H4), (1.32), and (3.72),

(3.85)
$$I_3 \leq \int_s^T \frac{C_\varepsilon}{(\sigma - \widehat{s})^{\alpha-\beta+\varepsilon}} |(I - e^{-A^*(\widehat{s}-s)})(A^*)^{-\varepsilon+\beta} Q \bar{y}_{t,x}(\sigma)| \, d\sigma$$
$$\leq \int_s^T \frac{C_\varepsilon |\widehat{s} - s|^\varepsilon}{(\sigma - \widehat{s})^{\alpha-\beta+\varepsilon}} |\bar{y}_{t,x}(\sigma)| \, d\sigma \leq C_\varepsilon |x| \, |\widehat{s} - s|^\varepsilon.$$

Combining (3.82)–(3.85), we obtain that $\psi_{t,x}(\cdot) \in C([t,T];X)$. Then, from relation (3.59), we see that $H(\cdot,t)x \in C([t,T];X)$. Consequently, it follows from (3.39) that $\bar{u}_{t,x}(\cdot) \in C([t,T];U)$.

Now, we show that $\bar{y}_{t,x}(\cdot)$ is continuous on $[t,T]$. In fact, for any $\varepsilon \in (0, 1-\alpha)$, $t \leq s \leq \widehat{s} < T$, we have (note (1.32) and (3.73))

(3.86)
$$|\bar{y}_{t,x}(\widehat{s}) - \bar{y}_{t,x}(s)|$$
$$\leq |e^{-A\widehat{s}}(I - e^{-A(\widehat{s}-s)})x| + \int_s^{\widehat{s}} |A^\alpha e^{-A(\widehat{s}-r)} B \bar{u}_{t,x}(r)| \, dr$$
$$+ \int_t^s |A^{\alpha+\varepsilon} e^{-A(s-r)}(I - e^{-A(\widehat{s}-s)}) A^{-\varepsilon} B \bar{u}_{t,x}(r)| \, dr$$
$$\leq C|(I - e^{-A(\widehat{s}-s)})x| + \frac{C|x|}{(T-\widehat{s})^\alpha} |\widehat{s} - s|^{1-\alpha} + \frac{C_\varepsilon |x|}{(T-s)^\alpha} |\widehat{s} - s|^\varepsilon.$$

This proves the continuity of $\bar{y}_{t,x}(\cdot)$ on $[t,T]$.

Then, our conclusion follows from the observation made before the statement of Theorem 3.9. □

The above says that under certain conditions, our optimal control has a *state feedback representation*. On the other hand, this gives a way of solving our Problem (LQ) as follows: First find a strong solution $H(s,t)$ of the Fredholm equation (3.42). Then, set $\bar{u}(\cdot)$ as in (3.71) to get an optimal control.

## §4. Riccati Integral Equation

From the previous section, we see that under certain conditions Problem (LQ) is uniquely solvable and, moreover, the optimal control is of linear state feedback form, in which the linear operator $H(s,t)$ of the formula (3.71) is a strong solution of the Fredholm integral equation (3.42). On the other hand, by a little closer observation of (3.71), we see that only the operator $H(t,t)$, $t \in [0,T]$ is needed (instead of $H(s,t)$, $0 \leq t \leq s < T$). In this section, we will try to find the equation that $H(t,t)$ satisfies. Let us again assume (H1), (H3), and (H4). Also, let $\Phi \gg 0$ and $R \gg 0$. Let $(\bar{y}_{t,x}(\cdot), \bar{u}_{t,x}(\cdot))$ be the optimal pair of Problem (LQ)$_t$. We define the operator $\mathcal{S}(\cdot,\cdot)$ as follows:

(4.1) $\qquad \mathcal{S}(s,t)x = \bar{y}_{t,x}(s), \qquad \forall 0 \leq t \leq s < T, \; x \in X.$

As $\bar{y}_{t,x}(\cdot)$ is continuous on $[t,T)$ and the optimal trajectory $\bar{y}_{t,x}(\cdot)$ is unique, we must have

(4.2) $\quad S(t,t) = I, \quad S(s,t) = S(s,r)S(r,t), \quad \forall 0 \le t \le r \le s < T.$

Moreover, we have the following lemma.

**Lemma 4.1.** *Let (H1), (H3), and (H4) hold. Let $\Phi \gg 0$ and $R \gg 0$. Then the operator $S$ defined by (4.1) has the following properties: For any $(s,x) \in [0,T) \times X$, the map $t \mapsto S(s,t)x$ is continuous on $[0,s]$; the map $t \mapsto Q_1 S(T,t)x$ is continuous on $[0,T)$; and for some absolute constant $C > 0$,*

(4.3) $\quad \begin{cases} \|S(s,t)\|_{\mathcal{L}(X)} \le \dfrac{C}{(T-s)^\alpha}, & \forall 0 \le t \le s < T, \\ \|Q_1 S(T,t)\|_{\mathcal{L}(X)} \le C, & \forall t \in [0,T). \end{cases}$

*Proof.* First of all, by (3.72) and (3.73), we see that (4.3) holds. We now prove the continuity of $S(s,\cdot)x$. To this end, let us take $0 \le t < \hat{t} \le s < T$ and $x \in X$. By (4.2) and (4.3),

(4.4) $\quad \begin{aligned} |S(s,\hat{t})x - S(s,t)x| &= |S(s,\hat{t})(I - S(\hat{t},t))x| \\ &\le \frac{C}{(T-s)^\alpha} \left| (e^{-A(\hat{t}-t)} - I)x + \int_t^{\hat{t}} A^\alpha e^{-A(\hat{t}-r)} B \bar{u}_{t,x}(r)\, dr \right| \\ &\le \frac{C}{(T-s)^\alpha} \left\{ |(e^{-A(\hat{t}-t)} - I)x| + \int_t^{\hat{t}} \frac{C|x|}{(\hat{t}-r)^\alpha (T-r)^\alpha}\, dr \right\} \\ &\le \frac{C}{(T-s)^\alpha} \left\{ |(e^{-A(\hat{t}-t)} - I)x| + \frac{C|x|(\hat{t}-t)^{1-\alpha}}{(T-s)^\alpha} \right\}. \end{aligned}$

This gives the continuity of $S(s,\cdot)x$ on $[0,s]$. Similarly, we observe that for any $0 \le t < \hat{t} < T$ and $x \in X$,

(4.5) $\quad \begin{aligned} |Q_1\{S(T,\hat{t})x - S(T,t)x\}| &= |Q_1 S(T,\hat{t})(I - S(\hat{t},t))x| \\ &\le \frac{C}{(T-s)^\alpha} \left\{ |(e^{-A(\hat{t}-t)} - I)x| + \frac{C|x|(\hat{t}-t)^{1-\alpha}}{(T-s)^\alpha} \right\}. \end{aligned}$

This proves our lemma. $\qquad\square$

Further, we have the following result concerning $H(s,t)$.

**Lemma 4.2.** *Under the assumption of Lemma 4.1, for any $x \in X$, the map $(s,t) \mapsto H(s,t)x$ is continuous on the set $\{0 \le t \le s < T\}$ and for some absolute constant $C > 0$, it holds that*

(4.6) $\quad \|H(s,t)\|_{\mathcal{L}(X)} \le \dfrac{C}{(T-s)^\alpha}, \quad \forall 0 \le t \le s < T,$

## §4. Riccati Integral Equation

(4.7) $$\|(A^*)^{-\alpha} H(s,t)\| \leq C, \qquad \forall 0 \leq t \leq s < T.$$

In particular, $t \mapsto H(t,t)x$ is continuous on $[0,T)$ and

(4.8) $$\|H(t,t)\|_{\mathcal{L}(X)} \leq \frac{C}{(T-t)^\alpha}, \qquad \forall t \in [0,T),$$

(4.9) $$\|(A^*)^{-\alpha} H(t,t)\| \leq C, \qquad \forall t \in [0,T).$$

*Proof.* First of all, by Theorem 3.9, we obtain (4.6). Next, by the relation (3.59) and Theorem 3.9, we have

(4.10)
$$\begin{aligned}|(A^*)^{-\alpha} H(s,t)x| &\equiv |(A^*)^{-\alpha}\psi_{t,x}(s)| \\ &\leq \left|e^{-A^*(T-s)}Q_1\bar{y}_{t,x}(T)\right| + \int_0^T \left|e^{-A^*(\sigma-s)}Q\bar{y}_{t,x}(\sigma)\right| d\sigma \\ &\leq C|x| + C|\bar{y}_{t,x}(\cdot)|_{\mathcal{X}[s,T]} \leq C|x|.\end{aligned}$$

Thus, (4.7) follows. Next, for any $x \in X$ and $0 \leq t \leq \bar{t} \leq s < T$, by (3.68),

(4.11)
$$\begin{aligned}|H(s,\bar{t})x - H(s,t)x| &= |\psi_{\bar{t},x}(s) - \psi_{t,x}(s)| \\ &= |\psi_{\bar{t},x}(s) - \psi_{\bar{t},\bar{y}_{t,x}(\bar{t})}(s)| = |H(s,\bar{t})x - H(s,\bar{t})\bar{y}_{t,x}(\bar{t})| \\ &\leq \|H(s,\bar{t})\| \, |(I - \mathcal{S}(\bar{t},t))x|.\end{aligned}$$

Then, similar to (4.4), we have the continuity of $H(s,\cdot)x$. Combining the above with Theorem 3.9, we obtain the continuity of $(s,t) \mapsto H(s,t)x$. □

Our main result of this section is the following.

**Theorem 4.3.** *Let (H1), (H3), and (H4) hold. Let $\Phi \gg 0$ and $R \gg 0$. Let*

(4.12) $$P(t) = -(A^*)^{-\alpha} H(t,t), \qquad \forall t \in [0,T).$$

*Then $P(t)^* = P(t) \in \mathcal{L}(X)$, for all $t \in [0,T)$; $P(\cdot) \in C([0,T);\mathcal{L}(X))$; and it satisfies*

(4.13)
$$\begin{aligned}P(t)x = {}& e^{-A^*(T-t)}Q_1 e^{-A(T-t)}x + \int_t^T e^{-A^*(\sigma-t)} Q e^{-A(\sigma-t)} x \, d\sigma \\ &- \int_t^T e^{-A^*(\sigma-t)} [(A^*)^\alpha P(\sigma)]^* BR^{-1} B^* (A^*)^\alpha P(\sigma) e^{-A(\sigma-t)} x \, d\sigma,\end{aligned}$$

$$x \in X, \ t \in [0,T),$$

(4.14) $$\lim_{t \uparrow T} \langle P(t)x, z \rangle = \langle Q_1 x, z \rangle, \qquad \forall x, z \in X,$$

(4.15) $$\langle P(t)x, x \rangle = J_{t,x}(\bar{u}_{t,x}(\cdot)), \qquad \forall t \in [0,T), \ x \in X.$$

398    Chapter 9. Linear Quadratic Optimal Control Problems

*Proof.* First of all, by the definition of $P(t)$ and Lemma 4.2, we see that $P(\cdot) \in C([0,T]; \mathcal{L}(X))$. Next, because

(4.16)
$$P(t)x \equiv -(A^*)^{-\alpha} H(t,t)x \equiv -(A^*)^{-\alpha} \psi_{t,x}(t)$$
$$= e^{-A^*(T-t)} Q_1 \bar{y}_{t,x}(T) + \int_t^T e^{-A^*(\sigma-t)} Q \bar{y}_{t,x}(\sigma) \, d\sigma,$$

then, for any $t \in [0,T)$ and $x, z \in X$,

(4.17)
$$\langle P(t)x, z \rangle = \langle Q_1 \bar{y}_{t,x}(T), e^{-A(T-t)} z \rangle$$
$$+ \int_t^T \langle Q \bar{y}_{t,x}(\sigma), e^{-A(\sigma-t)} z \rangle \, d\sigma$$
$$= \langle Q_1 \bar{y}_{t,x}(T), \bar{y}_{t,z}(T) \rangle + \int_t^T \langle Q \bar{y}_{t,x}(\sigma), \bar{y}_{t,z}(\sigma) \rangle \, d\sigma$$
$$- \int_t^T \langle Q_1 \bar{y}_{t,x}(T), A^\alpha e^{-A(T-r)} B \bar{u}_{t,z}(r) \rangle \, dr$$
$$- \int_t^T \langle Q \bar{y}_{t,x}(\sigma), \int_t^\sigma A^\alpha e^{-A(\sigma-r)} B \bar{u}_{t,z}(r) \, dr \rangle \, d\sigma$$
$$= \langle Q_1 \bar{y}_{t,x}(T), \bar{y}_{t,z}(T) \rangle + \int_t^T \langle Q \bar{y}_{t,x}(\sigma), \bar{y}_{t,z}(\sigma) \rangle \, d\sigma$$
$$+ \int_t^T \langle B^* \psi_{t,x}(r), \bar{u}_{t,z}(r) \rangle \, dr$$
$$= \langle Q_1 \bar{y}_{t,x}(T), \bar{y}_{t,z}(T) \rangle$$
$$+ \int_t^T \{ \langle Q \bar{y}_{t,x}(s), \bar{y}_{t,z}(s) \rangle + \langle R \bar{u}_{t,x}(s), \bar{u}_{t,z}(s) \rangle \} \, ds$$
$$= \langle x, P(t)z \rangle.$$

Hence, $P(t)^* = P(t)$. By taking $z = x$, we obtain (4.15). Now, from (4.16),

## §4. Riccati Integral Equation

we have the following estimate (note (3.24)):

(4.18)
$$\begin{aligned}
&|\langle P(t)x, z\rangle - \langle Q_1 x, z\rangle| \\
&\leq |\langle Q_1 \bar{y}_{t,x}(T), e^{-A(T-t)}z\rangle - \langle Q_1 x, z\rangle| \\
&\quad + C|z| \int_t^T |\bar{y}_{t,x}(\sigma)| \, d\sigma \\
&\leq |\langle Q_1 e^{-A(T-t)}x, e^{-A(T-t)}z\rangle - \langle Q_1 x, z\rangle| \\
&\quad + \Big|\Big\langle Q_1 \int_t^T A^\alpha e^{-A(T-r)} B\bar{u}_{t,x}(r) \, dr, e^{-A(T-t)}z\Big\rangle\Big| \\
&\quad + C\sqrt{T-t}|z| \, |\bar{y}_{t,x}(\cdot)|_{\mathcal{X}[t,T]} \\
&\leq o(1) + |\langle Q_1 L_1 K_t^* \bar{u}_{t,x}(\cdot), e^{-A(T-t)}z\rangle| \\
&\leq o(1) + |\bar{u}_{t,x}(\cdot)|_{\mathcal{U}[t,T]} |K_t(L_1^* Q_1)e^{-A(T-t)}z|_{\mathcal{U}[t,T]} \\
&\leq o(1) + C\Big(\int_t^T |(L_1^* Q_1 e^{-A(T-t)}z)(s)|^2 \, ds\Big)^{1/2} \to 0.
\end{aligned}$$

This gives (4.14). Finally, let us derive (4.13). To this end, we consider the following equation (denote $M = BR^{-1}B^*$):

(4.19) $$z_{t,\xi}(s) = e^{-A(s-t)}\xi - \int_t^s e^{-A(s-r)}M(A^*)^\alpha P(r)A^\alpha z_{t,\xi}(r) \, dr.$$

It is easy to show that for any $\xi \in \mathcal{D}(A^{\alpha/2})$, (4.19) admits a unique solution $z_{t,\xi}(\cdot) \in L^2(t,T;\mathcal{D}(A^\alpha)) \cap C([0,T];X)$. We define $G(s,t)\xi = z_{t,\xi}(s)$. Then, it is seen that

(4.20) $$\bar{y}_{t,x}(s) = A^\alpha G(s,t)A^{-\alpha}x.$$

Almost exact as in Chapter 2, Lemma 5.6, we can show that

(4.21) $$G(s,t)\xi = e^{-A(s-t)}\xi - \int_t^s G(s,r)M(A^*)^\alpha P(r)A^\alpha e^{-A(r-t)}\xi \, dr.$$

Also, from (4.16) and (4.20), we have that for any $x \in \mathcal{D}(A^\alpha)$,

(4.22)
$$\begin{aligned}
e^{-A^*(T-t)}Q_1 A^\alpha G(T,t)x &+ \int_t^T e^{-A^*(\sigma-t)}QA^\alpha G(\sigma,t)x \, d\sigma \\
&= P(t)A^\alpha x = \big((A^*)^\alpha P(t)\big)^* x, \qquad \forall t \in [0,T).
\end{aligned}$$

Hence, by (4.16), (4.21), and (4.22), we have
(4.23)
$$P(t)x = e^{-A^*(T-t)}Q_1 A^\alpha G(T,t)A^{-\alpha}x$$
$$+ \int_t^T e^{-A^*(\sigma-t)} Q A^\alpha G(\sigma,t) A^{-\alpha} x \, d\sigma$$
$$= e^{-A^*(T-t)} Q_1 A^\alpha \Big\{ e^{-A(T-t)} A^{-\alpha} x$$
$$- \int_t^T G(T,r) M(A^*)^\alpha P(r) A^\alpha e^{-A(r-t)} A^{-\alpha} x \, dr \Big\}$$
$$+ \int_t^T e^{-A^*(\sigma-t)} Q A^\alpha \Big\{ e^{-A(\sigma-t)} A^{-\alpha} x$$
$$- \int_t^\sigma G(\sigma,r) M(A^*)^\alpha P(r) A^\alpha e^{-A(r-t)} A^{-\alpha} x \, dr \Big\} d\sigma$$
$$= e^{-A^*(T-t)} Q_1 e^{-A(T-t)} x + \int_t^T e^{-A^*(\sigma-t)} Q e^{-A(\sigma-t)} x \, d\sigma$$
$$- \int_t^T e^{-A^*(T-t)} Q_1 A^\alpha G(T,r) M(A^*)^\alpha P(r) e^{-A(r-t)} x \, dr$$
$$- \int_t^T e^{-A^*(\sigma-t)} Q A^\alpha \int_t^\sigma G(\sigma,r) M(A^*)^\alpha P(r) e^{-A(r-t)} x \, dr \, d\sigma$$
$$= e^{-A^*(T-t)} Q_1 e^{-A(T-t)} x + \int_t^T e^{-A^*(\sigma-t)} Q e^{-A(\sigma-t)} x \, d\sigma$$
$$- \int_t^T e^{-A^*(r-t)} \Big\{ e^{-A^*(T-r)} Q_1 A^\alpha G(T,r)$$
$$+ \int_r^T e^{-A^*(\sigma-r)} Q A^\alpha G(\sigma,r) \, d\sigma \Big\} M(A^*)^\alpha P(r) e^{-A(r-t)} x \, dr.$$

By the density of $\mathcal{D}(A^\alpha)$ in $X$, we see that (4.13) follows from the above and (4.22). □

Equation (4.13) is referred to as the *integral Riccati equation*, (4.14) gives the terminal condition (in the weak sense), and (4.15) gives a representation for the optimal value of the cost (the value function). Operator valued function $P(\cdot)$ satisfying (4.13) with terminal condition (4.14) is called a *strong solution* of (4.13).

The result of this section tells us that under proper conditions, the integral Riccati equation (4.13) admits a strong solution $P(\cdot)$ and our optimal control $\bar{u}(\cdot)$ can also be written as

(4.24) $\qquad \bar{u}(s) = -R^{-1} B^* (A^*)^\alpha P(s) \bar{y}(s), \qquad s \in [0,T).$

Hence, this gives another way of solving our Problem (LQ).

We point out that under some further conditions, one can show that

## §5. Problem in Infinite Horizon

(4.13) admits a unique solution $P(\cdot)$. We omit the details here (see Lasiecka–Triggiani [1]).

## §5. Problem in Infinite Horizon

The problems studied in the previous sections are usually referred to as the *finite horizon LQ problems* because the time intervals we considered were finite. In this section, we consider the time-invariant linear quadratic optimal control problems in infinite horizons, namely, in the infinite time interval $[0, \infty)$. Such problems are referred to as *infinite horizon LQ problems*. We should point out that there are some significant differences between the finite and infinite horizon LQ problems. For the simplicity of presentation, in this section we only consider bounded control problems. Thus, we do not need the analyticity of the semigroup $e^{At}$.

### §5.1. Reduction of the problem

We consider the following control system:

$$(5.1) \quad \begin{cases} \dot{y}(t) = Ay(t) + Bu(t), & t \in [0, \infty), \\ y(0) = x, \end{cases}$$

with the cost functional

$$(5.2) \quad J(x; u(\cdot)) = \int_0^\infty \big\{ \langle Qy(t), y(t) \rangle + 2 \langle Sy(t), u(t) \rangle + \langle Ru(t), u(t) \rangle \big\} \, dt.$$

As before, we let $X$ and $U$ be two Hilbert spaces, $A$ generate a $C_0$ semigroup on $X$, $B \in \mathcal{L}(U, H)$, $Q \in \mathcal{L}(X)$, $S \in \mathcal{L}(X, U)$, and $R \in \mathcal{L}(U)$ with $Q$ and $R$ self-adjoint. Similar to the previous sections, for $x \in X$ and $u(\cdot) \in \mathcal{U} \equiv L^2(0, \infty; U)$, the corresponding state trajectory $y(\cdot; x, u(\cdot))$ is defined to be the mild solution of (5.1), i.e.,

$$(5.3) \quad y(t) \equiv y(t; x, u(\cdot)) = e^{At}x + \int_0^t e^{A(t-s)} Bu(s)\, ds, \quad t \in [0, \infty).$$

We will not distinguish (5.1) and (5.3) below. Our optimal control problem is that for given $x \in X$, minimizing the cost functional (5.2) subject to the state equation (5.1) (or (5.3)) with the control taken in $\mathcal{U}$. Such a problem is referred to as an *infinite horizon LQ problem*.

Now, let us make an observation. For the finite horizon problems discussed in previous sections, we know that for any $x \in X$ and $u(\cdot) \in \mathcal{U}$, the corresponding cost is always defined. However, this is not necessarily the case for the present infinite horizon problem as the integral in (5.2) is taken over the infinite time interval $[0, \infty)$. This is one of the most important differences between the finite and infinite horizon LQ problems.

Let us first look at the case where for any $x \in X$, the cost functional $J(x; u(\cdot))$ is well defined for all $u(\cdot) \in \mathcal{U}$. In order for $J(x; 0)$ to be defined, we need $\langle Qe^{A\cdot}x, e^{A\cdot}x \rangle$ to be integrable on $[0, \infty)$. This can be ensured if $e^{A\cdot}x \in L^2(0, \infty; X)$ because $Q \in \mathcal{L}(X)$. In particular, this will be the case if for some $M, \omega > 0$,

$$\|e^{At}\| \le Me^{-\omega t}, \qquad t \ge 0. \tag{5.4}$$

The above conditions about the $C_0$ semigroup $e^{At}$ lead to the following definitions.

**Definition 5.1.** (i) A $C_0$ semigroup $e^{At}$ on $X$ is said to be $L^2$-*stable* if for any $x \in X$, $e^{A\cdot}x \in L^2(0, \infty; X)$. In this case, we also say that $A$, the generator of $e^{At}$, is $L^2$-stable.

(ii) A $C_0$ semigroup $e^{At}$ is said to be *exponentially stable* if there exist constants $M, \omega > 0$, such that (5.4) holds. In this case, we also say that $A$ is exponentially stable.

It is clear that if the $C_0$ semigroup $e^{At}$ is exponentially stable, then it is $L^2$-stable. The following result is due to Datko [1], which gives the converse.

**Proposition 5.2.** *Let $e^{At}$ be an $L^2$-stable $C_0$ semigroup on the Hilbert space $X$. Then it is exponentially stable.*

*Proof.* First of all, we note that $\mathcal{X} \equiv L^2(0, \infty; X)$ is a Hilbert space. Let us define the operator $\mathcal{T} : X \to \mathcal{X}$ as follows

$$\mathcal{T}x = e^{A\cdot}x, \qquad x \in X. \tag{5.5}$$

By the $L^2$-stability of the semigroup $e^{At}$, this operator is well defined and is a linear operator. Regarding $\{\mathcal{T}x \mid |x| \le 1\}$ as a family of linear functionals on $\mathcal{X}$, we have

$$\begin{aligned} |\langle \mathcal{T}x, \varphi \rangle_{\mathcal{X}}| &\equiv \left| \int_0^\infty \langle e^{At}x, \varphi(t) \rangle_X \, dt \right| \\ &\le \left\{ \int_0^\infty |e^{At}x|^2 \, dt \right\}^{1/2} \|\varphi(\cdot)\|_{\mathcal{X}} < \infty, \qquad \forall \varphi \in \mathcal{X}, \ |x| \le 1. \end{aligned} \tag{5.6}$$

Thus, by the Principle of Uniform Boundedness (Chapter 2, Theorem 1.11) and the linearity of the operator $\mathcal{T}$, there exists a constant $L > 0$, such that

$$\int_0^\infty |e^{At}x|^2 \, dt \equiv \|\mathcal{T}x\|_{\mathcal{X}} \le L^2|x|^2, \qquad \forall x \in X. \tag{5.7}$$

Next, by the properties of semigroups, we know that there exist constants $\bar{M}, \bar{\omega} > 0$, such that

$$\|e^{At}\| \le \bar{M}e^{\bar{\omega}t}, \qquad t \ge 0. \tag{5.8}$$

## §5. Problem in Infinite Horizon

On the other hand,

$$
\begin{aligned}
(5.9) \quad \frac{1-e^{-2\bar{\omega}t}}{2\bar{\omega}}|e^{At}x|^2 &= \int_0^t e^{-2\bar{\omega}s}|e^{At}x|^2\,ds \\
&\leq \int_0^t e^{-2\bar{\omega}s}\|e^{As}\|^2|e^{A(t-s)}x|^2\,ds \\
&\leq \bar{M}^2 \int_0^t |e^{A(t-s)}x|^2\,ds \leq \bar{M}^2 \int_0^\infty |e^{As}x|^2\,ds \leq \bar{M}^2 L^2 |x|^2.
\end{aligned}
$$

Hence, there exists a constant $\gamma > 0$, such that

$$(5.10) \qquad |e^{At}x|^2 \leq \gamma^2 |x|^2, \qquad \forall x \in X,\ t \geq 0.$$

Consequently,

$$
\begin{aligned}
(5.11) \quad t|e^{At}x|^2 &= \int_0^t |e^{At}x|^2\,ds \leq \int_0^t |e^{As}e^{A(t-s)}x|^2\,ds \\
&\leq \gamma^2 \int_0^t |e^{A(t-s)}x|^2\,ds \leq \gamma^2 \int_0^\infty |e^{As}x|^2\,ds \leq L^2\gamma^2|x|^2.
\end{aligned}
$$

This yields

$$(5.12) \qquad \|e^{At}\| \leq \frac{L\gamma}{\sqrt{t}}, \qquad \forall t > 0.$$

Then $\|e^{At_0}\| < 1$ for $t_0 > 0$ large enough. Now, for any $t > 0$, there exists an integer $m \geq 0$, such that $mt_0 \leq t < (m+1)t_0$. Thus (note (5.10) and $\log\|e^{At_0}\| < 0$),

$$
\begin{aligned}
(5.13) \quad \frac{\log\|e^{At}\|}{t} &= \frac{\log\|e^{Amt_0}e^{A(t-mt_0)}\|}{t} \leq \frac{m\log\|e^{At_0}\|}{t} + \frac{\log\gamma}{t} \\
&\leq \frac{m}{m+1}\frac{\log\|e^{At_0}\|}{t_0} + \frac{\log\gamma}{t}.
\end{aligned}
$$

This implies that

$$(5.14) \qquad \varlimsup_{t\to\infty} \frac{\log\|e^{At}\|}{t} \leq \frac{\log\|e^{At_0}\|}{t_0} < 0.$$

Therefore, there exist $M, \omega > 0$, such that (5.4) holds. By Definition 5.1, the semigroup $e^{At}$ is exponentially stable. □

From the above result, we conclude that the exponential and the $L^2$-stabilities are equivalent. Thus, assuming the exponential stability is no more than assuming the $L^2$-stability (which looks weaker). For this reason, we will only use the notion of exponential stability in what follows.

The above analysis tells us that if $A$ is exponentially stable, then $J(x; u(\cdot))$ is well defined for all $(x, u(\cdot)) \in X \times \mathcal{U}$. However, if $A$ is not exponentially stable, the set (denoted by $\mathcal{U}(x)$) of all controls for which

$J(x; u(\cdot))$ is well-defined depends on $x$, in general. This can be seen from the following simple example.

*Example 5.3.* Let $X = \mathbb{R}$, $A = 0$, $B = Q = R = 1$, and $S = 0$. Then the state trajectory $y(\cdot)$ is given by

$$y(t; x, u(\cdot)) = x + \int_0^t u(s)\, ds, \qquad t \geq 0. \tag{5.15}$$

We take $u(t) = u_0(t) \equiv e^{-t}$. Then (5.15) becomes

$$y(t; x, u(\cdot)) = x + 1 - e^{-t}, \qquad t \geq 0. \tag{5.16}$$

Thus, we see immediately that $J(x; u_0(\cdot))$ is well defined only for $x = -1$. This means that $u_0(\cdot) \in \mathcal{U}(-1) \setminus \mathcal{U}(x)$, for all $x \neq -1$.

The dependence of the control set $\mathcal{U}(x)$ on $x \in X$ is not convenient for our later investigations. On the other hand, in many applications, if no control actions are applied at all, the control system is not necessarily stable. This is often the case if in the state equation (5.1), the generator $A$ is not necessarily exponentially stable. In such a case, we usually would like to use control actions to obtain some sort of stability for the systems. This leads to the following notion.

**Definition 5.4.** System (5.1) is said to be *exponentially stabilizable* if there exists a $K \in \mathcal{L}(X, U)$, such that $A - BK$ generates an exponentially stable $C_0$ semigroup. In this case, the operator $K$ is called a *stabilizing feedback operator* (note that such an operator is not necessarily unique).

Now, we let system (5.1) be exponentially stabilizable with the stabilizing feedback operator $K$. Set

$$u(t) = -Ky(t) + v(t), \qquad t \in [0, \infty), \qquad v(\cdot) \in \mathcal{U}. \tag{5.17}$$

Then system (5.1) becomes

$$\dot{y}(t) = (A - BK)y(t) + Bv(t), \qquad y(0) = x. \tag{5.18}$$

The cost functional (5.2) becomes

$$\widetilde{J}(x; v(\cdot)) = \int_0^\infty \big\{ \langle (Q - K^*S - SK + K^*RK)y(t), y(t) \rangle \\ + 2\langle (S - RK)y(t), v(t) \rangle + \langle Rv(t), v(t) \rangle \big\}\, dt. \tag{5.19}$$

Clearly, for any $x \in X$ and $v(\cdot) \in \mathcal{U}$, the corresponding state trajectory $y(\cdot) \in \mathcal{X} \equiv L^2(0, \infty; X)$ and the cost functional $\widetilde{J}(x; v(\cdot))$ is well defined. Consequently, $u(\cdot)$ given by (5.17) is in $\mathcal{U}$ and the original cost functional $J(x; u(\cdot))$ is well defined. We see that $J(x; u(\cdot))$ and $\widetilde{J}(x; v(\cdot))$ have the same form. The advantage of system (5.18) is that the operator $A - BK$ is exponentially stable and consequently the cost functional $\widetilde{J}(x; v(\cdot))$ is

## §5. Problem in Infinite Horizon

well defined for all $(x, v(\cdot)) \in X \times \mathcal{U}$. This will be convenient for further discussions.

In what follows, we only consider exponentially stabilizable systems. Such systems are general enough for many applications. From the above observation, it is seen that without loss of generality, we may assume the generator $A$ of the original system to be exponentially stable (otherwise, we may take the control $u(\cdot)$ of form (5.17) first and change the original system (5.1) to (5.18) with the cost functional (5.19), which is the same form as (5.2)). To conclude this subsection, we state our optimal control problem as follows.

**Problem (LQ)$_\infty$.** For given $x \in X$, find $\bar{u}(\cdot) \in \mathcal{U}$, such that

$$(5.20) \qquad J(x; \bar{u}(\cdot)) = \inf_{u(\cdot) \in \mathcal{U}} J(x; u(\cdot)).$$

### §5.2. Well posedness and solvability

Similar to the finite horizon LQ problem, we have the following notions.

**Definition 5.5.** (i) Problem (LQ)$_\infty$ is said to be *well posed* at $x \in X$ if

$$(5.21) \qquad \inf_{u(\cdot) \in \mathcal{U}} J(x; u(\cdot)) > -\infty.$$

In the case where Problem (LQ)$_\infty$ is well posed at all $x \in X$, we simply say that Problem (LQ)$_\infty$ is well posed.

(ii) Problem (LQ)$_\infty$ is said to be *solvable* at $x \in X$ if there exists a $\bar{u}(\cdot) \in \mathcal{U}$, such that (5.20) holds. If Problem (LQ)$_\infty$ is solvable at all $x \in X$, we simply say that Problem (LQ)$_\infty$ is solvable.

Now, we introduce the following notation: For all $y(\cdot) \in \mathcal{X}$ and $u(\cdot) \in \mathcal{U} \equiv L^2(0, \infty; U)$, $t \geq 0$,

$$(5.22) \qquad \begin{cases} (Lu(\cdot))(t) = \int_0^t e^{A(t-s)} Bu(s)\, ds, & (Qy(\cdot))(t) = Qy(t), \\ (Sy(\cdot))(t) = Sy(t), & (Ru(\cdot))(t) = Ru(t). \end{cases}$$

Then it is known that (note the exponential stability of $e^{At}$)

$$(5.23) \qquad \begin{cases} (L^*y(\cdot))(t) = \int_t^\infty B^* e^{A^*(\sigma-t)} y(\sigma)\, d\sigma, & t \geq 0, \\ (S^*u(\cdot))(t) = S^*u(t), & t \geq 0. \end{cases}$$

Next, we define

$$(5.24) \qquad \begin{cases} \Phi = R + L^* Q L + S L + L^* S^*, \\ \Theta = L^* Q e^{A \cdot} + S e^{A \cdot}, \qquad \Gamma = \int_0^\infty e^{A^* t} Q e^{At}\, dt. \end{cases}$$

Some simple calculations yield (compare (1.54))

(5.25) $\quad J(x;u(\cdot)) = \langle \Phi u(\cdot), u(\cdot) \rangle + 2\langle \Theta x, u(\cdot) \rangle + \langle \Gamma x, x \rangle.$

Hence, Problem $(LQ)_\infty$ becomes the minimization problem for a quadratic functional over $\mathcal{U}$ with $x \in X$ as a parameter.

**Proposition 5.6.** *Let $A$ be exponentially stable and Problem $(LQ)_\infty$ be well posed at some $x \in X$. Then*

(5.26) $\qquad\qquad J(0;u(\cdot)) \geq 0, \qquad \forall u(\cdot) \in \mathcal{U},$

(5.27) $\qquad\qquad\qquad\qquad R \geq 0.$

The proof is very similar to those of Propositions 2.1 and 3.1. Next, we have a result similar to Theorem 2.6. The proof is left to the interested readers.

**Theorem 5.7.** (i) *Let Problem $(LQ)_\infty$ be well posed at some $x \in X$. Then*

(5.28) $\qquad\qquad\qquad\qquad \Phi \geq 0.$

(ii) *Let (5.28) hold. Then Problem $(LQ)_\infty$ is solvable at $x \in X$ if and only if there exists a $u_0(\cdot) \in \mathcal{U}$, depending on $x$, in general, such that*

(5.29) $\qquad\qquad\qquad \Phi u_0(\cdot) + (\Theta x)(\cdot) = 0.$

(iii) *If $\Phi \gg 0$, i.e., there exists a $\delta > 0$, such that*

(5.30) $\qquad \langle \Phi u(\cdot), u(\cdot) \rangle \geq \delta \int_0^\infty |u(t)|^2\, dt, \qquad \forall u(\cdot) \in \mathcal{U}.$

*Then Problem $(LQ)_\infty$ is uniquely solvable on $X$, i.e., for each $x \in X$, Problem $(LQ)_\infty$ admits a unique optimal control, which is given by*

(5.31) $\qquad\qquad\qquad \bar{u}(\cdot) = -(\Phi^{-1}\Theta x)(\cdot).$

*Moreover, for some self-adjoint operator $P \in \mathcal{L}(X)$, it holds that*

(5.32) $\qquad V(x) \equiv \inf_{u(\cdot) \in \mathcal{U}} J(x;u(\cdot)) = \langle Px, x \rangle, \qquad \forall x \in X.$

The following result is comparable with Theorem 3.2 for the finite horizon LQ problem. The proof is similar to that for Theorem 3.2.

**Theorem 5.8.** *Let $\Phi \geq 0$ and $R \gg 0$. Then Problem $(LQ)_\infty$ is solvable at some $x \in X$ if and only if the following two-point boundary value problem admits a solution:*

(5.33)
$$\begin{cases} y(t) = e^{At}x + \int_0^t e^{A(t-s)}\{BR^{-1}B^*\psi(s) - BR^{-1}Sy(s)\}\, ds, \\ \psi(t) = -\int_t^\infty e^{A^*(\sigma-t)}\{(Q - S^*R^{-1}S)y(\sigma) \\ \qquad\qquad\qquad\qquad + S^*R^{-1}B^*\psi(\sigma)\}\, d\sigma. \end{cases}$$

## §5. Problem in Infinite Horizon

*In this case, the following gives an optimal control:*

(5.34) $$u(t) = R^{-1}B^*\psi(t) - R^{-1}Sy(t), \qquad t \in [0,\infty),$$

*and (5.32) holds for some $P \in \mathcal{L}(X)$. In addition, if $\Phi \gg 0$, then Problem $(LQ)_\infty$ admits a unique optimal control determined by (5.33) and (5.34).*

The differential form of (5.33) is the following:

(5.35) $$\begin{cases} \dot{y}(t) = (A - BR^{-1}S)y(t) + BR^{-1}B^*\psi(t), \\ \dot{\psi}(t) = (Q - S^*R^{-1}S)y(t) - (A - BR^{-1}S)^*\psi(t), \\ y(0) = x, \qquad \psi(\infty) = 0. \end{cases}$$

### §5.3. Algebraic Riccati equation

Unlike the finite horizon problem, in this section we are going to derive the algebraic equation satisfied by the operator $P \in \mathcal{L}(X)$ which appeared in Theorem 5.7, without the introduction of the Fredholm type integral equation. The method we will use is the dynamic programming method. We recall the definition of the value function $V(x)$ in (5.32). Similar to Chapter 6, §6, we denote $y_x(\cdot) = y(\cdot\,;x,u(\cdot))$. The following result is comparable with Chapter 6, Proposition 6.2.

**Proposition 5.9.** (Optimality Principle) *Let $A$ be exponentially stable, $R \gg 0$, and $\Phi \geq 0$. Then for any $t > 0$, it holds that*

(5.36) $$V(x) = \inf_{u(\cdot) \in \mathcal{U}} \left\{ \int_0^t \left\{ \langle Qy_x(s), y_x(s) \rangle + 2\langle Sy_x(s), u(s) \rangle + \langle Ru(s), u(s) \rangle \right\} ds + V(y_x(t)) \right\}.$$

By (5.32), we know that $V(x) = \langle Px, x \rangle$, which is smooth in $x$. Thus, similar to Chapter 6, §6, we have the following (see Chapter 6, (6.17) with $\lambda = 0$):

(5.37) $$\langle \nabla V(x), Ax \rangle + H(x, \nabla V(x)) = 0, \qquad \forall x \in \mathcal{D}(A),$$

where

(5.38) $$\begin{aligned} H(x,p) &= \inf_{u \in U} \left\{ \langle p, Bu \rangle + \langle Qx, x \rangle + 2\langle Sx, u \rangle + \langle Ru, u \rangle \right\} \\ &= \langle Qx, x \rangle + \inf_{u \in U} \left\{ \langle B^*p + 2Sx, u \rangle + \langle Ru, u \rangle \right\} \\ &= \langle Qx, x \rangle - \langle R^{-1}(\tfrac{1}{2}B^*p + Sx), \tfrac{1}{2}B^*p + Sx \rangle. \end{aligned}$$

Because $\nabla V(x) = 2Px$, we obtain from (5.37) and (5.38) that (note $X$ is a real Hilbert space)

(5.39) $$\begin{aligned} &\langle Px, Ax \rangle + \langle Ax, Px \rangle \\ &+ \langle \{Q - (PB + S^*)R^{-1}(B^*P + S)\}x, x \rangle = 0, \qquad \forall x \in \mathcal{D}(A). \end{aligned}$$

The above is called the *algebraic Riccati equation* for the operator $P$. We should note that in the case where $A$ is bounded, the term $\langle Px, Ax \rangle + \langle Ax, Px \rangle$ can be written as a symmetric form $\langle (PA + A^*P)x, x \rangle$.

Next, we would like to derive the integral form of the above equation. To this end, we recall that for any $x \in \mathcal{D}(A)$, $e^{As}x \in \mathcal{D}(A)$. Thus, for any $x \in \mathcal{D}(A)$, (5.39) implies

(5.40)
$$\begin{aligned}&\frac{d}{ds}\langle e^{A^*s}Pe^{As}x, x \rangle \\ &+ \langle e^{A^*s}(Q - (PB + S^*)R^{-1}(B^*P + S))e^{As}x, x \rangle \\ &= 2\langle Pe^{As}x, Ae^{As}x \rangle \\ &+ \langle \{Q - (PB + S^*)R^{-1}(B^*P + S)\}e^{As}x, e^{As}x \rangle = 0.\end{aligned}$$

Integrating the above from 0 to $\infty$ and noting the exponential stability of the semigroup $e^{At}$, we have

(5.41)
$$\langle Px, x \rangle = \int_0^\infty \langle e^{A^*s}[Q - (PB + S^*)R^{-1}(B^*P + S)]e^{As}x, x \rangle \, ds,$$
$$\forall x \in \mathcal{D}(A).$$

By the density of $\mathcal{D}(A)$ in $X$ and the self-adjointness of $P$, we obtain

(5.42)
$$P = \int_0^\infty e^{A^*s}(Q - (PB + S^*)R^{-1}(B^*P + S))e^{As} \, ds.$$

This is called the *integral form* of the algebraic Riccati equation. The following is the main result of this section.

**Theorem 5.10.** *Let $A$ be exponentially stable, $\Phi \geq 0$, and $R \gg 0$.*

*(i) If Problem $(LQ)_\infty$ is well posed, then the Riccati equation (5.42) admits a self-adjoint solution $P \in \mathcal{L}(X)$ and this $P$ is given through (5.32).*

*(ii) If the Riccati equation (5.42) admits a self-adjoint solution $P \in \mathcal{L}(X)$ and $A - BR^{-1}(B^*P + S)$ is exponentially stable, then Problem $(LQ)_\infty$ is solvable. In this case, the optimal pair $(\bar{y}(\cdot), \bar{u}(\cdot))$ satisfies the following relation:*

(5.43) $$\bar{u}(t) = -R^{-1}(B^*P + S)\bar{y}(t), \qquad t \geq 0;$$

*and the value function is represented by (5.32).*

*Proof.* We have proved (i) before the statement of the above theorem. The proof of (ii) is very simple and we leave the details to the readers. □

## §5.4. The positive real lemma

In this subsection, we would like to study some relations between the solvability of the algebraic Riccati equation (5.42) and certain characteristics of

## §5. Problem in Infinite Horizon

the control system. To this end, let us first make some observations. From Chapter 2, Proposition 4.14, we know that when $A$ is exponentially stable, it holds that

$$(5.44) \quad (i\omega - A)^{-1}x = \int_0^\infty e^{-i\omega t} e^{At} x \, dt, \qquad \forall \omega \in \mathbb{R}, \ x \in X.$$

Now, for any $u(\cdot) \in \mathcal{U}$, we define $u(t) = 0$ for $t < 0$. Then we have the natural embedding $\mathcal{U} \subset L^2(-\infty, \infty; U)$. Introduce the *Fourier transformation* of $u(\cdot)$ as follows:

$$(5.45) \quad \widetilde{u}(\omega) = \frac{1}{\sqrt{2\pi}} \int_{-\infty}^\infty e^{-i\omega t} u(t) \, dt \equiv \frac{1}{\sqrt{2\pi}} \int_0^\infty e^{-i\omega t} u(t) \, dt, \qquad \forall u \in \mathcal{U}.$$

It is standard that the following *Parseval equality* holds:

$$(5.46) \quad \int_{-\infty}^\infty \langle \widetilde{u}(\omega), \widetilde{v}(\omega) \rangle \, d\omega = \int_0^\infty \langle u(t), v(t) \rangle \, dt, \qquad \forall u, v \in \mathcal{U}.$$

Next, we observe the following (recall (5.22) for the operator $L$):

$$(5.47) \quad \begin{aligned} \sqrt{2\pi}(\widetilde{Lu})(\omega) &= \int_0^\infty e^{-i\omega t} (Lu(\cdot))(t) \, dt \\ &= \int_0^\infty e^{-i\omega t} \int_0^t e^{A(t-s)} Bu(s) \, ds \, dt \\ &= \int_0^\infty \left( \int_s^\infty e^{-i\omega(t-s)} e^{A(t-s)} Bu(s) \, dt \right) e^{-i\omega s} \, ds \\ &= (i\omega - A)^{-1} B \int_0^\infty e^{-i\omega s} u(s) \, ds = \sqrt{2\pi}(i\omega - A)^{-1} B \widetilde{u}(\omega). \end{aligned}$$

Thus, by the definition of $\Phi$ (see (5.24)) and the Parseval equality, we obtain

$$(5.48)$$
$$\int_0^\infty \langle (\Phi u(\cdot))(t), u(t) \rangle \, dt$$
$$= \int_0^\infty \Big\{ \langle Ru(t), u(t) \rangle + \langle Q(Lu(\cdot))(t), (Lu(\cdot))(t) \rangle$$
$$\quad + \langle S(Lu(\cdot))(t), u(t) \rangle + \langle u(t), S(Lu(\cdot))(t) \rangle \Big\} \, dt$$
$$= \int_{-\infty}^\infty \Big\{ \langle R\widetilde{u}(\omega), \widetilde{u}(\omega) \rangle + \langle Q(i\omega - A)^{-1} B\widetilde{u}(\omega), (i\omega - A)^{-1} B\widetilde{u}(\omega) \rangle$$
$$\quad + \langle S(i\omega - A)^{-1} B\widetilde{u}(\omega), \widetilde{u}(\omega) \rangle + \langle \widetilde{u}(\omega), S(i\omega - A)^{-1} B\widetilde{u}(\omega) \rangle \Big\} \, d\omega$$
$$= \int_{-\infty}^\infty \langle \widetilde{\Phi}(\omega)\widetilde{u}(\omega), \widetilde{u}(\omega) \rangle \, d\omega, \qquad \forall u \in \mathcal{U},$$

where

$$(5.49) \quad \begin{aligned} \widetilde{\Phi}(\omega) &= R + B^*(-i\omega - A^*)^{-1} Q(i\omega - A)^{-1} B \\ &\quad + S(i\omega - A)^{-1} B + B^*(-i\omega - A^*)^{-1} S^*. \end{aligned}$$

We call $\widetilde{\Phi}(\omega)$ the *frequency characteristics* associated with Problem (LQ)$_\infty$. We write $\widetilde{\Phi}(\cdot) \geq 0$ if

(5.50) $$\widetilde{\Phi}(\omega) \geq 0, \quad \forall \omega \in \mathbb{R},$$

and $\widetilde{\Phi}(\cdot) \gg 0$ if there exists a $\delta > 0$ such that

(5.51) $$\widetilde{\Phi}(\omega) \geq \delta I, \quad \forall \omega \in \mathbb{R}.$$

Thus, (5.48) implies

(5.52) $$\begin{cases} \widetilde{\Phi}(\cdot) \geq 0 & \Rightarrow \quad \Phi \geq 0, \\ \widetilde{\Phi}(\cdot) \gg 0 & \Rightarrow \quad \Phi \gg 0. \end{cases}$$

It is important to note that the converse of the above does not hold because the map defined by (5.45) is not onto $L^2(-\infty, \infty; U)$. To see this, we take $v(\cdot) = \chi_{[-1,0]}(\cdot)$. Clearly, $\widetilde{v}(\cdot) \neq 0$ and for any $u(\cdot) \in \mathcal{U}$, we have (note that $u(t) = 0$ for all $t < 0$)

(5.53) $$\int_{-\infty}^{\infty} \langle \widetilde{u}(\omega), \widetilde{v}(\omega) \rangle \, d\omega = 0.$$

This means that $\widetilde{v}(\cdot)$ is perpendicular to the image of the map defined by (5.45), which implies that this map is not onto.

On the other hand, from Theorems 5.7 and 5.10, we see that $\Phi$ is related to the solvability of the Problem (LQ)$_\infty$ as well as to the solvability of the Riccati equation (5.42). Thus, it is expected that there are some relations between the frequency characteristics $\widetilde{\Phi}(\cdot)$ and the solvability of the Riccati equation (5.42). This leads to the main result of this subsection, the *generalized positive real lemma*.

**Theorem 5.11.** *Let $A$ be exponentially stable, $\Phi \geq 0$, and $R \gg 0$.*

*(i) If Riccati equation (5.39) admits a self-adjoint solution $P \in \mathcal{L}(X)$, then $\widetilde{\Phi}(\cdot) \geq 0$.*

*(ii) If $\widetilde{\Phi}(\cdot) \gg 0$, then Riccati equation (5.39) admits a self-adjoint solution $P \in \mathcal{L}(X)$.*

*Proof.* (i) Let Riccati equation (5.39) admit a self-adjoint solution $P \in \mathcal{L}(X)$. Take any $y \in X$. We know that $x \equiv (i\omega - A)^{-1} y \in \mathcal{D}(A)$ and for any $\omega \in \mathbb{R}$,

(5.54) $$Ax = A(i\omega - A)^{-1} y = -y + i\omega(i\omega - A)^{-1} y = -y + i\omega x.$$

§5. Problem in Infinite Horizon

Thus, together with (5.39), we obtain

$$
\begin{aligned}
\langle (-i\omega - A^*)^{-1} Q(i\omega - A)^{-1} y, y \rangle &= \langle Qx, x \rangle \\
&= -\langle Px, -y + i\omega x \rangle - \langle -y + i\omega x, Px \rangle \\
&\quad + \langle (PB + S^*) R^{-1}(B^*P + S) x, x \rangle \\
&= \langle \{(-i\omega - A^*)^{-1} P + P(i\omega - A)^{-1}\} y, y \rangle \\
&\quad + \langle (-i\omega - A^*)^{-1}(PB + S^*) R^{-1}(B^*P + S)(i\omega - A)^{-1} y, y \rangle.
\end{aligned}
$$
(5.55)

Consequently,

$$
\begin{aligned}
(-i\omega - A^*)^{-1} Q(i\omega - A)^{-1} &= (-i\omega - A^*)^{-1} P + P(i\omega - A)^{-1} \\
&\quad + (-i\omega - A^*)^{-1}(PB + S^*) R^{-1}(B^*P + S)(i\omega - A)^{-1}.
\end{aligned}
$$
(5.56)

Hence, by (5.49),

$$
\begin{aligned}
\widetilde{\Phi}(\omega) &= R + B^*(-i\omega - A^*)^{-1}(PB + S^*) \\
&\quad + (B^*P + S)(i\omega - A)^{-1} B \\
&\quad + B^*(-i\omega - A^*)^{-1}(PB + S^*) R^{-1}(B^*P + S)(i\omega - A)^{-1} B \\
&= \{R^{1/2} + B^*(-i\omega - A^*)^{-1}(PB + S^*) R^{-1/2}\} \\
&\quad \cdot \{R^{1/2} + R^{-1/2}(B^*P + S)(i\omega - A)^{-1} B\} \geq 0.
\end{aligned}
$$
(5.57)

(ii) Let $\widetilde{\Phi}(\cdot) \gg 0$. Then, by (5.52), $\Phi \gg 0$. Thus, Theorem 5.7 implies that Problem (LQ)$_\infty$ is uniquely solvable and our conclusion follows from Theorem 5.10. □

The above proof for (ii) is based on the solvability of the LQ problem. It seems to us that such a result should have a direct algebraic proof. Interested readers are welcome to find such a proof.

The above result is a generalization of the usual *positive real lemma*, which gives the condition ensuring that the *transfer operator* of some linear system is positive and real. We now explain the lemma in detail. Let us consider the following control system:

(5.58)
$$
\begin{cases} \dot{y}(t) = Ay(t) + Bu(t), \\ z(t) = Sy(t) + Ju(t). \end{cases}
$$

The variable $z(\cdot)$ represents the *observation*. Sometimes, we also call it the *output*. We assume that $A$ is exponentially stable, $B \in \mathcal{L}(U, X)$, $S \in \mathcal{L}(X, U)$, and $J \in \mathcal{L}(U)$. Clearly, for any $u(\cdot) \in \mathcal{U}$, we must have $y(\cdot) \in L^2(0, \infty; X)$ and $z(\cdot) \in L^2(0, \infty; U)$. Now, take the Fourier transformation (see (5.45)) on both sides of (5.58). Assuming that all the required conditions are satisfied, we have

(5.59)
$$
\begin{cases} i\omega \widetilde{y}(\omega) = A\widetilde{y}(\omega) + B\widetilde{u}(\omega), \\ \widetilde{z}(\omega) = S\widetilde{y}(\omega) + J\widetilde{u}(\omega). \end{cases}
$$

Then it follows that

(5.60) $$\tilde{z}(\omega) = \{J + S(i\omega - A)^{-1}B\}\tilde{u}(\omega), \qquad \forall \omega \in \mathbb{R}.$$

The above gives a direct relation from the input $\tilde{u}(\cdot)$ to the output $\tilde{z}(\cdot)$ (in the so-called *frequency domain*). Thus, the operator

(5.61) $$W(s) = J + S(s - A)^{-1}B, \qquad s \in \mathbb{C}$$

plays a very important role in the study of the input-output behavior for the system (5.58). Operator $W(\cdot)$ is called the *transfer operator* of the system (5.48). This operator is said to be *positive real* if

(5.62) $$W(i\omega)^* + W(i\omega) \geq 0, \qquad \forall \omega \in \mathbb{R},$$

and it is said to be *strictly positive real* if there exists a $\delta > 0$, such that

(5.63) $$W(i\omega)^* + W(i\omega) \geq \delta I, \qquad \forall \omega \in \mathbb{R}.$$

The following result follows from Theorem 5.11 (with $Q = 0$).

**Corollary 5.12.** *Let $A$ be exponentially stable and $R \equiv J^* + J \gg 0$.*

(i) *Let the following Riccati equation*

(5.64) $$P = -\int_0^\infty e^{A^*t}(PB + S^*)R^{-1}(B^*P + S)e^{At}\,dt$$

*admit a self-adjoint solution $P \in \mathcal{L}(X)$. Then $W(\cdot)$ is positive real.*

(ii) *If $W(\cdot)$ is strictly positive real, then the Riccati equation (5.64) admits a self-adjoint solution $P \in \mathcal{L}(X)$.*

Usually, we write (5.64) as follows:

(5.65) $$\begin{cases} \text{Find } P \in \mathcal{L}(X),\ L \in \mathcal{L}(X,U),\ W_0 \in \mathcal{L}(U),\ \text{such that} \\ P = -\int_0^\infty e^{A^*t}L^*Le^{At}\,dt, \\ B^*P + S = W_0 L, \qquad W_0^*W_0 = J^* + J. \end{cases}$$

The most familiar form of the positive real lemma looks like the following.

**Corollary 5.13.** (Positive Real Lemma) *Let $A$ be exponentially stable and $J + J^* \gg 0$.*

(i) *If there exist $P$, $L$, and $W_0$ satisfying (5.65), then $W(\cdot)$ is positive real.*

(ii) *If $W(\cdot)$ is strictly positive real, then there exist $P$, $L$, and $W_0$ satisfying (5.65).*

## §5.5. Feedback stabilization

We have seen from Theorem 5.10 that under certain conditions, the optimal control for the Problem $(LQ)_\infty$ is given by the linear state feedback (5.43)

## §5. Problem in Infinite Horizon

where $P$ is a solution of the Riccati equation (5.39) (or (5.42)), and by denoting $K = R^{-1}(B^*P+S)$, the operator $A - BK$ is exponentially stable. It is natural to ask the following question: Suppose system (5.1) is stabilizable with the stabilizing feedback operator $K \in \mathcal{L}(X,U)$, i.e., $A - BK$ is exponentially stable (see Definition 5.4). Are there operators $Q, S, R$, so that some optimal control for the corresponding Problem (LQ)$_\infty$ is given by

$$\bar{u}(t) = -K\bar{y}(t), \qquad \forall t \geq 0? \tag{5.66}$$

This question is referred to as the *inverse problem of optimal regulators*. The following result gives a positive answer for this problem.

**Theorem 5.14.** *Let system (5.1) be stabilizable with the stabilizing feedback operator $K \in \mathcal{L}(X,U)$. Then Problem (LQ)$_\infty$ with $Q = K^*K$, $S = K$, and $R = I$ admits a unique optimal control that is given by (5.66).*

*Proof.* Consider the following system:

$$\begin{cases} \dot{y}(t) = (A - BK)y(t) + Bv(t), & t \in (0, \infty), \\ y(0) = x, \end{cases} \tag{5.67}$$

with the cost functional

$$J(x; v(\cdot)) = \int_0^\infty |v(t)|^2 \, dt. \tag{5.68}$$

In the present case, we have $\Phi = I \gg 0$. Thus, by Theorem 5.7, this problem admits a unique optimal control $\bar{u}(\cdot)$. The corresponding Riccati equation reads (see (5.42))

$$P = -\int_0^\infty e^{(A-BK)^*s} PBB^*P e^{(A-BK)s} \, ds. \tag{5.69}$$

On the other hand, from (5.68) and (5.32), we see that

$$\langle Px, x \rangle = V(x) \geq 0, \qquad \forall x \in X, \tag{5.70}$$

which implies that $P \geq 0$. Combining with (5.69), we obtain that $P = 0$ and the optimal control $\bar{v}(\cdot) = 0$. On the other hand, the algebraic form of (5.69) is of the form (see (5.39) with $Q = 0$, $S = 0$, $R = I$, and $A$ replaced by $A - BK$)

$$\begin{aligned}
0 &= \langle Px, (A - BK)x \rangle + \langle (A - BK)x, Px \rangle - \langle PBB^*Px, x \rangle \\
&= \langle Px, Ax \rangle + \langle Ax, Px \rangle \\
&\quad + \langle \{K^*K - (PB + K^*)(B^*P + K)\}x, x \rangle, \\
&\qquad \forall x \in \mathcal{D}(A).
\end{aligned} \tag{5.71}$$

Clearly, this Riccati equation corresponds to the Problem (LQ)$_\infty$ with $Q = K^*K$, $S = K$, and $R = I$. The optimal control for this problem has the form (5.43) with $P = 0$, $R = I$ and $S = K$, which gives (5.66). □

*Remark 5.15.* In the above, we can freely choose $Q$ and $S$. Such a freedom means that the inverse problem of optimal regulators always solvable. However, the problem is much harder if we require that $S = 0$. In the case where $X = \mathbb{R}^n$ and $U = \mathbb{R}$, Kalman [3] proved that if

(5.72) $\quad \{I + K(i\omega - A)^{-1}B\}^*\{I + K(i\omega - A)^{-1}B\} \geq I, \qquad \forall \omega \in \mathbb{R},$

then the inverse problem of optimal regulators is solvable. This result was generalized to the case $U = \mathbb{R}^m$ by Anderson (see Anderson–Moor [1]). It is not known if this is true for general infinite dimensional cases.

## §5.6. Fredholm integral equation and Riccati integral equation

In this subsection, we consider the case where $S = 0$ and $A$ is exponentially stable. In this case, we know that Problem (LQ)$_\infty$ is solvable if and only if the following two-point boundary value problem:

(5.73) $\quad \begin{cases} y(t; 0, x) = e^{At}x + \int_0^t e^{A(t-s)} M\psi(s; 0, x)\, ds, \\ \psi(t; 0, x) = -\int_t^\infty e^{A^*(\sigma-t)} Qy(\sigma; 0, x)\, d\sigma, \end{cases}$

admits a solution $(y(\cdot\,; 0, x), \psi(\cdot\,; 0, x)) \in L^2(0, \infty; X \times X)$, where $M = BR^{-1}B^*$. Substituting the first equation into the second, we obtain

(5.74) $\quad \psi(t; 0, x) = -V(t)x - \int_0^\infty V(t-s)M\psi(s; 0, x)\, ds,$

with

(5.75) $\quad V(t) = \begin{cases} \left(\int_0^\infty e^{A^*\sigma}Qe^{A\sigma}\, d\sigma\right)e^{At}, & \forall t \geq 0, \\ e^{A^*(-t)}\int_0^\infty e^{A^*\sigma}Qe^{A\sigma}\, d\sigma, & \forall t < 0. \end{cases}$

Equation (5.74) suggests that we consider the following Fredholm equation:

(5.76) $\quad H(t) = -V(t) - \int_0^\infty V(t-s)MH(s)\, ds.$

This is a Fredholm integral equation with unknown $H(\cdot)$ over time interval $[0, \infty)$. We have the following result whose proof is straightforward (compare with Theorem 3.8).

## §5. Problem in Infinite Horizon

**Theorem 5.16.** (i) *Let (5.76) admit a strongly continuous solution $H(\cdot) \in L^2(0, \infty; \mathcal{L}(X))$. Then (5.74) admits a strongly continuous solution $\psi(\cdot\,; 0, x) \in L^2(0, \infty; X)$ and*

(5.77) $$\psi(t; 0, x) = H(t)x, \qquad \forall t \in [0, \infty), \ x \in X.$$

(ii) *Problem (5.73) admits a unique strongly continuous solution $(y(\cdot\,; 0, x), \psi(\cdot\,; 0, x)) \in L^2(0, \infty; X \times X)$ if and only if (5.76) admits a unique strongly continuous solution $H(\cdot)$. In this case,*

(5.78) $$\psi(t; 0, x) = H(0) y(t; 0, x).$$

The next result is comparable with Theorem 4.3.

**Theorem 5.17.** *Let $A$ be exponentially stable.*

(i) *If the Riccati integral equation*

(5.79) $$P = \int_0^\infty e^{A^* t}(Q - PMP) e^{At}\, dt,$$

*admits a self-adjoint solution $P \in \mathcal{L}(X)$ such that $A - MP$ is exponentially stable, then the Fredholm integral equation (5.76) admits a strongly continuous solution $H(\cdot) \in L^2(0, \infty; \mathcal{L}(X))$. Moreover,*

(5.80) $$H(0) = -P.$$

(ii) *Let (5.76) admit a unique strongly continuous solution $H(\cdot) \in L^2(0, \infty; \mathcal{L}(X))$. Then $P = -H(0)$ is a solution of (5.79).*

*Proof.* Following the proof of Theorem 4.3, we can obtain (ii). We now prove (i). Let $P \in \mathcal{L}(X)$ be a self-adjoint solution of (5.79). We consider the Volterra integral equation on $[0, \infty)$:

(5.81) $$H(t) = -P e^{At} - \int_0^t P e^{A(t-s)} M H(s)\, ds.$$

This equation admits a unique strongly continuous solution $H(\cdot)$. Moreover, by Chapter 2, Lemma 5.6, we know that

(5.82) $$H(t) = -P e^{At} - \int_0^t H(t-s) M P e^{As}\, ds.$$

Hence,

(5.83) $$\begin{aligned} H(t)^* &= -e^{A^* t} P - \int_0^t e^{A^* s} P M^* H(t-s)^*\, ds \\ &= -e^{A^* t} P - \int_0^t e^{A^*(t-s)} P M^* H(s)^*\, ds. \end{aligned}$$

This gives $H(t)^* = -e^{(A^*-PM^*)t}P$. Consequently, $H(t) = -Pe^{(A-MP)t}$. By the exponential stability of $A - MP$, $H(\cdot) \in L^2(0,\infty;\mathcal{L}(X))$. Now, we substitute (5.79) into (5.81) to get

$$H(t) = -\int_0^\infty e^{A^*\sigma}(Q - PMP)e^{A\sigma}\, d\sigma\, e^{At}$$

(5.84)
$$- \int_0^t \Big(\int_0^\infty e^{A^*\sigma}(Q - PMP)e^{A\sigma}\, d\sigma\Big) e^{A(t-s)} MH(s)\, ds$$

$$= -V(t) - \int_0^t V(t-s)MH(s)\, ds + W(t),$$

where (note (5.79) and (5.81))

$$W(t) = \int_0^\infty e^{A^*\sigma} PMPe^{A\sigma}\, d\sigma\, e^{At}$$

$$+ \int_0^t \int_0^\infty e^{A^*\sigma} PMPe^{A(\sigma+t-s)}\, d\sigma\, MH(s)\, ds$$

$$= \int_0^\infty e^{A^*\sigma} PM\Big\{Pe^{A(\sigma+t)} + \int_0^t Pe^{A(\sigma+t-s)}MH(s)\, ds\Big\}\, d\sigma$$

$$= \int_0^\infty e^{A^*\sigma} PM\Big\{-H(\sigma+t) - \int_t^{t+\sigma} Pe^{A(\sigma+t-s)}MH(s)\, ds\Big\}\, d\sigma$$

$$= -\int_t^\infty e^{A^*(\sigma-t)} PMH(\sigma)\, d\sigma$$

(5.85)
$$- \int_t^\infty e^{A^*(\sigma-t)} PM \int_t^\sigma Pe^{A(\sigma-s)}MH(s)\, ds\, d\sigma$$

$$= -\int_t^\infty e^{A^*(s-t)} PMH(s)\, ds$$

$$- \int_t^\infty \Big\{\int_s^\infty e^{A^*(\sigma-t)} PMPe^{A(\sigma-s)}\, d\sigma\Big\} MH(s)\, ds$$

$$= -\int_t^\infty e^{A^*(s-t)} \Big\{P + \int_s^\infty e^{A^*(\sigma-s)} PMPe^{A(\sigma-s)}\, d\sigma\Big\} MH(s)\, ds$$

$$= -\int_t^\infty e^{A^*(s-t)} \Big\{\int_0^\infty e^{A^*\sigma} Qe^{A\sigma}\, d\sigma\Big\} MH(s)\, ds$$

$$= -\int_t^\infty V(t-s)MH(s)\, ds.$$

Hence, combining (5.84) and (5.85), we obtain (5.76). Then, from (5.81), we obtain (5.80). □

### Remarks

The finite dimensional linear quadratic optimal control problem was first studied by Bellman-Glicksberg-Gross [2, Chapter 4] in 1958. They reduced the problem to a minimization of quadratic functionals. In 1960, Kalman [1] found the relation between the LQ problem and optimal linear state feedback control. In the mid-1960s, Lions [1,2] discussed the LQ problem for partial differential equations. This general LQ problem for evolution equations with bounded controls was investigated by Lukes–Russell [1] in 1969 and by Curtain–Pritchard [1] in 1976. The LQ problem with unbounded controls was studied by Lions [2], Curtain–Pritchard [2] and Balakrishnan [3] in the 1970s. Since the mid-1980s, Lasiecka and Triggiani systematically studied the LQ problems for parabolic, hyperbolic, and other equations with boundary controls. The readers are encouraged to consult the book by Lasiecka–Triggiani [1] for an excellent survey in this direction.

At the same time, Lions [4] studied the LQ problem for parabolic equations in $\mathbb{R}^n$ ($n \leq 3$) with pointwise controls by introducing some suitable function spaces. Yao [4], Yao–You [1], and Chen [2] discussed the LQ problem for parabolic equations with pointwise controls in the domain or on the boundary. They reduced the problem to the minimization of densely defined quadratic functionals on some Hilbert space.

We note that in all the above results, the operator $Q$ appearing in the cost functional was assumed to be at least positive semidefinite. The case with $Q$ indefinite was first studied in Molinari [1] for finite dimensions in 1977. In the early 1980s, You [1] discussed the same problem in infinite dimensions, but with the control bounded.

In 1984, You [2] found a relation between the optimal feedback control and a Fredholm integral equation. A little later, in 1985, S. Chen [1] directly discussed an equivalent relation between the Riccati equation (which is nonlinear) and a linear Fredholm integral equation.

In this chapter, we have discussed the LQ problem with the operator $Q$ not necessarily positive semidefinite and with unbounded controls. Such a case has not appeared in the existing literature. We point out that our results here are complementary to those presented in Lasiecka–Triggiani [1].

It is interesting to point out that the discussion of minimization for densely defined quadratic functionals can be found in Friedrichs [1]. A later reference is Michlin [1]. In §2, we adopt the idea of Friedrichs to discuss the well posedness and solvability of our LQ problem. We have seen that the difficulty is the self-adjointness of the operator $\Phi$. Our assumptions look strong, but they are still very general, and they enable us to keep the results at a reasonable length. Theorem 2.9 is taken from S. Chen [2]. The idea of §3 is essentially taken from You [2]. The point is that the (linear) Fredholm integral equation is enough to solve our LQ problem and we do not even have to introduce the Riccati equation! §4 gives an equivalence relation of sorts between the Riccati equation and the Fredholm integral equation. This result is essentially due to S. Chen [1]. §5 is devoted to the infinite horizon problem. The stabilization of the system naturally comes into play. Proposition 5.2 is due to Datko [1]. We would like to point out here that there are some interesting results concerning the exact controllability and the exponential stabilizability. See Russell [2] and Lions

[6] for some relevant results. In §5, we first introduce the Riccati equation and then the Fredholm integral equation, which reverses the order of §§3–4 for the finite horizon problem.

Finally, we would like to mention other works on the LQ problem: S. Chen [3–5], Da Prato–Delfour [1], Da Prato–Ichikawa [1–3], Flandoli [1], Lasiecka–Triggiani [2], Lee–You [1], Pritchard–Salamon [1], You [3,4], and Zabczyk [1]. For some works on stabilizations, see G. Chen [1,2], Chen–Russell[1], Conrad [1], Hansen [1], Huang [1], Komornik-Russell-Zhang [1], Kwan–Wang [1], K.S. Liu [1], Liu-Huang-Chen [1], Sun [1], and Yong–Zheng [1,2].

# References

Abergel, F., and Casas, E.
[1] Some optimal control problems of multistate equations appearing in fluid mechanics, *Math. Model. & Numer. Anal.,* 27 (1993), 223–247.

Abergel, F., and Temam, R.
[1] Optimality conditions for some nonqualified problems of distributed control, *SIAM J. Control Optim.,* 27 (1989), 1–12.
[2] On some control problems in fluid mechanics, *Theoretical & Comp. Fluid Mech.,* 1 (1990), 303–325.

Acquistapace, P.; Flandoli, F.; and Terreni, B.
[1] Initial boundary value problems and optimal control for nonautonomous parabolic systems, *SIAM J. Control Optim.,* 29 (1991), 89–118.

Adams, A. R.
[1] *Sobolev Spaces,* Academic Press, New York, 1975.

Ahmed, N. U.
[1] Existence of optimal controls for a class of systems governed by differential inclusions on a Banach space, *J. Optim. Theory Appl.,* 50 (1986), 213–237.

Ahmed, N. U., and Teo, K. L.
[1] Necessary conditions for optimality of a Cauchy problem for parabolic partial differential systems, *SIAM J. Control,* 13 (1975), 981–993.
[2] Optimal control of systems governed by a class of nonlinear evolution equations in a reflexive Banach space, *J. Optim. Theory Appl.,* 25 (1978), 57–81.
[3] *Optimal Control of Distributed Parameter Systems,* North-Holland, Amsterdam, 1981.

Anderson, B. D. O., and Moore, J. B.
[1] *Linear Optimal Control,* Prentice-Hall, Englewood Cliffs, NJ., 1971.

Anita, S.
[1] Optimal control of a nonlinear population dynamics with diffusion, *J. Math. Anal. Appl.,* 152 (1990), 176–208.

Aubin, J. P., and Frankowska, H.
[1] *Set-valued Analysis,* Birkhäuser, Boston, 1990.

Balakrishnan, A. V.
[1] Optimal control problems in Banach spaces, *SIAM J. Control,* 3 (1965), 152–180.
[2] *Applied Functional Analysis,* Springer-Verlag, New York, 1976.
[3] Boundary control of parabolic equations L-Q-R theory, *Proc. V Intl.*

Summer School, Control Inst. Math. Mech., Acad. Sci., GDR, Berlin, 1977.

Ball, J. M.
[1] Strong continuous semigroups, weak solutions and the variation of constants formula, *Proc. Amer. Math. Soc.*, 63 (1977), 370–373.

Banks, H. T., and Kent, G. A.
[1] Control of functional equations of retarded and neutral type to target sets in function space, *SIAM J. Control*, 10 (1972), 567–593.

Barbu, V.
[1] *Optimal Control of Variational Inequalities*, Pitman, London, New York, 1984.
[2] Hamilton–Jacobi equations and nonlinear control problems, *J. Math. Anal. Appl.*, 120 (1986), 494–509.
[3] The dynamic programming equation for the time-optimal control problem in infinite dimensions, *SIAM J. Control Optim.*, 29 (1991), 445–456.
[4] Null controllability of first order quasilinear equations, *Diff. Int. Eqs.*, 4 (1991), 673–681.
[5] *Analysis and Control of Nonlinear Infinite Dimensional Systems*, Academic Press, Boston, 1993.

Barbu, V., and Da Prato, G.
[1] *Hamilton Jacobi Equations in Hilbert Spaces*, Pitman, London, New York, 1983.
[2] Hamilton-Jacobi equation in Hilbert spaces: variational and semigroup approach, *Ann. Mat. Pura. Appl.*, 142 (1985), 303–349.

Barbu, V.; Da Prato, G.; and Popa, C.
[1] Existence and uniqueness of the dynamic programming equation in Hilbert space, *Nonlinear Anal.*, 7 (1983), 283–299.

Barbu, V., and Friedman, A.
[1] Optimal design of domains with free-boundary problems, *SIAM J. Control Optim.*, 29 (1991), 623–637.

Barbu, V., and Precupanu, Th.
[1] *Convexity and Optimization in Banach Spaces*, D. Reidel Publishing, Dordrecht, 1986.

Barbu, V., and Tiba, D.
[1] Boundary controllability of the coincidence set in the obstacle problems, *SIAM J. Control Optim.*, 29 (1991), 1150–1159.

Barles, G.
[1] Deterministic impulse control problems, *SIAM J. Control Optim.*, 23 (1985), 419–432.

Barron, E. N.
[1] The Pontryagin maximum principle for minimax problems of optimal control, *Nonlinear Anal.*, 15 (1990), 1155–1165.

Basile, N., and Mininni, M.
[1] An extension of the maximum principle for a class of optimal control problems in infinite-dimensional spaces, *SIAM J. Control Optim.*, 28 (1990), 1113–1135.

Bellman, R.
[1] On the theory of dynamic programming, *Proc. Nat. Acad. Sci., USA*, 38 (1952), 716–719.
[2] *Dynamic Programming*, Princeton Univ. Press, Princeton, NJ, 1957.

Bellman, R.; Glicksberg, I.; and Gross, O.
[1] On some variational problems occurring in the theory of dynamic programing, *Rend. Circ. Mat. Palermo* (2), 3 (1954), 1–35.
[2] *Some Aspects of the Mathematical Theory of Control Processes*, Rand Co., Santa Monica, CA, 1958.

Bensoussan, A.; Da Prato, G.; Delfour, M. C.; and Mitter, S. K.
[1] *Representation and Control of Infinite Dimensional Systems, Vol.I*, Birkhäuser, Boston, 1992.

Bensoussan, A., and Lions, J. L.
[1] Contrôle impulsionnel et systèmes d'inéquations quasi variationnels, *C. R. Acad. Sci. Paris*, 278 (1974), 747–751.
[2] *Impulse Control and Quasi-Variational Inequalities*, Bordes, Paris, 1984.

Bergounioux, M.
[1] A penalization method for optimal control of elliptic problems with state constraints, *SIAM J. Control Optim.*, 30 (1992), 305–323.

Berkovitz, L. D.
[1] *Optimal Control Theory*, Springer-Verlag, New York, 1974.
[2] Existence and lower closure theorems for abstract control problems, *SIAM J. Control*, 12 (1974), 27–42.

Bermudez, A., and Saguez, C.
[1] Optimal control of a Signorini problem, *SIAM J. Control Optim.*, 25 (1987), 576–582.

Bhat, M. G.; Huffaker, R. G.; and Lenhart, S. M.
[1] Optimal trapping strategies for diffusive nuisance beaver populations, *Natural Resources Modeling J.*, 6 (1992), 71–97.
[2] Controlling forest damage by dispersive beaver populations: A spatiotemporal dynamic trapping strategy, *Ecological Appl.*, to appear.

Bonnans, J. F., and Casas, E.
[1] Optimal control of semilinear multistate systems with state constraints, *SIAM J Control Optim.*, 27 (1989), 446–455.
[2] Un principe de Pontryagin pour le contrôle des systemes semilinéaires elliptiques, *J. Diff. Eqs.*, 90 (1991), 288–303.
[3] A boundary Pontryagin's principle for the optimal control of state-constrained elliptic systems, *Int. Ser. Numer. Math.*, 107 (1992), 241–249.

Bonnans, J. F., and Tiba, D.
[1] Pontryagin's principle in the control of semilinear elliptic variational inequalities, *Appl. Math. Optim.*, 23 (1991), 299–312.

Bourbaki, N.
[1] *General Topology*, Addison-Wesley, Reading, MA, 1966.

Bourgin, R. D.
[1] *Geometric Aspects of Convex Sets with the Radon-Nikodým Property*, Lecture Notes in Math., Vol. 993, Springer-Verlag, Berlin, 1983.

Bradley, M. E., and Lenhart, S. M.
[1] Bilinear optimal control of a Kirchhoff plate, preprint.

Brokate, M.
[1] Necessary optimality conditions for the control of semilinear hyperbolic boundary value problems, *SIAM J. Control Optim.*, 25 (1987), 1353–1369.

Brokate, M., and Friedman, A.
[1] Optimal design for heat conduction problems with hysteresis, *SIAM J. Control Optim.*, 27 (1989), 697–717.

Bucci, F.
[1] A Dirichlet boundary control problem for the strongly damped wave equation, *SIAM J. Control Optim.*, 30 (1992), 1092–1100.

Butkovsky, A. G.
[1] Maximum principle of optimal control for distributed parameter systems, *Avtomatika & Telemekhanika*, 22 (1961), 1288–1301 (in Russian).
[2] *Distributed Control Systems*, Elsevier, New York, 1969.

Butkovsky, A. G.; Egorov, A. I.; and Lurie, K. A.
[1] Optimal control of distributed systems (a survey of Soviet publications), *SIAM J. Control*, 6 (1968), 437–476.

Butkovsky, A, G., and Lerner, A. Ja.
[1] Optimal control systems with distributed parameters, *Avtomatika & Telemekhanika*, 21 (1960), 682–691 (in Russian).

Cannarsa, P.
[1] Regularity properties of solutions to Hamilton Jacobi equations in infinite dimensions and nonlinear optimal control, *Diff. Int. Eqs.*, 2 (1989), 479–493.

Cannarsa, P., and Da Prato, G.
[1] Nonlinear optimal control with infinite horizon for distributed parameter systems and stationary Hamilton-Jacobi equations, *SIAM J. Control Optim.*, 27 (1989), 861–875.
[2] Some results on nonlinear optimal control problems and Hamilton-Jacobi equations in infinite dimensions, *J. Funct. Anal.*, 90 (1990), 27–47.

Cannarsa, P., and Frankowska, H.
[1] Value function and optimality conditions for semilinear control problems, *Appl. Math. Optim.*, 26 (1992), 139–169.

[2] Value function and optimality conditions for semilinear control problems, II: Parabolic, *Appl. Math. Optim.*, to appear.

Capuzzo-Dolcetta, I., and Evans, L. C.
[1] Optimal switching for ordinary differential equations, *SIAM J. Control Optim.*, 22 (1984), 143–161.

Cârjă, O.
[1] On the minimal time function for distributed control systems in Banach spaces, *J. Optim. Theory Appl.*, 44 (1984), 397–406.
[2] On continuity of the minimal time function for distributed control systems, *Boll. Un. Mat. Ital.*, 6 (1985), 293–302.

Carlson, D. A.; Haurie, A.; and Jabrane, A.
[1] Existence of overtaking solutions to infinite dimensional control problems on unbounded time intervals, *SIAM J. Control Optim.*, 25 (1987), 1517–1541.

Casas, E.
[1] Control of an elliptic problem with pointwise state constraints, *SIAM J. Control Optim.*, 24 (1986), 1309–1318.
[2] Optimal control in coefficients of elliptic equations with state constraints, *Appl. Math. Optim.*, 26 (1992), 21–37.
[3] Boundary control of semilinear elliptic equations with pointwise state constraints, *SIAM J. Control Optim.*, 31 (1993), 993–1006.

Casas, E., and Fernández, L. A.
[1] Optimal control of semilinear elliptic equations with pointwise constraints on the gradient of the state, *Appl. Math. Optim.*, 27 (1993), 35–56.
[2] Distributed control of systems governed by a general class of quasilinear elliptic equations, *J. Diff. Eqs.*, 104 (1993), 20–47.

Casas, E.; Fernández, L. A.; and Yong, J.
[1] Optimal control problems of quasilinear parabolic equations, *J. Royal Soc. Edinburgh*, to appear.

Casas, E., and Yong, J.
[1] Maximum principle for state-constrained optimal control problems governed by quasilinear elliptic equations, *Diff. Int. Eqs.*, to appear.

Cesari, L.
[1] Optimization with partial differential equations in Dieudonné-Rashevsky form and conjugate problems, *Arch. Rat. Mech. Anal.*, 33 (1969), 339–357.
[2] Existence theorems for abstract multidimensional control problems, *J. Optim. Theory Appl.*, 6 (1970), 210–236.
[3] Geometric and analytic views in existence theorems for optimal control in Banach spaces, I: Distributed parameters; II: Distributed and boundary controls; III: Weak solutions, *J. Optim. Theory Appl.*, 14 (1974), 505–520; 15 (1975), 467–497; 19 (1976), 185–214.

[4] Existence of solutions and existence of optimal solutions, *Lecture Notes in Math.*, Vol. 979, Springer-Verlag, Berlin, 1983, 88–107.

[5] *Optimization Theory and Applications, Problems with Ordinary Differential Equations,* Springer-Verlag, New York, 1983.

Chen, G.

[1] Energy decay estimates and exact boundary value controllability for the wave equation in a bounded domain, *J. Math., Pures Appl.,* 58 (1979), 249–273.

[2] Control and stabilization for the wave equation in a bounded domain, *SIAM J. Control Optim.,* 17 (1979), 66–81.

Chen, G., and Russell, D. L.

[1] A mathematical model for linear elastic systems with structural damping, *Quart. Appl. Math.,* 39 (1982), 433–454.

Chen, S.

[1] Riccati equations arising in infinite dimensional optimal control problems, *Control Theory Appl.,* 2 (1985), no.4, 64–72 (in Chinese).

[2] Optimal control and feedback synthesis for systems governed by parabolic differential equations with pointwise control on the boundary, *Control Theory Appl.,* 3 (1986), no.2, 20–41 (in Chinese).

[3] The existence of solutions to the infinite dimensional algebraic Riccati equations with indefinite coefficients, *Lecture Notes in Control & Inform. Sci.,* Vol.159, Springer-Verlag, 1991, 43–50.

[4] Structure of solutions to the algebraic Riccati equations and positive real lemma, *Chinese J. Contemporary Math.,* 11 (1991), 97–105.

[5] Necessary and sufficient conditions for the existence of positive solutions to algebraic Riccati equations with indefinite quadratic terms, *Appl. Math. Optim.,* 26 (1992), 95–110.

Choo, K. G.; Teo, K. L.; and Wu, Z. S.

[1] On an optimal control problem involving first order hyperbolic systems with boundary controls, *Numer. Funct. Anal. Optim.,* 4 (1981–82), 171–190.

[2] On an optimal control problem involving second order hyperbolic systems with boundary controls, *Bull. Austral. Math. Soc.,* 27 (1983), 139–148.

Clarke, F. H.

[1] *Optimization and Nonsmooth Analysis,* Wiley, New York, 1983.

Colonius, F.

[1] *Optimal Periodic Control, Lecture Notes in Math,* Vol.1313, Springer-Verlag, Berlin, 1988.

Conrad, F.

[1] Stabilization of beams by pointwise feedback control, *SIAM J. Control Optim.,* 28 (1990), 423–437.

Conti, R.

[1] Time-optimal solution of a linear evolution equation in Banach spaces,

*J. Optim. Theory Appl.,* 2 (1968), 277–284.

Conway, J. B.

[1] *A Course in Functional Analysis,* Springer-Verlag, New York, 1985.

Crandall, M. G., H. Ishii, and Lions, P. L.

[1] User's guide to viscosity solutions of second order partial differential equations, *Bull. Amer. Math. Soc. (NS),* 27 (1992), 1–67.

Crandall, M. G., and Lions, P. L.

[1] Viscosity solutions of Hamilton–Jacobi equations, *Trans. Amer. Math. Soc.,* 277 (1983), 1–42.

[2] Hamilton Jacobi equation in infinite dimensions I: Uniqueness of viscosity solutions; II: Existence of viscosity solutions; III; IV: Hamiltonians with unbounded linear terms; V: Uunbounded linear terms and $B$-continuous solutions, *J. Funct. Anal.,* 62 (1985), 379–396; 65 (1986), 368–425; 68 (1986), 214–247; 90 (1990), 237–283; 97 (1991), 417–465.

Cristescu, R.

[1] *Topological Vector Spaces,* Noordhoff, Leyden, 1977.

Curtain, R. F., and Pritchard, A. J.

[1] The infinite-dimensional Riccati equation for systems defined by evolution operators, *SIAM J. Control Optim.,* 14 (1976), 951–983.

[2] An abstract theory for unbounded control action for distributed parameter systems, *SIAM J. Control,* 15 (1977), 566–611.

[3] *Infinite Dimensional Linear Systems Theory, Lecture Notes in Control Inform. Sci.,* Vol.8, Springer-Verlag, New York, 1981.

Da Prato, G.

[1] Synthesis of optimal control for an infinite dimensional periodic problem, *SIAM J. Control Optim.,* 25 (1987), 706–714.

Da Prato, G., and Delfour, M. C.

[1] Unbounded solutions to the linear quadratic control problem, *SIAM J. Control Optim.,* 30 (1992), 31–48.

Da Prato, G., and Ichikawa, A.

[1] Optimal control of linear systems with almost periodic inputs, *SIAM J. Control Optim.,* 25 (1987), 1007–1019.

[2] Quadratic control for linear periodic systems, *Appl. Math. Optim.,* 18 (1988), 39–66.

[3] Quadratic control for linear time-varying systems, *SIAM J. Control Optim.,* 28 (1990), 359–381.

Datko, R.

[1] Extending a theorem of A. M. Lyapunov to Hilbert space, *J. Math. Anal. Appl.,* 32 (1970), 610–616.

Deimling, K.

[1] *Nonlinear Functional Analysis,* Springer-Verlag, New York, 1985.

Derzko, N., and Sethi, S. P.

[1] Distributed parameter systems approach to the optimal cattle ranching

problem, *Optim. Control Appl. Methods,* 1 (1980), 3–10.

Diaz, J. I.
[1] *Nonlinear Partial Differential Equations and Free Boundaries, Vol.I, Elliptic Equations,* Pitman, London, New York, 1985.

Di Blasio, G.
[1] Global solutions for a class of Hamilton-Jacobi equations in Hilbert spaces, *Numer. Funct. Anal. Optim.,* 8 (1985–86), 261–300.

Diestel, J.
[1] *Geometry of Banach spaces, Lecture Notes in Math.,* 485, Springer-Verlag, New York, 1975.
[2] *Sequences and Series in Banach Spaces,* Springer-Verlag, New York, 1984.

Diestel, J., and Uhl, J. J., Jr.
[1] *Vector Measures,* AMS, Providence, RI, 1977.

Dunford, N., and Schwartz, J. T.
[1] *Linear Operators, Part I: General Theory,* Intersciences Pub., New York, 1957.

Egorov, A. I.
[1] The maximum principle in the theory of optimal regulation, *Studies in Integro-Differential Equations in Kirghizia,* No.1 (1961), 213–242 (in Russian).
[2] Necessary optimality conditions for distributed parameter systems, *SIAM J. Control,* 5 (1967), 352–408.

Egorov, Yu. V.
[1] Optimal control in Banach spaces, *Dokl. Acad. Nauk SSSR,* 150 (1963), 241–244 (in Russian).
[2] Certain problems in optimal control theory, *USSR Comput. Math. & Math. Phys.,* 3 (1963), 1209–1232 (in Russian).
[3] Necessary conditions for optimal control in Banach spaces, *Mat. Sb.,* 64 (1964), 79–101 (in Russian).

Ekeland, I.
[1] On the variational principle, *J. Math. Anal. Appl.,* 47 (1974), 324–353.
[2] Nonconvex minimization problems, *Bull Amer. Math. Soc. (New Series),* 1 (1979), 443–474.

Ekeland, I., and G. Lebourg
[1] Generic Fréchet-differentiability and perturbed optimization problems in Banach spaces, *Trans. Amer. Math. Soc.,* 224 (1976), 193–216.

Elliott, C. M., and Ockendon, J. R.
[1] *Weak and Variational Method for Moving Boundary Problems,* Pitman, Boston, 1980.

Evans, L. C., and Friedman, A.
[1] Optimal stochastic switching and the Dirichlet problem for the Bellman equation, *Trans. Amer. Math. Soc.,* 253 (1979), 365–389.

Fabre, C.; Puel, J.-P.; and Zuazua, E.
[1] Approximate controllability of the semilinear heat equation, IMA preprint no. 1067.

Fattorini, H. O.
[1] Time-optimal control of solutions of operational differential equations, *SIAM J. Control*, 2 (1964), 54-59.
[2] Some remarks on complete controllability, *SIAM J. Control*, 4 (1966), 686–694.
[3] On complete controllability of linear systems, *J. Diff. Eqs.*, 3 (1967), 391–402.
[4] An observation on a paper of A. Friedman, *J. Math. Anal. Appl.*, 22 (1968), 382–384.
[5] The time optimal control problems in Banach spaces, *Appl. Math. Optim.*, 1 (1974), 163–188.
[6] The maximum principle for nonlinear noncovex systems in infinite dimensional spaces, *Lecture Notes in Control & Inform. Sci.*, Vol. 75, Springer-Verlag, 1985, 162–178.
[7] A unified theory of necessary conditions for nonlinear nonconvex control systems, *Appl. Math Optim.*, 15 (1987), 141–185.
[8] Optimal control problems for distributed parameter systems in Banach spaces, *Appl. Math. Optim.*, 28 (1993), 225–257.
[9] Existence theory and maximum principle for relaxed infinite dimensional optimal control problems, *SIAM J. Control Optim.*, to appear.

Fattorini, H. O., and Frankowska, H.
[1] Necessary conditions for infinite-dimensional control problems, *Math. Control Signal Systems*, 4 (1991), 41–67.

Fattorini, H. O., and Murphy, T.
[1] Optimal problems for nonlinear parabolic boundary control systems, *SIAM J. Control Optim.*, to appear.

Fattorini, H. O., and Sritharan, S. S.
[1] Necessary and sufficient conditions for optimal controls in viscous flow problems, *Proc. Royal Soc. Edinburgh, Ser.A*, to appear.
[2] Existence of optimal controls for viscous flow problems, *Proc. Royal Soc. London, Ser.A,* to appear.
[3] Optimal chattering controls for viscous flow, preprint.

Fife, P.
[1] Mathematical aspects of reacting and diffusing systems, *Lecture Notes in Biomathematics,* Vol.28, Springer-Verlag, New York, 1979.

Flandoli, F.
[1] Riccati equation arising in a boundary control problem with distributed parameters, *SIAM J. Control Optim.*, 22 (1984), 76–86.

Fleming, W. H.
[1] *Future Directions in Control Theory — A Mathematical Perspective,* SIAM, Philadelphia, 1988.

Frankowska, H.
[1] Some inverse mapping theorems, *Ann. Inst. Henri Poincaré, Anal. Non Linear,* 7 (1990), 183–234.
[2] A priori estimates for operational differential inclusions and necessary conditions for optimality, *Lecture Notes in Control & Inform. Sci.,* Vol.144, Springer-Verlag, 1990, 519–528.

Friedman, A.
[1] Optimal control for parabolic equations, *J. Math. Anal. Appl.,* 18 (1967), 479–491.
[2] Optimal control in Banach spaces, *J. Math. Anal. Appl.,* 19 (1967), 35–55.
[3] Optimal control in Banach spaces with fixed end-points, *J. Math. Anal. Appl.,* 24 (1968), 161–181.
[4] *Variational Principles and Free-boundary Problems,* John-Wiley & Sons, New York, 1982.
[5] Optimal control for variational inequalities, *SIAM J. Control Optim.,* 24 (1986), 439–451.
[6] Optimal control for parabolic variational inequalities, *SIAM J. Control Optim.,* 25 (1987), 482–497.

Friedman, A., and Hoffmann, K.-H.
[1] Control of free boundary problems with hysteresis, *SIAM J. Control Optim.,* 26 (1988), 42–55.

Friedman, A.; Huang, S; and Yong, J.
[1] Bang-bang optimal control for the dam problem, *Appl. Math. Optim.,* 15 (1987), 65–85.
[2] Optimal periodic control for the two-phase Stefan problem, *SIAM J. Control Optim.,* 26 (1988), 23–41.

Friedman, A., and Jiang, L.
[1] Nonlinear optimal control problem in heat conduction, *SIAM J. Control Optim.,* 21 (1983), 940–952.

Friedrichs, K.
[1] Spektraltheorie halbbeschränkter Operatoren und Anwendung auf die Spektralzerlegung von Differentialoperatoren, *Math. Ann.,* 109 (1934), 465–487.

Fujiwara, D.
[1] Concrete characterizations of the domains of fractional powers of source elliptic differential operators of the second order, *Proc. Japan Acad.,* 43 (1967), 82–86.

Fung, Y. C.
[1] *Foundations of Solid Mechanics,* Prentice-Hall, Englewood Cliffs, NJ, 1965.

Gal-el-Hak, M.
[1] Control of low-speed airfoil aerodynamics, *AIAA Journal,* 28 (1990), 1537–1552.

Gilbarg, D., and Trudinger, N. S.
[1] *Elliptic Partial Differential Equations of Second Order*, 2nd edition, Springer-Verlag, 1983.

Goebel, K., and Reich, S.
[1] *Uniform Convexity, Hyperbolic Geometry, and Nonexpansive Mappings*, Marcel Dekker, Inc., New York and Basel, 1984.

Gozzi, F.
[1] Some results for an optimal control problem with semilinear state equations, *SIAM J. Control Optim.*, 29 (1991), 751–768.

Granger, R. A.
[1] *Fluid Mechanics*, CBS College Publishing, New York, 1985.

Grisvard, P.
[1] *Elliptic Problems in Nonsmooth Domains*, Pitman, Boston, 1985.

Grusa, K.-U.
[1] *Mathematical Analysis of Nonlinear Dynamic Processes*, Longman Scientific & Technical, England, 1988.

Gunzburger, M. D.; Hou, L.; and Svobodny, T. P.
[1] Boundary velocity control of incompressible flow with an application to viscous drag reduction, *SIAM J. Control Optim.*, 30 (1992), 167–181.

Hale, J. K.
[1] *Theory of Functional Differential Equations*, Springer-Verlag, New York, 1977.

Halkin, H.
[1] Necessary conditions for optimal control problems with infinite horizon, *Econometrica*, 42 (1974), 267–273.

Hansen, S. W.
[1] Exponential energy decay in a linear thermoelastic rod, *J. Math. Anal. Appl.*, 167 (1992), 429–442.

Haurie, A.
[1] Stability and optimal exploitations over an infinite time horizon of interacting populations, *Optim. Control Appl. Methods*, 3 (1982), 241–256.

Haurie, A.; Sethi, S.; and Hartl, R.
[1] Optimal control of an age-structured population model with application to social services planning, *Large Scale Systems*, 6 (1984), 133–158.

He, Z.-X.
[1] State constrained control problems governed by variational inequalities, *SIAM J. Control Optim.*, 25 (1987), 1119–1144.

Hille, E., and Phillips, R. S.
[1] *Functional Analysis and Semi-groups*, AMS, Providence, RI, 1957.

Himmelberg, C. J.; Jacobs, M. Q.; and Van Vleck, F. S.
[1] Measurable multifunctions, selectors and Filippov's implicit functions lemma, *J. Math. Anal. Appl.*, 25 (1969), 276–284.

Hou, S. H.
[1] Existence theorems of optimal control problems in Banach spaces, *Nonlinear Anal.*, 7 (1983), 239–257.

Hu, B., and Yong, J.
[1] Pontryagin maximum principle for semilinear and quasilinear parabolic equations with pointwise state constraints, *SIAM J. Control Optim.*, to appear.

Huang, C., and Yong, J.
[1] Coupled parabolic and hyperbolic equations modeling age-dependent epidemic dynamics with nonlinear diffusion, *SIAM J. Math. Anal.*, to appear.

Huang, F. L.
[1] Characteristic condition for exponential stability of linear dynamical systems in Hilbert spaces, *Ann. Diff. Eqs.*, 1 (1985), 43–56.

Ishii, H.
[1] Uniqueness of unbounded viscosity solution of Hamilton-Jacobi equations, *Indiana Univ. Math. J.*, 33 (1984), 721–748.

Istratescu, V. I.
[1] *Strict Convexity and Complex Strict Convexity*, Marcel Dekker, New York and Basel, 1984.

Jin, F., and Li, X.
[1] On optimal control of functional differential systems, *Lecture Notes in Control & Inform. Sci.*, Vol. 102, Springer-Verlag, 1987, 112–119.

Kalman, R. E.
[1] Contributions to the theory of optimal control, *Bol. Soc. Math. Mexicana*, 5 (1960), 102–119.
[2] On the general theory of control systems, *Proc. 1st IFAC Congress, Moscow, 1960;* Butterworth, London, 1961, Vol.1, 481–492.
[3] When is a linear control system optimal? *Trans. ASME Ser.D: J. Basic Eng.*, 86 (1964), 1–10.

Kharatishvili, G. L.
[1] The maximum principle in the theory of optimal processes with a delay, *Soviet Math. Dokl.*, 2 (1961), 28–32.

Kime, K. A.
[1] Boundary controllability of Maxwell's equations in a spherical region, *SIAM J. Control Optim.*, 28 (1990), 294–319.

Kinderlehrer, D., and Stampacchia, G.
[1] *An Introduction to Variational Inequalities*, Academic Press, New York, 1981.

Komornik, V.; Russell, D. L.; and Zhang, B. Y.
[1] Control and stabilization of the Korteweg-de Vries equation on a periodic domain, *J. Diff. Eqs.*, to appear.

Korobov, V. I., and Šon, N. K.
[1] Controllability of linear systems in a Banach space in the presence of constraints on the control, I,II, *Diff. Eqs.*, 16 (1980), 505–513.

Kufner, A.; John, O.; and Fučík, S.
[1] *Function Spaces,* Noordhoff, Leyden, 1977.

Kwan, C. C., and Wang, K. N.
[1] Sur la stabilisation de la vibration élastique, *Scientia Sinica,* 17 (1974), 446–467.

Ladyzhenskaya, O. A., and Ural'tseva, N. N.
[1] *Linear and Quasilinear Elliptic Equations,* Academic Press, New York, 1968.

Lagnese, J. E.
[1] Exact boundary controllability of Maxwell's equations in a general region, *SIAM J. Control & Optim.*, 27 (1989), 374–388.
[2] *Boundary Stabilization of Thin Plates,* SIAM, Philadelphia, 1989.

Lagnese, J. E., and Lions, J. L.
[1] *Modelling Analysis and Control of Thin Plates,* Masson, Paris, 1988.

Langlais, M.
[1] A nonlinear problem in age dependent population diffusion, *SIAM J. Math. Anal.*, 16 (1985), 510–529.

Lasiecka, I.
[1] Unified theory for abstract parabolic boundary problems — a semigroup approach, *Appl. Math. Optim.*, 6 (1980), 287–333.

Lasiecka, I., and Triggiani. R.
[1] Differential and algebraic Riccati equations with applications to boundary/point controlproblems: continuous theory and approximation theory, Lecture Notes in Control & Inform. Sci., Vol.164, Springer-Verlag, Berlin, 1991.
[2] Riccati differential equations with unbounded coefficients and non-smooth terminal condition—the case of analytic semigroups, *SIAM J. Math. Anal.*, 23 (1992), 449–481.

Lee, E. B., and Markus, L.
[1] *Foundations of Optimal Control Theory,* John Wiley, New York, 1967.

Lee, E. B., and You. Y.
[1] Quadratic optimization for infinite-dimensional linear differential difference type systems: Syntheses via the Fredholm equation, *SIAM J. Control Optim.*, 28 (1990), 265–293.

Lenhart, S. M.
[1] Optimal control of a convective-diffusive fluid problem, preprint.

Lenhart, S. M., and Bhat, M.
[1] Application of distributed parameter control model in wildlife damage management, *Math. Model & Methods in Appl. Sci.*, 2 (1992), 423–439.

Lenhart, S. M., and Protopopopescu, V.
[1] Optimal control for parabolic systems with competitive interactions, *Math. Methods Appl. Sci.*, to appear.

Lenhart, S. M.; Protopopopescu, V; and Stojanovic, S.
[1] A two-sided game for non local competitive systems with control on source terms, preprint.
[2] A minimax problem for semilinear nonlocal competitive systems, *Appl. Math. Optim.*, to appear.

Lenhart, S. M., and Wilson, D. G.
[1] Optimal control of a heat transfer problem with convective boundary conditions, *J. Optim. Theory Appl.*, to appear.

Lenhart, S. M., and Yong, J.
[1] Optimal control for degenerate parabolic equations with logistic growth, *Nonlinear Anal.*, to appear.

Leung, A., and Stojanovic, S.
[1] Direct method for some distributed games, *Diff. Int. Eqs.*, 3 (1990), 1113–1125.
[2] Optimal control for elliptic Volterra-Lotka type equations, *J. Math. Anal. Appl.*, 173 (1993), 603–619.

Li, X.
[1] Vector-valued measure and the necessary conditions for the optimal control problems of linear systems, *Proc. IFAC 3rd Symposium on Control of Distributed Parameter Systems, Toulouse, France*, 1982, 503–506.
[2] Maximum principle of optimal periodic control for functional differential systems, *J. Optim. Theory & Appl.*, 50 (1986), 421–429.

Li, X., and Chow, S. N.
[1] Maximum principle of optimal control for functional differential systems, *J. Optim. Theory & Appl.*, 54 (1987), 335–360.

Li, X., and Yao, Y.
[1] Time optimal control of distributed parameter systems, *Scientia Sinica*, 24 (1981), 455–465.
[2] On optimal control for distributed parameter systems, *Proc. IFAC 8th Triennial World Congress, Kyoto, Japan*, 1981, 207–212.
[3] Maximum principle of distributed parameter systems with time lags, *Lecture Notes in Control & Inform. Sci.*, Vol.75, Springer-Verlag, 1985, 410–427.

Li, X., and Yong, J.
[1] Necessary conditions of optimal control for distributed parameter systems, *SIAM J. Control Optim.*, 29 (1991), 895–908.

Lieberman, G. M.
[1] Boundary regularity for solutions of degenerate elliptic equations, *Nonlinear Anal.*, 12 (1988), 1203–1219.

Lions, J. L.
[1] Sur le contrôle optimal de systèmes décrits par des équations aux dérivées partelles linéaires, Remarques générales; Équations elliptiques; Équations d'evolution, *C. R. Acad. Sci. Paris, Sér. A–B* 263 (1966), A661–A663; A713–A715; A776–A779.
[2] *Optimal Control of Systems Governed by Partial Differential Equations*, Springer-Verlag, New York, 1971
[3] *Some Aspects of the Optimal Control of Distributed Parameter Systems*, SIAM, Philadelphia, 1972.
[4] *Some Methods in the Mathematical Analysis of Systems and Their Control*, Science Press, Beijing, China, 1981.
[5] *Control of Distributed Singular Systems*, Gauthier-Villars, Paris, 1985.
[6] Exact controllability, stabilization and perturbations for distributed systems, *SIAM Review,* 30 (1988), 1–68.

Lions, J. L., and Magenes, E.
[1] *Non-Homogeneous Boundary Value Problems and Applications Vol.I*, Springer-Verlag, New York, 1971.

Littman, W.
[1] Boundary control theory for hyperbolic and parabolic partial differential equations with constant coefficients, *Ann. Scu. Norm. Sup. Pisa, Ser. IV,* 5 (1978), 567–580.

Liu, K. S.
[1] Energy decay problems in the design of a point stabilizer for coupled string vibrating systems, *SIAM J. Control Optim.,* 26 (1988), 1348–1356.

Liu, K. S.; Huang, F. L.; and Chen, G.
[1] Exponential stability analysis of a long chain of coupled vibrating strings with dissipative linkage, *SIAM J. Appl. Math.,* 49 (1989), 1694–1707.

Liu, Z.
[1] Approximation and control of a thermoviscoelastic system, *Ph.D. thesis,* Virginia Polytechnic Inst. and State Univ., Blacksburg, VA, 1989.

Lukes, D. L., and Russell, D. L.
[1] The quadratic criterion for distributed systems, *SIAM J. Control,* 7 (1969), 101–121.

MacCamy, R. C.
[1] A population model with nonlinear diffusion, *J. Diff. Eqs.,* 39 (1981), 52–72.

Mackenroth, U.
[1] Convex parabolic boundary control problems with pointwise state constraints, *J. Math. Anal. Appl.,* 87 (1982), 256–277.

Michlin, S. G.
[1] On convergence of Galerkin's method, *Dokl. Acad. Nauk SSSR,* 61

(1948), 197–199.

Mignot. F.
[1] *Contrôle dans les inequations variationelles elliptiques*, J. Funct. Anal., 22 (1976), 130–185.

Mignot, F., and Puel, J. P.
[1] Optimal control in some variational inequalities, *SIAM J. Control Optim.*, 22 (1984), 466–476.

Mizohata, S.
[1] Unicité du prolongement des solutions pour quelques opérateaurs différentiels paraboliques, *Mem. Coll. Sci. Univ. Kyoto, Ser. A31* (3), (1958), 219–239.

Molinari, B. P.
[1] The time-invariant linear-quadratic optimal control problem, *Automatica*, 13 (1977), 347–357.

Morrey, C. B., Jr.
[1] *Multiple Integrals in the Calculus of Variations*, Springer-Verlag, New York, 1966.

Mossino, J.
[1] An application of duality to distributed optimal control problems with constraints on the control and the state, *J. Math. Anal. Appl.*, 50 (1975), 223–242.

Murray, J. D.
[1] *Mathematical Biology*, Springer-Verlag, New York, 1989.

Nababan, S., and Teo, K. L.
[1] On the existence of optimal controls of the first boundary value problem for parabolic partial delay-differential equations in divergence form, *J. Math. Soc. Japan*, 32 (1980), 343–362.

Naito, K.
[1] Controllability of semilinear control systems dominated by the linear part, *SIAM J. Control Optim.*, 25 (1987), 715–722.

Naito, K., and Seidman, T. I.,
[1] Invariance of the approximately reachable set under nonlinear perturbations, *SIAM J. Control Optim.*, 29 (1991), 731–750.

Nakagiri, S.
[1] Optimal control of linear retarded systems in Banach spaces, *J. Math. Anal. Appl.*, 120 (1986), 169–210.

Neittaanmäki, P.; Sokolowski, J.; and Zolesio, J. P.
[1] Optimization of the domain in elliptic variational inequalities, *Appl. Math. Optim.*, 18 (1988), 85–98.

Neustadt, L.
[1] *Optimization*, Princeton Univ. Press, Princeton, NJ, 1976.

Pan, L., and Yong. J.
[1] Optimal control for quasilinear retarded parabolic systems, *Math. Systems, Estimation & Control,* to appear.

Papageorgiou, N. S.
[1] Existence of optimal controls for nonlinear systems in Banach spaces, *J. Optim. Theory Appl.,* 53 (1987), 451–459.
[2] Properties of the relaxed trajectories of evolution equations and optimal controls, *SIAM J. Control Optim.* 27 (1989), 267–288.
[3] Existence of optimal controls for nonlinear distributed parameter systems, *Funkcialaj Ekvacioj,* 32 (1989), 429–437.

Pazy, A.
[1] *Semigroups of Linear Operators and Applications to Partial Differential Equations,* Springer-Verlag, New York, 1983.

Peichl, G., and Schappacher, W.
[1] Constrained controllability in Banach spaces, *SIAM J. Control Optim.,* 24 (1986), 1261–1275.

Pelukhov, L. V.
[1] Optimal control of processes described by equations of hyperbolic type, *J. Appl. Math. Mech.,* 41 (1977), 385–397.

Pelukhov, L. V., and Troitskii, V. A.
[1] Variational optimization problems for equations of hyperbolic type, *J. Appl. Math., Mech.,* 36 (1972), 545–555.
[2] Some optimal problems of the theory of longitudinal vibrations of rods, *J. Appl. Math. Mech.,* 36 (1972), 842–851.
[3] Variational optimization problems for equations of the hyperbolic type in the presence of boundary controls, *J. Appl. Math. Mech.,* 39 (1975), 244-253.

Peng, S., and Yong, J.
[1] Determination of controllable set for a controlled dynamic system, *J. Austral. Math. Soc., Ser. B,* 33 (1991), 164–179.

Phillips, R. S.
[1] Perturbation theory for semi-groups of linear operators, *Trans. Amer. Math. Soc.,* 74 (1953), 199–221.

Pironneau, O.
[1] On optimum problems in Stokes flow, *J. Fluid Mech.,* 59 (1973), 117–128.
[2] On optimum design in fluid mechanics, *J. Fluid Mech.,* 64 (1974), 97–110.

Pontryagin, L. S.
[1] The maximum principle in the theory of optimal processes, *Proc. 1st Congress IFAC,* Moscow, 1960.

Pontryagin, L.S.; Boltyanskii, V.G.; Gamkrelidze, R.V.; and Mischenko,

E.F.
[1] *Mathematical Theory of Optimal Processes,* Wiley, New York, 1962.

Preiss, D.
[1] Differentiability of Lipschitz functions on Banach spaces, *J. Funct. Anal.,* 91 (1990), 312–345.

Pritchard, A. J., and Salamon, D.
[1] The linear quadratic control problem for infinite dimensional systems with unbounded input and output operators, *SIAM J. Control Optim.,* 25 (1987), 121–144.

Rakotoson, J. M.
[1] Réarrangement relatif dans les équations elliptiques quasi-linéaries avec un second membre distribution: Application à un théorème d'existence et de régularité, *J. Diff. Eqs.,* 66 (1987), 391–419.

Rodrigues, J.-F.
[1] *Obstacle Problems in Mathematical Physics,* North-Holland, Amsterdam, 1987.

Russell, D. L.
[1] Nonharmonic Fourier series in the control theory of distributed parameter systems, *J. Math. Anal. Appl.,* 18 (1967), 542–560.
[2] Controllability and stabilizability theory for linear partial differential equations: recent progress and open questions, *SIAM Rev.,* 20 (1978), 639–739.
[3] The Dirichlet-Neumann boundary control problem associated with Maxwell's equations in a cylindrical region, *SIAM J. Control Optim.,* 24 (1986), 199–229.
[4] Mathematical models for the elastic beam and their control-theoretic implications, *Semigroup Theory and Applications,* Brezis et al. eds, Longman, New York, 1985.

Russell, D. L., and Zhang, B. Y.
[1] Controllability and stabilizability of the third-order linear dispersion equation on a periodic domain, *SIAM J. Control Optim.,* 31 (1993), 659–676.

Sachs, E.
[1] A parabolic control problem of the Stefan-Boltzmann type, *Z. Angew. Math. Mech.,* 58 (1978), 443–449.

Saguez, C.
[1] Contrôle optimal d'un système gouverné par une inéquation variationnelle parabolique; observation du domaine de contact, *C. R. Acad. Sci. Paris,* 287 (1978), 957–959.

Sakawa, Y.
[1] Solution of an optimal control problem in a distributed parameter system, *IEEE Trans. Automat.,* 9 (1964), 420–426.

Saut, J. C., and Scheurer, B.
[1] Unique continuation for some evolution equations, *J. Diff. Eqs.*, 66 (1987), 118–139.

Schmidt, E. J. P. G.
[1] Boundary control problem for the heat equation with non-linear boundary condition, *J. Diff. Eqs.*, 78 (1989), 98–121.

Seidman, T. I.
[1] Time-invariance of the reachable set for linear control problems, *J. Math. Anal. Appl.*, 72 (1979), 17–20.
[2] Invariance of the reachable set under nonlinear perturbations, *SIAM J. Control Optim.*, 25 (1985), 1173–1191.

Shi, S.
[1] Optimal control of strongly monotone variational inequalities, *SIAM J. Control Optim.*, 26 (1988), 274–290.

Shiryaev, A. N.
[1] *Optimal Stopping Rules,* Springer-Verlag, New York, 1978.

Singer, I.
[1] *Bases in Banach Spaces, I,* Springer-Verlag, New York, 1970.

Song, J., and Yu, J. Y.
[1] On the theory of distributed parameter systems with ordinary feedback control, *Scientia Sinica,* 18 (1975), 281–310.
[2] Distributed parameter systems with pointwise observation and pointwise control, *Scientia Sinica,* 22 (1979), 131–141 (in Chinese).
[3] *Population System Control,* Springer-Verlag, Berlin, 1988.

Sritharan, S. S.
[1] *Optimal Control of Viscosity Flow,* SIAM, Philadelphia, 1993.

Stampacchia, G.
[1] Le problème de Dirichlet pour les equations elliptiques du second ordre à coefficients discontinus, *Ann. Inst. Fourier Grenoble,* 15 (1965), 189–258.

Stegall, C.
[1] Optimization of functions on certain subsets of Banach spaces, *Math. Ann.,* 236 (1978), 171–176.

Stojanovic, S.
[1] Optimal damping control and nonlinear parabolic systems, *Numer. Funct. Anal. Optim.,* 10 (1989), 573–591.
[2] Optimal damping control and nonlinear elliptic systems, *SIAM J. Control Optim.,* 29 (1991), 594–608.

Stojanovic, S., and Yong, J.
[1] Optimal switching for systems governed by nonlinear evolution equations, *Numer. Funct. Anal. Optim.,* 9 (1987), 995–1030.
[2] Optimal switching for partial differential equations I, II, *J. Math. Anal. Appl.,* 138 (1989), 418–438; 439–460.

Sun, S. H.
[1] On spectrum distribution of completely controllable linear systems, *SIAM J. Control Optim.*, 19 (1981), 730–743.

Suryanarayana, M. B.
[1] Necessary conditions for optimization problem with hyperbolic partial differential equations, *SIAM J. Control Optim.*, 11 (1973), 130–147.
[2] Existence theorems for optimization problem concerning linear, hyperbolic partial differential equations without convexity conditions, *J. Optim. Theory Appl.*, 19 (1976), 47–61.

Tahraoui, R.
[1] Contrôle optimal dans les équations elliptiques, *SIAM J. Control Optim.*, 30 (1992), 495–521.

Tataru, D.
[1] Viscosity solutions for the dynamic programming equations, *Appl. Math. Optim.*, 25 (1992), 109–126.
[2] Viscosity solutions of Hamilton-Jacobi equations with unbounded nonlinear terms, to appear.

Temam, R.
[1] *Navier–Stokes Equations: Theory and Numerical Analysis, 3rd edition,* North-Holland, Amsterdam, 1984.

Tiba, D.
[1] Boundary control for a Stefan problem, *Optimal Control of Partial Differential Equations, K.-H. Hoffmann and W. Krabs eds.,* Birkhäuser Verlag, Basel, 1983.
[2] *Optimal control of nonsmooth distributed parameter systems, Lecture Notes in Math,* Vol. 1495, Springer-Verlag, Berlin, 1990.

Torrejón, R., and Yong, J.
[1] A control problem for clamped extensible beams, *Appl. Anal.*, 28 (1988), 163–180.

Triggiani, R.
[1] Controllability and observability in Banach space with bounded operators, *SIAM J. Control*, 13 (1975), 462–491.

Troianiello, G. M.
[1] *Elliptic Differential Equations and Obstacle Problems,* Plenum Press, New York, 1987.

Trölzsch, F.
[1] *Optimal Conditions for Parabolic Control Problems and Applications,* Teubner Verlagsgesellschaft, Leipzig, 1984.

Tsien, H. S., and Song, J.
[1] *Engineering Cybernetics, 2nd edition,* Science Press, Beijing, 1980.

Wagner, D. H.
[1] Survey of measurable selection theorems, *SIAM J. Control Optim.*, 15 (1977), 859–903.

Wald, A.
[1] *Statistical Decision Functions,* John Wiley, New York, 1950.

Wang, L.
[1] The invariance of approximately reachable set for distributed parameter control systems under nonlinear perturbation, *J. Sys. Sci. & Math. Sci.,* 13 (1993), 193–200 (in Chinese).

Wang, P. K. C.
[1] Control of distributed parameter systems, *Advances in Control Systems, Theory & Applications, C .T. Leondes, eds.,* Vol. 1, Academic Press, New York, 1964, 75–172.
[2] Optimal control of parabolic system with boundary conditions involving time delays, *SIAM J. Control,* 13 (1975), 274–293.

Warga, J.
[1] *Optimal Control of Differential and Functional Equations,* Academic Press, New York, 1972.

Weidmann, J.
[1] *Linear Operators in Hilbert Spaces,* Springer-Varlag, New York, 1980.

Wolfersdorf, L. V.
[1] Optimal control for processes governed by mildly nonlinear differential equations of parabolic type I, II, *Z. Angew. Math. Mech.,* 56 (1976), 531–538; 57 (1977), 11–17.

Xu. H.
[1] Necessary conditions of optimal control for distributed parameter systems, Master thesis, Fudan University, 1988.

Yao, Y.
[1] A maximum principle for semilinear distributed systems I, *Chin. Ann. Math.,* 3 (1982), 679–690 (in Chinese).
[2] The maximum principle for semilinear distributed systems II, the time optimal control problem, *Chin. Ann. Math. Ser. A,* 4 (1983), 781–792 (in Chinese).
[3] Vector measure and maximum principle of distributed parameter systems, *Sci. Sinica Ser. A,* 6 (1983), 102–112.
[4] The regulator pointwise control for parabolic systems, *Proc. 9th IFAC Triennial World Congress, Budapest, Hungary,* (1985), 46–49.
[5] Optimal control for a class of elliptic equations, *Control Theory Appl.,* 1 (1984) No.3, 17–23 (in Chinese).

Yao, Y., and You, Y.
[1] Closed-loop optimal solutions for some parabolic systems with pointwise control ($n = 2, 3$), *J. Appl. Sci.,* 3 (1985), 223–232.

Yong, J.
[1] Systems governed by ordinary differential equations with continuous, switching and impulse controls, *Appl. Math. Optim.,* 20 (1989), 223–235.

[2] Optimal switching and impulse controls for distributed parameter systems, *System Sci. & Math. Sci.*, 2 (1989), 137–160.
[3] Maximum principle of the optimal controls for a nonsmooth semilinear evolution system, *Lecture Notes in Control & Inform. Sci.*, Springer-Verlag, 144 (1990), 559–569.
[4] Optimal control for distributed parameter systems with mixed constraints, *Colloquium Math.*, 60/61 (1990), 35–48.
[5] Time optimal control for semilinear distributed parameter systems—existence theory and necessary conditions, *Kodai Math. J.*, 14 (1991), 239–253.
[6] Existence of the value for a differential game with switching strategies in a Banach space, *System Sci. & Math. Sci.*, 4 (1991), 321–340.
[7] Existence theory of optimal control for distributed parameter systems, *Kodai Math. J.*, 15 (1992), 193–220.
[8] Pontryagin maximum principle for semilinear second order elliptic partial differential equations and variational inequalities with state constraints, *Diff. Int. Eqs.*, 5 (1992), 1307–1334.
[9] Infinite dimensional Volterra-Stieltjes evolution equations and related optimal control problems, *SIAM J. Control Optim.*, 31 (1993), 539–568.
[10] Optimal control for nonlinear abstract evolution systems, *Diff. Int. Eqs.*, 6 (1993), 1145–1159.
[11] A minimax control problem for second order elliptic partial differential equations, *Kodai Math. J.*, 16 (1993), 469–486.
[12] Zero-sum differential games involving impulse controls, *Appl. Math. Optim.*, 29 (1994), 243–261.

Yong, J., and Zhang, P.
[1] Necessary conditions of optimal impulse controls for distributed parameter systems, *Bull. Austral. Math. Soc.*, 45 (1992), 305–326.

Yong, J., and Zheng, S.
[1] Feedback stabilization and optimal control for Cahn–Hilliard equation, *Nonlinear Anal.*, 17 (1991), 431–444.
[2] Feedback stabilization for the phase field equations, *Appl. Anal.*, 42 (1991), 59–68.

Yosida, K.
[1] *Functional Analysis, 6th edition,* Springer-Verlag, Berlin, 1980.

You, Y.
[1] Optimal control for linear systems with quadratic indefinite criterion on Hilbert spaces, *Chin. Ann. Math. Ser. B*, 4 (1983), 21–32.
[2] On the solution of a class of operator Riccati equation, *Chin. Ann. Math. Ser. A*, 5 (1984), 219–227 (in Chinese).
[3] Closed-loop optimal solution to quadratic boundary control of parabolic systems, *Acta Math. Sinica*, 28 (1985), 809–816 (in Chinese).
[4] A nonquadratic Bolza problem and a quasi-Riccati equation for dis-

tributed parameter systems, *SIAM J. Control Optim.*, 25 (1987), 905–920.

Zabczyk, J.
[1] Remarks on the algebraic Riccati equations in Hilbert space, *Appl. Math. Optim.*, 2 (1975–76), 251–258.

Zeidler, E.
[1] *Nonlinear Functional Analysis and Its Applications, I*, Springer-Verlag, New York, 1986.

Zhang, X.; Li, X.; and Chen, Z.
[1] *Differential Equation Theory of Optimal Control Systems*, High Education Press, 1989 (in Chinese).

Zhou, H. X.
[1] A note on approximate controllability for semilinear one-dimensional heat equation, *Appl. Math. Optim.*, 8 (1982), 275–285.
[2] Approximate controllability for a class of semilinear abstract equations, *SIAM J. Control Optim.*, 21 (1983), 551–565.
[3] Controllability properties of linear and semilinear abstract control systems, *SIAM J. Control Optim.*, 22 (1984), 405–422.

Zhou, X. Y.
[1] Maximum principle, dynamic programming and their connection in deterministic control, *J. Optim. Theory Appl.*, 65 (1990), 363–373.

Zubarev, S. V.
[1] Necessary conditions for optimality for certain systems with distributed parameters, *Ukrainian Math. J.*, 31 (1979), 344–346.

# Index

adjoint system, 131,170
adjoint variable, 131
annihilator, 32

basis, 37
beam, 5
   Euler–Bernoulli, 6
   Rayleigh, 7
   Timoshenko, 8
boundary
   free, 17,18,21
   moving, 20

$C_0$ group, 52
capacitability, 86
capacity, 86
   carrying, 9
   right continuous, 86
closed loop, 264
codimension, 35
   finite, 35
   of subset, 134
   of subspace, 35
complete, 25
concavity, 234
   semi-, 234
   weak semi-, 235
condition
   boundary
      Dirichlet, 2
      Neumann, 2
      Robin, 2
   ellipticity, 76
   exact controllability, 294
   initial, 2
   maximum, 131,190
   transversality, 131,207
constraint
   pointwise state, 5
   state, 5,159,169
   terminal state, 5
continuity
   $B$-, 231
   equi , 48
   local modulus of, 227
   modulus of, 227
   of Banach space valued function
      in operator norm, 47
      strong, 46,47,53
      weak-, 47
      weak*-, 45,46
   of functions
      Hölder, 72
      Lipschitz, 72
      locally uniform, 98
      strongly (sequentially)
         lower semi-, 28
      weakly (sequentially)
         lower semi-, 31,239
      weakly* (sequentially)
         lower semi-, 31
   of linear operator, 27
   of multifunction, 89
      lower semi-, 89
      pseudo-, 91
      upper semi-, 89
control, 3,104,169
   admissible, 104,116,210
      impulse, 323,351
      switching, 323,351
   boundary, 3,208
      Dirichlet, 361
      Neumann, 364
   distributed, 3,208
   feasible, 104,169
   impulse, 320,321,323
   optimal, 104,127,131,169,210,370
   pointwise, 365
   state feedback, 256,264,392
   switching, 319
   time optimal, 294
controllability, 274
   approximate, 274,279,286
   approximate null, 274
   exact, 275,282
   exact null, 6,10,275
   infinitesimal time, 275
   local, 275
   small time local, 275

convergence
  strong, 25
  weak, 30
  weak*, 30
cost
  functional, 3
  impulse, 321,327
  running, 319,321,327
  switching, 319,327

derivative
  conormal, 77,209,364
  distributional, 72
  Fréchet, 257
  Gâteaux, 257
differentiability
  Fréchet, 45,257
  Gâteaux, 45,257
  strong, 48
differential inclusion, 264
discount factor, 264,319,327
distribution, 72
duality pairing, 28

eigenvalue, 33
eigenvector, 33
embedding
  compact, 73
  continuous, 73
  Sobolev, 73
energy
  kinematic, 6
  potential, 15
  strain, 6,12
  stress, 12
equation
  algebraic Riccati, 407
  beam, 6
  continuity, 12
  delay, 9,58
  elliptic, 4
    quasilinear, 191
    semilinear, 78,168
  Euler, 13
  evolution, 63
  Fredholm integral, 387,414
  Hamilton-Jacobi-Bellman,
    225,266,331
  heat, 2

  integral Riccati, 395
  Maxwell's, 21,23
  Navier–Stokes, 12
  parabolic, 4
  Riccati, 395,407,414
  state, ix,3
  wave, 5
estimate
  $L^p$-, 77
  Schauder, 77
exact penalization, 215
exponential stabilizability, 402
extension, 29,49,249

flow
  creep, 13
  steady, 14
  Stokes, 13
fluid
  dilatant, 192
  homogeneous, 12
  incompressible, 13
  inviscid, 13
  isotropic, 12
  pseudoplastic, 192
  Newtonian, 12,192
formula
  Green, 2
  Newton–Leibniz, 68
  variation of constants, 68
Fourier transformation, 409
frequency characteristics, 410
frequency domain, 412
function
  Banach space valued, 45
  convex, 29
  distance, 181
  $\delta$-, 173
  Fréchet differential bump, 45,246
  minimum time, 300
  multi-, 18,89
  Rademacher, 303
  simple, 45
  test, 239,267
  trace of, 75
  value, 227,264,300,328
functional
  linear bounded, 27
  linear continuous, 27

*Index* 445

graph
   of multifunction, 92
   of operator, 28
   maximal monotone, 122
   monotone, 122

Hamiltonian, 131,170
Hausdorff metric, 90,311

inner product, 24
integral
   Bochner, 45
   Lebesgue–Stieltjes, 58

law
   Fourier, 1
   Newton's second, 12
   parallelogram, 25
lemma
   Filippov, 102
   positive real, 408
      generalized, 410

map
   Dirichlet, 362
   Neumann, 364
measurability
   of Banach space value function
      strong, 46
      weak, 46
   of multifunction
      Borel, 94
      Lebesgue, 94
      Souslin, 94
measure
   Hausdorff, 217
   nonatomic, 143
   outer, 88
membrane, 15
method
   dynamic programming, 224
   Hilbert uniqueness, 285
model
   Malthus, 8
   logistic, 9
monotone
   maximal, 122
   strictly, 195

multi-index, 71

net, 31
norm, 24
   dual, 29
   equivalent, 42

observation, 411
obstacle, 16
   impulse, 331
   switching, 331
operator
   adjoint, 31,50
   bounded, 27
   compact, 33
   completely continuous, 33
   closable, 49
   closed, 49
   continuous, 27
   densely defined, 49
   domain of, 27
   embedding, 249
   evolution, 69
   fractional power of, 57
   graph of, 28
   kernel of, 28
   Laplacian, 59,76
   linear, 28
   nonclosable, 51
   orthogonal projection, 33
   range of, 28
   resolvent of, 33
   self-adjoint, 33,50
   spectrum of, 33
   symmetric, 50
   transfer, 411,412
   unbounded, 49
optimal
   control, 104,131,370
   pair, 104,131,370
   synthesis, 264
   trajectory, 104,131,370
output, 411

Parseval equality, 409
pair
   admissible, 104,116,130,125,169,195
   feasible, 104,116,130,125,169,195

optimal, 104,116,131,125,169,195
partial isometry, 51,377
perturbation
   needlelike, 143
   spike, 143
phase transition, 20
positive real, 412
   strict, 412
principle
   bang-bang, 164
   dynamic programming, 224
   Ekeland variational, 135
   Hamilton, 7,8
   (Pontryagin) maximum, 130,131
      132,165,168,189,199,202,211,295
     qualified, 210
   optimality, 216,331
   uniform boundedness, 28
problem
   boundary control, 208
   controllability, 10,274
   Dirichlet, 76
   distributed control, 366
   free boundary, 15,18,302
   minimax control, 4,197
   Neumann, 77,78
   obstacle, 16
     evolutionary, 20
   optimal control, 3
     impulse, 321
     linear-quadratic (LQ), 6,361,370
       classical, 370
         finite horizon, 401
         infinite horizon, 401
         regular, 370,381
         singular, 381
         standard, 381
     switching, 320
     time, 4,294
   patch control, 365
   Robin, 78
   Stefan, 20
     one-phase, 20
     two-phase, 20
   two-point boundary value, 381
property
   Cesari, 106
   Radon–Nikodým, 273
   semigroup, 52

unique continuation, 282,294

rate
   birth, 8,10
   death, 8,10
   growth, 8
restriction, 29,49
right-Lebesgue point, 54

selection, 100
   measurable, 100
semigroup
   analytic, 55
   $C_0$, 52
   compact, 55
   contraction, 55
   differentiable, 55
   generator of, 53
sequence
   Cauchy, 25
   minimizing, 112,120
set
   active, 121
   boundary of, 26
   closed, 26
   closure of, 26
   coincidence, 16
   compact, 26
   convex, 36
   convex hull of, 38,106
   exact controllable, 300
   first category, 26
   inactive, 171
   interior of, 26
   modified endpoint constraint, 134
   moving, 308
   noncoincidence, 16
   nowhere dense, 26
   open, 26
   reachable, 134,170,274,275
   relatively compact, 26
   second category, 26
   separable, 26
   Souslin, 84
   totally bounded, 27
sink, 2
solution
   classical, 76
   mild, 63,115

Index 447

  strong, 63,76,389,400
  weak, 63,64,76,77,79,115,194,213
solvable, 370,405
source, 2
space
  Banach, 25
    reflexive, 30
    separable, 26
    strictly convex, 41,135
    uniformly convex, 41
  dual, 28
  Hilbert, 25
  metric, 81
  normed linear, 24
  Polish, 81
  pivot, 114
  separable, 135
  Sobolev, 30,33,71
  Souslin, 84
  state, ix
  topological, 36,81
  topological vector, 36
  weakly compactly generated, 44
spectrum, 33
  approximate point, 33
  point, 33
  radius, 33
  residual, 33
stability
  exponential, 402
  $L^2$-, 402
  of optimal cost, 183,186,199
  strong, 210
stabilization, 6,10
  feedback, 412
stabilizing feedback operator, 404
state, ix,3
  co-, 131
  feasible, 169,195
  optimal, 127,170,210
support, 38
supporting point, 38,314
subdifferential, 146,257
superdifferential, 257
system
  adjoint, 131,170
  competition, 11
  control, 3
    unbounded, 366

  multispecies, 11
  predator-prey, 11
  variational, 134,170
  Walsh, 303

Theorem
  Alaoglu, 31
  Arzelà–Ascoli, 48
  Aubin–Lions, 118
  Baire Category, 26
  Ball, 63
  Banach–Steinhaus, 28
  closed graph, 28
  Eberlein–Shmul'yan, 31
  Eidelheit, 38
  Ekeland–Lebourg, 245
  Fredholm Alternative, 34
  Gagliardo, 75
  Goldstine, 41
  Hahn–Banach, 29
  Hille–Yosida, 53
  Himmelberg-Jacobs-Van Vleck, 94
  inverse mapping, 28
  Mazur, 39
  Mil'man–Pettis, 42
  open mapping, 28
  renorming, 42
  Riesz, 34
  Riesz Representation, 20,30,32
  Schauder, 34
  Sobolev Embedding, 73
  trace, 74
time
  approximate encounter, 309
  exact encounter, 309
  first approximate encounter, 309
  first exact encounter, 309
  first hitting, 295
  minimum approximate hitting, 309
  minimum hitting, 295
topology, 24
  strong, 30
  weak, 30
  weak*, 30
trajectory
  admissible, 104
  feasible, 104
  optimal, 104,118,131,370
  target, 274,315

triple
    admissible, 210
    feasible, 210
    optimal, 210

uniformly positive definite, 370

variational inequality, 16,183
    quasi-, 334
viscosity
    kinematic, 13
    solution, 223,239,267,344
    subsolution, 240,267,344
    supersolution, 240,267,344
vorticity, 15

well-posedness, 371,405

Yosida approximation, 53,54,65

# Systems & Control: Foundations & Applications

*Series Editor*
Christopher I. Byrnes
School of Engineering and Applied Science
Washington University
Campus P.O. 1040
One Brookings Drive
St. Louis, MO 63130-4899
U.S.A.

*Systems & Control: Foundations & Applications* publishes research monographs and
advanced graduate texts dealing with areas of current research in all areas of systems and
control theory and its applications to a wide variety of scientific disciplines.

We encourage the preparation of manuscripts in TEX, preferably in Plain or AMS TEX
LaTeX is also acceptable—for delivery as camera-ready hard copy which leads to rapid
publication, or on a diskette that can interface with laser printers or typesetters.

Proposals should be sent directly to the editor or to: Birkhäuser Boston,
675 Massachusetts Avenue, Cambridge, MA 02139, U.S.A.

Estimation Techniques for Distributed Parameter Systems
*H.T. Banks and K. Kunisch*

Set-Valued Analysis
*Jean-Pierre Aubin and Hélène Frankowska*

Weak Convergence Methods and Singularly Perturbed
Stochastic Control and Filtering Problems
*Harold J. Kushner*

Methods of Algebraic Geometry in Control Theory: Part I
Scalar Linear Systems and Affine Algebraic Geometry
*Peter Falb*

$H^\infty$-Optimal Control and Related Minimax Design Problems
*Tamer Başar and Pierre Bernhard*

Identification and Stochastic Adaptive Control
*Han-Fu Chen and Lei Guo*

Viability Theory
*Jean-Pierre Aubin*

Representation and Control of Infinite Dimensional Systems, Vol. I
*A. Bensoussan, G. Da Prato, M. C. Delfour and S. K. Mitter*

Representation and Control of Infinite Dimensional Systems, Vol. II
*A. Bensoussan, G. Da Prato, M. C. Delfour and S. K. Mitter*

Mathematical Control Theory: An Introduction
*Jerzy Zabczyk*

$H_\infty$-Control for Distributed Parameter Systems: A State-Space Approach
*Bert van Keulen*

Disease Dynamics
*Alexander Asachenkov, Guri Marchuk, Ronald Mohler, Serge Zuev*

Theory of Chattering Control with Applications to Astronautics, Robotics, Economics, and Engineering
*Michail I. Zelikin and Vladimir F. Borisov*

Modeling, Analysis and Control of Dynamic Elastic Multi-Link Structures
*J. E. Lagnese, Günter Leugering, E. J. P. G. Schmidt*

First Order Representations of Linear Systems
*Margreet Kuijper*

Hierarchical Decision Making in Stochastic Manufacturing Systems
*Suresh P. Sethi and Qing Zhang*

Optimal Control Theory for Infinite Dimensional Systems
*Xunjing Li and Jiongmin Yong*

Generalized Solutions of First-Order PDEs: The Dynamical Optimization Process
*Andreĭ I. Subbotin*